OBSERVATORIES AND TELESCOPES OF MODERN TIMES

Ground-Based Optical and Radio Astronomy Facilities since 1945

This volume gives an historical overview of the development of professional optical and radio observatories from 1945 to today. It covers the environment in which these facilities were developed by organisations in the United States, Europe and elsewhere, often led by larger-than-life individuals, as well as exploring the financial and political factors that both constrained and encouraged progress. As progressively more expensive optical facilities were built, they exploited new technologies to significantly improve their performance: CCDs, active and adaptive optics, and spun honeycomb and segmented mirrors. The second half of this volume turns to the parallel history of radio astronomy facilities throughout the world, finishing with the ALMA observatory in Chile. This is the ground-based companion to the author's previous work on space astronomy, *New Cosmic Horizons*; it is written for both technical and non-technical readers interested in the modern history of astronomy and its observational facilities.

DAVID LEVERINGTON received his first degree in Physics from Oxford University in 1963. Since then he has held a number of senior positions in the space industry, working for both the European Space Agency and British Aerospace (now BAE Systems) before taking early retirement in 1992. Subsequently he has written four books: *A History of Astronomy from 1890 to the Present* (Springer-Verlag, 1996), *New Cosmic Horizons: Space Astronomy from the V2 to the Hubble Space Telescope* (Cambridge University Press, 2000), *Babylon to Voyager and Beyond: A History of Planetary Astronomy* (Cambridge University Press, 2003) and *Encyclopedia of the History of Astronomy and Astrophysics* (Cambridge University Press, 2013). He was also technical consultant for ABC-CLIO's *Space Exploration and Humanity: A Historical Encyclopedia* (2010) supported by the History Committee of the American Astronautical Society. He was awarded his PhD in 1997.

OBSERVATORIES AND TELESCOPES OF MODERN TIMES

Ground-Based Optical and Radio Astronomy Facilities since 1945

DAVID LEVERINGTON

CAMBRIDGE
UNIVERSITY PRESS

CAMBRIDGE
UNIVERSITY PRESS

Shaftesbury Road, Cambridge CB2 8EA, United Kingdom

One Liberty Plaza, 20th Floor, New York, NY 10006, USA

477 Williamstown Road, Port Melbourne, VIC 3207, Australia

314–321, 3rd Floor, Plot 3, Splendor Forum, Jasola District Centre, New Delhi – 110025, India

103 Penang Road, #05–06/07, Visioncrest Commercial, Singapore 238467

Cambridge University Press is part of Cambridge University Press & Assessment, a department of the University of Cambridge.

We share the University's mission to contribute to society through the pursuit of education, learning and research at the highest international levels of excellence.

www.cambridge.org
Information on this title: www.cambridge.org/9780521899932

First published 2017

A catalogue record for this publication is available from the British Library

Library of Congress Cataloging-in-Publication data
Names: Leverington, David, 1941–
Title: Observatories and telescopes of modern times: ground-based optical and radio astronomy facilities since 1945 / by David Leverington.
Description: New York : Cambridge University Press, 2016. |
Includes bibliographical references and indexes.
Identifiers: LCCN 2016026406 | ISBN 9780521899932 (hardback: alk. paper)
Subjects: LCSH: Astronomical observatories. | Radio astronomy observatories.
| Observatories. | Telescopes. | Radio telescopes. |
Astronomy – History – 20th century.
Classification: LCC QB81 .L48 2016 | DDC 522/.1–dc23
LC record available at https://lccn.loc.gov/2016026406

ISBN 978-0-521-89993-2 Hardback

Contents

Preface

This book is a history of modern astronomical observatories and their telescopes. As such it covers the history of optical/infrared and radio/microwave observatories and telescopes that have been built since the Second World War. I have tried to cover the most innovative and trend-setting professional facilities and, as a result, I have excluded a discussion of most of the optical observatories established before the war even though some of them were still adding telescopes after 1945. This is because most of their new telescopes, with few exceptions, were not really innovative.

In many ways, the Palomar Observatory can be seen as the last of the major pre-war optical observatories as its first telescope, its 18/26 inch (46/66 cm) Schmidt, was built in the mid 1930s and its two main telescopes, the 200 inch (5.1 m) and the large Palomar (Oschin) Schmidt, were largely designed before the war. So maybe it should be excluded from this book, even though building of its two main telescopes was not completed until the late 1940s. But Palomar designs were innovative in many ways, setting the standards for large optical telescopes and observatories for some time to come. Consequently, I have begun my narrative with a brief outline of the design and development of the Palomar Observatory.

Ground-based astronomical observatories these days consist of more than just optical/infrared and radio/microwave facilities. For example, a number of neutrino, cosmic-ray, gamma-ray and gravity-wave observatories have been built, but I have excluded these to avoid complicating the book. Likewise, I have excluded small professional and large amateur facilities, interesting though many of them are, as this would take the book into completely new territory and make it considerably longer.

Discussing the history of professional astronomical observatories has its challenges as there are so many of them and their histories are often interconnected. It is theoretically possible to consider their history as one large interconnected story, but to produce a readable text I would have had to leave out interesting details which, in my opinion, would have made the book less interesting. So I have taken the alternative approach of considering their histories observatory by observatory whilst pointing out, where appropriate, how their histories are linked. Although this makes the book read a little more like an encyclopedia, I hope it is all the clearer for that enabling me to describe the unique circumstances that have led to the foundation and development of each observatory in turn.

Part 1

Optical Observatories

Part 1

Optical Observatories

1

Palomar Mountain Observatory

1.1 The 200 inch (5.1 m) Hale Telescope

The early twentieth century had seen the emergence of the United States as the world leader in the construction of large optical telescopes.[1] For example, two large solar telescopes had been built on Mount Wilson, California before the First World War. In addition, George W. Ritchey had also completed a 60 inch (1.5 m) reflector at the same observatory in 1908, and nine years later Ritchey and W. L. Kinney had completed the 100 inch (2.5 m) Hooker reflector there. As a result by the early 1920s Mount Wilson was also the premier observatory in the world.

No sooner had the 100 inch telescope been completed than George Ellery Hale, the director of the Mount Wilson Observatory, began to consider building an even larger instrument.(1) He mentioned his ideas to Francis Pease, who had recently joined the staff on Mount Wilson. By 1921 Pease, who by then had outlined the design for a 300 inch (7.5 m), was convinced that a 100 ft (30 m) telescope was feasible. But Hale was much more cautious, partly because of the difficulties that he had already experienced with building the 100 inch, and partly because of the difficulty he anticipated of raising the money to build such an enormous telescope. In fact, as he recognised, the time was not ripe for raising finances for even a 300 inch.

Nevertheless, Pease continued with designing his 300 inch. Then in 1926 he and Walter Adams took H. J. Thorkelson of the General Education Board of the Rockefeller Foundation on a tour of the Mount Wilson Observatory. During the tour, Pease showed him his design of the 300 inch. This design impressed Thorkelson so much that he mentioned it to Wickliffe Rose of the Rockefeller Foundation's International Education Board shortly afterwards.

[1] It could be argued that the United States had become dominant in the construction of large optical telescopes by the end of the nineteenth century. On the other hand, the Mount Wilson Observatory had had to purchase optical blanks for their 60 inch (1.5 m) and 100 inch (2.5 m) telescopes from France in 1895 and 1906 as they could not be produced in the United States at the time.

Two years later Hale wrote an article for Harper's Magazine on 'The Possibilities of Large Telescopes' in which he outlined their importance. He also floated the idea of finding a donor to back the financing of a new large telescope, following on the path already trodden by Messrs Lick, Yerkes, Hooker and Carnegie in funding telescopes. Hale arranged for a proof of his article to be sent to Rose, which he backed up with a long letter asking if the Rockefeller Foundation's General Education Board would consider making a grant to determine the maximum size of mirror that could be made. Rose, a supporter of big science, was interested in Hale's proposed project.

In March 1928, Hale followed up his letter with a visit to Rose where he explained the difficulties that he foresaw in building a 200 inch (5.1 m) telescope. This may have made Rose think twice about being involved in funding such a large telescope but, rather than being put off, Rose asked if a 300 inch was feasible. Hale suggested that the jump from 100 inch to 300 inch was probably too large and a 200 inch was a much safer alternative. The only disagreement between the two men at that time was which organisation should be given the telescope in the event that the Rockefeller Foundation agreed to fund it. Hale naturally thought that that should be the Carnegie Institution's Mount Wilson Observatory, which had experience of building and running their 60 inch and 100 inch telescopes. Rose disagreed and suggested that an education establishment like the California Institute of Technology (Caltech) should be the recipient. This ought not to have surprised Hale bearing in mind Rose's membership of Rockefeller's Education Board. But Rose then went on to suggest that the staff at Mount Wilson should, under Caltech's guidance, build the telescope and run its research programme.

The problem with Rose's idea was that John C. Merriam, the president of the Carnegie Institution, was not willing to enter into such an agreement with an organisation like Caltech who had no experience of running a great observatory. In fact Caltech did not even have a department of astrophysics.

In the end, after a convoluted set of negotiations between the Rockefeller Foundation, the Carnegie Institution and Caltech, the Rockefeller trustees agreed in 1928 to pay the estimated construction cost of $6 million for the 200 inch telescope and ancillary equipment. But they were not willing to finance the annual running costs and Caltech could not afford to pay for them. Fortunately Henry Robinson, a wealthy Los Angeles businessman and Caltech trustee, agreed to cover them. So the Rockefeller Foundation agreed to give the construction money to Caltech who were to build and operate the telescope, whilst the Carnegie Institution and the Mount Wilson Observatory agreed to provide assistance. Or as Elihu Root, chairman of the Carnegie trustees, put it – the Carnegie Institution would furnish the brains while the Rockefellers provided the funds.

A four-member Observatory Council was set up with Hale as chairman to direct the planning, construction and operation of the 200 inch. This was assisted by an Advisory Committee consisting of members of the observatory and institute staffs, as well as external astronomers, physicists, chemists and mathematicians. There were also committees set up to cover the mirror, the mounting and the site. In one way or another Hale managed to involve the best American scientists and engineers in this great undertaking.

Light pollution was already becoming a problem at Mount Wilson in the late 1920s and Rose's decision to give the telescope to Caltech also meant that a new location was required. So alternative sites in the south-western United States, as close as possible to Caltech in Pasadena, were tested for weather conditions and seeing starting in 1928.(2) At this stage, Hale already favoured Palomar Mountain as it had been recommended by the Lick astronomer William J. Hussey in 1903 as the most promising site in the San Diego area. But it was too inaccessible then and Hale had decided to locate the Carnegie observatory on Mount Wilson instead. Now in 1928 the inaccessibility problem was not as severe.

Following the site survey Palomar Mountain and Table Mountain behind Mount Wilson were considered to be the best sites with Palomar receiving the final decision in 1934 as the location for the 200 inch observatory. This was mainly because of Palomar's dark skies compared with those of the Mount Wilson area which was now too close to Los Angeles and its surrounding towns. The land to build and protect the new observatory consisted of 120 acres (49 hectares) purchased from private individuals and 40 acres which were transferred from the Cleveland National Forest to Caltech. Construction of a road to the observatory was begun by the County of San Diego the following year.

Initially it was thought that the equatorial mounting system for the primary mirror should be either an open-fork or an English yoke type.(3) At first, the fork design was preferred as it allowed access to the sky's polar regions, but in 1932 a new yoke design was proposed that also allowed such access. This was in the form of a horseshoe in which the telescope tube was lowered into the horseshoe far enough for it to be able to point to the pole. Unfortunately calculations indicated that the horseshoe was not sufficiently rigid for it to remain circular when the telescope tube moved over to the east or west. But this problem was solved by prestressing the horseshoe structure during manufacture.(4)

The 200 inch was large enough to allow an observer's cage to be fitted in the telescope tube at the primary focus (see Figure 1.1) without cutting off any more light than a Newtonian mirror. This 72 inch (1.8 m) diameter cage carried the secondary mirrors. But the weight of this system at the top of the tube and that of the primary mirror at the bottom caused the telescope tube to flex too much. This would have been a particular difficulty for the Ross corrector lens (see later) as it had to be collimated very accurately with the primary mirror. But that problem was

Figure 1.1. The 200 inch (5.1 m) Hale reflector which was completed on Palomar Mountain in the late 1940s. The observer is seen sitting in the prime focus cage. (Courtesy Palomar Observatory and California Institute of Technology.)

solved by Mark Serrurier who designed a parallelogram of trusses to connect the top and bottom rings of the tube to the section housing the bearings of the declination axis. This system allowed both ends of the telescope to move by virtually the same amount as the telescope moved. So the focus of the primary mirror did not deviate from its mean position on the plate in the observer's cage by more than about 0.01 inch (0.25 mm). Called the Serrurier truss, this solution has since been adopted for other large telescopes.

The most difficult problem facing Hale and the telescope team was how to make a 200 inch diameter mirror. After consulting experts worldwide Hale and the Observatory Council decided to ask the General Electric (GE) Company to make the mirror blank of fused quartz[2] at an estimated cost of about $250,000.[3] This

[2] Other materials that were considered at this time included Pyrex, stainless steel or some other metallic alloy, and metal with a glass surface of the same expansion coefficient. (See George E. Hale, The Astrophysical Observatory of the California Institute of Technology, *Astrophysical Journal*, 82, 1935, pp. 111–139.)

[3] The agreement with GE was that they would be paid at cost, excluding profit and overhead, so this $250,000 estimate was seen, at the time, to be on the high side of the possible cost of a 200 inch. See Donald E. Osterbrock, *Pauper and Prince; Ritchey, Hale and Big American Telescopes*, University of Arizona Press, 1993, p. 234.

was because fused quartz has an expansion coefficient of about one-fifteenth that of plate glass that had been used for the primary mirror of the Mount Wilson 100 inch.(5) But by late 1928 only small quartz discs had been produced and even in these it had proved impossible to eliminate bubbles from the molten silica. So GE decided to melt clear quartz plates onto the surface of the discs.(6) In the event that did not work for large diameter discs, so GE tried spraying the disc surfaces with molten quartz instead. By mid 1931, after a considerable effort, only two 60 inch discs had been produced. The first had a crack caused by an operative opening the furnace cover too early, and the other was unsuitable as pieces of fire brick had fallen from the cover of the furnace.(7) Yet over $600,000 had been spent. Possible timescales for production of the required 200 inch blank were anyone's guess, and the extra cost was estimated to be at least $1 million. Consequently Hale cancelled the contract.

In the meantime, Hale and Pease had decided to make the mirror a relatively fast *f*/3.3 because this allowed shorter exposures for nebula photography. It also enabled the tube to be made relatively short, allowing the dome to be made smaller also. The mirror was lightened by more than half by having an hexagonal pattern of ribs on its back to provide stiffening. This design also substantially reduced the annealing time of the mirror blank. The main problem caused by having such a relatively fast mirror was that the diameter of the field of good definition was only about 0.4 inch (1 cm) for the 200 inch because of coma. A similar but less severe problem had occurred with Mount Wilson's 60 inch f/5 and 100 inch f/5 telescopes. But this had been solved by Frank Ross of the Yerkes Observatory who designed corrector lenses for them. So Hale asked Ross to design similar corrector lenses for the 200 inch. This resulted in a series of two and three component lenses which produced well-corrected fields up to 15 cm in diameter. They also provided a choice of aperture ratios from f/3.6 to f/6.0. Which of these was used depended on the type of object being observed and the seeing conditions.

The primary mirror was not rigid enough to avoid bending under its own weight, even with the hexagonal pattern of ribs at the back. Consequently the mirror was carried on thirty-six support mechanisms inserted into the hexagonal holes. These elaborate lever-based mechanisms balanced the changing forces as the mirror was moved relative to gravity. But problems with friction within the mechanism caused its lever arm to be redesigned in 1948.(8)

Returning to the problem of making the 200 inch primary mirror, following the problems experienced by GE in producing the blank in fused quartz, Corning was contracted to produce one in Pyrex as, although the expansion coefficient of Pyrex was about five times that of fused quartz, it was only about one-third that of plate glass.(9) Corning produced a set of increasing diameter mirrors to adjust the quartz content of the Pyrex and to develop new methods of casting, preliminary cooling and annealing. First of all they produced a 26 inch (66 cm), followed by a 30 inch (76 cm) which was used for the Coudé flat, a 60 inch (1.5 m), and finally a 120 inch (3.0 m), all with ribbed backs. Unfortunately some of the cores required to produce the ribs at the back of the 60 inch mirror came loose during the mirror's production in July 1932. But the quartz content of the Pyrex was then changed to produce a material with a lower expansion coefficient,[4] and a 120 inch disc successfully cast in the summer of 1933.

The first attempt to produce a 200 inch disc was begun in March 1934 but, unfortunately, several cores came loose and rose to the surface as molten Pyrex was being ladled into the mirror mould. Attempts to break up the loose cores failed so a decision was taken to cool the blank relatively quickly to see if its quality was acceptable. Although this rapid cooling did not result in any cracks, polarised light tests indicated strains and the loose cores were embedded in the Pyrex.[5] Consequently it was decided to have another attempt to produce a satisfactory 200 inch blank, whilst retaining this first attempt as a spare.

The cores had come loose in producing the first 200 inch blank as the steel bolts that were designed to retain them had melted in the intense heat. So the bolts were replaced for the second attempt by those made of chrome-nickel steel, and a special air-cooling system was introduced beneath the mould to keep the interior of every core at an acceptable temperature. The new casting attempt took place in December 1934. After the glass had solidified the blank was moved to an electric annealing oven where it stayed at a steady temperature for two months. Its temperature was then gradually reduced at less than 1°C per day to produce a strain-free disc. But two-thirds of the way through this cooling process the Corning plant was flooded by its adjacent river and the electricity had to be cut off for three days. However, much to everyone's relief this had no adverse effect on the blank which turned out to be fine. Subsequently it was transported to Pasadena by train in April 1936, taking two weeks to cross the country.(10)

On receipt at the Pasadena optical shop, the front and back surfaces of the mirror blank were first ground flat. Then the front surface was ground to approximately the correct curvature by deepening the centre by nearly 4 inches (10 cm) and removing about 5 tons of glass. In total 31 tons of abrasive were used during this process and the subsequent polishing. The mirror was mounted on its 36-lever support system during figuring and testing. But this showed a problem with the outer 20 inches (50 cm) of the mirror that had been purposely produced with a slightly turned-up

[4] This Pyrex had an expansion coefficient of just three times that of fused quartz.

[5] Hale, who was not present as he was ill at the time, says in his 1935 Astrophysical Journal paper (7) that the floating cores were fished out. But both King (4) and Wright (1) say that it proved impossible to either break them up or fish them out.

edge as it was expected to sag slightly in operation. Unfortunately the mirror was too stiff and so it did not sag enough. But the error was only about one wavelength of green light. So the mirror was sent to the new observatory on Palomar Mountain in this condition, hoping to rectify the problem in situ.

In the meantime, Hale had died and the Second World War had interrupted work. So it was not until November 1947 that the mirror was finally transported to its observatory where it was placed in the telescope for the first time. The telescope was subsequently dedicated to Hale on June 3 of the following year in a ceremony presided over by Vannevar Bush, president of the Carnegie Institution, and Lee Du Bridge, president of Caltech. The Palomar Observatory was then administratively joined to Mount Wilson to become the Mount Wilson and Palomar Observatories. So Hale had got what he wanted in the end.[6]

It turned out to be impossible to rectify the problem of the raised edge of the mirror in situ, so in 1949 it was decided to remove the mirror from the telescope and repolish the offending edge on the dome floor. This delicate task was undertaken by Don Hendrix, the chief optician of the Mount Wilson observatory, assisted by Mel Johnson who had previously worked on the mirror in the Pasadena optical shop.(11) Following a successful operation, the telescope became available for regular scheduled observations in November 1949.

The Hale Telescope's design with its massive mirror and support system, low expansion glass, lensed field correctors, Serrurier truss design, equatorial mount, oil-pad bearings and hemispherical dome had a major effect on the design of other large telescopes for the next twenty years or so. In fact there was no major push to design a more mass-efficient large telescope for some time as the Hale telescope design was so successful with its oil-pad bearings and Serrurier truss design which were able to support such enormous weights.(12)

The Hale telescope remained the largest and most productive optical telescope in the world for many years. In the 1950s observations with it confirmed the linear relationship between redshift and distance to greater distances than previously possible. It also showed that the Andromeda nebula was twice as far away as previously thought, and that the value of the Hubble constant needed substantial revision. The Hale telescope also played an important role in understanding stellar evolution and in the discovery of quasars in the early 1960s.[7]

[6] In 1969, the Las Campanas Observatory in Chile became part of the group that was renamed the Hale Observatories. Then in 1980 it was decided to split the Palomar observatory from the Mount Wilson and Las Campanas observatories, with Caltech operating Palomar (except for the 60 inch [1.5 m] reflector which was still jointly operated) and the Carnegie Institution, the other two observatories. Subsequently Carnegie handed over the Mount Wilson Observatory to the non-profit making Mount Wilson Institute, and in 1989 the Las Campanas Observatory became part of the Observatories of the Carnegie Institution of Washington, see *Sky and Telescope*, October 1980, p. 280 and February 1996, p. 8.

[7] Sandage gives an excellent overview of the cosmological and astrophysical discoveries made with the 200 inch telescope in his Annual Reviews of Astronomy and Astrophysics article of 1999 (11).

Eventually, in 1976, the 5.1 m Hale telescope's position as the largest optical telescope in the world was finally lost to the 6.0 m (236 inch) Bolshoi Teleskop Altazimutalnyi[8] (BTA) that was built at the Special Astrophysical Observatory near Zelenchukskaya in the Caucasus. The BTA was the first modern large telescope to have an altazimuth mount (see Section 5.3), rather than the equatorial of the previous large reflectors. Unfortunately, the BTA's performance was poor because problems with its mirror were compounded by relatively poor seeing conditions at the observatory.

The design of the Hale telescope itself has been gradually improved over the years. For example in 1965 an observer's cage was fitted at the Cassegrain focus underneath the primary mirror's structure. This not only made it easier for astronomers to observe at the Cassegrain focus but also allowed the installation of larger and more complex instrumentation there. The Ross corrector lenses were also replaced by a Wynne corrector that provided a much wider usable field of view at the prime focus. But probably the largest change in using the 5 m Hale, like that with other large telescopes of the time, occurred in the recording and analysis of images. In the late 1940s, almost all observations had been made on the Hale using photographic plates. But by 1978, for example, photographic plates were used on only 15% of observing nights.(13) This was because they had been gradually replaced by image intensifiers, television cameras, imaging pulse-counting systems and finally CCDs (see Section 5.4), even though the photographic plates of the late 1970s were much better than those available in the late 1940s.[9]

In 1998 a Shack-Hartmann adaptive optics system[10] called PALAO (Palomar Adaptive Optics), developed by the Jet Propulsion Laboratory (JPL), was fitted to the Hale 5 m. It was produced as part of an agreement between Caltech and JPL which allowed JPL access to the telescope in return for their support with instrumentation.(14) PALAO was operated with PHARO (Palomar High Angular Resolution Observer) a dedicated near infrared science camera which had been designed and built by Cornell University under a cooperative agreement between Cornell and Caltech. The system produced near infrared images with a resolution of about 20 mas (milliarcsec).

Eight years later LuckyCam, which had been developed by Cambridge University in the UK, was installed on the telescope. It took images at a rate of about 20 per second, allowing the best to be added together to produce very sharp results. Unusually for modern high-resolution systems, LuckyCam operated in the

[8] Translates as 'Large Altazimuth Telescope'.

[9] The last photographic plate to be exposed on the 5 m Hale was a 1 hour exposure of the supernova remnant 3C 58 by Sidney van den Bergh on September 29 1989, see *Sky and Telescope*, February 1990, p. 134.

[10] Shack-Hartmann arrays are outlined in Section 5.2.

visible waveband. Operating with an adaptive optics system it produced images with a resolution of 35 mas.

1.2 Palomar Schmidt Telescopes

The Mount Wilson 60 and 100 inch f/5 reflecting telescopes had been built in the early twentieth century to photograph dim objects. But, as mentioned above, they initially had relatively narrow, corrected fields of view owing mainly to off-axis coma. In the late 1920s/early 1930s when the f/3.3 Palomar reflector was being planned it was clear that it would have an even worse problem as coma was known to vary inversely as the square of the focal ratio.(15) At about this time Ross was designing coma correctors for these large reflectors, but their corrected field of view would still be rather small.

But in 1931 John Anderson, the Mount Wilson instrumental physicist, became aware of a radical new design of telescope produced by Bernhard Schmidt of the Hamburg Observatory that could be used for very wide field photography. It was clear that such a telescope would be a valuable complement to the large reflectors on Mount Wilson and Palomar as it could find and accurately locate the faint stars, nebulae, and galaxies that these telescopes could then study in more detail.(16) Schmidt's design had a spherical rather than a parabolic primary mirror and an aspherical glass plate to correct for spherical aberration. This aspherical corrector plate was smaller than the primary mirror and was placed at its centre of curvature to eliminate coma. The main problem was that the focal plane was curved, so the photographic plate which was placed in the focal plane had to be curved to keep the image in focus.

John Anderson on hearing of Schmidt's radical design suggested to Hale that they should build a Schmidt telescope or camera[11] at Palomar. Shortly afterwards Russell Porter was commissioned to design such an instrument which was built in the Mount Wilson optical shop. It had a 26 inch (66 cm) *f*/2 primary mirror and an 18 inch (46 cm) corrector plate. With its 8° field of view, it was used by Fritz Zwicky and Joseph Johnson from 1936 to 1939 to produce 1625 photographs of 175 regions of the sky that contained nearby galaxies. They were looking for supernovae, twelve of which they discovered over that three year period.(17)

The early success of the 18 inch Schmidt, as it was called, persuaded Hale to ask the Rockefeller Foundation for funding to construct a much larger Palomar Schmidt. Corning made the blank for its 72 inch (1.8 m) *f*/2.4 spherical primary mirror. This was then ground and figured by Don Hendrix and Ralph Dietz in the Mount Wilson optical shop where it was completed in 1940. Hendrix then started

[11] Some authors use the term 'telescope' and some 'camera' when describing Schmidts.

work on its 48 inch (1.22 m)[12] corrector plate but had to interrupt this work when the USA entered the Second World War in December 1941.

After the war, Hendrix returned to figuring the 48 inch corrector assisted by Floyd Day who had joined Mount Wilson as a night assistant. The corrector was finished in 1948 and by January of the following year the large Schmidt (see Figure 1.2) was in full operation. The spherical primary mirror, which weighed over a ton, was supported by 36 weighted compound levers. It also had 18 equally spaced supports around the edge to help support the mirror when the telescope was tilted. The first official plate, taken by Hendrix in September 1948, was of M31. This was the first time that the galaxy had been imaged with such excellent resolution and such faint limiting magnitude on a single plate.

This 48 inch Schmidt was used for the National Geographic – Palomar Observatory Sky Survey (POSS I) which was carried out between 1949 and 1958 on both blue (103aO) and red (103aF) sensitive plates recording stars, galaxies and other objects down to magnitude 21. These 14 inch (35 cm) square plates each covered an area of 6.5° × 6.5°. They were specially made only 1 mm thick to enable them to be bent to match the curved Schmidt focal plane. This was achieved by holding them against a mandrill with a vacuum system. The survey covered the sky in 935 fields from the north celestial pole to –33° declination at 6° intervals.(18) Amongst other things POSS I was used by Peter Nilson to produce the Uppsala General Catalogue of Galaxies and by George Abell to produce the first large catalogue of galaxy clusters. Abell's catalogue allowed astronomers to produce a major new map of the distant universe. POSS I was also invaluable in identifying objects in the visible waveband that had been discovered in other wavebands such as radio galaxies, X-ray sources and quasars.

After completion of the POSS I survey the large Schmidt was used to continue the Palomar Supernova Search which had been started by Zwicky and Johnson in the 1930s. Three moonless nights per month were scheduled for the search which consisted of imaging 38 fields containing a total of over 3000 galaxies in clusters and groups. When the search was officially closed in 1975 it had discovered 183 supernovae. Then in 1982 the large Schmidt's Quick-V Survey was used to produce the northern half of the Hubble Space Telescope's (HST's) first Guide Star Catalog published in 1989, which included the position and intensity of about 19 million stars.

A new achromatic corrector plate had been installed on the UK Schmidt in 1977 to enable it to provide images over a wider wavelength range than with its old corrector (see Section 9.2). This new corrector was so successful that a similar achromatic corrector, produced by Grubb-Parsons in England, was also installed

[12] This was its nominal diameter. It was actually 49.5 inch (1.26 m) in diameter.

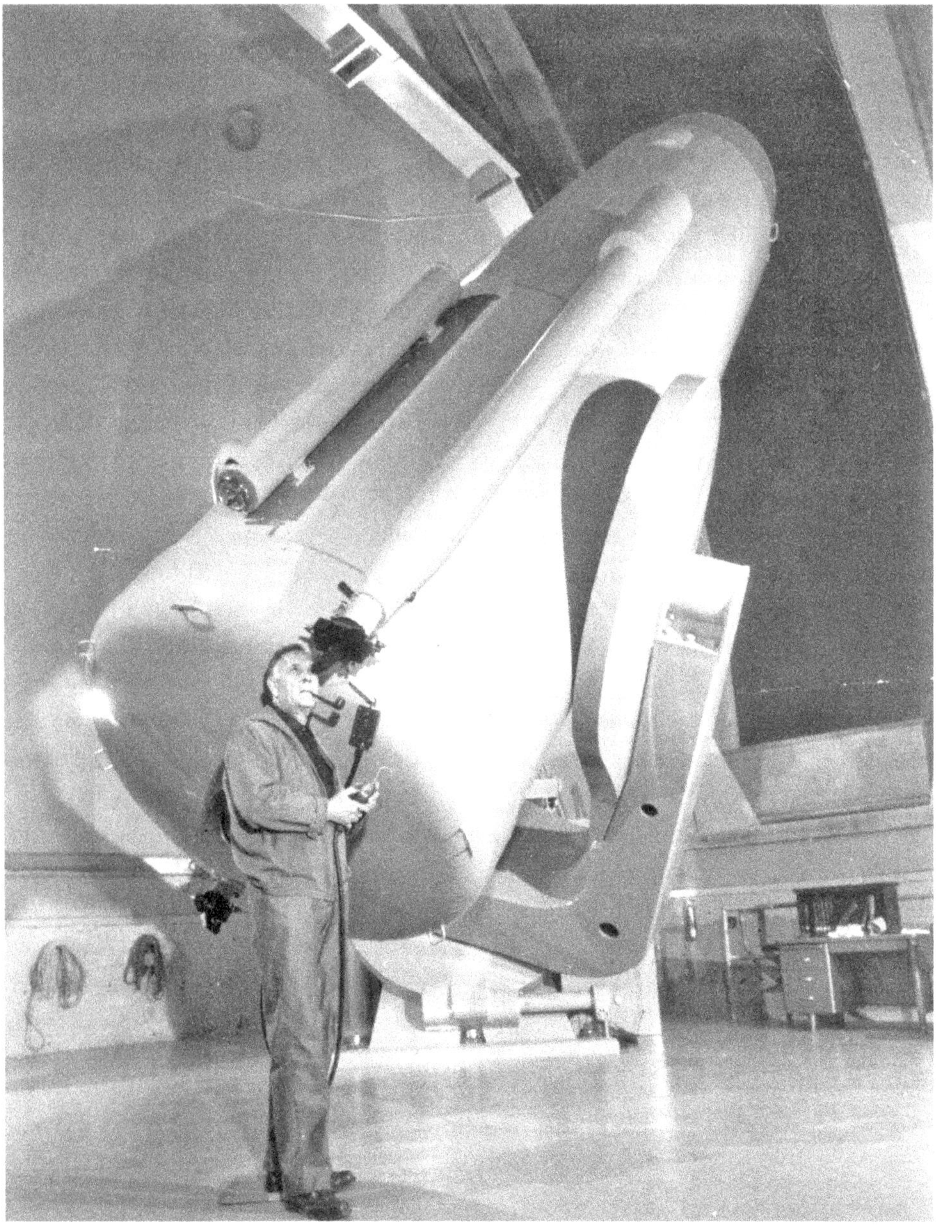

Figure 1.2. Edwin Hubble with the 48/72 inch (1.2/1.8 m) Palomar Oschin Schmidt. (Courtesy Palomar Observatory and California Institute of Technology.)

on the large Palomar Schmidt in the early 1980s. At about the same time an automatic guider and an internal calibration source were also installed on this Schmidt.

Subsequently, a new sky survey (POSS II) was undertaken at Palomar over the period 1985 to 2000 as a northern complement to the UK Schmidt survey. POSS II used finer grain and faster plates than had been available at the start of POSS I; these were IIIaJ blue, IIIaF red and IVN near-infrared sensitive plates.[13] Funding was provided for the Schmidt upgrade and execution of POSS II by the Alfred P. Sloan Foundation, Caltech, the National Geographic Society and the National Science Foundation, whilst Eastman Kodak provided all of the plates together with a cash grant. Then in 1986 the large Palomar Schmidt was renamed the Samuel Oschin Telescope following a donation of $1 million by the Samuel Oschin Family Foundation to help finance astronomical research at Palomar. A digitised version of POSS II was later produced as a collaborative project by the Space Telescope Science Institute (STScI) and Caltech to provide the northern half of the new HST guide star catalogue, Guide Star Catalog II (GSC-II).[14]

The Jet Propulsion Laboratory's NEAT (Near Earth Asteroid Tracker) CCD camera was installed on the Oschin Schmidt[15] in 2001. Two years later this was replaced by the 161 megapixel QUEST 2 (QUasar Equatorial Survey Team) CCD camera of Yale and Indiana University.(19) This camera was used for the Yale/Caltech[16] Palomar-Quest Sky Survey to observe transient sources which included quasars, blazars, supernovae and the afterglows of gamma-ray bursts. In addition, it was used to search for asteroids which the project team considered as 'contaminants' in their search for optical transients.(20) The Quest Sky Survey was completed in October 2008. Two months later the Palomar Transient Factory (PTF) Survey was begun which was a fully automated, wide-field survey to detect optical transients using an array of twelve 2K × 4K CCDs with a total light sensitive area of 7.26 square degrees. Photometric follow-up was undertaken by the automated Palomar 1.5 m telescope.[17]

Over its lifetime the Palomar (Oschin) Schmidt whilst imaging millions of stars has discovered thousands of supernovae and many thousands of asteroids.[18] More recently the Oschin Schmidt has discovered the large classical Kuiper belt object Quaoar (in 2002), the detached object Sedna (2003), the plutino Orcus (2004) and the dwarf planet Eris (2005).[19]

[13] The blue and red sensitive plates were hypersensitised by alternately placing them in separate atmospheres of hydrogen and nitrogen, whilst the near infrared plates were hypersensitised by soaking them in a silver nitrate bath shortly before use.

[14] This catalogue was also used by the Gemini, European VLT and TNG (Telescopio Nazionale Galileo) telescopes, see Barry M. Lasker et al., The Second-Generation Guide Star Catalog: Description and Properties, *Astronomical Journal*, 136, 2008, p. 735.

[15] It is often called the Oschin Schmidt rather than its official name of the Samuel Oschin Telescope mentioned previously.

[16] Yale and Caltech were the main collaborators but there were a number of others involved.

[17] This 60 inch telescope was installed at the Palomar observatory in the late 1960s to take some of the workload off the 200 inch. It was automated in 2004 to make observations of the afterglows of gamma-ray bursts and other transient events (see S. Bradley Cenko et al., The Automated Palomar 60 inch Telescope, *Publications of the Astronomical Society of the Pacific*, 118, 2006, pp. 1396–1406.

[18] For example, as long ago as 1960 it had discovered about 2,400 asteroids in just 11 nights of observations in the Palomar-Leiden Survey. This was followed in the 1970s by discovering more than 3,500 Trojans over 24 nights in the Palomar-Leiden Trojan Surveys.

[19] The years quoted are the discovery dates which were sometimes a year or so later than the telescope observation dates.

References

1. Wright, Helen, *Explorer of the Universe; A Biography of George Ellery Hale*, American Institute of Physics Press, 1994, p. 387.
2. Wright, Helen, *The Great Palomar Telescope*, Faber and Faber, 1953.
3. Van Helden, Albert, Building Large Telescopes, 1900–1950, in Gingerich, O., (ed.), *The General History of Astronomy Volume 4A; Astrophysics and Twentieth-century Astronomy to 1950: Part A*, Cambridge University Press, 1984.
4. King, Henry C., *The History of the Telescope*, Charles Griffin, 1955 (Dover reprint 1979), pp. 408–410.
5. Ibid., pp. 402–403.
6. Van Helden, Albert, Building Large Telescopes, 1900–1950, in Gingerich, O., (ed.), *The General History of Astronomy Volume 4A; Astrophysics and Twentieth-century Astronomy to 1950: Part A*, Cambridge University Press, 1984, p. 147.
7. Hale, George E., The Astrophysical Observatory of the California Institute of Technology, *Astrophysical Journal*, 82, 1935, p. 117.
8. Wilson, R. N., *Reflecting Telescope Optics II; Manufacture, Testing, Alignment, Modern Techniques*, Springer-Verlag, Corrected Second Printing, 2001, p. 260.
9. King, Henry C., *The History of the Telescope*, Charles Griffin, 1955 (Dover reprint 1979), p. 403.
10. Di Cicco, Dennis, The Journey of the 200-inch Mirror, *Sky and Telescope*, April 1986, pp. 347–348.
11. Sandage, Allan, The First 50 Years at Palomar: 1949–1999 The Early Years of Stellar Evolution, Cosmology, and High-Energy Astrophysics, *Annual Reviews of Astronomy and Astrophysics*, 37, 1999, pp. 445–486.
12. Learner, Richard, The Legacy of the 200-inch, *Sky and Telescope*, April 1986, pp. 349–353.
13. Oke, J. B., Palomar's Hale Telescope: The First 30 Years, *Sky and Telescope*, December 1979, pp. 505–509.
14. Hayward, T. L., PHARO: A Near-Infrared Camera for the Palomar Adaptive Optics System, *Publications of the Astronomical Society of the Pacific*, 113, 2001, pp. 105–108.
15. Ross, Frank E., *Publications of the Astronomical Society of the Pacific*, 7, 1933, pp. 98–99.
16. Osterbrock, Donald E., Don Hendrix, Master Mount Wilson and Palomar Observatories Optician, *Journal of Astronomical History and Heritage*, 6, No. 1, 2003, pp. 1–12.
17. Murdin, Paul and Lesley, *Supernovae*, Cambridge University Press, rev. ed. 1985.
18. Schombert, James, Surveying the Northern Sky, *Sky and Telescope*, August 1987, pp. 128–131.
19. Baltay, C., et al., The QUEST Large Area CCD Camera, *Publications of the Astronomical Society of the Pacific*, 119, 2007, pp. 1278–1294.
20. Djorgovski, S. G., The Palomar-Quest Digital Synoptic Sky Survey, *Astronomische Nachrichten*, 329, No. 3, 2008, pp. 288–291.

2

The United States Optical Observatory

2.1 Introduction

Although the Palomar Observatory had been very successful after the completion of its main instruments in the late 1940s, its existence had only exacerbated an underlying problem in the organisation of astronomy in the United States. This was because at that time the vast majority of telescopes were already owned by a limited number of prestigious universities or observatories. These had been paid for by private benefactors and their use had been generally limited to members of these institutions. As a result, astronomers at less wealthy universities had only access to instruments of more limited performance. There was no tradition of publicly funded observatories in the United States at that time, unlike that of many European countries.

The first major collaboration between two American universities in the provision of an optical observatory was instituted in 1932, when it was agreed that the new McDonald Observatory would be the joint responsibility of the universities of Texas and Chicago.[1] Then in 1940 Otto Struve, McDonald's director, published an article 'Cooperation in Astronomy' in *Scientific Monthly*.(1) In this, he suggested setting up a collaborative observatory between a larger number of astronomical institutions. This would have provided the wider access required, although Struve was assuming, at the time, that the initial finance would be provided by a private benefactor, in this case either the Carnegie Institution or the Rockefeller Foundation. Unfortunately, this project had to be put on hold at the time because of the imminent involvement of the United States in the Second World War.

The Second World War saw a turning point in the funding of technology in general by the United States government. In both the United States and Europe large sums of money were spent on nuclear, radio and radar research, and in Germany,

[1] The agreement, which was to run for thirty years, provided for the University of Texas to build and equip the McDonald Observatory and for the University of Chicago to operate it. See D. S. Evans and J. D. Mulholland *Big and Bright; A History of the McDonald Observatory*, University of Texas Press, 1986.

in particular, on the development of rockets like the V2. All of this was to have a major effect on the development of astronomy in the United States and elsewhere after the war. In the United States, in particular, it also had the effect of generating much more centralised government funding for all types of scientific research, both civil and military. This culminated, on the civil side, in the foundation of the National Science Foundation (NSF) in 1950.(2) Its first director, Alan T. Waterman, who was deputy director of the Office of Naval Research (ONR), was appointed the following year.

In 1949 the National Research Council (NRC) had asked the American Astronomical Society (AAS) to consider how the proposed National Science Foundation could best support astronomical research in the United States. In response, the AAS suggested in 1951 that the NSF should provide something of the order of \$200k to \$400k annually for new capital facilities.[2] But first, the fledgling NSF had to get its initial budget approved by Congress. This proved to be even more difficult than normal for a government-sponsored institution. The Bureau of the Budget approved a budget request for Fiscal Year (FY) 1952 of \$14 million, which the House Appropriations Committee reduced to just \$300k! The Senate then proposed a budget of \$6.3 million. Eventually a figure of \$3.5 million was approved one-third of the way into FY1952.(3) It is a miracle that any programmes could be planned with such an uncertainty in the annual budget.[3]

2.2 Founding of Association of Universities for Research in Astronomy (AURA)

Light pollution was becoming more and more of a problem in the United States after the Second World War for observatories located near to large cities. So much so that the University of Arizona began to consider moving its 36 inch (0.9 m) Steward reflector from its site at Tucson, Arizona to a less polluted site. At about the same time John B. Irwin, professor of astronomy at Indiana University, suggested(4) that a telescope dedicated to photoelectric photometry should be installed in the desert, and be available to astronomers from the eastern states, where the weather was poor. As a result, a meeting was held on April 23 1952 between representatives of the universities of Arizona, Ohio State and Indiana, and the NSF to discuss the possibility of the NSF funding a site survey for such a photoelectric telescope. This resulted in a formal proposal being made for a survey, followed, if successful, by the permanent move of the 36 inch Steward reflector to the selected

[2] Report of Committee: Draft of Proposal for the Support of Astronomy by the National Science Foundation, *Astronomical Journal*, 56, October 1951, pp. 147–148.

[3] This annual budget approval system still causes problems with major programmes in the United States today, especially if the President is not of the same political party as the majority party in both houses of Congress.

site. However, the proposal was formally rejected by the NSF in September 1952 as it was too expensive. Nevertheless, Raymond J. Seeger, programme director at the NSF, saw merit in the general idea, and this resulted in a conference held at the Lowell Observatory at Flagstaff, Arizona in the autumn of 1953 to discuss possible new facilities for photoelectric photometry.[4]

Over thirty astronomers attended the Flagstaff conference. But it was hijacked by Leo Goldberg of the University of Michigan who, in the closing session, raised the subject of a national observatory. He suggested that 'What this country needs is a truly National Observatory to which every astronomer with ability and a first class problem can come on leave from his university'.[5] In particular, he proposed that this national optical observatory should be funded by the NSF, and should have both a 100 to 120 inch (2.5 to 3.0 m) telescope, and a 36 inch (0.9 m) telescope. In addition, he suggested that the observatory, which should carry out all types of observing including photometric, spectroscopic, photographic and solar work, should be located in the south-western United States, where the weather cycle was out of phase with that of California.

Goldberg's suggestion came at an opportune time as, by now, the NSF was eager to develop large scale facilities in a number of scientific areas. So it appointed an advisory panel under the chairmanship of Robert R. McMath, the president of the AAS, to examine the need, location and possible organisation of Goldberg's suggested National Astronomical Observatory (NAO).(5)

McMath was an industrialist and civil engineer who had taken up astronomy as a hobby some years previously. By 1953 he was director of the University of Michigan's observatory and was a highly respected solar observer. He also moved in the right political circles, helping to achieve a substantial increase in the NSF budget from $4.75 million for FY1953 to $8.0 million in the following year. So he was the ideal choice to chair the NSF advisory panel.

This panel, chaired by McMath, endorsed the idea of a National Astronomical Observatory at its meeting in November 1954, and recommended that Aden B. Meinel, of the Yerkes Observatory,[6] should undertake a search for suitable sites for both stellar and solar work. They further recommended that a 30 inch (0.8 m) telescope should be installed as soon as possible, followed by an 80 inch (2.0 m) telescope, but a decision on a very large telescope should be deferred. Later Meinel suggested that the 30 inch should be increased to a 36 inch (0.9 m), as there was an

[4] Astronomical Photoelectric Conference Sponsored by the National Science Foundation and Lowell Observatory at Flagstaff, Arizona, August 31–September 1, 1953, *Astronomical Journal*, 60, January 1955, pp. 17–32.

[5] *Proceedings of the National Science Foundation Astronomical Photoelectric Conference*, published October 1955, pp. 107–108.

[6] Meinel had previously helped to design what would become the 1,000 ft (300 m) diameter Arecibo radio telescope. He had also tried, unsuccessfully, to interest astronomers in building a 500 inch (~12.5 m) diameter optical telescope (see Section 3.1).

unused 36 inch mirror blank at Yerkes, and a 36 inch mounting had recently been designed for the McDonald Observatory.

Meinel, assisted by Helmut Abt, started his survey of possible sites in the southwestern United States in early 1955.(6) Aerial reconnaissance produced about 150 possible sites, which were reduced to five by mid 1956 following ground-based surveys. Initial site testing was then undertaken. This resulted in the choice being narrowed down to Kitt Peak and Hualapai Mountain, both in Arizona, with most of the selection criteria favouring Kitt Peak. As a result, on March 1 1958 the Scientific Committee overseeing the site surveys unanimously voted for Kitt Peak at the north end of the Quinlan Mountains as the site for the National Astronomical Observatory.(7) Unfortunately, things were complicated by the fact that the 6,875 ft (2,100 m) high Kitt Peak was on a Papago Indian Reservation. But permission was eventually obtained from the Tribal Council to use the site, provided the land was only used for astronomical and related research. A lease was then signed in October 1958 granting 200 acres of the mountain top for scientific research, together with an additional 2,200 acres to be used as a buffer zone.

In the meantime institutional plans had been developed for the observatory whilst the site surveys were under way. By early 1957 the number of universities interested in working together at the NAO had stabilised at seven, namely California, Chicago, Harvard, Indiana, Michigan, Ohio State and Wisconsin. As a result, on October 28 1957 these seven universities became founder members of the 'Association of Universities for Research in Astronomy' or 'AURA' as it became known. Yale became the eighth member shortly afterwards. The NSF agreed to provide funds to AURA for observatory buildings and capital equipment. But, as was the case with all other NSF funding, AURA funding depended on an annual vote by Congress.

Aden Meinel had not been universally popular with McMath's panel whilst he had undertaken the site surveys, because he often operated outside his remit. There were also a number of criticisms that he had not carried out the site surveys in as methodical a manner as expected.[7] Nevertheless, in December 1957 the AURA Board appointed him the first Observatory Director. A year later the observatory was named the Kitt Peak National Observatory, which was abbreviated to KPNO.

In the meantime, the first artificial satellite, Sputnik I, had been launched by the Soviet Union on October 4 1957, followed by the launch only one month later by that of the much heavier Sputnik II. These launches created a degree of panic in the United States because of their military significance, which resulted in Congress agreeing on a rapid increase in funds for both military and civilian projects. For

[7] The problems that members of McMath's panel had with Meinel are outlined by Frank K. Edmondson in his *AURA and its US National Observatories*, Cambridge University Press, 1997. However, Edmondson was intimately involved with these matters at the time and it is unclear how objective is his version of events.

example, the NSF budget for FY1959 was increased almost overnight by about $30 million. This allowed preliminary work to start on a large solar telescope that had been previously proposed by McMath, but which had been deferred due to lack of funds.

Building an astronomical observatory on a virgin mountaintop is very challenging. In the case of Kitt Peak one rough track had been built up the mountain during site testing, but this was only really suitable for caterpillar vehicles, as it was extremely steep. So a second track had to be built for four-wheel drive vehicles during the observatory construction phase. Then in 1963 a twelve mile public road, State Route 386, was constructed at a cost of $2.5 million.

There was also a problem of providing an adequate water supply at the top of the mountain, which had a minimum annual rainfall of only 8 inches (20 cm).[8] The solution adopted was to build a three acre (1.2 hectare) concrete-lined, dish-shaped depression as a catchment area, and two 500,000 gallon (2 million litre) water storage tanks.

2.3 The National Observatory Telescopes on Kitt Peak

The first medium-sized telescope to be erected at the KPNO was a 36 inch (0.9 m) instrument built at a cost of about $230,000, including its observatory building. This 'Number One' 36 inch was completed on March 12 1960, just three days before the dedication of the observatory. Its Kanigen-coated,[9] aluminium primary mirror was replaced in 1963, and the original mirror was used in a second 36 inch telescope that was completed in 1966. The mirror of the first instrument was again replaced in both 1966 and 1970. In the latter case, the new mirror was made of Cer-Vit,[10] a new material of extremely low coefficient of expansion.

The 84 inch (2.1 m) Telescope

The next telescope to be built was an 84 inch (2.1 m) reflector. Aden Meinel's first design had been rejected by AURA as it was thought to have too many new features. But Meinel was still able to incorporate some non-standard concepts in the final telescope design. For example, most reflectors at that time used a standard Cassegrain arrangement with a parabolic mirror, whereas Meinel adopted a Ritchey – Chrétien design using a primary ground to a fourth order curve. This had

[8] This was the minimum annual rainfall recorded since 1930. The average rainfall was about 14 inches (35 cm) per year.

[9] Kanigen is a nickel-phosphorus alloy.

[10] *Sky and Telescope*, October 1967, p. 231.

the advantage of producing less off-axis distortion, and so made it more suitable for wider field photography.

AURA purchased an 84 inch Pyrex blank for $114,000 from Corning in 1958. At the time it was thought that the outer 2 inches (5 cm) may well be unusable after grinding and polishing, because of the mirror's large focal ratio (of f/2.6) and the complex shape of its surface.[11] But when it was completed it was found that the full 84 inches could be used. The mirror was ribbed and honeycombed on the back to reduce weight whilst retaining rigidity. It was produced by a relatively new process called 'sagging', which was claimed to reduce the number of bubbles in the glass. (8) Traditionally large glass blanks had been produced by ladling molten glass into the mirror mould, but in this new system nine chunks of glass were positioned in the mould and heated until molten.

During the last stages of grinding the mirror, a small piece of the grinding tool came loose, and created a series of 1/8 inch (3 mm) deep gouges in the mirror surface. It also chipped the edge of the mirror's central hole. This not only required extensive re-grinding of the mirror, but the central hole had to be increased in size slightly to eliminate the chipped edge. Eventually, however, the telescope was ready for use by visiting astronomers[12] in September 1964.(9) Appropriately, the first to use this 84 inch telescope was George van Biesbroeck of the University of Arizona on his 84th birthday. The telescope, which cost about $2.4 million(10), including its building, was completed within budget.

McMath-Pierce Solar Telescope

The NSF had insisted in 1958 that a condition of its funding for a large solar telescope was that it must be located on the same site as the National Astronomical Observatory, to minimise construction and operating costs. Outline drawings for this solar telescope had already been produced in 1957 by W. Zabriskie showing three possible structures. These and other configurations were examined in more detail by the Chicago firm of Skidmore, Owings and Merrill to enable a final design to be chosen.

In outline, the selected design consisted of an 80 inch (2.0 m) diameter optical flat operating as a heliostat mirror that followed the Sun. This mirror then reflected the Sun's light onto an image-forming, parabolic mirror 60 inches (1.5 m) in diameter some 500 ft (150 m) away along the polar axis of the telescope. The light was then reflected back to a 48 inch (1.2 m) diameter flat mirror, which deflected

[11] Kitt Peak's 80-inch Stellar Telescope, *Sky and Telescope*, January 1962, pp. 4–9.

[12] Unusually, its availability to visiting astronomers was advertised in the astronomical press (see, for example, N. U. Mayall, *Astronomical Journal*, 69, 1964, p. 527).

Figure 2.1. The McMath-Pierce Solar Telescope on Kitt Peak. The heliostat which is on the top of the vertical tower on the left reflected sunlight down the tunnel angled from the top left to the bottom right. (Courtesy NoobX via Wikimedia Commons.)

the light into an observing room, where the 34 inch (84 cm) diameter solar image could be viewed.(11)

The main problem in designing a solar telescope is caused by heat – both reflected off the ground and trapped within the telescope structure itself. In the first case, the Sun's light is distorted before it reaches the heliostat, and in the second it is distorted after leaving the heliostat and before it is imaged. To solve the first problem the heliostat was mounted 100 ft (30 m) above the ground (see Figure 2.1), whilst the last 300 ft (90 m) of the 500 ft light path between the heliostat and the 60 inch mirror was located underground in a tunnel, helping to solve the second problem.

The main 80 inch heliostat was mounted on a 27-ton base frame which, in turn, was mounted on a massive 26 ft (8 m) diameter concrete cylinder 100 ft (30 m) high, with steel-reinforced walls some 4 ft (1.2 m) thick. A water-cooled structure was built around this pier to keep it cool, along with another water-cooled structure surrounding the first 200 ft (60 m) of the light path from the heliostat. Interestingly, in the winter the tunnel below ground was warmer than the ambient air, and so, to avoid convection at that time of year, the tunnel was lined with pipes carrying refrigerant to cool it to the outside ambient temperature.

The 60 and 48 inch mirrors used in the initial construction of this solar telescope had had an interesting history. They had been produced as 60 inch quartz blanks by General Electric (GE) in 1931 as part of their contract to produce the 200 inch main Palomar mirror. However, they had lain unused since then at the GE plant after the GE Palomar contract had been cancelled. When the blanks had been made, George Ellery Hale had suggested that they may be useful for solar work, as quartz has a very low coefficient of expansion. Then in about 1950 GE notified Ira S. Bowen at Palomar Observatory that they wanted to dispose of them. Bowen, knowing that McMath was interested in building a large solar telescope, agreed to give them to him free of charge, provided McMath paid for their transportation, which he did. One of these 60 inch mirrors was then used in the Kitt Peak solar telescope until it was replaced in 1966. Another of the 60 inch mirrors was cut down to 48 inch diameter and used in this solar telescope until 1965, when it was replaced by a beryllium mirror. The solar telescope, which was named the McMath solar telescope at its dedication in November 1962, cost about $4 million to build. Thirty years later it was rededicated as the McMath-Pierce Solar Telescope in honour of Keith Pierce who had been responsible for its development. At about this time the McMath-Pierce was the only large solar telescope that could operate well beyond 2.5 microns in the infrared.

Kitt Peak Vacuum Telescope

This McMath-Pierce Solar Telescope was followed in the early 1970s by the 60 cm Kitt Peak Vacuum Telescope (KPVT). In this telescope, the problems of solar heating within the instrument were reduced by having the main optical path from the heliostat within a long vacuum tube. This KPVT was specifically built to provide ground-based support for the Skylab missions by generating magnetograms on an almost continuous basis during daylight hours. After the last Skylab mission in 1974, it was then used to study coronal holes which are linked to magnetic storms on Earth. This interested both NASA and the National Oceanic and Atmospheric Administration (NOAA) who joined the NSF to fund the telescope's operation.

Mayall 158 inch (4.0 m) Telescope

It had been envisaged since the mid 1950s that the 80 or 84 inch astronomical reflector[13] would be followed by a much larger reflector. By 1960 the 36 inch had been finished, and work was well underway in building both the 84 inch and the

[13] The estimated usable size changed over time, as explained earlier.

large solar telescope. So McMath suggested in early 1961 that AURA should start planning for an even larger reflector, which he suggested should have a mirror of about 150 inch (3.8 m) diameter. This was accepted in principle by AURA, and preliminary work was undertaken later that year, so that by January 1962 there was already an outline design available. The mirror was to have a focal ratio of about f/2.8, and the optical system was to provide a minimum distortion image field of about 45' to 1° diameter at its prime focus. It was also suggested that the telescope should be about 150 ft (45 m) above ground level to improve the seeing.[14] Total costs were estimated at about $10 million using a fused quartz primary mirror, or $9 million using Pyrex. Although fused quartz was more expensive, it has a lower coefficient of expansion than Pyrex, and so it was the preferred material for the large reflector.[15]

There were two bidders for the fused quartz primary mirror blank in early 1964, namely Corning and General Electric. Initially, the Corning price was much higher than that of GE, but Corning had apparently obtained a copy of the GE bid.(12) Consequently they were able to undercut the GE price just before the bid deadline. After some discussion, both AURA and the NSF decided that the last Corning bid should be disallowed, and decided to give the contract to GE at their original bid price of about $1.15 million. Their 158 inch (4.0 m) fused quartz mirror blank was delivered to Tucson, Arizona about three years later in October 1967, where its surface was ground and polished to the required shape in Kitt Peak's optical shop.

The 158 inch (4.0 m) telescope had a horseshoe mount similar to that of the 200 inch on Palomar Mountain. It was fixed on top of a 92 ft (28 m) high hollow, steel-reinforced, concrete cylinder, 37 ft (11 m) in diameter, completely isolated from the surrounding observatory building to reduce telescope vibration. The 15 ton, 2 ft (60 cm) thick mirror was supported by computer-controlled, pneumatic pads to maintain the mirror's shape as it moved to cover the sky. Its minimum distortion field of view of about 50 arcmin, which was produced at the prime focus with the aid of a triplet corrector lens was about six times that obtainable with the 200 inch (5.1 m) Hale reflector on Palomar Mountain.[16]

The concrete cylinder, observatory structure and dome were completed by the end of 1970. First light was achieved on February 27 1973, and the telescope (see Figure 2.2) was formally dedicated a few months later. It was officially named after Nicholas U. Mayall who had been director of KPNO from 1960 until he retired in 1971. When completed it was the second largest optical telescope in the United States (behind the 200 inch Hale reflector).

[14] Air turbulence from the ground was a major cause of poor seeing or image quality, even at night.

[15] Cer-Vit was not available at that time.

[16] First 4-meter Photographs from Kitt Peak, *Sky and Telescope*, July 1973, pp. 10–13.

Figure 2.2. The 158 inch (4.0 m) Mayall Telescope (centre) and the 90 inch (2.3 m) Bok Telescope (bottom left) on Kitt Peak. (Courtesy Montebest via Wikimedia Commons.)

Remote Control Telescope

One telescope that is often forgotten when discussing the Kitt Peak National Observatory is the 50 inch (1.3 m), which was designed in the early 1960s as a prototype for remotely controlled telescopes both on the ground and in space.[17] At that time computers were very large and rudimentary machines compared with those of today, and the idea of controlling a mountain-based telescope from some distance away using commercial telephone lines was very challenging, to say the least. This Remote Control Telescope[18] (RCT)(13), as it was called, had an aluminium primary mirror, which provided a considerable mass saving compared with a glass-based alternative. When first installed in the summer of 1965 at Tucson, the mirror worked perfectly, but during the subsequent winter on Kitt Peak its performance deteriorated badly.

Investigations showed that the main problem was caused by the outer parts of the mirror, as masking it down from 50 inches to 38 inches yielded a dramatic

[17] At the time KPNO were hoping to be involved in designing a 50 inch space telescope.
[18] Sometimes called the Remotely Controlled Telescope, with the same acronym.

improvement in image quality. Before figuring, the aluminium mirror had been coated on its front and back surfaces by an equal thickness of Kanigen. This Kanigen had been used as it provides a smooth surface suitable for optical finishing. After polishing, however, the Kanigen on the front surface, and particularly near the edge, was thinner than on the back. It was the differential expansion of these Kanigen layers and of the aluminium which was causing the problem.[19]

In 1969 the RCT was changed from remote to manual operation and, in the following year, its metal mirror was replaced with one made of Cer-Vit. The telescope has been used since then as both a test facility and an operational telescope. In the first case, it has been used to try out new instrumentation before installation on the KPNO 84 inch and 158 inch telescopes. As an operational telescope, it has been used at both optical and infrared wavelengths.(14)

Restructuring

KPNO had been the subject of some criticism towards the end of the 1960s – as the main building phase was coming to an end – of having very good facilities that produced relatively moderate results. The problem was not helped by the American astronomical fraternity having differing views as to the function of this national observatory. Should it just provide facilities for its member universities to use, or should it also have a quality research staff of its own? AURA gave the first indication of its intensions in resolving these issues when it decided in 1971 to appoint Leo Goldberg, a first-rate astronomer, as director when Mayall retired later that year. A special AURA Visiting Committee, created with Goldberg's blessing, recommended that he recruit a team of high-quality astronomers to not only undertake high-quality research but also develop first-class instrumentation.(15)

Consequently, with the approval of the AURA Organization Committee, Goldberg changed the administrative structure of KPNO to have a single scientific staff, making it much more like a university academic department. He formed a number of scientific programme groups each headed by a programme director, and expanded these groups by recruiting some promising young astronomers. This enabled KPNO to carry out quality research of its own, as well as providing facilities for its member universities. Goldberg also decided to put more emphasis on infrared astronomy, and reduced planetary research, whilst the relatively modest-sounding rocket programme was closed in 1973 to save money. So at that time Goldberg had both Kitt Peak's new 158 inch telescope, and a more scientifically oriented staff to use it effectively. Even so, this large national telescope was still

[19] American Astronomers Report, *Sky and Telescope*, October 1967, p. 213.

significantly smaller than the Palomar 200 inch, operated by Caltech and Carnegie as a private facility, which was out-of-bounds to most American astronomers.

2.4 Other Telescopes on Kitt Peak

Completion of the 158 inch in 1973 saw the end of the main construction phase of the Kitt Peak National Observatory. But this was not the end of construction of optical telescopes on the mountain, as those of the KPNO national observatory were joined by those of a number of other American institutions.

Steward Observatory's 36 inch (0.9 m) and the Spacewatch Project

The first of these telescopes was the 36 inch (0.9 m) Steward Observatory reflector of the University of Arizona, which had originally been located on the university's Tucson campus in 1921. Whilst at Tucson it had been used in 1955 to show the Moon and planets to representatives of the Papago Tribal Council, to help obtain permission to construct the National Astronomical Observatory on Kitt Peak. But light and atmospheric pollution at Tucson had reduced its usefulness there and so in 1962 it was moved to Kitt Peak and rebuilt on land leased from the National Science Foundation.

Twenty years later this 36 inch telescope had fallen into disuse. As a result, the Steward Observatory granted the Spacewatch Project exclusive access to it on condition that the project was responsible for the telescope's refurbishment and maintenance. Spacewatch then used it to perfect a sky-scanning system using CCDs to search for near-Earth asteroids and comets, whilst in 2002 the project installed a new primary mirror and CCD mosaic.

In the meantime, the Spacewatch Project had designed and built a 72 inch (1.8 m) telescope in-house to extend the work of the 36 inch. This 72 inch was initially envisaged as being a normal prime-focus f/2.7 reflector, but to save money the project decided to design a folding prime focus system with a flat secondary mirror to keep the telescope as short as possible. It saw first light at the Steward Observatory on Kitt Peak in 2000.

Steward Observatory's 90 inch (2.3 m) Bok Telescope

Returning to the Steward Observatory situation in the 1960s, the NSF provided a grant of $1.3 million to the University of Arizona in 1965 to enable them to build a 90 inch (2.3 m) reflector on Kitt Peak.(16) This was to pay for the telescope, mounting, drive and optics, whilst the State of Arizona agreed to pay $0.7 million for the building and dome. General Electric were chosen to supply the three-ton mirror blank, which was of fused quartz, whilst the blank's surface was figured at the University of Arizona's optical shop in Tucson. It was aluminised at the Lick Observatory.

The design of this Steward Observatory's 90 inch followed that of the National Observatory's 84 inch in using a Ritchey-Chrétien configuration because of its freedom from coma. The 90 inch primary mirror had a focal ratio of f/2.7, which was also almost exactly the same as that of the 84 inch. The disc of glass cut out from the centre of the primary was fashioned on both sides to create hyperbolic secondary surfaces. These provided an f/9 Cassegrain focus using one side of the mirror and an f/31 coudé using the other.[20] So a change from Cassegrain to coudé only required the secondary mirror to be flipped by 180°.

The start of this 90 inch telescope's construction slightly pre-dated that of the national observatory's 158 inch. Both telescopes were mounted some distance above the ground on high concrete piers to reduce air turbulence. The 158 inch's hollow cylindrical pier was made of steel-reinforced concrete formed by gradually raising a circular mould as the concrete set at the bottom. On the other hand, the Steward Observatory's 90 inch was a much smaller and lighter instrument. As a result, it was mounted at the top of a much smaller 37 ft high concrete pier 15 ft (5 m) wide simply made of two leaves of concrete blocks in-filled with concrete.

The support system for the 90 inch mirror was unusual. It was provided at the back of the mirror by air bags in which the pressure was controlled, and around the side by a mercury-filled tube. The telescope housing was also somewhat unusual as, instead of the normal dome, it was made of a revolving cylinder (see Figure 2.2) with shutters that moved horizontally onto the shoulders of the cylinder. This provided an easier method of construction.

The 90 inch was officially dedicated on June 23 1969, just a few years after funding had been approved. At first light, it was the fourth largest optical telescope in the United States. Some years later, on April 28 1996 it was officially named the Bok telescope in honour of Bart Bok who had been director of the Steward Observatory from 1966 to 1969.

In 1978 Hubbard, Angel and Gresham used the 90 inch to carry out the first experiments in astronomical spectroscopy using a fibre-fed spectrograph.(17) They used an optical fibre placed at the telescope's prime focus to feed an image tube spectrograph. This experiment was followed by the construction by John Hill and colleagues of the first fibre-fed instrument for multi-object spectroscopy, called Medusa, that was also fitted to the 90 inch. It used optical fibres glued to an aperture plate which fed light to a linear slit from where it entered a normal spectrograph. This system was used in 1980 to produce 26 simultaneous galaxy spectra.

[20] Steward 90-inch Telescope Dedicated, *Sky and Telescope*, September 1969, pp. 164–165.

Five years later the aperture plate was replaced by computer-controlled robotic arms to position the optical fibres.

University of Michigan's 52 inch (1.3 m) and the McGraw-Hill Observatory

In addition to the NSF grant to the Steward Observatory in the mid 1960s, the NSF had also agreed in 1967 to partially fund a new 52 inch (1.3 m) reflector of the University of Michigan.(18) It was built at the university's Portage Lake Observatory, only 16 miles from their campus at Ann Arbour, to enable easy access for both undergraduates and graduate students, who were to be its major users. Nearby was a 24 inch (60 cm) optical reflector plus two radio telescopes also belonging to the university.

This new 52 inch, which was constructed over the period 1968–1969 by Tinsley Laboratories, used a Cer-Vit primary blank provided by Owens-Illinois. The finished telescope had an f/2.6 primary mirror with an f/13.5 Cassegrain and an f/33.5 coudé system. An f/7.6 Cassegrain option was also available. The bolt configuration for the attachment of its accessories was made identical to that of the Kitt Peak 84 inch and Cerro Tololo 60 inch. This enabled the accessories developed on this Michigan 52 inch to be fitted to these two other telescopes.

Unfortunately the location at Portage Lake proved unsuitable for the 52 inch, as not only were the skies not clear enough, there was also a growing problem of light pollution. So the 52 inch was moved to Kitt Peak in 1975 and placed in a new purpose-built observatory, called the McGraw-Hill Observatory, with finance provided by McGraw-Hill and the Alfred P. Sloan Foundation.(19) Operation of the observatory was undertaken by the MDM consortium which consisted of the University of Michigan, Dartmouth College and MIT, all of whom provided instrumentation.[21] These institutions had, for a number of years, been collaborating on X-ray research, and this new telescope would be used to study X-ray sources in the visible waveband. In the event, the first observations with the 52 inch took place on May 4 1975, just four days before the launch of the X-ray satellite, SAS-3.

So there was by mid 1975 a significant number of optical telescopes on Kitt Peak. In particular, the Kitt Peak National Observatory had its 158 inch (4.0 m) Mayall, its 84 inch (2.1 m), and a number of smaller reflectors. There was also the McMath solar telescope, plus a vacuum solar telescope. Two other observatories had also been set up on land leased from the National Science Foundation, namely the Steward Observatory with its 90 inch (2.3 m) and 36 inch (0.9) reflectors, and the McGraw-Hill Observatory with its 52 inch (1.3 m) reflector.

[21] The 1.3 m telescope is now called the McGraw-Hill telescope, operated by the MDM observatory.

Move of the Burrell Schmidt

Four years later another institution moved an important telescope to Kitt Peak when Case Western Reserve University (CWRU) moved their highly successful Burrell Schmidt to the mountain. This telescope(20) that had a 36 inch (90 cm) primary mirror and a 24 inch (60 cm) corrector plate had been located at CWRU's Warner and Swasey Observatory at East Cleveland, Ohio in 1941. But sixteen years later light pollution had caused it to be moved 30 miles away to Montville, Ohio. Now it had to be moved again because of light pollution. Under the terms of this latest move, CWRU and KPNO shared operating expenses equally and, consequently, the telescope's use was shared equally also. Before this latest move, this Burrell Schmidt's primary mirror was refigured and a new corrector plate was installed to provide much sharper images over wavelengths up to 1.1 microns in the near infrared.

Hiltner 2.4 m Telescope

In 1982 the members of the MDM consortium decided to extend their collaboration by building a 2.4 m (95 inch) reflector on Kitt Peak.[22] The design was based on that of the 2.3 m (92 inch) University of Wyoming instrument that had been installed a few years earlier on Jelm Mountain, Wyoming. The f/2.07 primary Cer-Vit mirror of the MDM instrument was cast by Owens-Illinois as a 50 cm thick blank that was cut in half horizontally – one half went to the MDM Observatory, and the other was used in the Australian Advanced Technology Telescope at Siding Spring, Australia. The MDM telescope, which was dedicated in April 1989, was named the Hiltner telescope after William A. Hiltner, chairman and emeritus professor of Michigan's astronomy department for his role in conceiving the instrument and establishing the MDM consortium.[23]

This 2.4 m Hiltner telescope was of a very compact design, which allowed its 11 m diameter dome to be just over a metre larger than that for the 1.3 m McGraw-Hill telescope nearby. Both telescopes were situated on the southwest ridge of the mountain about 3 km away from, and 150 m below the main peak where most of the other telescopes were situated. The Hiltner telescope cost only about $2.0 million to build, and the moving mass of the telescope and mounting weighed only about 30 tons, which was some 35% less than that of the Wyoming instrument that had virtually the same size mirror.[24]

[22] New 2.4 Meter Telescope on Kitt Peak, *Sky and Telescope*, November 1982, pp. 410–411.

[23] *Sky and Telescope*, July 1989, p. 8.

[24] The MDM observatory is now owned and operated by the University of Michigan, Dartmouth College, Ohio State University, Columbia University and Ohio University.

2.5 National Observatory Funding Problems

AURA announced in 1983 that it had decided to consolidate its three observatories consisting of KPNO, the Cerro Tololo Inter-American Observatory in Chile (see Section 2.6), and the National Solar Observatory (NSO, see Section 15.1), as the National Optical Astronomy Observatory (NOAO). John T. Jefferies was appointed its first director, following his successful management of the University of Hawai'i's Institute for Astronomy. The NOAO was to have its headquarters at Tucson, Arizona, whilst AURA moved its corporate offices from Tucson to Washington, D.C., to be close to the NSF, NASA and other scientific institutions. In the following year Sidney Wolff, a previous colleague of Jefferies at the University of Hawai'i, was appointed director of KPNO, and in 1987 she was appointed to succeed him as the director of the NOAO.

NSF funding of NOAO was progressively reduced in the 1980s so that by 1988 KPNO, for example, was operating on only half the financial budget in real terms of ten years previously.(21) As a result, AURA was faced with the prospect of closing down or selling some NOAO facilities. To assist them in making their decision, Sidney Wolff and Goetz Oertel, the president of AURA, wrote to about 2,000 astronomers asking for their views. Three basic options were considered[25]: Option 1, emphasise constructing new telescopes and advanced instrumentation and close down or sell all but the most important telescopes at KPNO, Cerro Tololo and the NSO. Option 2, keep open as many of the existing telescopes as possible, so serving the maximum number of astronomers, and largely abandon the development of new facilities. Option 3, concentrate financial resources on those facilities that were unique as far as the American astronomical community was concerned. These latter facilities were mostly, but not solely, at Cerro Tololo (see next section) which had access to the far southern sky. This option also involved a substantial reduction in the development of new facilities.

In response to Wolff's and Oertel's letter the American astronomical community overwhelmingly opted for keeping open existing telescopes, at the expense of the development of new facilities.[26] As a result, AURA closed their Advanced Development Program, and stopped supporting Roger Angels' spinning oven laboratory in Tucson (see Section 4.1). They also eliminated financial support for an optical interferometer, and agreed to close the Number One 36 inch (0.9 m) telescope at KPNO (see Section 2.3). Maintenance would also be deferred and staff leaving for whatever reason would not be replaced.

Funding of pure research is always a thorny issue for any government, as its financial demands look like a bottomless pit with little in the way of financial

[25] Astronomy's Financial Crisis, *Sky and Telescope*, June 1988, p. 583.

[26] Budget Ax Falls, *Sky and Telescope*, September 1988, p. 239.

returns, as least in the short-to-medium term. Unfortunately, in the late 1980s the American government was also running a very large budget deficit, so there was a drive to restrict spending to the absolute minimum. There was also a perceived weakness in applied research in the United States at the time. As a result, the government decided to put more money into applied research at the expense of pure research, compounding the funding problem at the NOAO and in other areas of pure research.

NOAO's financial problems could be eased by undertaking more joint programmes with universities and by leasing or selling off some existing telescopes. These solutions were not ideal, as they restricted access of the general American astronomical community to these telescopes.

WIYN 3.5 m Telescope

The first significant example of a public-private partnership at Kitt Peak was begun in the late 1980s. This was the 3.5 m (138 inch) WIYN telescope of the universities of Wisconsin, Indiana and Yale[27], and NOAO.[28] NOAO's contribution included mirror assembly, provision of the site and a major share of the operating costs, whilst the universities paid for most of the capital cost items. As a result, the observing time on the telescope was split about 40%, 26%, 17% and 17% for NOAO, Wisconsin, Indiana and Yale, respectively.[29] NOAO considered the WIYN telescope as a technology development project to assist them in the design of the Gemini telescopes (see Section 3.3).

The Steward Observatory mirror laboratory, led by Roger Angel, had been developing a new technique for producing large telescope mirrors relatively cheaply using borosilicate glass in a spinning oven (see Section 4.1). The laboratory was gradually increasing the size of its mirrors, with the first of their 3.5 m mirrors going to the Astrophysical Research Consortium's (ARC's) telescope being built at Apache Point, New Mexico. The second 3.5 m mirror, which was spun in December 1988, was produced for the WIYN telescope.

The WIYN telescope was one of a new type of lightweight, compact, medium-sized instruments of the late 1980s and early 1990s. It had an altazimuth mount, instead of the usual equatorial, with a design based on that of the ARC mount. This made the telescope mounting system far simpler and compact, at the expense of field rotation of the image, which could be compensated by computer control.

[27] Yale joined the consortium a little after the other two universities.

[28] NOAO announced a similar joint venture with Columbia University and the University of North Carolina a little later. This was to build the 4 m SOAR (Southern Astrophysical Research) telescope, see Section 3.4.

[29] WIYN: A New Name atop Kitt Peak, *Sky and Telescope*, December 1990, pp. 586–587.

A key feature of the WIYN telescope was its highly reflective octagonal 'dome', which was designed to minimise solar heating during the day and optimise air flow over the telescope at night, to minimise atmospheric image distortions. In addition, it used active optics to control distortions of the honeycomb primary mirror, as the angle of the mirror changed relative to the local vertical and as the wind velocity varied. There was also a highly sophisticated thermal control system for the primary mirror which allowed individual parts to be cooled by passing thermally conditioned air through parts of its honeycomb substrate. This ensured that the temperature of the mirror surface was within 0.2°C of the ambient 'dome' air temperature. The WIYN telescope was formally dedicated on October 15 1994.

A comparison between the design and performance of the two largest instruments on Kitt Peak is highly instructive. The 4.0 Mayall, which achieved first light some 20 years before the 3.5 m WIYN, had a 15-ton mirror and a 375-ton moveable mass, compared with the WIYN's 2 and 46 tons, respectively. In addition, the top of the Mayall's dome was over three times the height of that of the WIYN, and the cost of the Mayall was over double that of the $14 million WIYN, in real terms. And yet the WIYN outperformed the Mayall by a large margin, with a wavefront error only about one-tenth that of the earlier telescope.(22)

SARA 0.9 m Telescope

As mentioned above, Sidney Wolff announced in 1988 that the Number One 36 inch would be decommissioned due to budget constraints. It would be given to that institution, on a permanent loan basis, that could best use it. The only provision was that the telescope must be moved from its site, as this was earmarked for the 3.5 m WIYN telescope. The winning proposal came from a new multi-university consortium called the Southeastern Association for Research in Astronomy (SARA) which had been specifically formed to produce the proposal. Consequently, the 36 inch was dismantled in 1990 and the original housing demolished. Construction of its new housing, designed for SARA by Stanley Black of Boston, was started in 1992 on a new site at Mercedes Point on Kitt Peak, near to the Burrell Schmidt of Case Western Reserve University (see Section 2.4). Reassembly of the telescope was undertaken the following year. In 2001 the WIYN consortium also led a group of universities to upgrade and operate the NOAO Number Two 0.9 m instrument.

Turn of the Century and Beyond

In the meantime, the 1.3 m RCT had been closed in 1995 as a KPNO facility, to save money. It has now been refurbished, automated and renamed the Robotically

Controlled Telescope (with the same acronym RCT) operated by the RCT consortium[30]. It began scientific observations again in 2004.

A number of medium and small telescopes have been located on Kitt Peak in recent years including the 1.2 m Calypso Telescope with a roll-off enclosure at the Edgar O. Smith Observatory. Design work on this optical telescope was begun in 1991 by a group of engineers who had worked on the Hubble Space Telescope. To produce the highest spatial resolution possible, they incorporated one of the few adaptive optics systems then operating in the optical waveband. The telescope was mainly used to study variable stars in globular clusters. In 2008 the Large Synoptic Survey Telescope (LSST) project purchased the telescope from Edgar Smith. It was used to conduct scientific investigations in support of the larger LSST project.[31]

Meanwhile in 1996 the specialist 0.6 m Wisconsin H-Alpha Mapper (WHAM) was moved from the Pine Buff Observatory of the University of Wisconsin-Madison to Kitt Peak. This instrument, which was normally operated remotely from Wisconsin, was specifically designed to study gas in the Milky Way. It was removed from Kitt Peak in 2008 and moved to the Cerro Tololo Inter-American Observatory in the following year.

Finally, the 0.6 m diameter Super-LOTIS (Livermore Optical Transient Imaging System) telescope has also been recently located on Kitt Peak designed to study the optical counterparts of gamma-ray bursts. It is also a robotic instrument housed in a roll-off-roof observatory, in this case located at the Steward Observatory on the mountain.(23)

On September 22 2003, the Kitt Peak Vacuum Telescope (KPVT) was closed after thirty years of observing the Sun. A new suite of instruments called SOLIS (Synoptic Optical Long-term investigations of the Sun) was then installed to monitor the Sun's long-term variability (over decades) and clarify its role in climate change. Designed and built by the National Solar Observatory it was funded by the National Science Foundation.

When the Kitt Peak National Observatory had been founded in the late 1950s it had been hoped that it would serve as the top quality optical observatory for those American observers who, at that time, did not have access to top quality instruments at a first-rate observing site. In this it has been very successful. But over the last 50 years or so other optical observatories have been developed outside of the continental United States, particularly on Mauna Kea in Hawaii, that have been found to have better 'seeing' than Kitt Peak. So it is to these that the best American instruments of today are being generally directed. These will be covered later in

[30] The RCT consortium consists of the universities of Western Kentucky, South Carolina State and Villanova.

[31] The LSST is an 8.4 m diameter very wide field telescope which, at the time of writing (in 2015), is expected to be completed on Cerro Pachón, Chile by about 2019. It was designed to have a field of view of 9.6 square degrees with a limiting magnitude of 24.5, or 27 for stacked images.

this book, but before turning to them, we need to consider the other major national optical observatory of NOAO, namely that of Cerro Tololo in Chile.

In 2012 the NSF's Division of Astronomical Sciences Portfolio Review Committee recommended that the 4.0 m Mayall, the 3.5 m WIYN and the 2.1 m Kitt Peak telescopes should lose their national funding to enable money to be diverted to more modern facilities elsewhere. Subsequently the Department of Energy's Lawrence Berkeley National Laboratory agreed to fund a radical modification to the top end of the Mayall to enable it to undertake a dark energy survey.[32] In addition, NASA agreed to fund a new spectrometer for the WIYN telescope to enable it to participate in NASA's exoplanet survey programme, whilst a consortium of universities is currently (in 2015) planning to take over responsibility for the 2.1 m telescope.

For ease of reference Table 2.1 gives a summary history of the optical telescopes on Kitt Peak that I have discussed above.

2.6 Cerro Tololo and its National Observatory Telescopes

Professor Federico Rutllant, director of the National Observatory of the University of Chile,[34] visited the United States in 1958 to attend the annual meeting of the Astronomical Society of the Pacific, a meeting of the American Astronomical Society, and a consulting committee on science called by the Organisation of American States.(24) During his stay he visited a number of the most important observatories in the USA seeking help from them in improving astronomical facilities in Chile. He asked, in particular, if they were interested in collaborating with the University of Chile in building and operating a moderate-sized observatory in Chile. However, the only people to indicate real interest were Gerard Kuiper and William Hiltner of Chicago's Yerkes Observatory. He also raised the possibility with Kuiper of collaboration between AURA and Chile. But, although the idea was of interest to AURA, they were unable to support it at that time as they were in the process of setting up their own observatory on Kitt Peak.

The collaborative agreement between the universities of Texas and Chicago to run the McDonald Observatory was due to come to an end in 1962, and both Kuiper and Hiltner saw the benefits of an extension of their collaboration into the southern hemisphere. Hiltner contacted the United States Air Force Geophysics Research Directorate (GRD), with which he had contacts, to ask if they were interested in such a facility, primarily for planetary research. They responded positively, and

[32] Additional funding will be provided by the Gordon and Betty Moore Foundation and the Heising-Simons Foundation.

[34] Historia del Observatorio Astronómico Nacional (www.oan.cl/historia).

Table 2.1. *Optical Telescopes Located on Kitt Peak*

Telescope	Year Construction Completed on Kitt Peak[a]	Comments
(a) Telescopes that were originally National Observatory telescopes		
16 inch (0.4 m) No. 1	1957	Used for site survey 1957–1958, housed in a trailer. Telescope then overhauled and sent to Cerro Tololo for site survey 1961.
16 inch No. 2	1959	Located in first dome to be assembled on Kitt Peak. Telescope moved to Tucson 1962. Then refurbished and sent to Cerro Tololo 1965.
36 inch (0.9 m) No. 1	1960	First medium-sized AURA telescope on Kitt Peak. Thirty years later responsibility passed to SARA consortium who were responsible for disassembly of telescope and reassembly in new location on Kitt Peak in 1993.
McMath-Pierce Solar Telescope	1962	
16 inch No. 3	1962	
16 inch No. 4	1963	
84 inch (2.1 m)	1964	Shut down in July 2014. Expected to reopen in 2015 under the control of a new university consortium.
50 inch (1.3 m) RCT	1965	Closed in 1995. Operated by RCT consortium since 2004.
36 inch No. 2	1966	Operation taken over by WIYN-led consortium 2001.
Kitt Peak Vacuum (Solar) Telescope	1973	Closed in 2003 for major modifications. with SOLIS instruments.
158 inch (4.0 m) Mayall Telescope	1973	Still operated by AURA. Currently (2015) it is planned to replace the telescope's top end with the massive $70 million Dark Energy Spectroscopic Instrument (DESI).
(b) Other telescopes		
36 inch (0.9 m) Steward Observatory	1962	Located at Steward Observatory, Tucson, Arizona 1921. Moved to Steward Observatory on Kitt Peak 1962. Operation taken over by Spacewatch group 1983.

(*continued*)

Telescope	Year Construction Completed on Kitt Peak[a]	Comments
90 inch (2.3 m) Bok Telescope, Steward Observatory	1969	
52 inch (1.3 m) McGraw-Hill Telescope	1975	Located at University of Michigan's Portage Lake Observatory 1969. Moved to McGraw-Hill Observatory Kitt Peak 1975, operated by MDM consortium.
24/36 inch (60/90 cm) Burrell Schmidt	1979	Located at Case Western Reserve University's Warner and Swasey Observatory 1941. Moved to Montville, Ohio 1957. Moved to Kitt Peak 1979 as a joint CWRU and AURA facility.
2.4 m (95 inch) Hiltner Telescope	1989	MDM consortium telescope.
3.5 m (138 inch) WIYN	1994	Built for the WIYN consortium telescope, of which NOAO (AURA) was a member. It is currently planned to add a NASA-funded spectrometer to enable the telescope to be used as part of the NASA exoplanet survey programme.
0.6 m WHAM	1996	Moved from University of Wisconsin-Madison's Pine Buff Observatory to Kitt Peak 1996. Removed from Kitt Peak 2008 and moved to Cerro Tololo.
72 inch (1.8 m) Spacewatch	2000	
0.6 m Super-LOTIS	2000	Located at Steward Observatory on Kitt Peak.
1.2 m Calypso Telescope	2001	Initially owned by Edgar Smith, completed on Kitt Peak 2001. LSST purchased 2008.

[a] Defining a year of completion is not as obvious as it sounds, as usually first light takes place before all the main elements of the telescope have been calibrated and commissioned.

this eventually led to an agreement between the universities of Chicago and Texas to set up a joint observatory in Chile with the University of Chile, with funding provided by the GRD. Formal approval of Air Force funding was received in December 1959 for the provision of a 60 inch (1.5 m) reflector.

Kuiper, without waiting for the formal approval of the Air Force, had asked Jurgen Stock, who had recently undertaken site surveys in South Africa for the

European Southern Observatory, to start site testing in Chile in April 1959. Because of funding limitations, Stock began by examining four sites in the Santiago area. He then extended his survey in the following year to include four sites further north on the southern end of the Atacama desert, where observing conditions seemed to be better. In the meantime, Kuiper had tried once more to interest AURA in an observatory in Chile, as he felt that this project, because of its international aspects, needed a broader political base as well as additional funding.

The NSF and AURA showed more interest this time, and the AURA Scientific Committee studied the design of the proposed 60 inch reflector. As a result in March 1960 they suggested a number of changes. These included the reduction in the focal ratio of the primary mirror from f/5 to f/3, and a change in the design of the mount from a fork to an asymmetric design that allowed more room for a Cassegrain focus. Two weeks later vice-president Warren Johnson of the University of Chicago proposed that AURA be asked to take over the running of the project, although the universities of Chicago and Texas still wished to be involved. AURA agreed in principle to the proposed change, and a formal agreement to this effect was negotiated over the next few months. The University of Chicago was already a member of AURA, and University of Texas' interests were covered by making Texas a joint member with Chicago of the AURA Board of Directors. The Air Force provided a grant of $100,000 in FY 1962 for the design of the telescope mounting, purchase of the 60 inch mirror blank, and general planning, whilst the NSF provided $50,000 for preliminary work.

Initially it had been suggested that a 24 inch (0.6 m) telescope should be located at the new observatory prior to the planned 60 inch. But the planned 24 inch was changed to a 36 inch at AURA's meeting in March 1962. In the meantime, Stock had continued his site surveys, with AURA finally settling on two alternative sites in the lower Andes. These were Cerro Tololo at an altitude of about 2,200 m (7,200 ft) and La Peineta, about 300 km north of Cerro Tololo, at an altitude of 3,100 m (10,200 ft). La Peineta had more hours of clear sky than Cerro Tololo, but it was subject to higher winds. In addition, Cerro Tololo was on private land that could be purchased, whereas La Peineta was on government-owned land that was only available on a 20 year renewable lease. As a result, on November 23 1962 Cerro Tololo was chosen as the site for the AURA southern observatory, which became known as the Cerro Tololo Inter-American Observatory (CTIO). With its location near the southern end of the Atacama desert it was ideal for infrared observations. As part of the agreement up to 25% of observing time was made available to astronomers from Central and South America.(25)

AURA then purchased 120 square miles (310 sq. km.) of land including and surrounding the observatory site for just $13,000. This purchased area included four other mountains. AURA also purchased a plot of land on the Pacific coast at La

Serena to build the observatory's headquarters. This was about 250 miles (400 km) north of Santiago and about 30 miles (50 km), as the crow flies, from Cerro Tololo.

Stock was appointed the first director of CTIO in April 1963. A provisional 38 km long access road was built up the mountain in the second half of that year, and work was started on flattening the top of the mountain to 14 m below the original top in January 1964. Mayall had suggested that this flattening should be limited to providing just enough room for the 60 inch and for a possible 150 inch (3.8 m) later. Fortunately for future users of the site, Stock ignored this advice and produced a flat area 190 × 110 yards (210 × 120 m) in size, which was enough for the location of a few more telescopes.

Two 16 inch (40 cm) photometric telescopes had been used for site testing at Kitt Peak in the late 1950s. One of these was hauled up Cerro Tololo on mule back and used for site testing there in 1961.[35] Four years later these two 16 inch telescopes were the first telescopes to be installed in their own permanent observatories on Cerro Tololo. They were later replaced by new ones of the same size.[36]

The first medium-sized telescope to be erected at the CTIO was, as at Kitt Peak, a 36 inch (0.9 m) reflector. In fact the first mirror for the Cerro Tololo instrument came from the first Kitt Peak 36 inch telescope. Initially the CTIO 36 inch, which had been made by Boller and Chivens, was set up on a concrete apron in the open in early 1967, as it never rained at Cerro Tololo at that time of year. The telescope's observatory was completed shortly afterwards, and first light was achieved with the 36 inch in its observatory in April 1967. Three months later, Victor Blanco became director of CTIO.

The 36 inch was followed by the 60 inch, which had a modified Ritchey-Chrétien design with a Pyrex primary mirror. This was polished at the KPNO optical shop at Tucson. The telescope's mounting, which had been made by the Westinghouse Electric Corporation, was of a similar asymmetric design to the 36 inch. The 60 inch saw first light on November 6 1967.

Blanco 158 inch (4.0 m) Telescope

Meanwhile a twin to the KPNO 150 inch (3.8 m) telescope had been included in AURA's five year master plan of March 1964, although in the following year it was dropped from the approved budget. Subsequently the AURA Scientific Committee examined the relative merits of a large Schmidt telescope and the 150 inch, coming out in favour of the latter. Shortly afterwards the Ford Foundation, who had been looking at the possibility of subsidising a large telescope in the southern

[35] Victor Blanco, *Brief History of the Cerro Tololo Inter-American Observatory*, February 1993. (www.ctio.noao.edu/diroff/ctio_history.htm)

[36] Both 16 inch telescopes have since been removed.

hemisphere, expressed an interest in the CTIO 150 inch. Finally in April 1967 the Foundation formally agreed to fund half the cost of this 150 inch up to a maximum of $5 million, provided the National Science Foundation provided matching funds.(26) Interestingly, although President Johnson and President Frei of Chile announced in April 1967 that the telescope would be built, it took another seven months before the US Congress agreed to include the CTIO 150 inch in the NSF budget.

The timing of the Ford grant in early 1967 was excellent as AURA were planning to invite bids for the construction of the dome and observatory building for the KPNO 150 inch in the middle of that year. So they were able to add the CTIO dome to the bid request. (The CTIO observatory building was due to be built by the AURA team.) The Sundt Construction Company of Tucson were the lowest bidders, even though their bid was $600,000 over the AURA budget. AURA also put out a bid invitation for both the Kitt Peak and Cerro Tololo telescope horseshoe mountings. Western Gear from near Seattle were the successful bidders, with a bid well within the AURA budget, which more than compensated for the Sundt price.

The mirror blank for the KPNO 150 inch (actually 158 inch or 4.0 m) had been ordered three years earlier, so there was no chance of doubling up this order for CTIO. However, the Anglo-Australian Observatory and the French National Observatory wanted to order similar-sized blanks for their new telescopes.[37] So AURA joined with them to produce a multiple purchase bid invitation. General Electric, Corning and Owens-Illinois submitted bids in October 1967, with the latter winning. They proposed using Cer-Vit at a price of about $660,000 each for the three blanks, which was about half the price of fused silica.

Owens-Illinois poured the CTIO blank in June 1969, but there was a power cut during the cooling period, and the blank cracked. However, a second attempt was successful six weeks later. It was shipped to Tucson in early 1970 where it was put into storage whilst the KPNO mirror was being finished.

Skidmore, Owings and Merrill of Chicago, who had worked on the design of the McMath-Pierce Solar Telescope, designed the observatory buildings for both Kitt Peak and Cerro Tololo 158 inch (4.0 m) telescopes. The building at Cerro Tololo (see Figure 2.3) had a total height of 135 ft (41 m) to the top of the dome. This was somewhat shorter than that on Kitt Peak, as the thermal conditions near the ground were more favourable at Cerro Tololo. There was also another difference in that the CTIO building was constructed using a bolted steel framework mounted on reinforced concrete pillars, because of frequent earthquakes in Chile, whilst the KPNO building had used a welded steel framework. The Cerro Tololo building and

[37] The French order was for what turned out to be the Canada-France-Hawaii Telescope on Mauna Kea (see Section 8.2).

Figure 2.3. The 158 inch (4.0 m) Blanco Telescope is seen (right) on Cerro Tololo. The main optical telescopes moving leftwards from the Blanco are the 60 inch (1.5 m), the Curtis Schmidt, the 40 inch (1.0 m) Yale and the 36 inch (0.9 m). (Courtesy David Walker via Wikimedia Commons.)

dome were completed in 1972, and the telescope mounting was shipped to Chile in February 1973.

In the meantime, grinding and polishing of the 158 inch Cer-Vit mirror blank had been started in early 1972 in the KPNO optical shop at Tucson. After completion, the mirror was sent to Chile, where it arrived in September 1974 after an 84- day journey by rail, ship and lorry. Once on the mountain, the mirror was coated with aluminium and installed on its massive 350-ton mount. Shortly afterwards the first photographs were taken at the f/2.6 prime focus by Victor Blanco, the observatory director. On September 8 1995, this telescope was named after him as the Victor M. Blanco Telescope.

2.7 Other Telescopes on Cerro Tololo

The University of Michigan had located its 24/36 inch (0.6/0.9 m) f/3.5 Curtis Schmidt(27), a twin of the Burrell Schmidt (see Section 2.4), at its Portage

Lake Observatory in 1950. But in 1967 it was moved to Cerro Tololo, initially on a ten year loan basis, where it was put into a new observatory building. The University of Michigan was allocated one-third of the available observing time. The Curtis Schmidt had been designed to be used with an objective prism that fitted over its 24 inch diameter corrector plate to provide spectra of all the bright stars in the field. It had been provided with 4° and 6° prisms at Portage Lake, but Victor Blanco designed a 1° 50' prism to be used after its move to Cerro Tololo. Although this prism naturally provided less spectral dispersion than the others, it permitted spectral classification of fainter stars because the spectrum was more concentrated.(28)

Kodak IIa-O plates had generally been used with the 4° prism, but its grain was too course to be used satisfactorily with the very short spectra produced by the lower dispersion 1° 50' prism. Fortunately, Kodak had recently introduced a very fine grain IIIa-J emulsion, which by hypersensitisation[38] had an equivalent speed to the IIa-O. This very new fine grain emulsion had enough resolution to allow the spectral classification of stars down to magnitude 16 with the 1° 50' prism. In 1975 the loan agreement for the Curtis Schmidt was extended by 25 years.

The Curtis Schmidt was used by the University of Chile and CTIO in the Calán/Tololo Supernova Survey from 1989 to 1995. It is currently used by the Michigan Orbital Debris Survey Telescope (MODEST) project in support of NASA's Orbital Debris Program.

In 1968 AURA built an observatory for the Lowell Observatory's 24 inch (0.6 m) reflector on a mountain spur about half a mile (800 m) from the main CTIO site. The Lowell Observatory used this telescope for planetary observations as part of the International Planetary Patrol Program. It was later handed over to AURA and then mothballed in 1996, as by then it was little used. In 2010 SARA spent about $250,000 to upgrade this telescope and make it remotely accessible on the Internet.

As mentioned previously, financial pressure had caused NOAO to concentrate its resources in later years on maintaining and operating the largest telescopes on Kitt Peak, and closing or handing over operational responsibility of the smaller telescopes to university consortia. NOAO adopted a similar approach on Cerro Tololo, with operation of the 36 inch (0.9 m) and 60 inch (1.5 m) being handed over to the SMARTS 1 consortium[39] in February 2003. NOAO is a member of this consortium, allowing NOAO users to still have access to these telescopes, although they now had to share access with other members of the consortium.

[38] The plates were hypersensitised by being baked in nitrogen gas at 65°C for four hours before use.

[39] SMARTS is an acronym of 'Small and Moderate Aperture Research Telescope System'. SMARTS 1 ran from 2003 to 2006 when a modified consortium, SMARTS 2 took over. Both institutional and private persons can be members. Currently (2015) the institutional members are the Universities of Georgia State, Ohio State, SUNY-Stony Brook, Wesleyan University and Yale, in addition to NOAO/CTIO.

In 1973 Yale University moved their 40 inch (1.0 m) Boller and Chivens reflector from Bethany, Connecticut to Cerro Tololo. From 1998 to 2002 this telescope was used by members of the YALO[40] consortium, and from 2004 by members of the SMARTS 1 consortium.

One of the most unusual telescopes at Cerro Tololo is the 1.3 m (50 inch) reflector. It was specifically optimised for the infrared, as it was used for the two Micron All Sky Survey (2MASS) starting in 1998.(29) This Cerro Tololo 2MASS telescope, as it was called, covered the southern part of the sky, and an identical telescope, located on Mt Hopkins, Arizona, covered the northern part. Each telescope housed three infrared arrays, enabling the sky to be surveyed simultaneously in the J (1.25 microns), H (1.65 microns) and K_s (2.16 microns) photometric bands.(30)

The infrared detectors of the 2MASS telescope were cooled down to 77K by liquid nitrogen to minimise thermal noise. But the most unusual part of the telescope was the Cassegrain secondary mirror support, which moved during each 1.3- second exposure. This was to compensate for the telescope movement as it smoothly and continuously scanned the sky at about 1 arcmin/sec. After each exposure the Cassegrain returned to its starting position, ready for the next 1.3 second exposure. All the operations were automatic. After the 2MASS survey had been completed in February 2001 this telescope was given to CTIO. Two years later it was handed over to the SMARTS 1 consortium.

So, at the time of writing (2015), only the 2.1 and 4.0 m telescopes on Kitt Peak and the 4.0 m telescope on Cerro Tololo are NOAO facilities, whilst a number of other organisations operate smaller telescopes at both observatories. NOAO also have access to some of these other telescopes for some of their observing time. That includes the WIYN and 0.9 m telescope on Kitt Peak and the SMARTS telescopes on Cerro Tololo.

The CTIO observatory has been very popular as one of the few astronomical observatories located at altitude in the southern hemisphere with a good range of telescopes (see Table 2.2). As a result in recent times a number of small telescopes have been located on Cerro Tololo (but not at the main CTIO observatory site, as it was full, see Figure 2.3) including six 0.4 m (16 inch) PROMPT (Panchromatic Robotic Optical Monitoring and Polarimetry Telescopes) telescopes of the University of North Carolina at Chapel Hill. These were designed to undertake follow-on observations of gamma-ray burst sources in the visible waveband. Other small telescopes on Cerro Tololo included the 0.6 m WHAM telescope (see Section 2.5) which had been relocated from Kitt Peak in 2009, and the 0.5 m

[40] YALO is an acronym of 'Yale, AURA, University of Lisbon and Ohio State University'. In 2003 it was used as the basis of the SMARTS consortium.

Table 2.2. *Optical Telescopes Located on Cerro Tololo*

(a) Telescopes that were originally National Observatory telescopes

Telescope	Year Construction Completed on Cerro Tololo[a]	Comments
16 inch (0.4 m) No. 1	1965	Initially used for site testing on Kitt Peak and then in 1961 on Cerro Tololo. Installed in permanent observatory on Cerro Tololo 1965. Now closed.
16 inch No. 2	1965	Initially used for site testing on Kitt Peak. Installed in permanent observatory on Cerro Tololo 1965. Now closed.
36 inch (0.9 m)	1967	Operated by SMARTS 1 consortium from 2003.
60 inch (1.5 m)	1967	Operated by SMARTS 1 consortium from 2003.
158 inch (4.0 m) Victor M. Blanco Telescope	1974	Still operated by AURA.

(b) Other telescopes

Telescope	Year Construction Completed on Cerro Tololo	Comments
24/36 inch (60/90 cm) Curtis Schmidt	1967	Located at University of Michigan's Portage Lake Observatory 1950. Moved to Cerro Tololo in 1967 as a joint University of Michigan/ AURA facility. Currently used by University of Michigan.
24 inch (0.6 m) Lowell Telescope	1968	Lowell Observatory telescope initially. 1973 operation passed to AURA. 2010 operation passed to SARA.
40 inch (1.0 m) Yale Telescope	1973	1966 at Yale Observatory, Bethany, Connecticut. Moved to Cerro Tololo 1973. Operated by YALO consortium from 1998, and SMARTS 1 consortium from 2004.
1.3 m (50 inch) Cerro Tololo 2MASS Telescope	1998	2MASS project telescope 1998. Given to AURA 2001, and handed over to SMARTS 1 consortium 2003.

(*continued*)

Telescope	Year Construction Completed on Cerro Tololo	Comments
Six 0.4 m PROMPT telescopes	2006	First 5 telescopes operational by 2006. Owned by University of North Carolina at Chapel Hill.
0.6 m WHAM	2009	Removed from Kitt Peak 2008 and moved to Cerro Tololo 2009.
0.5 m CHASE telescope	2010	Used by Chilean CHASE project.

[a] Defining a year of completion is not as obvious as it sounds, as usually first light takes place before all the main elements of the telescope have been calibrated and commissioned.

which was installed near the PROMPT telescopes in 2010 as part of the Chilean Automatic Supernova Search (CHASE).

But the flattened top of the mountain, where the main CTIO telescopes were located, had been completely full since the mid 1970s. So the 8.1 m Gemini South (see Section 3.3) and 4.1 m SOAR[41] telescopes (see Section 3.4) were built on the nearby Cerro Pachón in the 1990s and 2000s, respectively.

References

1. Struve, Otto, Cooperation in Astronomy, *Scientific Monthly*, 50, 1940, pp. 142–147.
2. England, J. Merton, *A Patron for Pure Science. The National Science Foundation's Formative Years, 1945–57*, National Science Foundation, NSF 82–24, 1982.
3. Edmondson, Frank K., *AURA and its US National Observatories*, Cambridge University Press, 1997.
4. Irwin, John B., Optimum Location of a Photoelectric Observatory, *Science*, 115, February 29 1952, pp. 223–226.
5. Goldberg, Leo, The Founding of Kitt Peak, *Sky and Telescope*, March 1983, pp. 228–232.
6. Kloeppel, James E., *Realm of the Long Eyes; A Brief History of the Kitt Peak National Observatory*, Univelt, California, 1983, reprinted 1996.
7. Meinel, Aden B., The National Observatory at Kitt Peak, *Sky and Telescope*, August 1958, pp. 492–499.
8. Kloeppel, James E., *Realm of the Long Eyes; A Brief History of the Kitt Peak National Observatory*, Univelt, California, 1983, reprinted 1996.
9. Mayall, Nicholas U., Observatory Report, *Astronomical Journal*, 71, 1966, p. 233.
10. Abt, Helmut A., The Cost-effectiveness in Terms of Publications and Citations of Various Optical Telescopes at the Kitt Peak National Observatory, *Publications of the Astronomical Society of the Pacific*, 92, 1980, pp. 249–254.
11. McMath, Robert R., and Pierce, A. Keith, The Large Solar Telescope at Kitt Peak, *Sky and Telescope*, August 1960, pp. 64–67, and September 1960, pp. 132–135.

[41] SOAR is an acronym of 'SOuthern Astrophysical Research'.

12. Edmondson, Frank K., *AURA and its US National Observatories*, Cambridge University Press, 1997, p. 172.
13. Mayall, Nicholas U., Observatory Report, *Astronomical Journal*, 71, 1966, pp. 238–239.
14. Sage, Leslie and Aschenbrenner, Gail, *A Visitor's Guide to the Kitt Peak Observatories*, Cambridge University Press, 2004.
15. McCray, W. Patrick, *Giant Telescopes: Astronomical Ambition and the Promise of Technology*, Harvard University Press, 2004, pp. 84–87.
16. Hilliard, R. L., Steward Observatory's New 90-inch Reflector, *Sky and Telescope*, August 1967, pp. 79–81.
17. Hearnshaw, John, *Astronomical Spectrographs and their History*, Cambridge University Press, 2009.
18. Wehinger, Peter A., and Mohler, Orren C., The New Michigan 52-inch Reflector, *Sky and Telescope*, February 1971, pp. 72–75.
19. Mook, Delo E., The New McGraw-Hill Observatory on Kitt Peak, *Sky and Telescope*, January 1976, pp. 23–25.
20. King, Henry C., *The History of the Telescope*, Charles Griffin and Co., 1955, Dover reprint, 1979, p. 370.
21. MacRobert, Alan, U.S. Astronomy in Crisis, *Sky and Telescope*, May 1988, pp. 469–470.
22. Murray, Jack, A Classic vs. High-Tech Telescope, *Sky and Telescope*, April 1995, pp. 32–34.
23. Williams, G. G., et al., The Robotic Super-LOTIS Telescope: Results and Future Plans, arXiv:0803.002v1, 29 February, 2008.
24. Edmondson, Frank K., *AURA and its US National Observatories*, Cambridge University Press, 1997.
25. Blanco, Victor M., The Inter-American Observatory in Chile, *Sky and Telescope*, February 1968, pp. 72–76.
26. Edmondson, Frank K., *AURA and its US National Observatories*, Cambridge University Press, 1997.
27. Miller, Freeman D., The Portage Lake Observatory of the University of Michigan, *Sky and Telescope*, February 1950, pp. 79–80.
28. Philip, A. G. Davis, Thin-Prism Spectra of Messier 4 in Scorpius, *Sky and Telescope*, April 1976, pp. 244–246.
29. Skrutskie, Michael, 2MASS: Unveiling the Infrared Universe, *Sky and Telescope*, July 2001, pp. 34–42.
30. Skrutskie, M. F., et al., The Two Micron All Sky Survey (2MASS), *Astronomical Journal*, 131, 2006, pp. 1163–1183.

3

From the Next Generation Telescope to Gemini and SOAR

3.1 Next Generation Telescope (NGT)

No sooner had the 200 inch (5.0 m) Palomar Telescope been completed in 1948, than some astronomers started to consider whether it would be possible to build an even larger instrument. One of the most determined was Aden Meinel who, in the early 1950s, produced the outline design of an optical reflector of about 500 inch (~12.5 m) diameter. In this the primary mirror was not to be made of one piece of glass, but of several hundred smaller mirrors. But he received little support for his ideas, particularly when the question of funding was raised. It was also not clear at that time whether it was feasible to build such a machine.

Twenty years later, the design of a very large telescope was readdressed by Leo Goldberg the director of KPNO. Telescope design had moved on since Meinel's work, and by 1974 the main building phase of telescopes at KPNO had been completed, leaving a number of talented engineers with relatively little to do.(1) So Goldberg asked these engineers to examine possible designs of an optical telescope with a collecting area equivalent to a 25 m (~1000 inch) diameter mirror.

KPNO's first design concept for this very large telescope, called PALANTIR[1] or rotating shoe telescope, consisted of a 75 m long by 25 m wide segment of the primary mirror which could be rotated in azimuth. This primary was composed of hundreds of polished aluminium segments. The secondary mirror assembly, which included a 3 m secondary mirror, was mounted on the elevation axle, which was parallel to the shortest side of the primary. By moving about this axle, the secondary could view the sky from the zenith down to about 30° above the horizontal. The main advantage of this design, over a conventional one, was that the primary mirror could be rigidly held and did not suffer from variable gravitational forces as the telescope looked at various parts of the sky.

[1] PALANTIR was an acronym of 'Program for A Large Aperture Novel Thousand Inch Reflector'.

Unfortunately, the proposed resolution of PALANTIR was rather poor, its secondary mirror could only see a fraction of its primary at any one time, and its capital cost of about $160 million was completely out of the question at that time. Nevertheless, the NSF were prepared to give KPNO $200,000 to enable them to further study alternative designs of this so-called Next Generation Telescope (NGT).

Over the next two years, KPNO developed a number of concepts for a 25 m telescope, including a large dish with a segmented mirror that looked more like a radio telescope, a much enlarged version of the Multiple Mirror Telescope (MMT) where a number of telescopes were fitted to a common mount (see Chapter 10), an array of six 10 m telescopes, and a much larger array of 2.4 m telescopes. These and other designs were discussed at an international conference[2] held at the University of Arizona in January 1980.(2) By then KPNO had spent almost $1 million on their NGT programme without any clear indication as to the best design of a 25 m telescope. The thinking was really a bit too 'blue sky' and so, to sharpen up the focus, the target size was reduced shortly afterwards to 15 m.

3.2 National New Technology Telescope (NNTT)

This 15 m telescope was variously called the National New Technology Telescope (NNTT) or the New Technology Telescope (NTT), and was a recommended new start in the 1982 Field Report.(3) It was to be designed for both optical and infrared wavelengths at an estimated cost of about $100 million. By 1982 the choice of designs had been reduced to two; either a larger MMT consisting of four 7.5 m mirrors, or a telescope with a segmented 15 m mirror. Both designs used an altazimuth mount to reduce the size and complexity of the instrument and its observatory. Although altazimuth designs produced field rotation of the image, this was now no longer a problem as it could be compensated by computer control. The University of Arizona were joint owners of the MMT (with the Harvard-Smithsonian Center for Astrophysics), so it was no surprise that they favoured the larger MMT approach. On the other hand, the segmented mirror design was favoured by Jerry Nelson and others at the University of California, as they had been working for some time on a segmented mirror design for a 10 m telescope for their observatory.

In March 1983 Geoffrey Burbidge, KPNO's director, asked a committee of astronomers from Caltech and the Universities of Arizona, California, and Texas to join some of his own staff to examine the two competing designs of the proposed NNTT. They were not to try to design the telescope but look at the two

[2] Adelaide Hewitt (ed.), *Kitt Peak National Observatory Conference, Optical and Infrared Telescopes for the 1990s, Tucson, Arizona, 7–12 January 1980, Proceedings Vols. I and II*, Kitt Peak National Observatory, Tucson, Arizona, May 1980; see, for example, L. D. Barr, The Kitt Peak NGT Project, pp. 23–53.

competing designs from an astronomer's point of view, and make a recommendation as to which design to adopt. Burbidge invited Robert Gehrz, of the University of Wyoming, to chair this Scientific Advisory Committee (SAC). The panel's first recommendation was that the NNTT should be designed to undertake spectroscopy and infrared imaging as its main objectives. From this, they produced a set of performance requirements to evaluate the two telescope designs.

In May 1984, whilst Gehrz's panel was still evaluating the NNTT designs, the University of California announced that Marion Hoffman, the widow of a wealthy businessman, had given the university \$36 million to help it to build the largest telescope in the world. But although large, this donation was not enough on its own to enable the university's proposed 10 m segmented mirror telescope to be built. So the University of California set about trying to find partners to help finance their project.

Two months later, Gehrz's panel met to make their final recommendation on the design of the NNTT. Although, at this time, the extra funding for the University of California's 10 m telescope had not been found, it looked likely that it would be. So, if a segmented mirror design was also chosen for the 15 m NNTT, the University of California's 10 m could act as a pipe cleaner for the larger telescope, which would clearly be of benefit to the NNTT. But waiting for the results of the 10 m would delay the start of the NNTT, and any delays in building the 10 m would have a direct knock-on effect on the NNTT schedule. On the other hand, although first light on the MMT had already been achieved, it was still not working as expected, leading some committee members to doubt the suitability of the MMT concept. In the end, the decision between the two alternative designs of NNTT was very close, with Gehrz's panel recommending in July 1984, by a small margin of votes, that the larger-size MMT(4) should be the NNTT. On balance, the committee felt more comfortable dealing with a design based on relatively conventional mirrors, than one in which the primary mirror relied on the high accuracy control of a large number of mirror segments.

The University of California did not have long to wait for a positive response to their request for partners for their 10 m segmented mirror telescope. In August 1984, they were informed by Caltech that Howard Keck, the son of the Standard Oil tycoon William Keck, was in discussions with them about a possible large donation which would probably be enough to pay for virtually the whole of the 10 m telescope, estimated to cost about \$85 million. Although, Caltech were prepared to go ahead on their own with a segmented mirror design, eventually they agreed to form a joint enterprise with the University of California to manage the project. Caltech would pay most of the construction costs, and so own the telescope, whilst the University of California would pay for its running costs over a period of 25 years. This became the 10 m segmented mirror Keck telescope (see Section 4.2).

Things now got complicated as the NNTT was to be basically a larger MMT. But the existing MMT was partially owned by the University of Arizona, and it was assumed that the mirrors of the larger MMT would also be moulded in the University of Arizona's rotating furnace (see Section 4.1). So it was beginning to look as though the University of Arizona was taking over the new national telescope in competition with Caltech's 10 m Keck telescope. This issue of who really ran NOAO's NNTT project produced a great deal of friction between the University of Arizona's Steward Observatory and the NOAO, whose offices were just across the road from the Steward Observatory's Tucson campus. Then in 1985 Ohio State University and the University of Arizona announced that they were considering building a Large Binocular Telescope (then called the Columbus Project, see Section 13.2) with two 8 m telescopes on a common mount. So many people also asked whether there was now a real requirement for the NNTT with both the Keck, and now this new large binocular telescope on the stocks. But the NNTT project still carried on with modest funding.

The next question was where the NNTT should be built. By this time a number of telescopes had been located at Mauna Kea in Hawaii (see Chapter 8) which provided a better viewing site than Kitt Peak. So should the NNTT be built on Mauna Kea, in spite of its distance from the USA mainland, or was there an equivalent site on the United States mainland? At this time, the University of Arizona was planning to build its next generation of astronomical facilities, including the Large Binocular Telescope, on the 3,250 m (10,700 ft) high Mount Graham in Arizona. So they favoured putting the NNTT on Mount Graham,[3] whereas John Jefferies, NOAO's new director, favoured Mauna Kea, a location he had helped to develop. Eventually in early 1987, after three years of site surveys, NOAO announced that they had chosen Mauna Kea as their preferred site for the NNTT, with Mount Graham as a backup.[4]

Meanwhile, the design of the NNTT had progressed sufficiently by 1985 to give a better estimate of its likely cost, and this did not make happy reading. The basic capital cost was about $125 million, plus about $20 million for instrumentation. This was at a time when the National Science Foundation (NSF) was in the process of paying for the Very Long Baseline Array radio telescope (see Section 20.10), and money was very tight generally (see Section 2.5).

[3] Sage suggested that Peter Strittmatter, the director of the University of Arizona's Steward Observatory, tried to use the NNTT to solve the problem they were having from environmentalists who were objecting to their building the Large Binocular Telescope on Mount Graham. See L. Sage, A Brief History of the Controversy Surrounding the Mount Graham International Observatory, in André Heck (ed.), *Organizations and Strategies in Astronomy,* Volume 4, Kluwer Academic Publishers, 2003, p. 84.

[4] News Notes, *Sky and Telescope*, July 1987, p. 8.

Goetz Oertel, the new president of AURA, NOAO's parent body, set up a committee in 1987 chaired by Steve Strom, from the University of Massachusetts, to examine the future role of NOAO. This committee recommended in September 1987 that NOAO build two 8 metre class telescopes, one in each hemisphere, instead of the 15 m NNTT.(5,6) Shortly afterwards AURA accepted this advice, leaving building the NNTT as a long term goal to help to mollify NNTT advocates. Subsequently in September 1989 AURA submitted a proposal for two 8 m telescopes to the NSF. At about the same time the NOAO decided to join the universities of Wisconsin and Indiana in developing a medium-sized telescope, as a prototype for the national 8 m instrument. This was to become the 3.5 m WIYN telescope (see Section 2.5) located on Kitt Peak. A little later NOAO also joined with Columbia University and the University of North Carolina in building the 4.1 m SOAR telescope in Chile (see Section 3.4), also as a prototype for the large national instrument.

The decision to scale back the American national large telescope project first from 25 m to 15 m, and then to two 8 m telescopes, together with the closure of NOAO's Advanced Development Program (see Section 2.5), caused many people to leave NOAO in disgust. Not only was Caltech's 10 m Keck under construction on Mauna Kea by this time, but the Carnegie Institution of Washington was planning an 8 m telescope of their own at the Las Campanas Observatory (see Chapter 12) and the University of Arizona and others had already started designing the Large Binocular Telescope. But, worse than that for American prestige, the European Southern Observatory (ESO) were planning to build the Very Large Telescope (see Section 7.2) consisting of an array of no less than four 8 m telescopes in Chile. And the Japanese were planning to build an 8 m telescope on Mauna Kea (see Section 8.5). Nevertheless, two 8 m telescopes were better than nothing, so Goetz Oertel and Sidney Wolff, NOAO's new director, set out to sell the project to American astronomers, the NSF and Congress.

The telescope design in the AURA proposal to the NSF of September 1989 used a primary mirror produced by the University of Arizona's spun honeycomb mirror facility developed by Roger Angel. This was because it was thought that this mirror would provide better image quality than a segmented mirror like that of the Keck. But there was still some doubt as to whether the University of Arizona's facility could produce such large mirrors of the required quality. So meniscus mirrors were considered as a backup solution, albeit at increased cost. Adaptive optics (see Section 5.2) were also incorporated into the 8 m telescope design from the start, rather than retrospectively as with many other telescopes, which should also help to produce better images. However, the capital cost of the two 8 m telescopes, which was estimated at about $144 million, was considerable.

3.3 Gemini

One way of saving money on these two telescopes was to find international partners to share both the capital funding and annual costs. In fact informal talks had already taken place in mid 1989 between representatives of the United States, the UK and Canada about a possible joint project with two 8 m telescopes. A tentative agreement had been reached assuming that the US would pay 50% of the capital costs, with the UK and Canada contributing 25% each. So it was not really surprising that, when the US Congress agreed to fund the two-telescope programme in September 1990, they should require the NSF to find international partners to help to fund the programme whose capital cost had, by then, reached an estimated $176 million. But it was still a shock to many American astronomers to see how Congress had constrained their room for manoeuvre by limiting the US contribution to 50% of the capital cost, or $88 million, whichever was the least. As a result, this became a design-to-cost programme with financial considerations constraining many of the technical decisions as the programme developed. This was the start of what was to be called the Gemini programme.

Many in America argued that it would be better to build just one 8 m telescope in Hawaii, and fund it 100%, than two telescopes funded at 50% each, as the latter gave them only limited control over their use. At this time American astronomers were not used to sharing major ground-based astronomical facilities with astronomers from other countries. But such arrangements were becoming more normal for American astronomers and scientists participating in spacecraft programmes. Although, even here, Americans were often taking some time to realise that they could not unilaterally make major changes to a project without at least consulting their international partners, even if the Americans were providing more than 50% of the funding.[5]

A similar discussion had been going on about possible ground-based 8 m telescope programmes amongst UK astronomers who were more used, than their American counterparts, to working on jointly funded and operated international programmes. But the question in the UK was whether they should work with the Americans and get something like 25% of the time on two 8 m telescopes, one in each hemisphere, or whether they should build an 8 m telescope on La Palma in the Canaries, in collaboration with Spain, where the UK would have more control. Then in June 1989, the Canadian Astronomical Society voted unanimously to join the two telescope American project. Finally, after much discussion, the UK's Science and Engineering Research Council (SERC) decided in December 1990 to partner the United States and Canada in the twin 8 m telescope project.

[5] As in the case of the ISPM spacecraft programme, for example, see David Leverington, *New Cosmic Horizons; Space Astronomy from the V2 to the Hubble Space Telescope*, Cambridge University Press, 2000, pp. 273–276.

Since 1964 the American Academy of Sciences has organised a decadal review of astronomy, to survey the current state of the subject in the United States and to make recommendations for expenditure over the next ten years. Mention has already been made of the 1982 review, chaired by George Field. The next decadal review, to look at possible investments in the 1990s, was chaired by John Bahcall of Princeton's Institute for Advanced Study starting in 1989.

John Bahcall suspected that, if it were put to the vote, the American astronomical community would not support the Gemini programme, but would rather the money was spent on providing one modest-sized telescope for every astronomy department.(7) But he regarded it as his job, and that of his review committee, to lead the astronomical community, not just follow it. So they analysed the role of the planned 8 m Gemini telescopes in comparison with other possible investments in ground- and space-based facilities, taking it as a given that large facilities were required to facilitate cutting-edge astronomical research.

In making their recommendations, Bahcall and his committee were influenced, amongst other things, by major technological developments over the preceding ten years or so by the US military in adaptive optics and infrared detectors. Adaptive optics were about to be declassified, whilst some infrared detector technology had already been made available. There were more large telescopes existing and planned in the northern hemisphere by various countries than in the southern, so Bahcall thought that the northern Gemini telescope should be more of a cutting edge instrument than its southern counterpart. As a result his committee recommended, largely following advice from the respected infrared astronomers Fred Gillett[6] and Frank Low[7], that the northern telescope on Mauna Kea should be an infrared telescope optimised to operate between 2 and 10 microns. On the other hand, the committee recommended that the sensitivity of the southern hemisphere telescope should be centred in the visible, with extensions into the near ultraviolet and near infrared. The northern hemisphere infrared telescope, with an estimated capital cost of about $80 million, was given a higher priority than its southern hemisphere counterpart, that was estimated to cost about $55 million. No mention was made in the report of international collaboration, as Bahcall, in particular, thought that both telescopes should be American. He was particularly keen that, if this was not possible, at least the northern telescope, optimised for the infrared, should be an American-only instrument.

[6] Gillett was a pioneer in infrared spectroscopy. At the time of the Bahcall report he worked at the NOAO.

[7] Low had built a cryogenic bolometer that revolutionised infrared detection in the early 1960s. He had installed infrared detectors in aircraft in 1965 to undertake astronomical observations, and had helped to persuade NASA to become involved, with the Netherlands and the UK, in the IRAS infrared spacecraft that was launched in 1983. He had worked at the University of Arizona since 1965.

Bahcall was aware that many people would point out that the 10 m Keck telescope under construction on Mauna Kea would provide American astronomers with a world-class facility, so there was no need to provide another large telescope to do the same thing at the same observatory. But, as he said in his report(8), 'The proposed national infrared-optimized 8-m telescope will differ from the private 10-m Keck telescope in three ways: (1) it will be available to all US astronomers, whereas the Keck telescope will be available to only about 3 percent of the national astronomical community; (2) it will be the only large telescope in the world optimized for performance in the infrared; and (3) it will use adaptive optics to achieve the maximum possible spatial resolution in the near-infrared.'

Bahcall's report, published in 1991, had already been overtaken by events to a large extent, as the twin 8 m telescope Gemini project had already been started as an international project as required by Congress. But the Bahcall report's key requirement that the northern telescope should be optimised for the infrared could not be ignored.

The Gemini project was run by the Gemini Board consisting of 4 American, 2 British and 2 Canadian members, based on an agreed 50/25/25 percentage split of the project's costs. The board was responsible for approving the telescopes' technical requirements, agreeing the project's annual budget, and deciding on workshare between the countries involved. The board's chairman was Robert Bless of the University of Wisconsin.

The early days of the Gemini project were full of problems, however, most of them people-oriented. Many of the Americans in the project team resented the fact that Congress had insisted that Gemini be an international project, when they saw it as their project.[8] After all, Gemini was the result of work done at the NOAO over many years. This problem was not helped by the fact that the Gemini project was based at NOAO's headquarters at Tucson, Arizona, and that its project manager, Larry Randall, was an ex-Kitt Peak employee. It would probably have been better if the project team, many of whom were NOAO employees, had been located elsewhere, but there were benefits in its being located in NOAO facilities. Then in early 1992 the Gemini Board, which incidentally was not allowed to have anyone associated with AURA on it, formally stopped NOAO management exercising any direct control over the project.

In the meantime, matters had been made worse when Canada's National Research Council announced in June 1991 that it was pulling out of Gemini, although a little later Canada said that it would participate at 15%, rather than the 25% initially agreed. Unfortunately, this left NSF with little time to find the missing 10% from

[8] This issue was similar to the one, mentioned previously, with the 'ownership' of the design of the NNTT between members of the University of Arizona's Steward Observatory and those of the NOAO.

other countries before Congress' deadline for contributions of January 1 1992 was reached. Congress had said that, unless foreign financing was fully committed by then, the NSF could build only the infrared-optimised telescope on Mauna Kea.[9] Unfortunately, that would have cost more than 50% of the cost of the two telescope programme approved by Congress – not something that Congress wanted to hear. So the deadline was extended, and the NSF was given until May 1993 to find the missing 10% for the two telescope programme, although the American share of Gemini's funding was frozen in the meantime.

Chile was an obvious candidate to provide some or all of the missing 10% because of AURA's long association with the country and the likelihood that the southern telescope would be built there. Other possible candidates for funding were Argentina, Brazil and Australia. In fact, some Australian astronomers had shown an interest in participating in Gemini in early 1992, but that interest did not result in a positive commitment. However, in early 1993 Chile, Argentina and Brazil agreed to participate with contributions of 5/2.5/2.5%, respectively. At about the same time the 2,700 m (8,900 ft) high Cerro Pachón[10] located near Cerro Tololo in Chile was chosen as the site for Gemini's southern telescope, in return for Chilean astronomers getting an extra 10% of observing time. Then in 1998 the Australian Research Council agreed to pay $9 million towards Gemini's capital costs (equivalent to about 5%), plus about $1 million/year for operational costs, which gave Gemini additional funding stability. This was just as well as, by then, it had become clear that the $17 million budget for instrumentation was far too small.

Gemini's original design had been based on using spun honeycomb mirrors produced by the University of Arizona's Steward Observatory's Mirror Laboratory (see Section 4.1). But by the end of 1991, the Gemini Board were becoming increasingly concerned about the Mirror Labs' slow progress in the manufacture of mirrors for other telescopes. As a result, in January 1992 the Steward Observatory appointed its first professionally trained manager, Stephen Hinman, to sort things out. He found that the laboratory operated, as one may have anticipated, as a typical university department, which is what it was, lacking financial discipline. For example, financial estimates were often produced for complex jobs without a proper analysis of what the job entailed. This naturally led to programme slippages with consequent cost increases.

In the early 1990s, there were basically three design options for the primary mirror of very large telescopes. They were spun honeycomb mirrors produced by

[9] Bahcall was quite prepared to see the southern telescope cancelled if that meant that the northern telescope could be 100% American. But not all American astronomers agreed, as this would severely compromise the access of most American astronomers to large southern hemisphere instruments.

[10] A site survey carried out in 1989 had resulted in Cerro Pachón being chosen for the location of the 4 m SOAR telescope in which NOAO also had an interest (see Section 3.4). So this mountain was to be the site of two large telescopes.

the Steward Mirror Lab, thin meniscus mirrors[11] of the sort produced by Schott in Germany and Corning in the United States[12], and segmented mirrors like that used in the 10 m Keck. At that time all three mirror types were being used in the design of various large telescopes. For example, the Keck telescope had been dedicated in November 1991 with 9 of its 36 mirror segments installed. Then in February 1992, Schott had successfully made its first 8.2 m diameter meniscus blank for the European Southern Observatory's Very Large Telescope in Chile. Finally in April 1992 the Steward Mirror Lab had produced the 6.5 m mirror blank for the MMT[13] on Mount Hopkins. In addition, a little earlier Corning had been chosen to produce the 8.2 m meniscus mirror for the Japanese Subaru telescope on Mauna Kea. So when, in May 1992, Gemini's management issued a formal request for quotations for their 8.1 m mirrors, all three competing technologies were available, although none, at that time, had yet been used in a complete large telescope.

Gemini sent out their request for quotations (RFQs) to Corning, Schott and the University of Arizona's Mirror Lab in May 1992 for fixed-price bids for two 8 m mirror blanks. In reply, Corning bid for two thin meniscus mirror blanks made of ultra low expansion (ULE) glass at about $11 million. As a publicly funded university, the University of Arizona had difficulty in accepting significant financial risks associated with a fixed-price bid.(9) Nevertheless, they eventually felt able to bid $18.5 million, fixed-price, for two blanks made of their normal borosilicate glass. Their relatively high price was partly due to the fact that, if their mirror casting was a failure, the borosilicate glass could not be reused, where Corning's ULE glass could be.

The Gemini Board appointed two committees to review both bids. They concluded that, not only was the Corning bid much cheaper than that of the Mirror Lab, but Corning's schedule was more believable. In addition, the ULE glass used by Corning had a much lower thermal expansion coefficient than the borosilicate glass used by the Mirror Lab.(10) On the other hand, it was thought that a thin meniscus mirror made from the Corning blank may not perform as well as the thicker mirror from the Mirror Lab, and, consequently, the thin meniscus mirror would need a more sophisticated active optics support system. On balance, however, both committees favoured the Corning bid. So in September 1992 the Gemini

[11] A thin meniscus mirror is basically a very thin monolithic piece of glass, with an aspect ratio (of diameter to thickness) of typically about 40.

[12] Early views on the relative merits of spun honeycomb and thin meniscus mirrors are outlined in L. D. Barr, The U.S. National Large Telescope Project: The NNTT and Others, in Marie-Helene Ulrich (ed.), *Proceedings of a ESO Conference on Very Large Telescopes and Their Instrumentation, held in Garching, March 21–24, 1988*, European Southern Observatory, 1988, pp. 73–92.

[13] This 6.5 m mirror was to replace the six 1.8 m mirrors originally used in the MMT. The change from six small mirrors to one large mirror was made because of problems with making the six co-aligned telescopes work satisfactorily (see Chapter 10) and because it was now more feasible to make one large mirror than when the MMT had been designed originally. So, it was just as well that AURA had not proceeded with their larger MMT design for the NNTT.

Board awarded the contract to Corning to supply one mirror blank, with an option for a second.

Peter Strittmatter, the director of the Steward Observatory, and Roger Angel were naturally disappointed by the Gemini Board's decision and cried foul, as they believed that the bidding process was unfair. They had, apparently, not expected to be in competition with a commercial company[14] and to have to quote a fixed-price, with the associated difficulty in taking a significant financial risk, as mentioned above.

The Gemini mirror decision caused a number of disgruntled American astronomers to join forces and push for various changes to the project. Not only were Peter Strittmatter and Roger Angel concerned about their failure to win the mirror contract, but John Bahcall felt that the design of the northern Gemini telescope on Mauna Kea, including that of the mirror, had not been properly optimised for infrared astronomy.[15] In this he was backed up by the respected infrared astronomer Frank Low of the University of Arizona. Bahcall had also never accepted that the northern telescope was an international instrument, and in this he was supported by Donald Hall, director of the University of Hawai'i's Institute of Astronomy. Unfortunately, Strittmatter, Angel, Low and Hall all had institutional reasons to press their point of view. What was clearly required was an independent review of the project. As a result, in December 1992 the NSF set up an independent committee to review the Gemini project, including its mirror choice, chaired by James Houck of Cornell University, a respected infrared scientist.

Although the Houck committee was a purely American committee, its findings could well have derailed the Gemini project, and that looked very likely when it condemned the mirror choice as too risky.[16] (11) However, as Frank Kapper, a Corning executive observed, it was somewhat surprising, in view of its mandate, that this committee was composed of astronomers and scientists and had no mechanical and structural engineers included.

The main problem that the committee highlighted was that of supporting the very thin meniscus mirror in the telescope. Although smaller diameter thin meniscus mirrors had been used satisfactorily, no one at that time had solved the problem for a thin 8 m mirror. It was true that the ESO and Japan had opted for a similar sized thin meniscus mirror in their new large telescopes, but their imaging requirements were considered to be less onerous than those of Gemini, and their

[14] It seems incredible that the Steward Observatory expected to be given a contract of this magnitude without competition. They still appeared to think that they were dealing with the NSF, which had funded their development work, rather than with an international organisation like Gemini, where 50% of the money was coming from outside the USA.

[15] Bahcall later admitted that he had formed his views on Gemini, especially those connected with the choice of mirror, on faulty technical advice, see W. Patrick McCray, *Giant Telescopes: Astronomical Ambition and the Promise of Technology*, Harvard University Press, 2004, p. 235.

[16] M. Mountain et al., Letters, *Sky and Telescope*, September 1993, pp. 6–7.

telescopes had not yet been completed. Another problem highlighted by Houck's committee was that of controlling the temperature of the meniscus mirror so that it was the same as that of the observatory's air to avoid creating air turbulence. It was thought that it would be easier to accomplish this with a Steward Observatory borosilicate mirror as cooling air could, in principle, be blown through its honeycomb backing. But such a system had not then been proven in a large telescope.[17]

The Gemini Board addressed the Houck committee's concerns, and said that all the technical issues that they had raised would be considered at Gemini's Preliminary Design Review (PDR) in late 1993. Then, if problems continued with using the thin meniscus mirror, the project would change to the Steward Observatory's mirror for the northern telescope, provided it was clearly shown to be the better solution. In this the Gemini Board were crucially supported by William Harris, the NSF's assistant director for Mathematical and Physical Sciences. In the event, Gemini's PDR which was held in December 1993[18] showed that the existing telescope design was sound.

Modern large telescopes, like Gemini, with altazimuth mounts usually have both a Cassegrain and Nasmyth focus. In the case of the Cassegrain focus, light enters the telescope and is reflected off the primary mirror to a secondary mirror. This secondary then reflects the light back through a hole in the primary to the Cassegrain focus. In the Nasmyth configuration, another mirror intercepts the light coming from the secondary and reflects it off to one side of the telescope along the altitude axis. There it is focused at the Nasmyth focus. This arrangement has the advantage that heavy and bulky instruments can be placed there on a permanent platform, as it only has to rotate in azimuth.

Initially, Gemini was designed to have an f/1.8 primary mirror, with both an f/16 Cassegrain focus, and two f/19.6 Nasmyth foci on either side of the telescope. But by 1993 the estimated cost of the Gemini programme had reached about \$250 million.(12) This cost figure was unacceptable and there were also technical problems in incorporating the Nasmyth platforms. As a result, the platforms were eliminated. This simplified the design of the area around the primary mirror and reduced the amount of heat-emitting hardware there, so improving Gemini's infrared design. But this was at the expense of the telescope's scientific performance and operational flexibility.

Gemini had also been originally designed with interchangeable secondary mirrors. One was a 1.0 m diameter lightweight secondary for infrared observing, and the other was a larger conventional mirror to produce a wider field of view. But

[17] In addition, as the thermal expansion coefficient for borosilicate glass was much higher than that of Corning's ULE glass, the Steward Observatory's mirror would have had to have its temperature controlled much more rigorously than that of the Corning mirror.

[18] *Sky and Telescope*, March 1994, pp. 10–11.

the high programme cost just mentioned meant that the wide-field mirror had to be eliminated, at least in the telescope's initial configuration. This just left the small infrared secondary, which considerably simplified the secondary mirror support structure. In total the elimination of the Nasmyth focus and large secondary mirror were thought to have saved about $24 million.(13)

The design requirements for the small infrared secondary mirror were quite severe as it had to be mounted on a rapid tip-tilt and chopping mechanism operating at up to 200 Hz and 3 Hz, respectively. It had originally been hoped to use a silicon carbide design, because of its stiffness and lightweight characteristics, but, unfortunately, problems in producing such mirrors with a diameter of 1 m resulted in the project using Schott Zerodur mirrors instead.

Corning produced the blank for the primary mirror in the same way as they had produced the 8.2 m blank for the Japanese Subaru telescope fusing together a number of large hexagons of glass into a single blank. Prior to fusing, the thermal expansion coefficient had been measured for each of the hexagons, and their positions carefully optimised to produce a minimum thermal distortion of the completed mirror. After fusing, the mirror blank was rough ground and then heated over a mould, until it sagged and took up a meniscus shape with the required radius of curvature.

Corning finished its first 8.1 m mirror blank for Gemini in October 1995 in their facility in Canton, New York, and then shipped it to REOSC[19] Optique, near Paris to be polished. REOSC had been selected to undertake the polishing as they had previously been chosen by the ESO to polish the four 8 m thin meniscus blanks for their VLT telescope. Polishing of Gemini's first 20 cm thick primary mirror was completed in January 1998, and installed in the Mauna Kea on what is now called the Gemini North telescope in December. A month later the secondary mirror was in place, allowing first light to be achieved at the end of January 1999. The Gemini project then installed a small adaptive optics unit called Hokupa'a (Hawaiian for the North Star) and an infrared imager on loan from the University of Hawai'i in May 1999. These items enabled the Gemini North telescope to achieve star images of just 0.08 arcsec in diameter at 2.2 microns.[20] The second primary mirror arrived at the Gemini South observatory on Cerro Pachón, Chile in early 2000 just under two years after the first had arrived on Mauna Kea.

As with all thin primary mirrors, the shape of Gemini's primary mirror was maintained by an active optics system, in this case using 120 actuators per mirror. The coating plants at both the north and south observatories were designed to coat the mirrors with either silver, for infrared observations, or aluminium. The

[19] REOSC is an acronym of 'Recherche et Étude en Optique et Sciences Connexes'.
[20] News Notes, *Sky and Telescope*, September 1999, pp. 18–19.

Figure 3.1. Gemini South on Cerro Pachón which was only about 10 km (6 miles) from the Inter-American Observatory on Cerro Tololo. (Courtesy Denys via Wikimedia Commons.)

silver was then to be coated by a very thin layer of silicon nitride(14) to protect it from tarnishing. But problems with the deposition process resulted in both primary mirrors initially being coated with aluminium, instead. Eventually the protected silver coating was applied on Gemini South's primary mirror in May 2004 and on Gemini North's primary about six months later.

In the meantime, Gemini had opened its Northern Operations Center in Hilo, Hawaii in August 1998, which, amongst other things, allowed remote operation of the telescope from a sea level site. The Southern Operations Center was opened later on the CTIO campus in La Serena, Chile.

Gemini North was dedicated on June 25 1999 and Gemini South (see Figure 3.1) on January 18 2002. At another ceremony on November 13 2002, the Gemini North telescope was named after Frederick C. Gillett, Gemini project scientist, who had died prematurely in April 2001. Among his many contributions to the programme were the specification and development of infrared-optimised, protected silver coatings for the primary mirrors.

The design of the 8 m Gemini, and similar large telescopes, around the turn of the twentieth century took place at a time of major developments in computers and telecommunications. These developments, together with those of CCDs, active optics and adaptive optics (see Sections 5.4, 5.1 and 5.2, respectively) had a fundamental effect on the way that the Gemini, in particular, was designed, integrated and operated. In fact the Gemini project was eventually managed and operated more like that of a spacecraft than of earlier telescopes.[21] For example, the Gemini telescope was designed with its instruments as a complete system to optimise the observational results, and not as a telescope to which observers would generally add their own instruments. Gemini's dedicated so-called 'facility instruments' typically weighed ~1 ton and cost ~$5 million to $10 million each, indicating not only their size but their complexity.

The facility instruments were arranged around the Cassegrain focus, with some of them being virtually permanent fixtures. Access to the various instruments installed on the telescope for a night run was effected by a movable tertiary mirror at the Cassegrain focus that directed the light to whichever instrument was required. This meant that changing from one instrument to another was very rapid. Gemini's science plan called for most of these instruments to be shuttled between Hawai'i and Chile to cover different observing periods, with the exception of Gemini's Multi-Object Spectrograph (GMOS). There were two versions of this GMOS produced, one for Mauna Kea optimised for red wavelengths and one for Chile optimised for blue.[22]

In 1994 the Canadian Herzberg Institute for Astrophysics won the contract to supply the ALTAIR (ALTitude-conjugate Adaptive optics for the InfraRed) adaptive optics system for Gemini North. They developed this using experience gained with an adaptive optics system on the Canada-France-Hawaii Telescope also on Mauna Kea. Three years later the contract was extended to allow the ALTAIR design to be modified to operate with laser-generated guide stars. ALTAIR NGS (Natural Guide Star), which used a 177 actuator deformable mirror, was delivered to Gemini North in 2002 and was operational from 2004, whilst ALTAIR LGS (Laser Guide Star) was delivered to Gemini North in 2005. These adaptive optics units intercepted light from the tertiary mirror and fed a corrected beam to the Gemini instruments. In 2007 another adaptive optics unit called NICI (Near Infrared Coronagraphic Imager) saw first light at Gemini South, becoming operational two years later. This 85-element adaptive optics system was designed

[21] Interestingly, NASA had taken over from the NSF as the main source of federal funding for US astronomers by the early 1990s. This was to some extent due to NASA's policy of funding astronomers to analyse their data produced by the Hubble Space Telescope (HST), which naturally helped to make the HST popular amongst professional astronomers.

[22] It had originally been envisaged that Gemini North would differ significantly from Gemini South in being optimised for the infrared, rather than for the visible waveband. However, as the programme progressed it became clear that it would be cheaper if two identical telescopes were produced. (See Lauren Gravitz, Matt Mountain: A Clear Vision of the Future, *Gemini Focus*, No. 38, June 2009, pp. 12–15.) But the instruments put on Gemini North have tended to be more infrared-oriented than those put on the southern telescope.

by MKIR (Mauna Kea Infrared) of Hilo, Hawai'i to detect large planets orbiting other stars. Then in 2011 the Gemini Multi-conjugate adaptive optics system, called GeMS (see Section 5.2), saw first light on Gemini South. Two years later it produced an image resolution of ~ 0.075 arcsec at a wavelength of about 1.65 microns.

With so many astronomers from Gemini partner countries wanting to make observations it was essential that these observations were made as efficiently as possible. The method of changing from one instrument to another via the tertiary mirror certainly helped. But the project also introduced 'queue observing'[23] on some nights, which in many ways mimicked the way observations were undertaken with spacecraft. In this system, introduced in spite of much scepticism amongst the Gemini team, observers specified what observations they wanted to make with what instruments. The observatory staff then decided how best to fit in these observations with those requested by other astronomers, taking into account the suitability of the seeing conditions for the various types of observations requested. The observatory staff, not the astronomers themselves, then carried out the observations. Alternatively astronomers could ask to carry out the observations themselves, but these so-called 'classical observations' were not reimbursed for weather or technical losses. More recently, Gemini has introduced 'target of opportunity' observing which can interrupt classical observing nights. In this case, the lost time is reimbursed to the classical observers. Interestingly, in spite of the initial scepticism, most observations are now carried out using queue observing.

Funding problems in the UK caused it to notify its Gemini partners in November 2007 that it would have to negotiate a phased withdrawal from the Gemini project. Both the Gemini Board and the astronomical community in the UK reacted with horror at this announcement which came out of the blue to both of them. In fact, the Gemini Board were so agitated that they decided at their meeting on January 25 2008 to determine that the UK had withdrawn unilaterally from the project on December 12 2007. Clearly it was in no-one's interest to have such a sudden withdrawal. On the one hand, the UK, according to the Gemini Agreement, would have to continue to pay its $4 million/annum contribution to Gemini's operational costs for the next two years, receiving no observing time in return. On the other hand, the Gemini project would instantly lose the benefit of the UK's technical expertise, and after two years the project would have to exist on a much smaller budget. So when the UK reversed its decision and agreed to continue its contributions, it was reinstated as a full member of Gemini. However, the UK did eventually withdraw from the Gemini project at the end of 2012.

[23] Also called 'service observing' or 'queue scheduling'.

3.4 SOAR

In the late 1980s the University of North Carolina at Chapel Hill and Columbia University, New York examined the possibility of building a state-of-the-art, 4 m-class telescope.(15) They wanted to locate it in the southern hemisphere, far from the poor observing conditions of the eastern United States, as there it would be able to observe the centre of the Milky Way, the Magellanic Clouds and other important regions of the southern sky unobservable from their home location. The universities thought that they would be able to raise sufficient funds to build such a telescope from private donations and/or institutional sources. But they anticipated having a problem paying for the observatory's infrastructure and annual operating costs, which would be relatively high because of its distance from its home institutions.

It so happened that at that time CTIO, which already had telescopes on Cerro Tololo in Chile, was heavily constrained in their capital budget. As a result, they would not have been able to raise funds to build such a large new telescope there. But they would not have found it nearly so difficult to fund such a telescope's infrastructure and operating costs, particularly if the telescope was located near to their existing observatory. As a result, the University of North Carolina, Columbia University and CTIO decided to get together to build and operate a new-technology 4 m telescope, which is now called the Southern Astrophysical Research (SOAR) Telescope.

The existing CTIO 4.0 m telescope on Cerro Tololo had a good wide-field capability. As a result, the SOAR partners agreed early on that their new telescope should focus on image quality. A site survey, carried out in 1989, found a suitable site on the 2,700 m Cerro Pachón, not far from Cerro Tololo, which had even better seeing than that of the CTIO itself. Shortly afterwards the Gemini project also decided to build their new 8 m telescope on Cerro Pachón.

The first and most important technical question to be resolved in the SOAR project was whether the primary mirror should be a thin meniscus type or be one made by the Steward Observatory's Mirror Lab of borosilicate honeycomb. There was some concern at the time that the Steward Observatory would be unable to meet its existing commitments without taking on an additional one. On the other hand, the European Southern Observatory's (ESO's) 3.6 m diameter New Technology Telescope (NTT) (see Section 7.1), which used a very thin meniscus type primary mirror with active optics, was working well. So when the SOAR project received a very competitive quote from Corning for a thin meniscus mirror made of ULE material they felt that this was the best solution. Shortly afterwards the Gemini project made the same decision.

The SOAR designers were considerably influenced by the design of ESO's 3.6 m NTT in more ways than just in the choice of mirror type. Having decided on a thin primary mirror, the first question the SOAR designers had to answer was how thin should it be? ESO had been cautious and had chosen a mirror with an aspect ratio (of diameter to thickness) of 15 to avoid being too dependent on the performance of its active optics system. But their active optics system had been successful, so the SOAR designers decided to be bolder and chose an aspect ratio of 40, giving a mirror thickness of just 10 cm which made the mirror completely dependent on active optics to achieve its target resolution. The ESO had a fast primary mirror with an *f* ratio of 2.2. Again this was successful, so SOAR went one step further and chose an *f* ratio of 2.0. It is far easier to design an active optics system for a mirror in an altazimuth mount than in an equatorial, as the mirror deflections due to gravity are in one dimension rather than two. In fact the ESO had used an altazimuth mount for their NTT, as this also allowed the telescope enclosure to be much smaller than if they had chosen an equatorial. So the SOAR designers followed suit.

There was a change in the SOAR project's partners in the mid 1990s, with Columbia University pulling out, and Brazil's Ministry of Science and Michigan State University joining the team.[24] Shortly afterwards, in August 1997 the SOAR partners agreed to build the telescope at a capital cost of $28 million, whilst the telescope would be operated by the CTIO.[25] The partners agreed that the SOAR telescope would not duplicate the capabilities of the existing 4 m Blanco telescope at CTIO, but would complement it. As a result, the two telescopes would together complement the capabilities of Gemini South.

SOAR was designed to operate between the atmospheric cut-off at 320 nm up to about 5 microns in the infrared. Its secondary mirror, like its primary, was made of ULE glass. An instrument support box (ISB) at each Nasmyth focus, on the altazimuth mount's altitude axis, could carry a cluster of three heavy instruments, whilst lighter instruments could be mounted at two folded Cassegrain ports on the elevation ring. A tertiary mirror rotated to select the focal station to be used, whilst beam steering optics in each ISB directed light to the chosen instrument. This allowed the observer to switch between instruments in a few minutes. The tertiary mirror was also designed to vibrate in two axes at up to 50 Hz to partially compensate for atmospheric turbulence.

The telescope site was levelled in 1998 and major contracts let in the following year for the mirror blanks, the active optics system, the telescope mount, and the telescope enclosure. Corning produced the 4.3 m diameter meniscus mirror

[24] The observing time allocations were Brazil 31%, Michigan State 13%, North Carolina 16% and NOAO 30%, with Chile receiving 10% as host country, see Jack Baldwin, SOAR Telescope Project Starts Construction, *NOAO Newsletter*, No. 59, September 1999.

[25] Thomas A. Sebring, Gerald Cecil and Gilberto Moretto, *The Soar Telescope Project: A Four-Meter Telescope Focused on Image Quality*, accessed via www.noao.edu/noao/meetings/spie98/gcecil.pdf.

blank[26] with the primary mirror's required radius of curvature which was then sent to Raytheon in Danbury, Connecticut for grinding and polishing. On completion, it was despatched in December 2003 via New York and the Panama Canal to Chile, arriving on Cerro Pachón about a month later. There it was aluminised using the Gemini South vacuum chamber.

SOAR's mirrors were installed in February 2004, and the telescope was dedicated in April of that year. But early scientific observations were severely compromised by a serious problem with the primary mirror's lateral support system. New lateral supports were installed in 2006 to enable the telescope to become fully operational.

In the meantime, the SOAR Optical Imager (SOI) and the OSIRIS[27] near infrared imager and spectrograph had been fitted to the telescope. Then in 2008 the Goodman high throughput optical imager and spectrograph (GHTS) was commissioned followed in 2010 by the SPARTAN near infrared camera, which had been built by Michigan State University. The SOAR Adaptive Optics Module (SAM), built by CTIO, was first used for scientific observations in early 2014.

References

1. McCray, W. Patrick, *Giant Telescopes; Astronomical Ambition and the Promise of Technology*, Harvard University Press, 2004.
2. Robinson, Leif J., Monster Mirrors and Telescopes, *Sky and Telescope*, June 1980, pp. 469–477.
3. McCray, W. Patrick, *Giant Telescopes; Astronomical Ambition and the Promise of Technology*, Harvard University Press, 2004, pp. 168–169.
4. Barr, L. D., The U.S. National Large Telescope Project: The NNTT and Others, in Ulrich, Marie-Helene, (ed.), *Proceedings of a ESO Conference on Very Large Telescopes and Their Instrumentation, held in Garching, March 21–24, 1988*, European Southern Observatory, 1988, pp 73–92.
5. Field, G. B., et al., *Astronomy and Astrophysics for the 1980s*, Vol. 1 (1982), Vol. 2 (1983), National Research Council, National Academy Press, Washington DC.
6. Robinson, Leif J., Update: Telescopes of the Future, *Sky and Telescope*, July 1986, pp. 23–24.
7. McCray, W. Patrick, *Giant Telescopes: Astronomical Ambition and the Promise of Technology*, Harvard University Press, 2004, p. 201.
8. Bahcall, et al., *The Decade of Discovery in Astronomy and Astrophysics*, National Research Council, National Academy Press, Washington, DC, 1991, pp. 76 and 78.
9. McCray, W. Patrick, *Giant Telescopes: Astronomical Ambition and the Promise of Technology*, Harvard University Press, 2004, p. 222.
10. Wilson, R. N., *Reflecting Telescope Optics II; Manufacture, Testing, Alignment, Modern Techniques*, Springer-Verlag, Corrected Second Printing, 2001, p. 221.

[26] The ULE material for the mirror blank had been purchased as long ago as 1990, remaining in storage for almost a decade whilst finance was raised for the project, see Steve Heathcote, and Tom Sebring, SOAR Construction Progressing on Schedule, *NOAO Newsletter*, No. 62, June 2000.

[27] OSIRIS is an acronym of 'Ohio State Infra-Red Imager and Spectrograph'.

11. Robinson, Leif J., and Murray, Jack, The Gemini Project: Twins in Trouble?, *Sky and Telescope*, May 1993, pp. 26–32.
12. Isbell, Douglas, Sidney Wolff: Call me the First Director, *Gemini Focus*, Issue 38, June 2009, pp. 20–23.
13. McCray, W. Patrick, *Giant Telescopes: Astronomical Ambition and the Promise of Technology*, Harvard University Press, 2004, p. 253.
14. Boccas, Maxime, Coating Gemini's Mirrors with Protected Silver, *Gemini Newsletter*, Issue 29, December 2004, pp. 9–13.
15. Carney, Bruce W., *The SOAR Telescope*, ASP Conference Series, 55, (1994), pp. 45–55.

4

Competing Primary Mirror Designs

4.1 Spun Honeycomb Mirrors

The second half of the twentieth century saw the development of three different designs of primary mirror for large optical reflectors. These were spun honeycomb mirrors as used in the Vatican Advanced Technology Telescope on Mount Graham, segmented mirrors (see Section 4.2) as used in the Keck Telescope on Mauna Kea, and thin meniscus mirrors (see Section 4.3) of the sort used in ESO's New Technology Telescope on La Silla. A fourth type, a liquid mirror (see Section 4.5), was also developed, but this had more limited applications. This present section outlines the development of spun honeycomb mirrors as produced at the University of Arizona's Steward Observatory Mirror Laboratory.

The Multiple Mirror Telescope (MMT) had been completed at the Whipple Observatory on Mount Hopkins, Arizona in 1979 with six 72 inch (1.8 m) honeycomb mirrors. These mirrors, made of low expansion, fused silica glass, had been obtained as Air Force surplus equipment. The design of this MMT (see Chapter 10) was innovative in many ways. For example, the observatory building had been designed to allow rapid air movement across both the mirror and telescope structure, allowing them to achieve ambient temperatures quickly and stay there. The honeycomb structure of the mirrors also assisted this.

Neville Woolf of the University of Arizona suggested to his colleague Roger Angel that he look into the possibility of building a larger telescope than the MMT of the same general design.(1) Angel was impressed by the performance of the MMT's honeycomb mirrors and decided to look into the possibility of building larger ones without using expensive low-expansion glass. Instead, Angel and his graduate student John Hill decided to use borosilicate glass, as not only was it cheaper but it had a much lower melting temperature than the low-expansion alternatives, making it easier to cast.

Honeycomb backing for mirrors had been around for some time. The design, which was then called cellar or ribbed, had first been suggested by George Ritchey in the 1920s,(2) and had been used in a modified form for the 200 inch (5 m) Palomar Telescope that saw first light in the late 1940s. This mirror was much lighter than a solid monolithic mirror of the same diameter, and it reached ambient temperature much more quickly. But the Palomar mirror had been produced by ladling buckets of molten glass into a mould, whereas Angel and Hill decided to melt the glass in the mirror mould itself.

Angel's University of Arizona mirror group first produced a 60 cm (24 inch) honeycomb casting in 1981 by heating the borosilicate glass well above its fusing temperature of 850°C to its working temperature of 1,250°C.(3) This experiment, which was successful, was undertaken to get a feel of the problems in handling, melting and cooling the glass. Angel then proposed to take a radical step and make larger honeycomb mirrors by melting the glass in a spinning furnace large enough to make mirrors of about 2 m diameter. The mould would have an array of ceramic cores, held in place by silicon carbide pins, to produce the honeycomb substrate.

If a bowl of liquid is rotated at constant velocity about an axis perpendicular to its surface, that surface will take up a parabolic shape. In fact, in 1908 R. W. Wood had used this effect to produce a telescope with a rotating container of liquid mercury as its primary mirror (see Section 4.5). But his system was not practical at the time, as variations in angular velocity caused ripples in the liquid mercury, and it was limited to looking more or less vertically upwards.

Angel decided to use the same principle to produce a parabolic glass mirror blank by rotating the molten glass, and continuing to rotate it until it set. This had the big advantage over a flat, parallel-sided blank in that it required much less glass to be ground out to produce the final mirror, saving in both time and the amount of raw glass used. In addition, the curvature of the mirror, and hence its *f* ratio, could be adjusted just by changing the rotation speed of the molten glass. Initial funding for Angel's experiments came from the University of Arizona. Shortly afterwards they were supported by the NSF as part of their early funding of technology development work for the NNTT.

In 1983 the Angel's mirror group (later called the Steward Observatory Mirror Lab) moved to a new site in an abandoned temple where their prototype spinning furnace, which spun between 8 and 15 revolutions per minute, could make mirrors of up to about 1.8 m diameter. With this system, on June 8 1984 Angel's team produced a 0.75 m blank using 70 kg (150 lb) of broken borosilicate glass heated up to about 1,200°C. Three days later it had cooled, leaving an excellent f/2 mirror blank.(4) Then in March 1985 the team produced a 1.8 m, f/1.0 mirror blank for the Vatican Advanced Technology Telescope (VATT, see Section 13.1).

At this time, the Mirror Lab was in the process of setting up a much larger facility under a stand at the University of Arizona's football stadium, which would be capable of producing mirror blanks of up to about 8 m diameter. This would allow it to produced mirrors for the NNTT (Section 3.2) which, at that time, was expected to have four 7.5 m mirrors, and the University of Arizona's Large Binocular Telescope (see Section 13.2) with its two 8 m mirrors.

So by 1985 the University of Arizona's Steward Observatory Mirror Lab had grown from a small experimental facility to an industrial style facility. Yet it was still run by astronomers with no formal system for estimating programme timescales and costs. This could not continue if they were to compete for multi-million dollar contracts against the likes of Corning.

One indication of the problem caused by expecting a university department, like the Steward Observatory's Mirror Lab, to operate commercially is given by Roger Angel's estimate in 1985 that the Lab could produce an 8 m mirror blank for $2.5 million each ($3.3 million at 1992 prices).(5) Three years later he was still quoting $2 million to $3 million for an 8 m blank,(6) whereas in 1992 the Mirror Lab's bid for two 8 m mirror blanks for the Gemini project was $12.8 million (assuming that Gemini would cover some of the risk that the mouldings would fail)[1], a unit price of about twice as much. But Gemini wanted a fixed price quotation, with the Mirror Lab or any other bidder covering the cost of failures. This resulted in the University of Arizona quoting $18.5 million for two blanks, which was about three times Angel's earlier estimate of unit prices.

The Arizona University Mirror Lab, as a university department, was not allowed to make a loss on a contract. So, not only did they find it difficult to quote a fixed price, they were unable to take the sort of commercial decision that Corning was thought to have taken in bidding for the Gemini mirrors of quoting a loss-leader price for strategic reasons.

As far as meeting timescales was concerned, in 1985 the Mirror Lab said that it would cast its first 3.5 m mirror blank in September 1986, whereas they only achieved that in April 1988, 19 months late. This was the latest in a series of delays. Of course, commercial undertakings were not immune from cost and schedule overruns, but the Mirror Lab's overruns were rather excessive and due, to a large extent, to their inexperience in running large projects.[2] Eventually, the University of Arizona made a change to a more structured project but this proved to be both painful and prolonged. In fact, it was not until 1992 that the Steward Observatory,

[1] Unfortunately, in the event of a moulding failure the glass and ceramic cores could not be reused. Corning did not have that problem as their ULE glass could be remoulded in the event of a failure.

[2] The fact that their technology was not mature did not help but, as Angel later admitted, their timescale estimates were unrealistically optimistic, see Jan McCoy, Angel Builds 'em Bigger, *Sky and Telescope*, August 1988, pp. 128–129.

pushed by the NSF, finally appointed its first professionally trained manager. At the same time the NSF stopped funding the Mirror Lab, leaving the facility to obtain funds from the telescope projects for which they made the mirrors.

In the meantime, the Mirror Lab had commissioned its new large spinning furnace under the stand of the university's football stadium, and in April 1988 it produced its first 3.5 m mirror blank from two tons of borosilicate glass. The glass, which had been heated to 1,170°C, had remained at that temperature for three hours to allow the mould to fill, and any bubbles to come to the top. The mould was then allowed to cool and continued spinning for 33 hours before the glass had set sufficiently to retain its shape. The maximum speed used was about 8.5 revolutions per minute to produce the f/1.75 blank. Then followed a period of six weeks of further cooling and annealing before the disc was delivered to the Astrophysical Research Consortium's (ARC's) contractor, Norman Cole's Arizona Technologies, for grinding and polishing for their 3.5 m telescope.[3] Later that year the Mirror Lab produced another 3.5 m, f/1.75 blank, this time for the WIYN telescope on Kitt Peak (see Section 2.5). Then in June 1989 they produced a 3.5 m, f/1.5 blank for the US Air Force Phillips Laboratory in New Mexico.

The contract from the Air Force was very important for the Mirror Lab as, not only did it pay for their 3.5 m mirror blank, but they also wanted the Lab to grind and polish their mirror.(7) To facilitate this the Air Force funded the laboratory's development of their stressed-lap polishing system, which the laboratory was developing to polish steep curves for fast mirrors.(8) This Air Force money came at a difficult time for the Mirror Lab and helped to keep it in operation. Their contract, typical of defence contracts, also required the Mirror Lab to keep much more accurate documentation than they were used to. This helped to transform the laboratory from a university research organisation to something more like an industrial one.

The first operational mirror to be polished using this stressed-lap polishing technique was the 1.8 m, f/1.0 mirror for the Vatican Advanced Technology Telescope. The mirror's quality, following its completion in 1990, was excellent with 80% of a star's light (including the effect of diffraction) being concentrated into a circle of just 0.28 arcsec diameter.[4] This was followed by stressed-lapping of the Air Force's 3.5 m, f/1.5 mirror which was completed in just 13 months to produce a mirror with even better optical performance.[5]

The Mirror Lab's next big challenge was to produce its first 6.5 m mirror blank, which it achieved in the spring of 1992. This was an f/1.25 blank for a mirror to

[3] The blank was later returned to the Mirror Lab for polishing owing to the slow progress of Arizona Technologies in undertaking this task.

[4] Assuming perfect seeing and a perfect secondary.

[5] In this case 80% of a star's light (including the effect of diffraction) was concentrated into a circle of just 0.17 arcsec diameter.

replace the six 1.8 m mirrors of the MMT on Mount Hopkins. It was at this stage that the Gemini project sent out its Request for Quotation for two 8 m diameter mirror blanks (see Section 3.3).

It is probably worth quoting a few figures here to give an idea of the size of the facility at the Mirror Lab in 1992, and what it had achieved in producing the 6.5 m blank.(9) The furnace, which spun at up to 7.4 revolutions per minute, was about 12 m (39 ft) in diameter and 6.5 m (22 ft) high, with a rotating mass of 70 tons[6]. At peak power it consumed 700 kilowatts of electricity, using 10 tons of glass to produce the 6.5 m diameter blank. Its spin rate required to produce the blank with an f/1.25 focal ratio generated a 33 cm (13 inch) central depression, with a total shape accurate to about 1 mm. Although the blank had been deliberately made about 1 cm too thick to allow for variations in the mould, only 1 ton of glass had to be ground away. This compares to about 12 tons that would have had to have been removed if the mirror had been made flat.

In early 1994 the Mirror Lab produced another 6.5 m, f/1.25 blank, this time for the Magellan project at Las Campanas Observatory in Chile. Then in 1997 it finally produced its first 8 m class blank. This was the first of two 8.4 m, f/1.14 blanks for the Large Binocular Telescope on Mount Graham. Since then it has clearly moved from a development to a production phase, producing (up to early 2015) five more 8 m class mirror blanks and three more 6.5 m mirror blanks.

4.2 Segmented Mirrors

William Parsons, the third Earl of Rosse, built the first segmented telescope mirror as long ago as 1828. His 15 cm (6 inch) mirror consisted of a central disc, which could be axially adjusted, surrounded by a torus 4 cm wide. The whole mirror, which was made of speculum metal, was then polished to produce a spherical shape and the central disc moved backwards until the images from both parts of the mirror coincided. This reduced the spherical aberration by about 50%.(10)

Just over a century later Guido Horn d'Arturo, of the University of Bologna, also experimented with the production of a segmented mirror. By 1953 he had completed a 1.8 m (6 ft) diameter primary mirror consisting of 61 hexagonal, spherically worked mirrors, each 20 cm across, with a focal length of 10.4 m. Each mirror segment was adjustable in depth and tilt to produce an optimum focus. This segmented primary was fixed on a marble slab and used as part of a zenith telescope with the stellar images recorded on a movable plateholder.[7]

[6] Including the mass of glass.

[7] www.europa.com/~telscope/tsitaly.txt.

In the 1970s PALANTIR or the rotating shoe telescope was considered as a possible design for a very large telescope by engineers at KPNO. Its primary mirror, which was made of hundreds of polished aluminium segments, was 75 m long by 25 m wide. It could be rotated in azimuth whilst the telescope's movable secondary enabled it to view the sky from the zenith down to about 30° above the horizon. The main advantage of this design, over a conventional one, was that the primary mirror could be made rigid and did not suffer from variable gravitational forces as it only rotated in azimuth. But its estimated resolution was found to be rather poor and the telescope would be very expensive to build, so the design was abandoned.

Keck Telescopes

In 1977 Jerry Nelson of the University of California's Lawrence Berkeley Laboratory was asked to join a committee that was considering the future of astronomy at the university.(11) When the 3 m Shane telescope had been completed at the university's Lick Observatory in 1959 it had been the second largest optical telescope in the world. But since then the observatory on Mount Hamilton had been gradually affected by light pollution. So the question was what size telescope should the university now build, and where should it be built?

Some astronomers favoured building a conventional 3 m, 5 m or maybe 7.5 m telescope, whereas others opted for a telescope of around 10 m diameter, but that would require a completely different design to the conventional one. The 4 m Mayall had been completed on Kitt Peak in 1973, and similar-sized telescopes had been completed in the following year on Cerro Tololo in Chile and at the Anglo-Australian Observatory in Australia. In addition a number of other institutions were already building or considering telescopes at the lower end of the 3 m to 7.5 m range. As a result, it was thought that it would be easier to raise funds to build a 10 m telescope as it would clearly stand out from the crowd and have far more PR appeal.

Consequently, the University of California's committee asked Nelson to look into the problem of building a 10 m telescope.(12) Was it feasible and, if so, how should it be built? There were clearly two basic alternatives, either a telescope with a single monolithic mirror or one with a segmented mirror built of many reflecting elements like a radio telescope dish. Nelson recognised that a single monolithic mirror presented a number of problems. The facilities to polish and aluminise it would be very expensive. If there was a fault with the mirror, the whole mirror may have to be rejected and a new one built. And there was the problem of how to transport such a large mirror. On the other hand polishing, aluminising and transporting a mirror made up of several small segments seemed to be much easier. So Nelson chose a segmented mirror as his preferred option, with each of the lightweight

glass segments being controlled by a system of sensors and pistons. At that time KPNO were also looking at something similar for their 25 m diameter NGT following the rejection of their PALANTIR design. As a result Nelson spent some time with their engineers exchanging ideas and setting up a small design team back in California. This team included Terry Mast, a physicist who he recruited from the Lawrence Berkeley Laboratory, and George Gabor, an engineer with experience in hardware and electronics, as key members who worked mainly on the active mirror control system.

The University of California's committee also asked Joe Wampler, who was a member of the university's astronomy department, to examine the possibility of building a new large telescope with a monolithic mirror. Wampler, who had been a leading voice in pushing for the largest mirror possible, thought that it would be much easier to make a thin meniscus mirror, than Nelson's proposed segmented mirror which relied on too many new developments.

Nelson now started to develop his design of 10 m telescope. To save space and money he proposed an altazimuth mounting (which required a much smaller enclosure than an equatorial) and a telescope with as short a focal length as possible. He suggested that this should be about 17 m to produce an f/1.7 instrument. The 60 mirror segments would be hexagonal, each 1.4 m in diameter (corner to corner) and just 10 cm thick,(13) aligned by a system of active optics. These thin mirror segments would not only reduce the mass of the mirror, but would also enable the mirror to respond more rapidly to changes in its thermal environment. They would be made of Schott Zerodur ultralow expansion glass ceramic. Nelson pointed out that the segmented mirror should not only be easier to build than one large mirror, but it should also be easier to produce one with a very short focal length. At the end of 1979 the University of California's so-called 'greybeards committee' of senior astronomers voted for Nelson's design, rather than Wampler's, in a close vote. A little later, the University of California decided to build the telescope on the 4,150 m (13,600 ft) high Mauna Kea after a site testing programme confirmed the excellent seeing there, particularly in the infrared.

The university then created an Executive Management Committee (EMC) to oversee a technology development programme to demonstrate the feasibility of Nelson's design. Harold Ticho, the chairman of the EMC, approached Gerald Smith of JPL in 1981 to see if he was interested in being the project manager for the 10 m telescope project.(14) Smith had assisted the University of Hawai'i in managing the design and construction of the NASA infrared telescope facility (IRTF) on Mauna Kea. So he was familiar with the problems of building a large telescope on the mountain chosen for the 10 m telescope. Unfortunately, at the time of Ticho's approach, Smith was still committed to NASA's IRAS (Infrared Astronomical Satellite) project. But the EMC asked if he could work, for the moment, with Jerry

Nelson part-time to outline the likely schedule and costs of the telescope, which was then called the Ten-Meter Telescope (TMT) Project. He did this until IRAS was launched in 1983. Then he worked full time to produce detailed schedule and cost estimates for the project, and to assist in the management of the technology development work.

It had been obvious from quite early on that one of the most difficult problems with a segmented mirror was how to polish the surface of the individual glass elements so that they collectively made a parabolic surface. The diameter of the segments had now been increased from 1.4 m to 1.8 m, so reducing their total number from 60 to 36, with a thickness of 7.5 cm rather than 10 cm, which would help them to reach ambient temperature more quickly. Nelson wondered if these segments could be stress-polished in a similar way to that that had been used by Bernhard Schmidt in the 1930s to produce non-spherical surfaces in his early Schmidt telescopes.[8] So Nelson approached Jacob Lubliner, one of the most imaginative of the university's civil engineers, who helped him to define and apply the stresses that would need to be applied to each individual off-axis mirror segment.(15) Initial tests on a circular blank showed that stress polishing produced the correct surface figure right to the edge. Cutting after polishing was preferred because it was expected that this would ensure that the correct figure would be maintained to within a few millimetres of the cut edge. But, unfortunately, problems appeared after the circular blank had holes drilled into its back for the mirror supports and after it was cut to its final hexagonal shape. These operations had caused the mirror segment to relax slightly. This problem was eventually solved by modifying the active optics system using a warping harness based on leaf springs near the top of the mirror-positioning actuators. This produced a small permanent force to correct for the segment's relaxation. Final correction was achieved by the application of ion beam figuring(16) to the surface of the mirror segments.[9]

In the meantime, the University of California had been trying to raise funds to enable them to build their proposed 10 m segmented mirror telescope. In May 1984 David Gardner, the president of the university, announced that Marion Hoffman,(17) the widow of a wealthy businessman, had given the university assets valued at $36 million to assist it to build the largest telescope in the world, provided it was named after her late husband, Maximilian E. Hoffman. Unfortunately, this was not enough money for the university to build their proposed 10 m telescope.

[8] Schmidt had invented the process of stress polishing in about 1930. In this, he placed a thin optical flat over the end of a tube and partially evacuated the tube to distort the flat by a specified amount. He then repolished the flat to make it flat again on one face. When this had been done he let the air back into the tube, causing the glass to take up its required shape. The actual shape produced depended on the type of glass, its thickness and the level of vacuum applied.

[9] This ion beam figuring was mainly undertaken by Kodak, following initial research work published in 1988 by Wilson, Reicher and McNeil of the University of New Mexico.

So the University of California, whilst looking for other partners, asked Caltech if they were interested in contributing $25 million for a quarter share in the project. If they had accepted they would have been the junior partner, which naturally did not appeal to them. So Marvin Goldberger, the president of Caltech, was delighted when, three months later, he was able to tell the University of California that they were in discussions with Howard Keck, the son of the oil tycoon William Keck, about providing enough to pay for most if not the whole of the 10 m telescope. Unfortunately for the University of California, however, the conditions of this $70 million Keck donation required that the telescope be named after William Keck, not Maximilian Hoffman, and be owned by Caltech, not the University of California.

Gardner suggested that maybe the two university's could build two 10 m telescopes side-by-side on Mauna Kea, one named after Hoffman and the other after Keck. They could be either used separately or together. But there were problems with the Hoffman donation, as Marion Hoffman had died the day after she had made a verbal agreement with Gardner to transfer the $36 million of assets. Although her secretary had, in accordance with Marion Hoffman's wishes, transferred these after her death, Marion's sister, who was one of the two directors of the Hoffman Foundation disputed the donation. There was also the problem that the donation had been made to build the largest telescope in the world, not one of two.

Caltech was not prepared to wait for these legal problems to be sorted out, which could take years. So Goldberger told the University of California that Caltech was willing, if necessary, to go ahead and build a 10 m segmented mirror telescope on their own. This really left the University of California with no option but to quietly return the Hoffman gift and agree to found a joint corporation with Caltech, called the California Association for Research in Astronomy (CARA), to manage the 10 m telescope project at the so-called W. M. Keck Observatory. Caltech would pay the construction costs, which were then estimated at about $85 million, and own the telescope, whilst the University of California would pay its operating costs, then estimated at about $3.5 million per year, for 25 years. The University of Hawai'i were guaranteed 10% of the telescope's observing time in exchange for providing a site on Mauna Kea, with the other 90% split equally between the two Californian universities. In September 1985, Howard Keck joined officials from the two universities for the ground-breaking ceremony on Mauna Kea.

As mentioned above (Section 4.1), the University of Arizona had had to change their management approach when their Mirror Lab was required to change from being a research and development (R & D) organisation to one required build a number of large, expensive mirrors to a price and timescale. Similarly, the management of the Keck project had to change when it moved from an R & D project, supervised by Jerry Nelson, to one spending the best part of $100 million to build the largest telescope in the world. The university partners appointed a board of directors for overall supervision of the Keck project, and a Science Steering Committee responsible for developing the telescope's scientific instruments. Jerry Nelson was appointed as chief scientist and Gerald Smith as project manager.

The design of the 670-ton steel and aluminium dome was begun in August 1985. It was manufactured by Coast Steel of Vancouver, Canada and in May 1987 it was shipped in parts to Hawaii where assembly was completed a few months later. The Keck team went to great lengths to try to ensure that the interior of the dome was as close in temperature to its external environment as possible. This involved providing an air-conditioning system to keep the interior cool during the day, as well as massive fans which could replace all the interior air in just five minutes, if required. In addition, the adjacent control building was covered with volcanic ash to mimic the reflectivity of its surroundings. Manufacture of the telescope structure had been started in 1987 by TIW Systems in Spain. It was partially assembled and tested there before being shipped to Hawaii where it arrived in August 1989.

The Keck project had decided in late 1985 to award a $10.8 million contract for making 42 mirror segments (36 + 6 spares) to Itek Optical Systems of Massachusetts. But Itek were more used to working on large defence contracts than on a relatively small R & D contract for university staff who knew more than they did about manufacturing and polishing the mirror segments. This resulted in frustration on both sides, exacerbated by timescale and cost overruns. As a result, Tinsley Laboratories of California were contracted to make some of the segments to try to keep the project on schedule. But eventually the Itek contract was cancelled and the remaining work transferred to Tinsley Laboratories.

The first production 2.1 m diameter mirror disc was produced and polished to shape in 1987. It was then cut into its final hexagonal shape at the end of the year. First light of the telescope with just 9 of the 36 segments installed was achieved on November 24 1990. At that stage there was no secondary mirror available, so the first light image was recorded by a CCD array at the mirror's prime focus. Dedication of the telescope took place on November 7 1991 with half of the 36 segments were in place.

The Keck telescope had been designed to use either of two secondary mirrors, a 1.45 m diameter, f/15 mirror made of Zerodur for observations in the visible waveband, and a 0.51 m diameter, f/25 chopping mirror(18) of gold-plated beryllium for infrared observations. The f/15 secondary arrived in Hawaii in mid 1991 and the infrared secondary was installed on the telescope about a year later. Light from the secondary mirror could either pass straight through a hole in the primary or be reflected by a tertiary between the secondary and primary mirrors to either of two Nasmyth platforms where heavy instruments could be located.

The success of the first light image of November 1990 encouraged the Keck Foundation to provide a grant for a second 10 m telescope, Keck II, 85 m away from the first telescope. As a result they announced on April 26 1991 that they would provide $74.6 million towards the construction of this second telescope, which was estimated to cost about $93.3 m. This second telescope could be used either as a stand-alone instrument or with its twin as an interferometer which would provide an angular resolution of ~3 mas (milliarcsecond) at 2.2 microns and ~12 mas at 10 microns.

First light on Keck I with all 36 mirror segments in place, but no scientific instruments, took place on April 14 1992. In the following year the Near-Infrared Camera (NIRC) was the first new Keck-specific scientific instrument to be used on Keck I, allowing the start of scientific observations. It had been designed to produce infrared images and low resolution spectra from about 1 to 5 microns. It had an indium antimonide 256 × 256 pixel array that was cooled to near absolute zero by liquid helium.[10] Shortly afterwards two other instruments were installed on Keck I, the Low Resolution Imaging Spectrometer (LRIS) and the High Resolution Echelle Spectrometer (HIRES),(19) both of which operated from about 0.3 to 1.1 microns with 2,048 × 2,048 CCDs. HIRES weighed about 8 tons so it was located on one of the telescope's two Nasmyth foci. Over the last twenty years or so the instrument suites on both Keck telescopes have been expanded as indicated in Table 4.1.

Scientific observations started using Keck II (see Figure 4.1) in 1996. At about the same time NASA joined as a one-sixth partner in the Keck Observatory mainly because of their interest in observing extrasolar planets with both the individual Keck telescopes and with the anticipated Keck interferometer. Subsequently NASA selected JPL to build the Keck interferometer in collaboration with CARA.

After a number of unsuccessful attempts the two telescopes of the Keck interferometer (see Figure 4.2) were first combined optically in March 2001 when engineering tests produced fringes for a number of stars. Scientific observations started in the following year using adaptive optics on both telescopes. This enabled the interferometer to measure much dimmer objects than previously possible. The first object to be observed was the T Tauri object DG Tau, which the interferometer found was surrounded by a hot dust disc with an inner edge ~ 0.15 AU from the star.

NASA had hoped to extend the two-telescope Keck interferometer by adding four 1.8 m outrigger telescopes. In 1998 they contracted EOS Technologies in Tucson, Arizona to build these telescopes but, although they were built at a cost of about $15 million, they were never located on the summit. Some years earlier an agreement had been made with the Hawaiian community, who regarded

[10] S and T Newswire, *Sky and Telescope*, July 1993, p. 9.

Table 4.1. *Keck Instruments*[a]

Instrument		Year Commissioned	Approx. Waveband (microns)	Comments
NIRC	Near-Infrared Camera	1993	0.8 to 5.3	Retired in 2010.
LRIS Red Side	Low Resolution Imaging Spectrometer	1993	0.38 to 1.1	Upgraded in 2009 with state-of-the-art detectors, electronics and software.
HIRES	High Resolution Echelle Spectrometer	1994	0.3 to 1.1	First light detector upgrade August 2004.
LWS	Long Wavelength Spectrometer	1999	3 to 25	
ESI	Echellette Spectrograph and Imager	1999	0.39 to 1.1	
NIRSPEC	Near-Infrared Echelle Spectrograph	1999	0.95 to 5.5	
LRIS Blue Side		2000	0.32 to 0.9	
NIRC – 2	Near-Infrared Camera 2	2001	1.2 to 4.8	A second-generation NIRC designed to work with Keck adaptive optics.
DEIMOS	Deep Imaging Multi-Object Spectrograph	2002	0.41 to 1.1	
OSIRIS	OH Suppressing Infrared Imaging Spectrograph	2005	1 to 2.4	Designed for use with Keck adaptive optics system.
MOSFIRE	Multi-Object Spectrograph for Infrared Exploration	2012	0.97 to 2.45	

[a] An excellent summary of the specifications of the instruments as of 2002 is given in *Keck Telescope and Facility Instrument Guide*, August 2002, accessed from www2.keck.hawaii.edu/observing/kecktelgde/ktelinstupdate.pdf.

Figure 4.1. The segmented mirror of the Keck II telescope. (Courtesy SiOwl via Wikimedia Commons.)

the mountain as sacred, to limit the number of domes built on the mountain. And although this agreement was modified in 2000, which would have allowed the construction of up to six outriggers, opponents of the construction challenged NASA's environmental assessment. Then in 2006 NASA withdrew finance from the outrigger project because of internal funding problems, although they continued to support the Keck I/II interferometer. Finally in 2011 NASA announced that they would stop funding the Keck interferometer completely. Officially that was because its primary task of imaging dusty discs surrounding nearby stars had been completed.[11] But the project had suffered from its technical complexity, the fact

[11] News Notes, *Sky and Telescope*, May 2012, p. 16.

Figure 4.2. The two Keck telescopes that could be used either individually or as an interferometer. (Courtesy NASA/JPL.)

that it had tied up both telescopes, and the cancellation of the outriggers which had eliminated the interferometric imaging capability.

The first adaptive optics system was fitted to Keck II, rather than Keck I. Its first light was in February 1999, followed by first light of the adaptive optics system on Keck I in December 2000. These systems each had a 6 inch (15 cm) deformable mirror that changed shape at a frequency of up to 660 Hz. In 2003 the Keck Foundation provided a $2 million grant to improve the system.[12] This resulted in the provision of the Next Generation Wavefront Controller which replaced the original ten-year old computers with new computer hardware and real-time software which could provide corrections at up to 2.4 kHz. It was saw first light in February 2007.

First light of the laser guide star (LGS) adaptive optics system which had been installed on Keck II was achieved in September 2003. It became operational in 2005.[13] Unfortunately the 13 watt pulsed dye laser system used was rather complex and required regular maintenance. In addition the coupling efficiency of the laser light with atmospheric sodium atoms was rather low. As a result, the Keck Observatory joined with the Very Large Telescope (VLT) project of the European Southern Observatory and a consortium of US observatories in looking for a

[12] *W. M. Keck Observatory Annual Report 2006*, p. 18.
[13] *W. M. Keck Observatory Annual Report 2006*, p. 17.

replacement laser. Narrowband continuous wave (CW) lasers had been found to have a much higher coupling efficiency than the existing laser, so the observatories looked to develop a 20 watt commercial CW laser. The new laser, which was produced by a joint team from the laser suppliers Toptica Photonics of Germany and the Canadian company MPB Communications, formed part of the Next Generation LGS which was commissioned on Keck II in 2012. Funding came from the Gordon and Betty Moore Foundation, which provided $2 million, and the Keck Foundation which provided $1.5 million.[14] In the meantime a 20 watt solid state laser designed and built by Lockheed Martin and Keck's development team had been installed on Keck I.[15]

The 4,150 m (13,600 ft) altitude of Mauna Kea was a very challenging environment for human observers. As a result there was a major push from early on in the Keck programme to make remote observations possible from an environment much nearer sea level. Consequently, a facility for remote operations of Keck I was provided from the 2,500 ft (750 m) altitude Keck headquarters in Waimea, Hawaii just before Keck II became operational in 1996. By the following year over 50% of Keck I operations were being carried out from Waimea. Subsequently, the bandwidth of the link from Waimea to the summit was increased, so that in 1999 more than 90% of Keck I/II operations were being undertaken from the Keck HQ.(20)

This solved the problem of high-altitude observing, but there was still a great amount of time and money being spent in travelling from the US mainland to Hawai'i. So the next step was to try to provide remote access to the Keck telescopes from the US mainland. If this could be provided, not only would travel costs and time be saved but more people would be able to be involved in the observations themselves.

Although it was feasible to use the Internet to enable most of these observations to be undertaken remotely, it was necessary to ensure that there was an adequate backup in case of problems with the link. As about a dozen different networks were involved in Internet communications from the mainland to Hawai'i, such a backup was clearly necessary. In the event the first mainland control rooms were set up at the University of California, Santa Cruz and Caltech and linked to Hawai'i in 2001. The peak Internet bandwidth available was 35 Mbit/sec, with backup ISDN lines providing 128 Kbit/sec per line. Although the latter bit rate was low, it was adequate to cover critical traffic until normal communications were resumed. Since the first sites became operational a number of other mainland sites have been connected.[16]

[14] *Keck Observers' Newsletter*, Summer 2012.

[15] *W. M. Keck Observatory Annual Report 2010*, p. 14.

[16] Gregory D. Wirth et al., *Operating a Wide-area Remote Observing System for the W. M. Keck Observatory*, accessed via www2.keck.hawaii.edu/inst/people/wirth/spie2008.pdf.

Hobby-Eberly Telescope

In 1983 Lawrence Ramsey and Daniel Weedman of Pennsylvania State University proposed an innovative design for a large, low-cost telescope optimised for optical spectroscopy. Although the primary mirror of their proposed Penn State Spectroscopic Survey Telescope (PSSST), as it was then called, was segmented, it was made of a large number of circular mirrors rather than hexagonal mirrors of the Keck type.(21) Using circular mirrors avoided the problems of cutting hexagonal segments, whilst the gaps between the circular mirrors were not too serious. This was because the telescope was not intended for infrared use, where the heat emitted from the gaps would have been of concern.

Each of the PSSST's seventy-three 0.9 m diameter mirrors had a spherical rather than the usual parabolic surface figure, so they were easy to make and they were all identical. The primary mirror, which had an effective aperture of about 7 m, was fixed in elevation, whilst it and the entire telescope and dome were rotatable in azimuth. So, although the primary mirror could be moved to observe different parts of the sky, the angle of the mirror to the Earth's gravitational vector was constant, again simplifying the design. Clearly the primary mirror could not track celestial objects, but the focal plane detector system could do so for ~40 minutes by moving along a rail system at the telescope's primary focus.[17] This focal plane system had a two-element reflecting arrangement to correct for spherical aberration over a ~1 arcmin diameter field, which was thought to be adequate. The focal plane detectors were linked to spectrographs in a room near the base of the telescope by fibre optics, thus minimising the weight carried on the focal plane rail system.

The University of Texas at Austin joined the project, renamed the SST, in 1985 and offered a site for the telescope at their McDonald Observatory. Five years later the University of Munich agreed in principle to join the project, followed shortly afterwards by the University of Göttingen and Stanford University. The telescope was then named the Hobby-Eberly Telescope (HET) after Bill Hobby, a previous Lieutenant-Governor of Texas, and Robert E. Eberly, a Pennsylvania State benefactor who donated money for the telescope.

The detailed design of the primary mirror changed a number of times during the ten year design phase of the project. Finally in 1994 the circular mirror elements were replaced by hexagonal ones in the design to produce an 11 × 9.8 m spherical primary mirror.(22) These 91 hexagonal mirror segments of Schott Zerodur glass provided a primary mirror with a surface area of 77.6 m², corresponding to a circular mirror of diameter 9.9 m. But the telescope's effective

[17] Consequently, it was described as an 'optical Arecibo' as the Arecibo radio telescope (see Section 22.6) also tracked objects by moving its focal plane detector system over its spherical focal surface.

aperture of 9.2 m was determined by the entrance pupil of the correction optics. To further complicate matters, the actual mirror area used depended on the angle between the primary mirror normal and the direction of the object being observed and, consequently, it varied as the object traversed the mirror. The telescope's fixed tilt of 35° limited sky coverage to about 70% of the sky visible from the observatory. But by accepting such a limitation, the cost of the telescope was very much reduced. Similarly observations were limited to those at the mirror's prime focus, as there was no Nasmyth or Cassegrain focus, again keeping the costs down.

The science instruments consisted of a low-resolution spectrograph contained within the prime focus instrument package (PFIP) on the prime focus tracker, and medium- and high-resolution spectrographs in the basement fed by optical fibres from the PFIP.(23) The spherical aberration corrector (SAC) also contained in the PFIP had a four-mirror double-Gregorian type corrector designed to produce images with a FWHM[18] resolution of better than 0.6 arcsec (ignoring 'seeing'[19]) over a 4 arcmin diameter field of view. Objects could be tracked from between 39 minutes and 2.8 hours, depending on their celestial coordinates.

The telescope was located at the McDonald Observatory on the 6,660 ft (2,030 m) Mount Fowlkes in the Davis mountains of western Texas. Ground breaking took place in March 1994, followed by first light in December 1996 when seven mirror segments had been installed. The telescope was dedicated in October 1997 with all mirrors present, becoming operational two years later. It had been built at a cost of $16 million,(24) or about 15 to 20% of that of a normal 9 m class telescope.

Unfortunately it was clear from its first observations that the telescope's resolution was poor and that it was almost impossible to retain focus. In addition the coatings on the primary mirror segments deteriorated more rapidly than expected. It also transpired that not only were the mirror segments difficult to align but they also lost their alignment rather quickly. This was not helped by the high temperature in the telescope's dome which caused mirror distortions.

It took the best part of ten years to disentangle the various problems and to effect the required improvements. The thermal environment in the dome was improved and the equipment that was used to monitor and correct the positioning of the mirror segments was significantly upgraded. In addition new coatings were applied to the mirror segments and a more thorough and regular cleaning system was introduced.

Even with these improvements the HET telescope did not fulfil the hopes of its original designers. Consequently it was decided in 2009 to change its mission

[18] Full Width Half Maximum.
[19] Median seeing at the observatory was about 0.8 arcsec FWHM.

from that of an all-purpose spectroscopic telescope to an instrument designed to research one topic, namely dark energy. As a result a further $40 million was spent in installing an unprecedented suite of multi-object spectrographs and in increasing the telescope's field of view from 4 to 22 arcmin. These changes are currently (2015) being implemented.

SALT

Meanwhile in the late 1980s astronomers in South Africa were looking to extend their observational facilities at the South African Astronomical Observatory (SAAO). This was located at an altitude of about 5,900 ft (1,800 m) near Sutherland, about 230 miles (370 km) from Cape Town. For some time the South Africans had envisaged building a 4 m class telescope like the European Southern Observatory's 3.5 m New Technology Telescope in Chile or the 3.5 m WIYN telescope on Kitt Peak. That was still their intention when in 1996 senior members of the HET project visited Australia, Chile and South Africa to see if any of these countries were interested in joining them in building a similar telescope to the HET in the southern hemisphere.(25) This approach was greeted with enthusiasm by senior members of the South African astronomical community who realised that it was possible to have a 10 m class telescope for about the same cost as a 4 m class instrument, provided they could accept the performance limitations inherent in the design of the larger instrument.

This period was an opportune one in South African history following the General Election in 1994, which had seen the end of the Apartheid era and the beginning of the reintegration of the country into the international arena. The new administration was keen on developing new projects, especially those with a scientific/technological element. So things moved rather rapidly after the HET visit and in July 1998 the government of South Africa announced that it would fund half the cost of the new telescope, now called Southern African Large Telescope (SALT). The government could see that it would not only help South African astronomers to integrate with their colleagues elsewhere in the worldwide astronomical community, but it would also provide opportunities for young African scientists, engineers and technicians in a high-technology scientific enterprise. As a flagship project, the government was especially keen to use it to stimulate an interest in science and technology education in the country.

Following the agreement in principle of the South African government to fund 50% of the project, Khotso Mokhele, president of the South African National Research Foundation (NRF), and Robert Stobie, director of the SAAO, then set out to try to find international partners to fund the missing 50%. Their efforts were rewarded in October 2000 by the signature of the Memorandum of Understanding between the NRF and organisations in the USA, Poland, Germany, New Zealand and the UK.[20]

Early in the project it was decided to change the fixed inclination from the 35° of the HET to 37° to enable SALT to observe the whole of the Small Magellanic Cloud from its observatory at the SAAO near Sutherland. In addition other design changes were made following the experience gained with the HET. For example the primary mirror's support system was modified by the addition of edge sensors and actuators for each of the hexagonal segments. But one of the most significant changes was to the SAC (spherical aberration corrector) which increased the telescope's field of view by a factor of four, improved the sharpness of the images, and together with newly developed mirror coatings provided better sensitivity at shorter wavelengths down to 320 nm. The new SAC also allowed a much larger instrument package to be mounted on the tracker.

The first light image was taken in October 2003 using seven mirrors and an SAC borrowed from the HET. Then just under two years later the first light image was taken with all 91 mirrors and the SALT SAC in place. The telescope was inaugurated in November 2005 by the country's president Thabo Mbeki. But, unfortunately, two serious problems were found during commissioning. (26) One problem with the Robert Stobie Spectrograph,[21] which was removed from SALT at the end of 2006, was solved relatively easily. But the problem with the telescope's focus which would vary from perfect to dreadful took a number of years to resolve. It turned out to be due to the use of dissimilar materials in the design of the SAC. The SAC's steel frame was bolted to an aluminium ring which in turn was glued into the carbon composite structure of the instrument payload. All three materials behaved differently under thermal load causing exceptionally high stress in the SAC. As a result the SAC was removed from the telescope in April 2009 for modifications and repair. Images taken in August 2010, after the modified SAC had been returned to SALT, showed that the problem had been solved.[22]

[20] In addition to the NRF who contributed 34%, the organisations who collaborated in building SALT were the University of Wisconsin-Madison (15%), Dartmouth College, New Hampshire (11%), Rutgers University (10%), Carnegie Mellon University (3%), and University of North Carolina at Chapel Hill (3%), all of the USA, the Nicolaus Copernicus Astronomical Centre, Poland (11%), the Georg-August University of Göttingen, Germany (5%), the University of Canterbury, New Zealand (4%), and the United Kingdom SALT Consortium (4%), representing a number of UK universities and institutions. In 2007, the American Museum of Natural History and the Inter-University Centre for Astronomy and Astrophysics in India joined the SALT consortium.

[21] Named after Robert Stobie who had been appointed director of the SAAO in 1992. He died in 2002 at the early age of 60.

[22] At the time of writing (2015) SALT still lacks an active alignment system for its mirror segments, requiring operators to undertake the task manually twice per night.

Figure 4.3. The Chinese LAMOST telescope. Light entered the telescope at the left hand dome and was reflected by a Schmidt corrector plate to the telescope's primary mirror at the top of the large slanted tunnel on the left. From there it was reflected back down the tunnel to numerous fibre optics linked to a number of spectrographs. (Courtesy Sheliak and Paul Hilscher via Wikimedia Commons.)

LAMOST

Meanwhile in 1997 the Chinese had begun the detailed design of a unique Schmidt telescope called LAMOST[23] with segmented mirrors to be used for wide field spectroscopy. To reduce costs the LAMOST optics were contained in a fixed structure aligned north–south.

Light from an astronomical source first hit the ~5.7 × 4.4 m active deformable mirror acting as a Schmidt corrector plate that could track the source. This mirror consisted of 24 hexagonal segments in a dome (the left hand dome in Figure 4.3) at ground level. The light was then reflected to the ~6.7 × 6.1 m fixed primary mirror, with a spherical surface figure, consisting of 37 hexagonal mirror segments at the top of a large slanted tunnel. From there the light was reflected back down the tunnel to the primary mirror's 1.75 m wide focal plane where fibre positioning units positioned 4,000 fibres to transfer the light to sixteen spectrographs located in a separate room. This system allowed the spectroscopy of up to

[23] For 'Large Sky Area Multi-Object Fiber Spectroscopic Telescope'. It was also known as the Guoshoujing Telescope after a thirteenth century Chinese scientist.

4,000 objects simultaneously. LAMOST had an effective aperture of 4 m and a field of view of 5°.

The LAMOST project had been approved by the National Development and Reform Commission in October 1997. Construction was begun at Beijing Astronomical Observatory's Xinglong station in September 2001 where the median seeing was found to be about 1.1 arcsec. But since then observation conditions have worsened significantly in terms of both atmospheric and light pollution as the observing site is only about 115 km (70 miles) from Beijing.

First light was achieved in August 2008, and the telescope, which cost 235 million yuan (about $US 34 million) to build, was inaugurated on October 16 of the same year. It was designed to operate over the spectral range 370 to 900 nm and go down to ~20 magnitude after an exposure of 1.5 hours. It is currently (2015) being used to conduct a spectroscopic survey of both stars and galaxies.

4.3 Thin Meniscus Mirrors

Large glass primary mirrors were made very thick in the early twentieth century to ensure that they did not distort as their orientation was changed. But these mirrors were not only very heavy, requiring massive support systems, but they also cooled down to ambient temperatures only very slowly. One solution, which was adopted for the mirror of the 200 inch (5.1 m) Palomar telescope, was to have a hexagonal pattern of ribs on its back which provided stiffness at reduced mass. But the ridged backing was not sufficient to provide support on its own. So in addition the Palomar mirror was carried by 36 support mechanisms inserted into its back. Elaborate lever arm mechanisms balanced the changing forces on the mirror as its orientation was varied. This design of the Palomar mirror gave some idea of what could be done in the 1930s to support a large mirror that was not thick enough to support itself.

The development of computer and more accurate sensor and actuator systems after the Second World War made much more sophisticated mirror support systems possible allowing much thinner mirrors to be used. These so-called active optics systems (see Section 5.1) were designed to maintain the shape of a mirror as its orientation was changed.

In the late 1970s the University of California asked Joe Wampler to examine the possibility of building a large telescope with a monolithic mirror. Wampler concluded that it was feasible to build a 7 m diameter telescope with a thin meniscus mirror, but this solution was rejected in favour of Nelson's segmented mirror design for what came to be called the 10 m Keck telescope. So the idea of a thin meniscus mirror supported by active optics was not proceeded with for the Keck telescope.

In fact the first large telescope to use a thin meniscus mirror with an active optics system was the European Southern Observatory's (ESO's) 3.6 m diameter New Technology Telescope (NTT)(27) which was completed at La Silla in 1989.[24] When this telescope was being designed it was thought to be feasible to build an active optics system to correct a 3.6 m mirror with an aspect ratio (of diameter to thickness) of 18. But this active optics system was unproven, so ESO management insisted that the telescope should be able to produce acceptable results in the passive mode, that is even if the active system failed. As a result a conservative aspect ratio (in today's terms) of 15 was chosen.

Much to everyone's surprise, technical first light with the ESO NTT in January 1989 revealed strong spherical aberration due to an error in the curvature of the primary mirror. Apparently there had been a spacer error made during testing at the manufacturer Carl Zeiss,[25] but Zeiss had realised this before the mirror had been delivered. Unfortunately they had made a mistake in quoting the sign of the error, and so in trying to correct it had doubled the error rather than eliminated it! Fortunately, as it turned out the active optics system was able to correct the double error, but at the expense of using 80% of its dynamic range.(28) Two months later astronomical first light produced the sharpest images yet seen by a ground-based telescope of 0.33 arcsec FWHM.

This was welcome news as the ESO had already decided in 1987 to build a Very Large Telescope (VLT) consisting of four individual 8.2 m telescopes using technology tried out on the NTT. The Schott Zerodur thin meniscus primary mirror of the NTT, in particular, with its active optics system, was to be used to help with the design of the larger telescope. But the primary mirrors of the 8.2 m telescopes were to be only 17.5 cm thick, producing an aspect ratio of an incredible 46.9. This required a much more sophisticated active optics system than that of the NTT controlling 16 natural flexure modes instead of seven. Schott made the first 8.2 m blank for the VLT in 1992. Six years later it went into operation in the first VLT. At about the same time two other 8 m telescopes with thin meniscus mirrors also went into operation. These were the Japanese Subaru and Gemini North telescopes both with an aspect ratio of 40. But both of these telescopes used Corning's ULE fused silica mirrors instead of Schott's Zerodur glass ceramic.[26]

[24] The 2.6 m Nordic Optical Telescope (NOT) completed on La Palma in 1989 (see Section 6.1) also had a thin mirror, but it was significantly smaller than ESO's NTT. In addition, although the NOT mirror was initially supported by a pneumatic system, this support was only modified in 1992 to be a fully functional active optics system.

[25] In the following year, a similar error had been made at Perkin-Elmer in testing the Hubble Space Telescope that had also resulted in strong spherical aberration.

[26] Both Corning's ULE and Schott's Zerodur had virtually zero thermal expansion coefficients.

4.4 Metal Mirrors

André Couder and Dmitri Maksutov independently investigated the possible use of metal mirrors in reflecting telescopes in the 1930s. Couder showed that metal was theoretically superior to glass because of the former's higher thermal conductivity, whilst Maksutov went so far as to produce a number of metal mirrors, including a stainless steel mirror for a 0.7 m telescope at Pulkovo. Maksutov investigated the use of both solid and ribbed metal mirrors, and suggested that copper, bronze and aluminium could be used with chromium coats to take a good polish, or stainless steel which could be polished directly. In all cases they could be used with reflective coats of aluminium.(29) In the early 1950s Maksutov went so far as to suggest that the new Russian 6 m BTA telescope should use a metal primary mirror, but his suggestion was rejected in favour of a mirror of low expansion glass.(30)

Shortly afterwards, Kanigen-coated aluminium primaries were installed on 0.9 m (36 inch) and 1.3 m (50 inch) telescopes on Kitt Peak (see Section 2.3). Unfortunately, the performance of the 1.3 m mirror deteriorated rapidly, causing it to be replaced by a Cer-Vit mirror in 1970. But in 1965 Harold Johnson had used a 1.5 m (60 inch) Kanigen-coated aluminium primary for photometric work at the Catalina Station of the Lunar and Planetary Laboratory (LPL). It was, apparently, much more stable than an adjacent telescope with a Pyrex mirror and produced images of 3 to 4 arcsec diameter.(31) Later, this was reduced to 1 to 2 arcsec. At about the same time Fred Forbes, also of the LPL, demonstrated the thermal stability of a 40 cm (16 inch) nickel-plated aluminium alloy (Tenzaloy) mirror produced by welding together six individually cast segments.(32) It later transpired that these metal mirrors exhibited considerable warping in the astigmatic mode, which gave metal mirrors a bad name in the USA.(33)

In Europe, however, things were much more healthy as in 1968 a highly successful 1.37 m (54 inch) telescope with a Kanigen-coated pure aluminium primary went into operation at Milan Observatory's Merate station in northern Italy. (34) The telescope was tested by ESO after 14 years of continuous use and found to be as good as that of a similar-sized conventional telescope with a glass mirror. Consequently, ESO considered using a nickel-coated aluminium or aluminium alloy primary for their 3.6 m NTT, using either a chemical process (Kanigen) or electrolytic process applied to the nickel,(35) but they were forced to reject these designs for timescale reasons. ESO also considered using aluminium as a backup to glass for their VLT's 8.2 m primary mirrors, but eventually settled on glass.[27]

[27] Claus Madsen, *The Jewel on the Mountaintop; The European Southern Observatory through Fifty Years,* ESO and Wiley-VCH, 2012, p. 205.

However ESO did decide to use a nickel-coated beryllium secondary for their VLT because of its lower mass compared with glass. With a diameter of 1.1 m this beryllium mirror weighed just 50 kg.(36)

4.5 Liquid Mirror Telescopes

The surface of a rotating liquid assumes the shape of a paraboloid with its focal length dependent solely on its angular velocity. This well-known relationship encouraged a number of physicists and astronomers in the nineteenth and early twentieth centuries to try to produce telescope mirrors of liquid mercury.[28] Robert Wood of Johns Hopkins University, who in 1908 undertook the most extensive early investigations, found that the biggest problem was caused by surface ripples. These were caused by problems with the bearings of his liquid container, an inaccurately levelled container, and fluctuations in the container's angular velocity.(37)

Wood was able to minimise the ripples as a result of his experiments and successfully use a rotating 20 inch (50 cm) mirror of liquid mercury as the primary mirror of a reflecting telescope. With this he resolved stars with separations as low as 2.3 arcsec. He also successfully investigated how to eliminate the residual ripples by covering the mercury surface with a thin layer of transparent glycerine. But although such mirrors were much cheaper to make than conventional mirrors, they could only be used to observe the zenith, and so Wood abandoned his investigations.

There were no further attempts to build a Liquid Mirror Telescope (LMT) until Ermanno Borra and colleagues of Laval University, Quebec began experiments in 1982 to extend Wood's work and to eventually build a liquid mirror telescope. After all, although LMTs were limited to zenith observations, they had the advantage that pointing, tracking and telescope flexure problems were eliminated, and atmospheric extinction and refraction were minimised. In addition the general progress of technology in the eighty odd years since the turn of the century made the Laval group's task far easier than that of Wood. For example air bearings, not available to Wood, significantly reduced vibrations, and the use of synchronous motors, also not available to Wood, got rid of variations in spin rate. Consequently, the Laval group was able to build and successfully test a 1 m diameter, f/1.6 and

[28] Mercury does not have as high a reflectivity as aluminium, for example, but it is still pretty high and it can be easily kept in good condition by frequent filtering. (Reflectivities of mercury are 79% at 310 nm and 90% at 10 microns, whilst those for freshly evaporated aluminium are 92% at 320 nm and 98% at 10 microns, see Ronald A. Schorn, Liquid-Mirror Telescopes, *Sky and Telescope*, September 1984, pp. 266–268. Freshly evaporated silver, on the other hand, has a very low reflectivity of just 9% at 320 nm, although it rises to 88% at 360 nm and 99.5% at 10 microns, see R. N. Wilson, *Reflecting Telescope Optics II; Manufacture, Testing, Alignment, Modern Techniques*, Springer-Verlag, Corrected Second Printing, 2001, p. 426.)

a 1.65 m diameter, f/0.9 liquid mercury mirror. These mirrors each consisted of a spinning film of mercury 4 to 7 mm thick on a spuncast polyester resin surface. The experience gained using these mirrors in an experimental set-up were extended in 1986 to using a 1 m, f/4.7 mirror to observe stars. This yielded a FWHM value of about 2 arcsec, which was an excellent result for their sea level environment. But it was noted that turbulent winds in excess of about 10 km/hour caused problems with the surface of the mercury. To solve this the Laval group introduced a transparent Mylar cover, although this inevitably introduced some scattered light. Then over the period 1989–1990 the group produced a 1.5 m, f/2 mirror for laboratory testing. In the process they modified the design of the surface on which the mercury was placed and found that they could reduce the thickness of the mercury to 2.5 mm without it puddling. This reduced the sensitivity of the mercury surface to vibration and wind disturbance, creating a mirror which was diffraction limited.

A joint programme was begun in late 1988 to develop a 2.65 m, f/1.89 LMT by a team led by Paul Hickson at the University of British Columbia (UBC) and by Borra at Laval University (LU).(38) The mirror consisted of a core of aluminium and styrofoam surrounded by a Kevlar skin. The upper surface of the Kevlar was covered by a layer of polyurethane resin formed by spin-casting which supported a 2 mm thick layer of mercury. The whole mirror, including the mercury, weighed only 270 kg. The CCD at the mirror's focus observed a 21 arcsec-wide strip of sky in a time-delay-integration (drift-scan) mode with continuous data readout. The integration time, which was the time taken for an object to drift the length of the CCD, was 130 s. Regular observations at an essentially sea level site in 1994 resulted in astronomical images with an FWHM resolution of about 2 arcsec indicating that they were limited only by atmospheric turbulence.

The Institut d'Astrophysique de Paris then joined the UBC/LU team which was hoping to build an LMT with a primary mirror of at least 5 metres diameter to conduct astronomical surveys. This was to become the Large Zenith Telescope (LZT) project(39) which has since included astronomers from the State University of New York at Stony Brook and Columbia University. In the event the team built a 6 m diameter telescope with a liquid mercury mirror that rotated at 8.5 revolutions per minute. Construction was completed in 2003 at the UBC Liquid-Mirror Observatory. This was at an altitude of about 400 m in the UBC Malcolm Knapp Research Forest about 60 km east of Vancouver. The telescope cost was an order of magnitude less than the cost of a conventional telescope of the same size.

Problems were experienced with the rotation speed of the LZT's mirror (see Figure 4.4), which seemed to be affected by local gusts of wind, and with the large velocity of the outer parts of the mirror, which caused ripples on the surface of the mercury, even though the mercury was only ~1 mm thick. The variations in

Figure 4.4. The 6 m diameter liquid mercury mirror of the Large Zenith Telescope. (Courtesy P. Hickson, University of British Columbia.)

rotation speed were eliminated by the application of a closed-loop servo control system. Whilst the ripple problem, which was found to be caused by atmospheric turbulence just above the mercury's surface, was solved by suspending a thin film of transparent Mylar a few centimetres above that surface.(40) The Mylar, which rotated at the same speed as the mercury, trapped the air between the mercury's surface and the Mylar, forcing that air to rotate at the same speed as the mercury. The air turbulence was still present, but it was now above the Mylar where it could no longer affect the mirror's mercury surface. The Mylar also protected that surface from wind disturbance.

Currently (2015) a team consisting of the University of Liège in Belgium, LU and UBC in Canada, the Aryabhatta Research Institute of Observational Sciences (ARIES) in India, and the Adam Mickiewicz University in Poland are building a 4 m diameter International Liquid Mirror Telescope (ILMT) at the ARIES Observatory at Devasthal in India.

Large telescopes using the new optical technology outlined in this chapter are listed in Table 4.2.

Table 4.2. *Operational Optical Telescopes at least 3.5 m Diameter Using New Optical Technology Listed by Type of Primary Mirror*[a]

Year Completed[b]	Telescope	Type of Primary Mirror	Diameter[c]	Type of Glass	Comments
1994	Astrophysical Research Consortium	Spun honeycomb	3.5 m	Borosilicate	
1994	WIYN	Spun honeycomb	3.5 m	Borosilicate	
2000	MMT upgrade	Spun honeycomb	6.5 m	Borosilicate	
2000 and 2002	Magellan I and II, Baade and Clay Telescopes	Spun honeycomb	Two 6.5 m telescopes	Borosilicate	Two separate telescopes
2008[d]	Large Binocular Telescope	Spun honeycomb	2 × 8.4 m	Borosilicate	Two telescopes on a common mount
1992 and 1996	Keck I and II	Segmented	Two 10 m telescopes	Schott Zerodur	Two separate telescopes
1997	Hobby-Eberly	Segmented Spherical figure	9.2 m	Schott Zerodur	Fixed in elevation
2005	SALT	Segmented Spherical figure	9.2 m	AstroSitall	Fixed in elevation
2008	LAMOST	Segmented Spherical figure	4 m	Schott Zerodur	Fixed primary mirror
2009	Gran Telescopio Canarias	Segmented	10.4 m	Schott Zerodur	Operates into the mid infrared
1989	ESO's NTT	Thin meniscus AR[e] 15	3.6 m	Schott Zerodur	
1998	Telescopio Nazionale Galileo	Thin meniscus	3.6 m	Schott Zerodur	
1998	Subaru	Thin meniscus AR 40	8.2 m	Corning ULE	
1998–2000	VLT	Thin meniscus AR 46.9	Four 8.2 m telescopes	Schott Zerodur	
1999 and 2002	Gemini North and South	Thin meniscus AR 40	Two 8.1 m telescopes	Corning ULE	
2004	SOAR	Thin meniscus AR 40	4.1 m	Corning ULE	
2008	VISTA[f]	Thin meniscus AR 24	4.1 m	Schott Zerodur	Near-infrared telescope
2012	Discovery Channel Telescope	Thin meniscus AR 43	4.3 m	Corning ULE	
2003	LZT	Liquid mercury	6 m		

[a] Valid 2015.

[b] The definition of when a telescope is completed varies from one telescope observatory to another. First light is sometimes declared for a telescope by owners or operators even when the secondary mirror is not present. There is also no consistent build standard for when a telescope is dedicated. The situation is even more unclear for telescopes with segmented mirrors as first light is often announced for a telescope without its full complement of mirror segments present. Even when all the telescope pieces are present there is a shake-down period required whilst the telescope is optimised and problems resolved before it is available to the scientific community for research. So the dates quoted should only be taken as indicative.

[c] The diameter quoted for a telescope is not always clear and consistent from one telescope observatory to another. Sometimes it is quoted as the physical diameter of the mirror, but sometimes as the useful diameter as seen through the telescope optics. The problem is particularly difficult for segmented mirrors that are not even circular and for those telescopes, like the Hobby-Eberly, that do not have full altazimuth tracking. So the diameters quoted in this table should only be taken as indicative.

[d] 2005 with one mirror and 2008 with two mirrors.

[e] Aspect Ratio

[f] VISTA is an acronym of 'Visible and Infrared Survey Telescope for Astronomy'.

References

1. McCray, W. Patrick, *Giant Telescopes: Astronomical Ambition and the Promise of Technology*, Harvard University Press, 2004, p. 121.
2. Ritchey, G. W., *L'Evolution de l'Astrophotographie et les Grandes Telescopes de l'Avenir*, Société Astronomique de France, 1929.
3. Angel, J. R. P., and Hill, J. M., Honeycomb Mirrors of Borosilicate Glass, in *Proceedings of the ESO Conference on Scientific Importance of High Angular Resolution at Infrared and Optical Wavelengths*, Garching, March 24–27 1981, pp. 61–65.
4. Mammana, Dennis L., The Incredible Spinning Oven, *Sky and Telescope*, July 1985, pp. 7–9.
5. Roger Angel: The Glass Giant of Arizona, *Sky and Telescope*, July 1985, pp. 10–11.
6. McCoy, Jan, Angel Builds 'em Bigger, *Sky and Telescope*, August 1988, pp. 128–129.
7. McCray, W. Patrick, *Giant Telescopes: Astronomical Ambition and the Promise of Technology*, Harvard University Press, 2004, p. 158.
8. Wilson, R. N., *Reflecting Telescope Optics II; Manufacture, Testing, Alignment, Modern Techniques*, Springer-Verlag, Corrected Second Printing, 2001, pp. 19–23.
9. Robinson, Leif J., Spinning a Giant Success, *Sky and Telescope*, July 1992, pp. 26–29 and 31.
10. King, Henry C., *The History of the Telescope*, Charles Griffin 1955 (Dover reprint 1979), p. 206.
11. McCray, W. Patrick, *Giant Telescopes: Astronomical Ambition and the Promise of Technology*, Harvard University Press, 2004, pp. 102–104.
12. Parker, Barry, *Stairway to the Stars; The Story of the World's Largest Observatory*, Perseus Publishing, 1994, pp. 68–83.
13. Nelson, Jerry, The University of California Ten Meter Telescope Project: The Segmented Design Hewitt, pp. 11–22, in Adelaide, (ed.), *Kitt Peak National Observatory Conference, Optical and Infrared Telescopes for the 1990s, Tucson, Arizona, 7–12 January 1980, Proceedings Vol. I*, Kitt Peak National Observatory, Tucson, Arizona, May 1980.
14. Smith, Jerry, Keck Profile: Jerry Smith: The Early Years, *Cosmic Matters*, W. M. Keck Observatory, November 2007.
15. Wilson, R. N., *Reflecting Telescope Optics II; Manufacture, Testing, Alignment, Modern Techniques*, Springer-Verlag, Corrected Second Printing, 2001, pp. 28–33.
16. Wilson, R. N., *Reflecting Telescope Optics II; Manufacture, Testing, Alignment, Modern Techniques*, Springer-Verlag, Corrected Second Printing, 2001, pp. 34–37 and p. 181.
17. Ciotti, Paul, Mr Keck's Bequest, *Los Angeles Times Magazine*, May 24 1987, pp. 17–20.
18. Lorell, Kenneth R., et al., Design and operation of the infrared chopping secondary mirror for the Keck 10-m telescope, *Proc. SPIE*, 2201, *Adaptive Optics in Astronomy*, 821, May 31 1994.
19. Krisciunas, Kevin, Science with the Keck Telescope, *Sky and Telescope*, September 1994, pp. 20–26.
20. Shopbell, Patrick L., et al., Remote Observing on the Keck Telescopes, *Astronomical Data Analysis Software and Systems XII*, ASP Conference Series, Vol. 295, 2003, Payne, H. E., Jedrzejewski, R. I., and Hook, R. N., (eds.), pp. 170–173.
21. Ramsey, L. W., and Weedman, D. W., The Penn State spectroscopic survey telescope in *Very large telescopes, their instrumentation and programs*, Proceedings of IAU Colloquium No. 79, Garching, West Germany, April 9–12, 1984, European Southern Observatory, 1984, pp. 851–860.

22. Ramsey, Lawrence W., The Spectroscopic Survey Telescope: Concept and Performance, in *Optical Astronomy from the Earth and Moon*, ASP Conference Series, Vol. 55, 1994, pp. 35–44.
23. Savage, Richard D., et al., *Current Status of the Hobby-Eberly Telescope Wide Field Upgrade and VIRUS*, in Stepp, Larry M., and Gilmozzi, Roberto, Ground-based and Airborne Telescopes II, *Proc. SPIE*, Vol. 7012, August 2008.
24. Zimmerman, Robert, Flawed Giants, *Sky and Telescope*, January 2015, pp. 60–67.
25. Buckley, David, et al., *Africa's Giant Eye; Building the Southern African Large Telescope*, SALT Foundation, 2005, p. 21.
26. Charles, P. A., Status of SALT, and Joint Science Projects with the GTC, *RevMexAA (Serie de Conferencias)*, 42, 2013, pp. 96–98.
27. Wilson, R. N., *Reflecting Telescope Optics II; Manufacture, Testing, Alignment, Modern Techniques*, Springer-Verlag, Corrected Second Printing, 2001, pp. 199–204.
28. Wilson, R. N., The History and Development of the ESO Active Optics System, *The Messenger*, No. 113, September 2003, p. 5.
29. Wilson, R. N., *Reflecting Telescope Optics II; Manufacture, Testing, Alignment, Modern Techniques*, Springer-Verlag, Corrected Second Printing, 2001, p. 219.
30. Wilson, R. N., *Reflecting Telescope Optics I*, Springer-Verlag, Second Edition, 2007, p. 450.
31. Johnson, Harold L., The Design of Low-Cost Photometric Telescopes, *Vistas in Astronomy*, 10, 1968, pp. 149–158.
32. Forbes, Fred F., A 40-cm Welded-Segment Lightweight Aluminium Alloy Telescope Mirror, *Applied Optics*, 8, 1969, pp. 1361–1363.
33. Wilson, R. N., *Reflecting Telescope Optics II; Manufacture, Testing, Alignment, Modern Techniques*, Springer-Verlag, Corrected Second Printing, 2001, p. 232.
34. Mottoni, G. de, An Italian 54-inch Reflector of Unusual Design, *Sky and Telescope*, May 1972, pp. 296–297.
35. Mischung, K. N., ESO's New Technology Telescope (NTT) metallic primary mirror project, *Proceedings of IAU's Colloquium Number 79 on Very large telescopes, their instrumentation and programs, Garching, West Germany*, European Southern Observatory, 1984, pp. 57–64.
36. Stanghellini, S., and Michel, A., Performance of the First Two Beryllium Secondary Mirrors of the VLT, *ESO Messenger* 94, December 1998, pp. 10–11.
37. Gibson, Brad K., Liquid Mirror Telescopes: History, *Journal of the Royal Astronomical Society of Canada*, 85, No. 4, (1991), 158–171.
38. Hickson, Paul, et al., UBC/LAVAL 2.7 Meter Liquid Mirror Telescope, *Astrophysical Journal*, 436, L201–L204, 1994.
39. Gamma, Eagle, Liquid Astronomy; The Large Zenith Telescope's Mercury Mirror Provides Critical Insights for Next-Gen Instruments, *Sky and Telescope*, April 2013, pp. 26–30.
40. Hickson, Paul, Liquid-Mirror Telescopes; An Old Idea for Astronomical Imaging is Undergoing a Technology-Driven Renaissance, *American Scientist*, 95, No. 3, May–June 2007, p. 216.

5

Active Optics, Adaptive Optics and Other Technical Innovations

5.1 Active Optics

Active optics is usually understood to refer to a low frequency control system applied to the primary mirror of large reflecting telescopes. It corrects for optical aberrations like spherical aberration and astigmatism either detected in the initial set-up of the telescope or caused by flexure of the mirror as the telescope's orientation is changed or its temperature varies. To make the corrections an active optics system detects errors in the mirror's surface shape either directly or via its effect on astronomical images and corrects them using a motorised support system under the mirror. The mirror can be either monolithic, in which case the flexure function has no discontinuities, or segmented where the flexure function is discontinuous.[1]

The first significant step on the road to designing an active optics system appears to have been taken by André Couder in 1931 when he suggested that the astigmatism of inadequately supported mirrors could be corrected by applying appropriate forces to the back of the mirror. Astigmatism left over after mirror manufacture could also be corrected in this way. But he could only measure the astigmatism qualitatively off-line making such corrections cumbersome and slow. As a result his suggested procedure was limited to the initial setting up of the telescope, rather than being applied continuously as in a modern active optics system. Independently Dmitri Maksutov appears to have come up with a broadly similar proposal in 1948, but it had the same limitations.

Technological developments in computer and sensor systems after the Second World War began to make it possible to design much more sophisticated active optics systems. In the event the first major telescope to use a form of active optics was the Multiple Mirror Telescope (MMT) which consisted of six identical 1.8 m

[1] R. N. Wilson, *Reflecting Telescope Optics II; Manufacture, Testing, Alignment, Modern Techniques*, Springer-Verlag, Corrected Second Printing, 2001. A simplified version of the active optics discussion in his book is contained in R. N. Wilson, The History and Development of the ESO Active Optics System, *The Messenger*, No. 113, September 2003, pp. 2–9.

telescopes on a common mount. The MMT was designed to bring all six images to a common focus on its central axis. It was planned to continuously monitor the flexure of the structures of the six individual telescopes using an internal laser system, and to correct them in real time to ensure that all six images coincided. The telescope was dedicated in 1979 but it took quite a number of years to get the active optics system to work. In the process the laser system had to be abandoned and the six images had to be digitally recorded separately.

It was decided to abandon the MMT concept in 1987 as by then it was clear that it was possible to build very large monolithic mirrors, which had been a very doubtful proposition when the MMT had originally been designed. Instead the MMT's six telescopes were replaced by a single 6.5 m diameter primary of lightweight construction in the same housing.

Active optics had been tried with segmented primary mirror telescopes in the early 1970s when a relatively simple 4.2 m infrared spectroscopic telescope had been built at the Meudon Observatory in France. It consisted of 36 square segments with a closed-loop feedback control system based on measurements of a natural star image. Unfortunately the quality of the individual segments was poor, and so the image had a resolution of only about 10 to 20 arcsec. The telescope was abandoned in 1975 because of a general lack of interest.

A few years later Jerry Nelson designed what was to become the 10 m Keck telescope (see Section 4.2) with 36 lightweight glass segments controlled by an active optics system. That system used a natural star to initially adjust the mirror segments. But in routine operations the system detected the movement of each mirror segment relative to its neighbour and made corrections twice a second.(1) It was, and still is highly successful.

Before the Second World War monolithic telescope mirrors had been made as thick and stiff as possible to avoid flexure. But this had put a limit on the size of telescopes that could reasonably be built as these thick heavy mirrors also required massive mounting systems. But technological developments in the 1960s and 1970s meant that much more sophisticated mirror support systems became possible for monolithic mirrors allowing much thinner mirrors to be used.

Interestingly Ray Wilson came up with the idea of an operational active optics system for a telescope with a large monolithic mirror when he was investigating another telescopic optical effect. In 1968 he was working at Zeiss on a study for the optics of the European Southern Observatory's 3.6 m telescope when he examined the problem of maintaining the alignment of the axis of the secondary mirror to that of the primary in Cassegrain telescopes during tracking. Misalignment produced what was called decentering coma. He concluded that it was virtually impossible to solve the misalignment problem structurally by making the system rigid, as the lateral offset of one mirror relative to the other had to be maintained to much better than 1 mm. Instead he suggested observing the aberration of a star's image in real time and moving the secondary laterally to correct it. This eventually led Wilson[2] to propose in 1977 that a similar approach could be applied to correcting image aberrations caused by flexure of the primary mirror. He suggested that an analysis of image distortions could be used to determine the forces needed to be applied to the primary mirror to correct them. This closed-loop system using mechanical actuators acting on the primary mirror would operate automatically in real-time.

A few years later the European Southern Observatory decided to build a New Technology Telescope (NTT) with a thin monolithic mirror of about 3.5 m diameter as a first step to building a very large telescope. By 1982 they had decided to incorporate an active optics system(2) into this NTT based on Wilson's design. But to check that the system worked correctly ESO built a prototype with 75 actuators operating on a thin 1.05 m diameter test mirror. The mirror's thickness of only 18.9 mm was calculated by scaling the NTT primary (diameter 3.58 m, thickness 240 mm) according to the gravity flexure law. In the event the experiment was a resounding success.(3, 4)

When the ESO's 3.6 m diameter NTT was completed in 1989 it was the first large telescope to use a thin monolithic mirror with an active optics system (see Figure 5.1). But to be on the safe side ESO's Director General Lodewijk Woltjer had insisted that the telescope should be able to produce acceptable results in the passive mode, even if the active system failed. As a result a rather thicker mirror had been chosen than was strictly necessary assuming that the active optics system worked.

In the event, due to an error made in testing, the curvature of the primary mirror of the NTT was built to the wrong figure, like that of the Hubble Space Telescope. But, unlike the Hubble Space Telescope, the NTT's active optics system was able to correct the error, although at the expense of using 80% of its dynamic range. Two months later astronomical first light produced the sharpest images yet produced by a ground-based telescope of 0.33 arcsec FWHM. But this was achieved manually as the NTT software had no provision to make the active optics correction automatically.

This excellent resolution came as something of a relief as the ESO had already decided in 1987 to build the Very Large Telescope (VLT) consisting of four individual 8.2 m telescopes using technology tried out on the NTT. The thin meniscus primary mirror of the NTT, in particular, with its active optics system, was to be used to help the design of the larger telescope. As a result the VLT active optics software was installed on the NTT in 1996, thus making the corrections automatic on the NTT for the first time. But the primary mirrors of the 8.2 m telescopes were to be only 17.5 cm thick. This meant that they were about 50 times more flexible

[2] Wilson, in the meantime, had joined ESO in 1972.

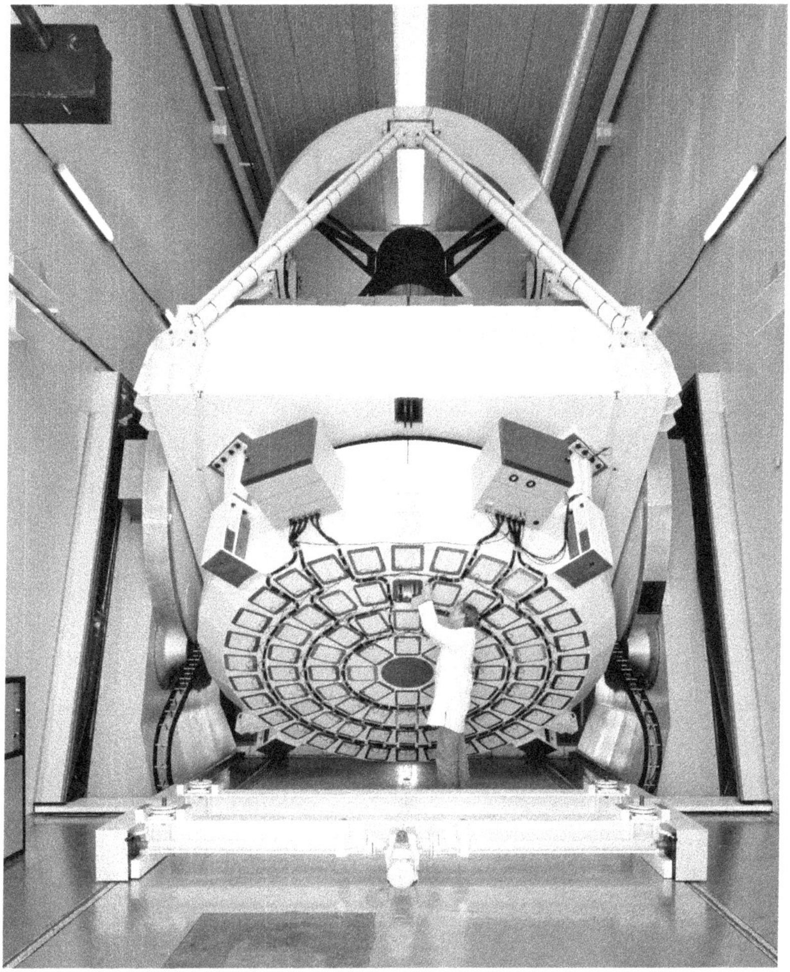

Figure 5.1. The underside of the primary mirror cell of ESO's New Technology Telescope showing the location of four rings of axial supports which formed its active optics system. The 78 axial supports are in recesses under the covers shown. (Courtesy ESO/C. Madsen.)

than the NTT primary so their active optics systems had to control 16 natural flexure modes instead of six.

Initial optical tests were carried out with the first 8.2 m VLT telescope with an uncoated primary mirror in early 1998. They resulted in a resolution of 0.4 arcsec. At about the same time two other 8 m telescopes with thin monolithic mirrors and active optics also went into operation. These were the Japanese Subaru and Gemini North telescopes both of which had mirrors almost as flexible as those of the VLT (see Table 4.2). By the end of 2000 first light had been achieved by all four of the ESO's VLTs.

5.2 Adaptive Optics

In 1953 Horace Babcock of the Mount Wilson Observatory suggested a method of eliminating the problem of 'seeing' in optical telescopes by using a thin film of oil deposited on a small mirror inserted into a telescope's light path. A control system would sense how atmospheric turbulence was modifying the astronomical image, and would adjust the shape of the oil-film mirror electrostatically to correct the image distortion.(5) Unfortunately the Carnegie Institution that supported research at Mount Wilson did not feel able to fund such a development programme at that time.

David Fried of Rockwell International found in the mid 1960s that atmospheric turbulence had less effect at infrared wavelengths than at optical wavelengths, so an adaptive optics system should be easier to build at these longer wavelengths. The isoplanatic angle, within which the phase of the wavefront was nearly uniform, was later estimated to be about 2 arcsec at 0.5 microns, 10 arcsec at 2.2 microns and 30 arcsec at 5.0 microns.

A small research team at Itek of Lexington, Massachusetts led by John Hardy produced a laboratory demonstration of an adaptive optics system in 1973.(6) This was under contract to the US military who were interested in imaging Soviet spacecraft in orbit. Hardy's system used 21 actuators that deformed a 5 cm diameter monolithic piezoelectric mirror. At the time a number of research institutes had been trying, with limited success, to improve images after they had been recorded, whereas Hardy's system corrected the images before they were recorded. In June 1974 a joint US Air Force/Itek team showed that Hardy's system could compensate for the atmospheric turbulence experienced by a helium neon laser beam transmitted along a horizontal path about 1 m above the ground over a length of 300 m. In spite of such an encouraging start, however, it took another eight years before a good quality spacecraft image could be produced from the ground. This was using a 10 cm diameter deformable mirror, with 168 actuators, attached to a 1.6 m military telescope on the 3,050 m (10,000 ft) summit of Haleakala in Hawaii. But this work was unknown to the astronomical community at the time as it was classified.

In parallel a number of civilian research groups were trying to build an adaptive optics system. For example, Robert Smithson and colleagues at the Lockheed Research Laboratories built a system using a deformable mirror of nineteen hexagonal segments arranged in a circular array 15 cm in diameter.(7) The angle of each mirror segment could be adjusted using three piezoelectric pistons or actuators. The light from each segment was focused by one of nineteen small lenses in what was called a Shack-Hartmann array, one lens for each segment, onto a CCD. This allowed the displacement of the light coming from each mirror segment to be determined, enabling correction signals to be applied. In 1983 Smithson and colleagues successfully tested the system by imaging sunspots and solar granulation using the 0.76 m Vacuum Tower Telescope (now called the Dunn Solar Telescope) at the US National Solar Observatory on Sacramento Peak. It achieved a resolution of about 2 to 3 arcsec in 6 arcsec seeing conditions.

Gerard Rousset, Fritz Merkle and colleagues produced a new adaptive optics system in the late 1980s as an early prototype for the system to be installed on ESO's VLT due to be built in Chile. Called COME-ON,[3] it was similar to Smithson's system with a Shack-Hartmann wavefront sensor and a deformable mirror, but its deformable mirror had a continuous reflecting surface instead of discrete segments. This continuous surface of silver-coated silicon was bonded to an array of nineteen piezoelectric actuators, each of which changed the local surface curvature when a voltage was applied. The wavefront changes were determined by observing a bright star close to the target object so that it suffered the same atmospheric distortion as the target. Unlike Smithson's system, however, COME-ON operated in the infrared, taking advantage of Fried's discovery.

In 1989 Rousset and Merkle's COME-ON system was tested on the 1.5 m telescope at the Haute-Provence Observatory in France, where it produced the first diffraction-limited images ever made of a faint astronomical object.(8, 9) It resolved a binary star system with an angular separation of only 0.5 arcsec at a wavelength of 2.2 microns, in seeing of ~1 arcsec in the visible waveband. The FWHM of each stellar component was 0.32 arcsec. Not only did the adaptive optics system produce a sharper image than the uncorrected one, but it also produced star images an order of magnitude brighter. In the following year the COME-ON system was installed on ESO's 3.6 m telescope on La Silla where it clearly resolved a binary star of 0.38 arcsec separation at a wavelength of 3.6 microns. This adaptive optics system produced a FWHM value down to 0.17 arcsec at 2.2 microns and a Strehl ratio[4] of better than 0.60 at 3.8 microns.(10)

ESO also developed an improved version of COME-ON called COME-ON+ that had a deformable mirror with 52 actuators and a better Shack-Hartmann wavefront sensor. Although it could be used on dimmer objects than its predecessor, light loss in the system was still a problem. In addition both systems suffered from requiring a relatively large team of specialists to optimise and operate them. So ESO developed a second generation system called ADONIS[5] which was designed to have improved performance, versatility and operational efficiency. As a result its computer system incorporated an artificial intelligence system to handle overall instrument control and the interface with the user. ADONIS was first installed on

[3] From 'Cge, Observatoire de Meudon, Eso, ONera'!

[4] The Strehl ratio is the ratio of the measured central intensity of the image of a point source to that of its theoretical intensity (i.e. the central intensity of the Airy disc).

[5] ADONIS is an acronym of 'ADaptive Optics Near-Infrared System'. For a description of the system see J.-L. Beuzit, et al., Adaptive Optics on a 3.6 m Telescope; The ADONIS System, *Experimental Astronomy*, 7, 1997, pp. 285–292.

ESO's 3.6 m telescope in late 1994 to assist, like its predecessors, in the development of an adaptive optics system for the VLT. In the following year it achieved diffraction-limited performance at 2.2 microns.

Curvature Sensor and Bimorph Systems

The adaptive optics systems described above required a computer to calculate the magnitude of the signals to be applied to each of the actuators attached to the deformable mirror. In the late 1980s, however, François Roddier proposed an alternative type of adaptive optics system. It used a curvature sensor that measured the brightness of the wavefront on two parallel planes, rather than a Shack-Hartmann wavefront sensor. The curvature sensor was then directly linked to a bimorph deformable mirror which consisted of a sandwich of two sheets of piezoelectric materials. No intervening computer was required. In 1995 a thirteen electrode prototype was fitted to the 3.6 m Canada-France-Hawaii telescope on Mauna Kea, achieving a resolution of 0.15 arcsec in the near infrared. In the following year this prototype was replaced by a nineteen element system named Pueo[6] which achieved a resolution of 0.10 arcsec at about 1.25 microns.(11) In 1997 yet another system, called Hokupa'a (immovable star), was installed on the 3.6 m telescope with 36 segments in its sensor and bimorph. Two years later it was moved to the 8 m Gemini North telescope also on Mauna Kea. At that time Hokupa'a was the only adaptive optics system available to guide an 8 m telescope using faint stars. In 2003 a 60 actuator curvature sensor/bimorph system called MACAO[7] was also installed on ESO's VLT where it achieved a resolution of 0.06 arcsec at about 2.2 microns.

Altitude-Conjugate Systems

In 1994 the Gemini consortium awarded the contract to build an adaptive optics system for Gemini North to the Canadian Herzburg Institute for Astrophysics, who had worked on the Pueo system. Called ALTAIR[8] the 177 actuator Gemini North system intercepted light from the telescope's tertiary mirror and fed a corrected beam to the telescope's instruments. Unlike most earlier instruments ALTAIR was an altitude-conjugate system which could correct for distortion by atmospheric turbulence at any given altitude. It was fitted to Gemini North in late 2002 and achieved a resolution of 0.06 arcsec at 1.6 microns in the following year.[9]

[6] Pueo is the name of a little Hawaiian owl with exceptionally sharp night vision.

[7] For 'Multi-Application Curvature Adaptive Optics'.

[8] ALTAIR is derived from 'ALTitude conjugate Adaptive optics for the InfraRed'.

[9] ALTAIR was built assuming that the dominant turbulent layer at Mauna Kea was at 6 km altitude, but this turned out to be incorrect. So a field lens had to be added to tune the instrument to the correct altitude, see Brent Ellerbroek, Blazing Gemini's Path to Multi-conjugate Adaptive Optics, *Gemini Focus*, Issue 38, June 2009, p. 26.

Laser Guide Star Systems

One problem with these adaptive optics systems was that they needed a nearby, reasonably bright star to enable the wavefront distortion of the target astronomical object to be determined. This was a particular problem when using wavelengths of less than about 3 microns. Renaud Foy and Antoine Labeyrie suggested in 1985 that, instead of being limited to observing objects near to relatively bright stars, it may be possible to produce an artificial star adjacent to the target astronomical object using a laser.(12) One solution involved using a sodium laser to excite the resonance fluorescence of sodium atoms concentrated in a layer about 90 km above the earth. Alternatively a laser could be used to produce Rayleigh scattering by oxygen and nitrogen molecules in the dense regions of the atmosphere up to about 30 km altitude.

In 1991, when the military work on adaptive optics was declassified, it was found that the military had been experimenting with systems to observe military targets using Rayleigh scattering to produce the laser guide 'stars'.(13) In fact the breakthrough had been achieved in February 1989 when a team led by Robert Fugate at the Sandia Optical Range[10] on Kirtland Air Force Base, New Mexico had managed to improve the resolution achieved by a 1.5 m telescope from 2 to 0.18 arcsec using a laser guide star.(14)

The first attempt to use an adaptive optics *astronomical* system with an artificial guide star was made in 1995 at the Lick Observatory. Instead of using Rayleigh scattering, however, astronomers used a pulsed sodium laser to excite sodium atoms in an atmospheric layer high above the earth. Initially the pulsed sodium laser facility was situated outside the dome of the 3 m Shane telescope. But the tests went so well that, in the following year, the Shane telescope was used to both transmit and receive the laser pulses, and also to observe the astronomical objects. The peak intensity of the corrected image was found to have increased by a factor of 2.4 compared with that of the uncorrected image, and the width of the image to have reduced from 0.75 to 0.31 arcsec. The Lick facility was later used as a testbed for the laser guide star (LGS) adaptive optics system that was fitted to the Keck II telescope.

To produce a laser guide star the laser light must traverse the earth's atmosphere once, and to observe it by the telescope the light must traverse the atmosphere again in the opposite direction. As a result the laser guide star could not provide information on the tilt of the wavefront from the adjacent astronomical source because the tilt of the laser guide star's wavefront cancels out in going backwards and forwards through the same part of the earth's atmosphere. So a natural guide star was still required, even with a laser guide star system, to provide information on the tilt of

[10] Later called the Starfire Optical Range.

Figure 5.2. A laser beam is emitted by ESO's VLT Yepun (Unit 4) telescope as part of its adaptive optics system. (Courtesy G. Hüdepohl/ESO.)

the wavefront from the astronomical source. But even though a natural guide star was still required, this laser guide star system had reduced demands on the brightness and distance from the astronomical target of the natural guide star used, leading to a ten-fold increase in the area of sky covered.

First light of the Keck II LGS adaptive optics system was achieved in September 2003, with the system becoming operational in 2005.[11] In the following year the European Southern Observatory also installed a laser guide star system on one of their VLT telescopes (see Figure 5.2). Unfortunately the 13 watt pulsed dye laser system used by the Keck was rather complex and required regular maintenance. In addition the coupling efficiency of its laser light with atmospheric sodium atoms was rather low. As a result, the Keck Observatory joined with the European Southern Observatory and a consortium of US observatories to produce a more efficient laser system. Narrowband continuous wave (CW) lasers had been found to have a much higher coupling efficiency than their existing laser, so the observatories looked to develop a 20 watt commercial CW laser. The new laser eventually formed part of the Next Generation LGS which was commissioned on Keck II in 2012.

[11] *W. M. Keck Observatory Annual Report 2006*, p. 17.

Multi-Conjugate Systems

Both the natural guide star and laser guide star systems discussed so far were only able to correct atmospheric turbulence for stars very close to the guide star on the celestial sphere. But in 1987 Jacques Beckers suggested that a new technique called 'Multi-Conjugate Adaptive Optics' could be used to correct images over a much larger area. This MCAO system was based on atmospheric tomography in which wavefront distortions would be observed in slightly different directions to enable the three-dimensional behaviour of the wavefront to be inferred and corrected.(15) In 2000 Ragazzoni, Marchetti and Valente reported that they had demonstrated such a system using three off-axis stars about 15 arcsec from the central star.(16) Further developments were reported by Maud Langlois and colleagues in 2004 who used such a system to observe the Sun with the 76 cm Dunn Solar Telescope at Sacramento Peak(17) and by Berkefeld, Soltau and von der Luehe in 2005 who also observed the Sun with the German 70 cm Vacuum Tower Telescope on Tenerife.(18) This work at both solar observatories increased the diameter of the corrected field to about 30 arcsec or so, enabling the whole of the visible disc of the Sun to be imaged at once.

Around the turn of the century a number of organisations were looking to build telescopes of 8 m or more in diameter. ESO, in particular, examined the possibility of building telescopes some tens of metres in diameter for which adaptive optics would be essential. As a result they decided to build a multi-conjugate demonstrator called MAD (for Multi-conjugate Adaptive optics Demonstrator) optimised for the K band (2.2 microns), with a field of view of 2 arcmin. It had three Shack-Hartmann wavefront sensors and two deformable mirrors, one deformable mirror conjugated at 8.5 km above the telescope and the other at the telescope pupil.(19) MAD was first successfully tested in ESO's laboratory at Garching, Germany in 2005. Then two years later it was attached to one of the VLT telescopes in Chile where it used three natural guide stars. In 0.7 arcsec seeing MAD produced images with FWHM values of from 0.087 to 0.107 arcsec over an area of 20 × 20 arcsec near the centre of the globular cluster Omega Centauri. The nearest guide star was ~1 arcmin away.

In the meantime the Gemini project had set out to design a multi-conjugate adaptive optics system using a number of laser guide stars, rather than natural guide stars, and deformable mirrors to estimate and correct atmospheric turbulence over a larger area. This GeMS system[12] saw first light on Gemini South in December 2011 using a constellation of five laser guide stars produced by splitting the light from a 50 Watt laser. The system also used three natural guide stars

[12] GeMS is an acronym of 'Gemini Multi-conjugate System'.

and three deformable mirrors optically conjugated at 0, 4.5 and 9 km, and one tip-tilt mirror.(20) It achieved a FWHM resolution of 0.08 arcsec in H band (about 1.65 microns) over a field 87 arcsec square.(21) By 2013 it was found that GeMS produced image resolutions of 0.075 arcsec or better 50% of the time, even though one of the deformable mirrors had failed.

Adaptive Secondary Mirrors

Although the performance of adaptive optics systems has gradually been improved over time, this has been at the expense of design complications. This had often led to crowding in or near the focal plane, and more components producing higher emissivity in the infrared. A possible solution to these problems was to make the telescope's secondary mirror fully deformable, so reducing the number of components in the system. But this was clearly not a simple job, considering the size and mass of secondary mirrors, particularly for large telescopes.

The first fully deformable secondary mirror made for a large telescope was built by the University of Arizona and the Osservatorio Astrofisico di Arcetri in Italy for the new MMT on Mount Hopkins, Arizona. Attached to the telescope in 2003,(22) this deformable secondary mirror consisted of a 1.6 mm thick glass shell, 64 cm in diameter, that floated about 30 microns in front of its backing. The shell was deformed by 336 voice coils pushing on magnets glued to its back surface. Experience with this system, which operated with a Shack-Hartmann wavefront sensor, was used to assist in the design of a similar mirror that was installed on the Large Binocular Telescope (LBT) on Mount Graham, Arizona in 2010.

Each of the two 91 cm diameter deformable secondary mirrors of the LBT had a 1.6 mm thick Zerodur shell controlled by 672 computer-controlled actuators. A pyramid sensor(23) was used to detect atmospheric distortions. The system produced a resolution of 0.040 arcsec and Strehl ratios of better than 80% in H band (1.65 microns) with a natural guide star system operating at 1 kHz.(24)

In the meantime, planning for an adaptive secondary mirror to be installed on ESO's VLT was started in 2004. Eight years later the mirror was delivered to ESO's Garching laboratory for testing prior to its expected installation on one of the four VLTs. The 112 cm diameter mirror consisted of a 2 mm thick Zerodur shell floating 65 microns above the supporting substrate with 1170 actuators operating at 1.2 kHz (see Figure 5.3). It is intended to operate this system with a four-laser guide star system.(25)

Today adaptive optics are being used on large telescopes all over the world, generally operating in the near infrared between 1 and 5 microns, both with and without artificial guide stars. Those described above are the ones that have been in the forefront of adaptive optics developments.

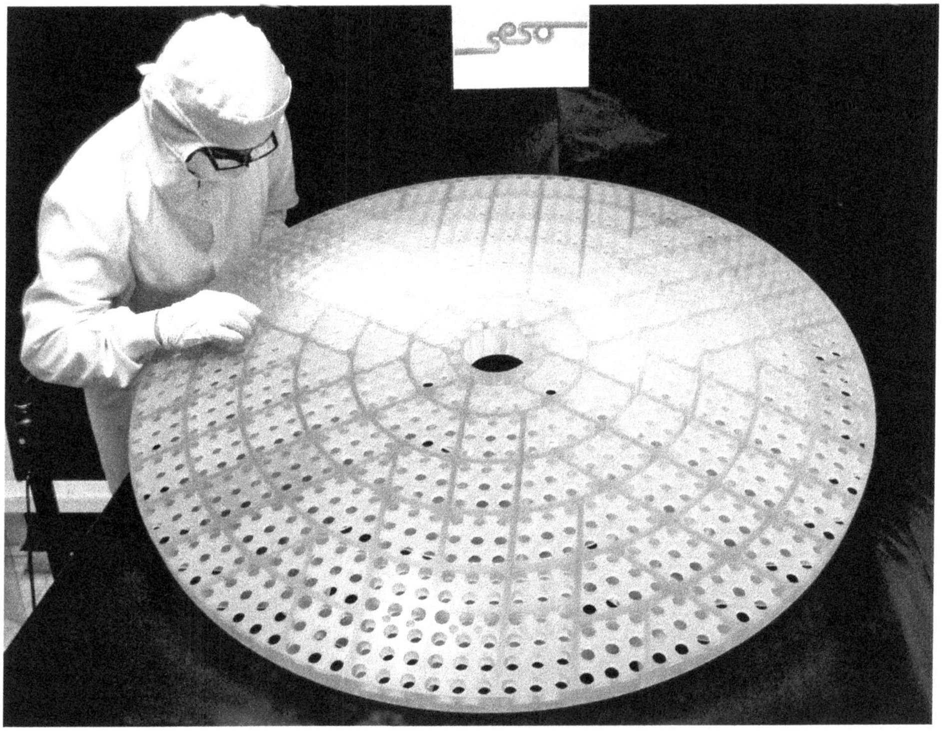

Figure 5.3. The reference optic for the deformable secondary mirror system of ESO's VLT showing the location holes for its 1170 actuators. (Courtesy ESO/SESO.)

5.3 The Change to Altazimuth Mounts

The development of computers in the West was much more advanced than that in the Eastern Bloc in the decades immediately after the Second World War. So it was a surprise when astronomers in the West found, in the 1960s, that the Russians were designing the largest telescope in the world with an altazimuth mount that required a computer to drive it. This was the 6.0 m (236 inch) Bolshoi Teleskop Altazimutalnyi (BTA) or Large Altazimuth Telescope which was eventually built at the Special Astrophysical Observatory near Zelenchukskaya in the Caucasus.(26)

Initially in the 1950s the design team under Dmitri Maksutov at the Pulkovo Observatory near Leningrad (now St Petersburg) considered an equatorially mounted instrument for their large telescope, like that of all other large optical telescopes of the period. But an equatorial mount is asymmetrical, causing serious loading and flexure problems with such a very large instrument, whereas an altazimuth design is symmetrical and basically rigid. Altazimuth mounts are also much smaller and easier to design and build than equatorials. So the decision was made to change to an altazimuth design. The main operational disadvantage of such a

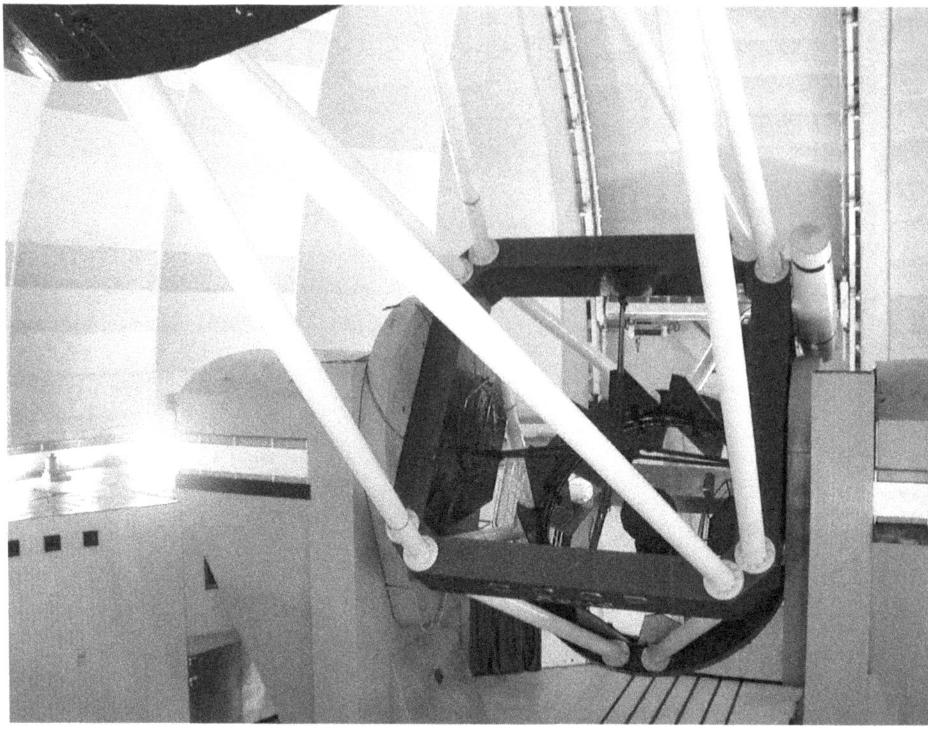

Figure 5.4. The Soviet Union's 6 m (236 inch) Bolshoi Teleskop Altazimutalnyi (BTA) at the Special Astrophysical Observatory near Zelenchukskaya. It was the world's first large optical telescope to have an altazimuth mount. (Courtesy DieterS. via Wikimedia Commons.)

mount was that correction rates were too fast near the zenith, so the BTA telescope would not be able to observe within a few degrees of the zenith.

This change to an altazimuth design resulted in a lighter and cheaper structure. But it also involved rotating the telescope simultaneously about two perpendicular axes at nonuniform rates to track objects. In addition the field of view rotated during an observation, and that needed compensation. Although computers in Russia at the time were very rudimentary, they were sufficiently powerful to undertake these tasks. Not only that, but the BTA computers would allow automatic photoelectric guiding using an attached 70 cm diameter telescope. The BTA telescope (see Figure 5.4) was completed in 1976. But, unfortunately, although the altazimuth mounting was a big success, the BTA's performance as a whole was poor. This was because problems with its primary mirror were made worse by the relatively poor seeing conditions at the observatory.(27)

It seems incredible now that large optical telescopes in the West continued to use equatorial mounts until the early 1970s, even though computer-controlled altazimuth mounts had been used for radio telescope dishes for some time by then.

But as Richard Learner persuasively pointed out(28) the designers of large optical telescopes after the Second World War had been too influenced by the design of the Palomar 200 inch (5 m). As a result they had generally stuck with the designs they knew and did not seriously examine the benefits of new technology in many aspects of telescope design.

The Multiple Mirror Telescope (MMT) (29), which was completed at the Whipple Observatory on Mount Hopkins, Arizona in 1979(30), was the first large modern optical telescope in the West to use an altazimuth mount. The use of such an altazimuth mount enabled the size of the MMT telescope building to be much reduced compared with that for a conventional equatorial. This allowed the building's temperature, and that of the telescope mirrors in particular, to reach that of the outside environment faster than with a much larger traditional dome.

The next large optical telescope to use an altazimuth mount was the 4.2 m William Herschel Telescope (WHT) (31) that was completed on La Palma in the Canary Islands in 1987. This altazimuth design reduced the weight of the mount by 120 tons, compared to that of an equatorial, and reduced the height of the dome by about 50%. To counteract the image rotation, instruments at the Cassegrain and prime foci were mounted on counterrotating turntables. But instruments were not rotated at the Nasmyth focus. Instead optical derotators were used there consisting of fused silica lenses cemented to fused silica prisms.

Since the development of these three telescopes, altazimuth mounts have become the mounts of choice for large optical reflectors.

5.4 Charge-Coupled Devices

Telescopes have benefited over the years from major developments in materials, instruments and detectors, like glass ceramics for mirrors, multi-object spectrographs using optical fibres and aperture plates, and so on, which are outside the scope of this book. But the development of CCDs, which have now taken over from photographic plates in astronomy, has had such a profound impact on the performance of telescopes that it requires a short review.

Charge-Coupled Devices, or CCDs, were invented by Willard Boyle and George Smith at the Bell Telephone Laboratories in New Jersey in 1969. At the time the laboratory was working on semiconductor bubble memories and the CCD, then called a 'Charge Bubble Device', was seen as a memory device. But Michael Tompsett, also of Bell Labs, realised that as silicon is sensitive to light, CCDs could also be used for all sorts of imaging. Subsequently Tompsett produced the first CCD camera.[13]

[13] Although Boyle and Smith came up with the idea behind CCDs, Tompsett appears to have produced the first CCD camera. Apparently in error, the Nobel Committee credited Boyle and Smith with that as well, see www.eng.cam.ac.uk/news/stories/2012/alumni_feature_Michael_Tompsett/.

Boyle and Smith's initial CCDs were surface-channel devices in which charge was stored and transferred at the semiconductor surface. But in 1973 they also invented the buried-channel CCD in which charge was stored and transferred in the bulk semiconductor below the surface. In 2009 Boyle and Smith were awarded the Nobel Prize for Physics for their work on CCDs.

Fairchild Semiconductor produced a buried-channel CCD that became available commercially as a 100 × 100 pixel array in 1974. It was the first CCD to be tested by the Jet Propulsion Laboratory (JPL), which found it clearly better than the earlier surface-channel CCDs. Unfortunately the Fairchild device used interline transfer architecture which could not achieve high quantum efficiency.[14] Meanwhile RCA was developing a full-frame, back-illuminated 512 × 320 frame transfer array for television applications using surface-channel technology. JPL tests showed that the RCA device had poor charge transfer efficiency and high read noise compared with the Fairchild device.(32)

Fairchild and RCA were trying to develop commercial CCDs no larger than 512 × 320, which was really too small for many astronomical purposes. It rapidly became clear that, for scientific applications, it would be necessary to work with a commercial organisation to specifically develop devices with all the best attributes then known for such use. As a result JPL worked with Texas Instruments (TI), under contract from NASA, to develop a CCD sensor for scientific applications based on full-frame, back-illuminated, buried-channel technology. In 1975 JPL successfully tested a 100 × 160 back-illuminated TI device. In the following year they tested a 400 × 400 TI device, which had been designed for the Mariner Jupiter-Uranus mission. Its performance was excellent. Unfortunately the process yield was very low, so TI changed their design from using aluminium gates to polysilicon gates which were much less prone to short-circuiting. JPL planned on producing a 800 × 800 device with polysilicon gates that could be used on the Jupiter Orbiter and Probe mission (now called Galileo) and possibly on the Large Space Telescope (later called the Hubble Space Telescope) if the CCD could be cooled sufficiently.

JPL then produced a 400 × 400 back-illuminated CCD camera system that could be taken to various observatories to demonstrate its capabilities. At that time astronomers were basically satisfied with the photographic and electronic systems that were available, but the JPL demonstrations were to change that. The performance of the CCD system was a revelation and, as a result, there was increased support for the use of CCDs in both ground- and space-based applications.

In 1976 Landauer, Hovland and Janesick of JPL and Bradford Smith of the University of Arizona produced the first high-resolution, astronomical images with

[14] This was because in interline transfer devices shift registers that are insensitive to light cover part of the imaging area.

a CCD. These were of Jupiter, Saturn and Uranus using a prototype TI array on the 1.5 m (60 inch) Mount Lemmon telescope in the near infrared. When Smith and his colleagues looked at the 890 nm image of Uranus they were surprised to find, for the first time, evidence of particles high above the methane layer in its atmosphere.

An upgraded CCD camera system was later taken to the Lick, Kitt Peak and Palomar observatories. The visit was so successful that these observatories made their own systems immediately afterwards using spare detectors given to them by the manufacturer.

NASA had previously decided in the early 1970s that the SEC Vidicon should be the detector used for the Wide Field Camera of the Hubble Space Telescope (HST). But a few years later they were beginning to have second thoughts because of the rapid improvements in CCDs which were much more sensitive than vidicons in the visible and infrared. CCDs also have a linear response to light and an excellent dynamic range.(33) However, one of the big problems with CCDs at that time was their very poor sensitivity to the ultraviolet, which was a key waveband for the HST. James Westphal suggested that this could be solved by coating the CCD with a fluorescent material called coronene which would convert the incident ultraviolet light to visible light, which the CCD could detect. This proved successful on CCD tests with the 200 inch (5 m) Hale telescope on Mount Palomar.(34)

Westphal proposed using two sets of four CCDs to provide a full image array for the first HST Wide Field/Planetary Camera (WFPC) of 1,600 × 1,600 pixels operating in either the wide field or planetary camera mode. About twenty years later a new Wide Field Camera was attached to the HST. It had two ultraviolet/visible CCDs, each of 4,096 × 2,048 pixels, and a separate infrared sensitive CCD of 1,024 × 1,024 pixels operating up to 1.7 microns.

Modifications in the design of CCDs over the years have resulted in back-thinned, back-illuminated CCDs now having a quantum efficiency from about 80% to more than 90% over the visible waveband, compared with a few percent for the fastest photographic emulsions. In addition, not only have pixel numbers and physical array sizes increased over the last few decades, but noise levels have also been reduced substantially. So, today, CCDs have taken over from photographic emulsions for astronomical imaging in both ground- and space-based telescopes.

References

1. Parker, Barry, *Stairway to the Stars; The Story of the World's Largest Observatory*, Perseus Publishing, 1994, pp. 79–81.
2. Wilson, R., Progress on the 3.5 m 'New Technology Telescope' (NTT), *ESO Messenger*, No. 29, September 1982, pp. 24–27.
3. Wilson, R. N., et al., Active Optics: the NTT and the Future, *ESO Messenger*, No. 53, September 1988, pp. 1–7.

4. Noethe, L., Active optics II. Results of an experiment with a thin 1 m test mirror, *Journal of Modern Optics*, 35, No. 9, 1988, pp. 1427–1457.
5. Babcock, H. W., The Possibility of Compensating Astronomical Seeing, *Publications of the Astronomical Society of the Pacific*, 65, October 1953, pp. 229–236.
6. Duffner, Robert W., *The Adaptive Optics Revolution; A History*, University of New Mexico Press, 2009.
7. Zirker, J. B., *An Acre of Glass; A History and Forecast of the Telescope*, Johns Hopkins University Press, 2005, pp. 202–203.
8. Merkle, F., et al., *The Messenger*, No. 58, December 1989, pp. 1–4.
9. Rousset, G., et al., First Diffraction-limited Astronomical Images with Adaptive Optics, *Astronomy and Astrophysics*, 230, 1990, pp. L29–L32.
10. Rigaut, F., et al., Adaptive Optics on a 3.6-m telescope: Results and Performance, *Astronomy and Astrophysics*, 250, 1991, pp. 280–290.
11. Rigaut, F., et al., Performance of the Canada-France-Hawaii Telescope Adaptive Optics Bonnette, *Publications of the Astronomical Society of the Pacific*, 110, No. 744, February 1998, pp. 152–164.
12. Foy, R., and Labeyrie, A., Feasibility of Adaptive Telescope with Laser Probe, *Astronomy and Astrophysics*, 152, 1985, pp. L29–L31.
13. Wilson, R. N., *Reflecting Telescope Optics II; Manufacture, Testing, Alignment, Modern Techniques*, Springer-Verlag, Corrected Second Printing, 2001, p. 416.
14. McCray, W. Patrick, *Giant Telescopes: Astronomical Ambition and the Promise of Technology*, Harvard University Press, 2004, p. 158.
15. Beckers, Jacques M., Multiconjugate Adaptive Optics: Experiments in Atmospheric Tomography, *Proc. SPIE* 4007, 1056, March 2000.
16. Ragazzoni, Roberto, et al., Adaptive-optics Corrections Available for the Whole Sky, *Nature*, 403, January 2000, pp. 54–56.
17. Langlois, Maud, et al., Solar Multiconjugate Adaptive Optics at the Dunn Solar Telescope: Preliminary Results, *Proc. SPIE*, 5490, October 2004, p. 59.
18. Berkefeld, Thomas, Soltau, Dirk, and von der Luehe, Oskar, Results of the Multi-conjugate Adaptive Optics System at the German Solar Telescope, Tenerife, *Proc. SPIE*, 5903, September 2005.
19. Marchetti, Enrico, et al., On-sky Testing of the Multi-Conjugate Adaptive Optics Demonstrator, *The Messenger*, No. 129, September 2007, pp. 8–13.
20. Neichel, B., et al., The Gemini MCAO System GeMS: Nearing the End of the Lab-story, *Proc. SPIE*, 7736, July 2010.
21. Neichel, Benoit, et al., Science Readiness of the Gemini MCAO system: GeMS, *Proc. SPIE*, 8447, September 2012.
22. Brusa, Guido, et al., MMT Adaptive Secondary: First AO Closed-loop Results, *Proc. SPIE*, 5169, December 2003.
23. Davies, Richard, and Kasper, Markus, Adaptive Optics for Astronomy, *Annual Reviews of Astronomy and Astrophysics*, 50, 2012, pp. 309–310.
24. Esposito, S., et al., Large Binocular Telescope Adaptive Optics System: New Achievements and Perspectives in Adaptive Optics, *Proc. SPIE*, 8149, September 2011.
25. Arsenault, Robin, et al., Delivery of the Second Generation VLT Secondary Mirror (M2) Unit to ESO, *The Messenger*, No. 151, March 2013, pp. 14–19.
26. Ioannisiani, Bagrat K. (Rodman, Richard, B., trans.), The Soviet 6-meter Altazimuth Reflector, *Sky and Telescope*, November 1977, pp. 356–362.
27. Keel, William C., Galaxies Though a Red Giant, *Sky and Telescope*, June 1992, pp. 626–627 and 630–632.
28. Learner, Richard, The Legacy of the 200-inch, *Sky and Telescope*, April 1986, pp. 349–353.

29. Weymann, R. J., and Carlton, N. P., The Multiple-Mirror Telescope Project, *Sky and Telescope*, September 1972, pp. 159–163.
30. Weekes, T. C. (ed.), *The MMT and the Future of Ground-Based Astronomy*, Smithsonian Astrophysical Observatory Special Report 385, 1979.
31. Boksenberg, A., The William Herschel Telescope, *Vistas in Astronomy*, 28, 1985, pp. 531–553.
32. Janesick, James R., *Scientific Charge-Coupled Devices*, SPIE Publications, 2001.
33. Janesick, James, Sky on a Chip: The Fabulous CCD, *Sky and Telescope*, September 1987, pp. 238–242.
34. Smith, Robert W., *The Space Telescope: A Study of NASA, Science, Technology, and Politics*, Cambridge University Press, 1989, p. 251.

6

European Northern Observatory and Calar Alto

6.1 European Northern Observatory, Canary Islands

Night-time Telescopes on Tenerife

The Astronomer Royal for Scotland, Charles Piazzi Smyth, set up two small astronomical stations at high altitude on Tenerife in the Spanish Canary Islands in 1856. Although they were not there for very long, these observatories were the first to show the great advantage of observing at high altitude.(1) No one took much notice, however, and it was not until over fifty years later that Jean Mascart of the Paris Observatory suggested that an international astronomical observatory should be established on Mount Guajara on Tenerife. Discussions then started between the Spanish, French and German governments, but they were abandoned at the start of the First World War.

There was no progress with the idea of setting up an observatory at high altitude in the Canaries for a number of years. But astronomers observed a total solar eclipse from the Canaries in 1959. In the same year Spain founded the Observatorio del Teide at Izana, on Tenerife, at an altitude of 2,380 m (7,250 ft). A 30 cm (12 inch) French photopolarimeter was installed there in 1964 to study the zodiacal light, followed by a Spanish 25 cm heliographic telescope in 1969. The largest telescope at El Teide (now part of the European Northern Observatory) is the UK's 1.5 m Infrared Flux Collector (IRFC) that was built in 1971 as the prototype for the 3.8 m United Kingdom Infrared Telescope that was completed on Mauna Kea eight years later.[1]

Site surveys carried out by JOSO (Joint Organization for Solar Observations) in the 1970s indicated that although El Teide on Tenerife was the better Canary Islands site for solar observations (see *Solar telescopes* later), the Roque de los

[1] The IRFC was transferred to Spanish control in 1983. Two years later its name was changed to the Carlos Sanchez Telescope.

Muchachos on La Palma was better for night-time observations. So although night-time telescopes continued to be built at El Teide (see Table 6.1), the largest such telescopes were built on La Palma (see Table 6.2).

Table 6.1. *Optical Telescopes at El Teide Observatory, Izana, Tenerife*

(a) Night-time telescopes

Telescope Name	Telescope Diameter	Year Construction Completed in Canaries	Comments
Photopolarimeter	30 cm	1964	French telescope to observe zodiacal light.
Telescopio Carlo Sánchez	1.52 m	1972	Initially called the UK's Infrared Flux Collector (IRFC). Transferred to Spanish control 1983.
Mons photometric telescope	50 cm	1972	Operated by the University of Mons (Belgium).
Six-channel photopolarimeter	6 × 20 cm	1981	Spanish facility.
Instituto de Astrofisica de Canarias IAC-80	82 cm	1991	
OGS (Optical Ground Station) Telescope	1.0 m	1998	European Space Agency telescope.
STARE (STellar Astrophysics and Research on Exoplanets) Telescope	10 cm	2001	Removed 2010.
Bradford Robotic Telescope	35 cm	2004	For educational use.
STELLA I and II	1.2 m each	2006	Robotic telescopes.
SONG	1.0 m	2013	Robotic Telescope Node of the Stellar Observation Network Group (SONG).

(b) Solar telescopes

Telescope Name	Telescope Diameter	Year Construction Completed in Canaries	Comments
Spanish heliographic telescope	25 cm	1969	
West German evacuated tower solar telescope	40 cm	1972	

(*continued*)

Table 6.1. *(continued)*

Telescope Name	Telescope Diameter	Year Construction Completed in Canaries	Comments
Node of BiSON (Birmingham Solar Oscillations Network)		1975	
Vacuum Gregory-Coudé Telescope (GCT)	45 cm	1985	Moved from Locarno Observatory. Dismantled in 2002 to make room for the GREGOR telescope.
Vacuum Tower Telescope (VTT)	70 cm	1987	Built by the German Kiepenheuer Institute.
THÉMIS	90 cm	1996	Joint French/Italian telescope.
GREGOR Telescope	1.5 m	2012	German telescope with adaptive optics.

Table 6.2. *Optical Telescopes at Roque de los Muchachos Observatory, La Palma*

(a) Night-time telescopes

Telescope name	Telescope Diameter	Year Construction Completed in Canaries	Comments
Swedish general purpose reflector	60 cm	1981	
Isaac Newton Telescope	2.54 m	1984	1967 first light in England with 98 inch (2.5 m) mirror. At that time the fifth largest optical telescope in the world. 1979 closed down in England and later moved to Canaries after installation of new mirror.
Jacobus Kapteyn Telescope	1.0 m	1984	Astrometric reflector. Now owned by the IAC.

(continued)

Table 6.2. *(continued)*

Telescope name	Telescope Diameter	Year Construction Completed in Canaries	Comments
Carlsberg Meridian Telescope	18 cm	1984	Refractor. Fully remote operation from 1997.
William Herschel Telescope	4.2 m	1987	When completed was third largest optical telescope in the world to have a single primary mirror.
Nordic Optical Telescope	2.56 m	1989	Funded by Denmark, Sweden, Norway, Finland and Iceland. First major telescope facility to use active optics.
Telescopio Nationale Galileo	3.58 m	1998	Time allocations 34% Spain, 33% UK, 28% Netherlands, 5% free.
Mercator Telescope	1.2 m	2000	Part of the southern sky extrasolar planet research programme.
Liverpool Telescope	2.0 m	2003	On first light was largest fully robotic telescope in the world used primarily for astronomical research.
Super WASP-N	8 × 200 mm Canon lenses	2003	Partners Super WASP-S at the South African Astronomical Observatory. Designed to detect exoplanets from variations in the brightness of stars.
Gran Telescopio Canarias	10.4 m	2009	Preliminary observations with 12 of 36 mirror segments in place in 2007. When completed was the largest optical telescope in the world. Time allocations 90% Spain, 5% Mexico, 5% University of Florida.

(continued)

Table 6.2. *(continued)*

(b) Solar telescopes

Telescope Name	Telescope Diameter	Year Construction Completed in Canaries	Comments
Swedish vacuum solar telescope	44 cm	1978	Moved from Capri to La Palma.
Swedish Vacuum Solar Telescope (SVST)	50 cm	1985	Replaced the Swedish vacuum solar telescope. Closed 2000.
Dutch Open Telescope	45 cm	1997	No telescope tube leaving mirror open to the air.
Swedish 1-m Solar Telescope	1.0 m	2002	Replaced 50 cm solar telescope.

Night-time Telescopes on La Palma

H. H. (Harry) Plaskett, then president of the Royal Astronomical Society (RAS), had made the case as long ago as February 1946 for a new large optical telescope to be built in the UK. He argued that the weather was, contrary to the general view, acceptable for such a telescope. Plaskett suggested building a 49/74 inch (125/ 188 cm) Schmidt similar to the large Palomar Schmidt (see Section 1.2). He used the 74 inch Radcliffe telescope at Pretoria as a base to estimate the costs of his proposed telescope as about £100,000.(2) With amazing speed the RAS sent a request on May 14 1946 to the Royal Society asking for the funding of a telescope of at least 72 inches (183 cm) aperture[2] hoping to take advantage of the delayed Newton tercentenary celebration, due to be held in July to gain support for their request.[3] At this stage the RAS envisaged either a 72 inch Cassegrain reflector plus a 27/40 inch (69/102 cm) Schmidt, or Plaskett's proposed 49/74 inch Schmidt. The latter was to have a removable corrector plate to allow the telescope to be used in a Cassegrain configuration for spectroscopy when the sky was too bright for Schmidt photography. Following their meeting on May 16 the Royal Society sent a request to the Treasury for £150,000 for an 84 inch reflector, which they subsequently increased to £200,000 for a 100 inch.(3) The reason given for the greater aperture was to reduce exposure

[2] It is probably no coincidence that this was also the diameter of the Dominion Astrophysical Observatory's reflector in Canada, which had been designed by Plaskett's father, John S. Plaskett.

[3] During the war it had been decided to delay the full tercentenary celebration of Newton's birth, which had been due in January 1943 (December 1942 Old Calendar), until after the end of hostilities.

times to make maximum use of short periods of good seeing in the variable British weather. Amazingly the Royal Society was able to announce just two months later, at the opening ceremony of the Newton Tercentenary in July 1946, that the Chancellor of the Exchequer had agreed in principle to provide funding for a 100 inch telescope and building, to be called the Isaac Newton Observatory.[4]

It had also been decided early in 1946 to move the Greenwich Royal Observatory (RGO) from Greenwich to Herstmonceux Castle in Sussex to escape London's light polluted skies. So the newly approved Isaac Newton Observatory would be built there. Then in 1949 a 98 inch (2.5 m) Pyrex mirror blank was donated by the Tracy McGregor fund and the University of Michigan[5] to the UK for this new telescope.(4) But progress was painfully slow over the next few years with the Isaac Newton Telescope (INT), not helped by cracks, bubbles and striations found in the donated mirror blank which caused difficulties in grinding it to the required f/ 3 shape. There were also extensive arguments over whether the telescope should be a conventional Cassegrain reflector or be a Cassegrain with a removable Schmidt corrector plate as advocated by Plaskett. In fact it was not until March 1956 that the new Astronomer Royal, Richard Woolley, decided that it should be a simple Cassegrain. Woolley had also considered, for a short while, abandoning the 98 inch telescope and purchasing a 74 inch from Grubb Parsons similar in design to the four that it had already built for other customers.[6] But he dropped that idea in April 1956.[7]

There had also been problems of finance starting in October 1954 when the Treasury announced that they would have to suspend funding for a time due to national budgetary restrictions.(5) The move of the RGO from Greenwich to Herstmonceux was completed in 1957, but in the same year the government temporarily halted funding for the Isaac Newton Telescope (INT) once more. This was because of serious financial problems in the UK caused by the invasion of Suez in 1956. In the event funding for the INT was finally restored in January 1959 which allowed the contract to build the telescope to be let to Grubb Parsons at the end of the year.

But problems continued to dog the INT project as in 1960 a number of astronomers raised the old question as to whether it made sense to locate such a large telescope in England, with its poor weather conditions, or whether it should be

[4] The Isaac Newton Observatory: Memoranda addressed to the President and Council of the Royal Society, *Monthly Notices of the Royal Astronomical Society*, 107, 1947, pp. 11–19.

[5] This mirror blank had been purchased from Corning in 1936 for a telescope at the University of Michigan that was never built because of the Great Depression in the United States.

[6] These were Cairo University, the Radcliffe Observatory, Mount Stromlo and the David Dunlap Observatory. Woolley had been the Director of the Mount Stromlo Observatory until just before he took over as Astronomer Royal, so he had a detailed knowledge of their 74 inch.

[7] The idea resurfaced for a time in 1958 as a cost-cutting idea. But it was finally abandoned in January of the following year.

located abroad. This question was only finally kicked into touch in 1962 when the decision to locate it at Herstmonceux was confirmed. In the meantime construction of the telescope had been proceeding with most of the mechanical parts completed by 1964 and the mirror ground to its final paraboloid shape two years later. Finally in 1967 first light was achieved on the INT, some 21 years after the programme had been approved, at a cost of almost £1 million. At that time it was the fifth largest optical telescope in the world.

Unfortunately it turned out that Plaskett had been unduly optimistic as the seeing and weather conditions in England proved to be unsatisfactory for such a large telescope. In fact an analysis of papers published in the astronomical press showed that the number using observations with the INT was about one sixth that for similar telescopes elsewhere. Inexpensive air fares had also arrived by the time that the INT had become available, making a foreign-based telescope much more attractive in 1967 than when the telescope had been proposed in the mid 1940s. Consequently Hermann Brück, Astronomer Royal for Scotland, suggested in January 1967 that a 150 inch (3.8 m) telescope should be built at a new UK northern hemisphere observatory.(6) A Northern Hemisphere Review Committee was set up as a result, and site testing of a number of sites in the North Atlantic area undertaken in the early 1970s. The results were compared with those of Mauna Kea, Hawaii to find a suitable location for this new northern hemisphere observatory to complement the UK's shared southern hemisphere observatory in Australia (see Section 9.2). The UK eventually chose the site at the Roque de los Muchachos, at an altitude of 2,300 m (7,600 ft), on La Palma. Although this observatory was not as high as that on Mauna Kea, it was much closer geographically to the UK, which was still important at the time, while being generally above the normal cloud ceiling of about 1,500 m.

The Canary Islands are part of Spain, and so agreement was required from the Spanish authorities to construct an observatory on La Palma. Eventually an agreement was reached in 1979 between Spain, on the one hand, and the UK, Sweden and Denmark, on the other, to allow the latter three countries to construct an observatory (now also part of the European Northern Observatory) at the Roque de los Muchachos. In return for giving their permission, Spain would receive 20% of observing time on all telescopes. Four years later West Germany became a signatory to the agreement when it was extended to include the El Teide observatory on Tenerife.

Under the terms of this agreement the UK were to build three main telescopes at the Roque de los Muchachos observatory. These were the 4.2 m William Herschel Telescope (WHT), the 2.5 m INT, and a 1.0 m astrometric reflector (later called the Jacobus Kapteyn Telescope or JKT) (see Figure 6.1). Sweden was to build a 44 cm solar tower telescope, and a general purpose 60 cm reflector, whilst Denmark was to

Figure 6.1. Panorama of the Roque de los Muchachos Observatory on La Palma in 2001. From left to right the main telescopes seen are the 4.2 m William Herschel Telescope, the 45 cm Dutch Open Telescope, the 44 cm Swedish Vacuum Solar Telescope, the 2.5 m Isaac Newton Telescope, and the 1.0 m Jacobus Kapteyn Telescope. (Courtesy Bobb Tubbs via Wikimedia Commons.)

provide an automatic transit circle (later called the Carlsberg Meridian Telescope) in a joint project with the UK. This transit circle, which was located at Brorfelde observatory in Denmark, was moved to La Palma in 1984. At that time it was one of the first fully automatic telescopes in the world. In 1997 it was modified to enable it to be operated remotely over the Internet from the UK, Denmark or Spain.

In the meantime the UK had decided to replace the 98 inch Pyrex primary mirror of the INT with a 100 inch (2.54 m) Zerodur mirror from Schott of West Germany. The INT was closed down in 1979 and returned to Grubb Parsons, the original manufacturer, to be refurbished and the new mirror installed in preparation for its move to this proposed Northern Hemisphere Observatory on La Palma. Its equatorial mounting had to be modified to take account of the change in latitude and the old prime focus assembly, which included an observing cage, was replaced with one operated by remote control. The old dome was left at Herstmonceux and a new one built at La Palma.

The JKT and INT both saw first light at the Roque de los Muchachos observatory in 1984. But the largest telescope on La Palma was, for many years, the 4.2 m (165 inch) WHT (see Figure 6.2).(7) This had been designed to complement

Figure 6.2. Dome of the 4.2 m William Herschel Telescope on La Palma. (Courtesy H. Raab via Wikimedia Commons.)

the 3.9 m Anglo-Australian Telescope in the southern hemisphere which had been completed in 1974. Initially conceived as a 4.5 m telescope, the 4.2 m WHT was almost cancelled in 1979 because of escalating costs. But a radical design change, and the agreement of the Netherlands to contribute financially, enabled the project to be given the go-ahead two years later. Studies of atmospheric turbulence close to the ground had shown that the height of the telescope altitude axis could be reduced, thus reducing the height of the telescope pier and of the telescope enclosure. In addition reducing the f ratio of the primary mirror from f/3.2 to f/2.5 allowed the diameter of the telescope dome to be significantly reduced. These and other design changes reduced the estimated cost of the telescope, building and instrumentation from £25 million to £15 million (in 1984 prices).

The WHT was built by Grubb Parsons of Newcastle, UK as their last telescope, using a Cer-Vit mirror blank which was one of four produced ten years earlier by Owens-Illinois. (The other three had been used on the Anglo-Australian Telescope, the Canada-France-Hawaii Telescope, and the Blanco Telescope on Cerro Tololo.) The WHT, which saw first light in June 1987, was one of the first large modern telescopes to use an altazimuth mount. A 1.0 m diameter convex hyperloidal secondary mirror, made of Zerodur, was used to produce Cassegrain and Nasmyth foci with unvignetted field diameters of 15 arcmin and 5 arcmin, respectively. A three-element field lens corrected off-axis coma and provided a field of view of about 50 arcmin at prime focus.(8) These optical configurations resulted in an f/2.8 prime focus system and f/11 Cassegrain and Nasmyth systems. At that time the WHT was the third largest optical telescope in the world to have a single primary mirror.

In 1988 the Spanish authorities passed the Sky Law to control light pollution in the Canaries. Then in the following year the Nordic Optical Telescope (NOT),[8] jointly owned by Sweden, Finland, Denmark and Norway, came into operation on La Palma. Although it was not in the original design, it had a form of active optics(9) which allowed the use of the thinnest single 2.5 m diameter primary mirror up to that time which was only 18 cm (7 inches) thick. This f/2.0 Zerodur mirror was ground and polished at the Tuorla Observatory of the University of Turku in Finland. The telescope, which was mounted in a co-rotating building, was designed to complement the Danish 1.5 m instrument located at the European Southern Observatory (ESO). About ten years later an Italian telescope was also built on La Palma. This was the 3.6 m Telescopio Nazionale Galileo, whose design, including that of its active optics system, was based on that of ESO's New Technology Telescope at La Silla.

Planning for the construction of the largest telescope, the Gran Telescopio Canarias, on La Palma was begun in 1987 following the completion of the WHT. At that time Spain and the United Kingdom considered building a telescope with a monolithic 8 m mirror, but in 1990 the UK decided to join the American-led Gemini project instead. Five years later the Spanish abandoned the design with a monolithic mirror and decided to build a telescope with a 10 m diameter segmented mirror instead, like that of the two Keck telescopes on Mauna Kea. The Canarian government set up the Gran Telescopio de Canarias company, in which the Spanish and Canarian governments each had a 50% share, to build the telescope. Formal Spanish government financial approval was given in 1998. Then in 2001 both Mexico and the University of Florida agreed to contribute 5% each to the costs of the telescope in return for each using 5% of its observing time.

This 10.4 m Gran Telescopio Canarias (see Figure 6.3) was designed to operate in the visible and infrared wavebands up to 25 microns. It had a altazimuth design with a primary mirror of 36 Schott Zerodur segments and a beryllium secondary. First light was achieved in 2007 with 12 mirror segments in place. Two years later, when all 36 segments had been fitted, it was the largest single-aperture optical telescope in the world.

The first two instruments to be fitted to the telescope, OSIRIS[9] and CanariCam, were located at the Nasmyth-B and -A focal stations, respectively. OSIRIS, which

[8] Arne Ardeberg, *Some Properties of the Nordic Optical Telescope*, www.not.iac.es/telescope/tti/proptxt.html.
[9] OSIRIS is an acronym of 'Optical System for Imaging and low Resolution Integrated Spectroscopy'.

Figure 6.3. Dome of the 10.4 m Gran Telescopio Canarias at the Roque de los Muchachos Observatory on La Palma. (Courtesy Pachango via Wikimedia Commons.)

was an imaging and low resolution spectrograph, operated in the optical and near infrared bands from 0.36 to 1.05 microns. It was first used for scientific observations in 2009. CanariCam, which followed three years later, was a diffraction-limited infrared imager, with spectroscopic, coronagraphic and polarimetric capabilities, operating from 7.5 to 25 microns.

Unfortunately it was soon found that OSIRIS had a high dark current and was operating at a much warmer temperature than expected. As a temporary fix a new Dewar was installed whilst the existing cryostat was dismantled. This showed that there were a number of conduction routes from the outside to the inside owing to a design fault. Once this was corrected in 2011 the cryostat worked correctly.

A cooling problem was also found with the CanariCam instrument. Here a two-stage closed cycle cooler was used instead of a bulky Dewar to reduce the detector temperature to about 9 K. Unfortunately the cooler kept loosing efficiency, resulting in the detector temperature exceeding about 12 K, which essentially made the instrument unusable. The operations team have kept repairing the cooler but this has caused considerable disruption to the observing schedule. A more permanent solution is currently (in 2015) under consideration.

In 2003 the Roque de los Muchachos observatory was the site for an innovative 2 m telescope built by Telescope Technologies Ltd. (TTL) of Liverpool, UK. This

Liverpool Telescope, as it was called, was contained in a clamshell enclosure that could be opened to uncover the telescope completely. TTL had been founded some years previously as a joint enterprise between Liverpool John Moores University and the Royal Greenwich Observatory (RGO), but the RGO had subsequently withdrawn from the partnership following changes in government priorities. On first light in July 2003 the Liverpool Telescope was the largest fully robotic telescope in the world. It could be operated either autonomously, carrying out pre-programmed observations, or be operated by remote control. Five percent of observing time was reserved for school students under the UK's National Schools' Observatory project.

Over the years the countries involved in the various Canary Island telescopes have changed. For example, in 1997 Iceland became the fifth member of the Nordic Optical Telescope Scientific Association (NOTSA) which ran the NOT. In addition, the financial contribution of various counties to the telescope projects has changed. So in the early twenty-first century, for example, the UK contribution to the Isaac Newton Group of telescopes, which initially comprised the WHT, INT and JKT, was reduced and balanced by both efficiency savings and additional contributions from the UK's partners. In addition the ownership of JKT, which had been withdrawn from service in 2003 as a common-user facility, was transferred to the Instituto de Astrofisica de Canarias in 2014. It is now operated by the Southeastern Association for Research in Astronomy (SARA).

Solar Telescopes

European astronomers, led by Karl-Otto Kiepenheuer, began discussions at the 1967 General Assembly of the International Astronomical Union (IAU) in Prague about the possibility of establishing a modern solar observatory at a superior site in the northern hemisphere. Further discussions over the next two years led to the establishment of the Joint Organization for Solar Observations (JOSO) in December 1969. Initial attendees included solar physicists from France, Germany, Italy, the Netherlands, Norway, Sweden and Switzerland. The agreed plan had three phases: undertake a comprehensive site survey campaign, install new or existing medium sized instruments at the chosen site, and design and install a state-of-the art Large European Solar Telescope (LEST) to study the fine structure of the solar atmosphere in the visible and near infrared.[10](10)

The search for an ideal site for a solar observatory concentrated on the southern Mediterranean and on the Atlantic coast of Europe.(11) The searchers were trying

[10] The word 'European' was dropped later from the definition of what 'LEST' stood for and LEST was redefined as 'Large Earth-based Solar Telescope' when Australia, China and the USA joined the project in the mid 1980s. In the event LEST was never built.

to find a site that would allow a spatial resolution ~0.1 arcsec for long periods of time over the whole year. Meteorological, topographic, orographic and accessibility information was examined on nearly 40 possible sites, and a short list of 15 sites prepared for more thorough examination using telescopes, microthermal sensors, radiosondes and aircraft.(12) Initially it was thought that flat coastal areas would be the best, but it gradually became clear by the early 1970s that mountain sites in the Canary Islands produced the best results. By 1976 just two of these locations had been shortlisted, namely El Teide on Tenerife and the Roque de los Muchachos on La Palma. Finally an extensive set of observations at both of these was carried out in 1979 by 53 observers from eight European countries. It was concluded that El Teide generally provided the better environment during daylight hours and the Roque de los Muchachos during night-time.

Whilst all this testing was going on Spain had built a 25 cm refractor for white light and hydrogen alpha heliography at the Observatorio del Teide on Tenerife in 1969. Three years later the German Fraunhofer Institute built a 40 cm evacuated Newtonian type solar tower telescope with a Zerodur mirror also at El Teide. Then in 1979 Spain signed an agreement with the UK, Sweden and Denmark to allow these three countries to build astronomical facilities at the Roque de los Muchachos on La Palma. Four years later West Germany signed the agreement when it was extended to include El Teide.

A number of solar instruments had already been transferred from Capri to La Palma by Swedish astronomers in 1978, in anticipation of relocating them to the Roque de los Muchachos Observatory. These included a vacuum solar tower telescope with a 60 cm heliostat that fed a 44 cm Cassegrain, an echelle spectrograph and double dispersion spectrograph. Then in 1985 Sweden upgraded their Roque de los Muchachos facility by building the Swedish Vacuum Solar Telescope (SVST) which used a 50 cm achromatic doublet as the entrance window of their vacuum system. At about the same time the German Göttingen Observatory removed their 45 cm Vacuum Gregory-Coudé Telescope (GCT)[11] from their solar observatory at Locarno, Switzerland and rebuilt it at El Teide. Then in 1987 the Kiepenheuer Institute built a Vacuum Tower Telescope (VTT), also at El Teide, with a Zerodur primary mirror of 70 cm diameter and a tower height of 33 m. An adaptive optics system was permanently installed on the VTT in 2003, producing an image resolution of about 0.2 arcseconds.

THEMIS[12] was a French/Italian telescope built at El Teide Observatory in the 1990s as a 90 cm Ritchey-Chrétien telescope dedicated to the investigation of solar magnetism. Its telescope tube was filled with helium to reduce thermal effects.

[11] Also called the Vacuum Gregory Telescope (VGT).

[12] THEMIS is an acronym of 'Télescope Héliographique pour l'Étude du Magnétisme et des Instabilités Solaires'.

A key element of its design was its ability to image the Sun in up to ten wavelengths simultaneously from about 400 to 1100 nm. This allowed the surface magnetic fields of the Sun to be determined in three dimensions. THEMIS achieved first light in 1996.

The 45 cm Dutch Open Telescope[13] was built at the Roque de los Muchachos Observatory on La Palma in 1997. As its name implies the telescope was used with an open tube when observing the Sun. This reduced local thermal irregularities, usually caused by a telescope's structure, whilst allowing a free flow of air across its primary mirror. The relatively constant trade winds at the site were an ideal environment for this sort of design. Post image processing with a despeckle algorithm enabled the telescope to achieve a resolution of 0.2 arcsec for sustained periods.

It became clear in 1998 that LEST would not be built. As a result Swedish astronomers examined whether it would be possible to replace their 50 cm SVST vacuum solar telescope at Roque de los Muchachos by a new solar telescope of 1 m diameter. They quickly concluded that it would. As a result the SVST was removed in 2000 and replaced with what came to be called the Swedish 1-m Solar Telescope (SST), which was completed just two years later.(13)

The SST was a vacuum telescope with a single element fused silica lens of 110 cm diameter and 97 cm clear aperture, corrected for coma, which also acted as a vacuum window. Fused silica was used for the lens because of its relatively low thermal expansion coefficient. All the mirrors were within the vacuum system, so eliminating mirror seeing and mirror dust problems. The SST had two secondary optical systems. One, with a minimum of optical surfaces, was used for narrow-band images. The other was a sophisticated Schupmann system to correct the chromatic aberration of the single element lens for wide-band imaging over a small part of the image. The SST produced diffraction-limited images with its adaptive optics when it saw first light in 2002. At that time it was the second largest solar telescope in the world.

The 45 cm Vacuum Gregory-Coudé Telescope was dismantled in 2002 to make room for the new 1.5 m GREGOR telescope which was completed at El Teide in 2012. It was thought that the GREGOR telescope would be too large, complicated and expensive if it were to be designed as a vacuum instrument. So it was designed with an open tube and retractable dome to allow a free flow of air across the primary mirror. It had initially been intended to make the first three mirrors, including the 1.5 m diameter primary, of Cesic, a lightweight silicon carbide material, as silicon carbide has a much higher thermal conductivity than Zerodur. This

[13] DOT Team, *Dutch Open Telescope, Status and Prospects, Report for the DOT Evaluation Committee, Utrecht, 2001*. www.staff.science.uu.nl/~rutte101/rrweb/dot-documents/decreport.pdf.

would have allowed the primary mirror in particular to be more easily cooled. In the event, although two of the mirrors were made of Cesic, problems in making the primary of this material resulted in it being made of Zerodur after all.[14]

Five minute vertical oscillations of the Sun's photosphere had been discovered by Robert Leighton and colleagues at the Mount Wilson Observatory in 1961. It was clearly important to observe the Sun 24 hours/day to get a full understanding of these and other possible oscillations and avoid false periodicities. So the Birmingham (UK) Solar Oscillations Network (BiSON) was gradually built up from the late 1970s to measure shifts in a chosen solar Fraunhofer line averaged over the whole disc. The first prototype solar instrument, of what was to become the BiSON network, was taken to the Observatorio del Teide in 1975 where it was developed over the next few years. This led in 1978 to the discovery of global, acoustic solar oscillations of low degree. This instrument remained at El Teide, whilst improved instruments were located at Haleakala, Hawaii; Carnarvon, Australia; Sutherland, South Africa; Las Campanas, Chile and Narrabri, Australia.[15](14)

6.2 Calar Alto

Germany had been a leader in observational astronomy in the early years of the twentieth century. But the after-effects of two world wars and the intervening recession had caused a major disruption to observational astronomy there. The climatic conditions in Germany like those in the UK, the Netherlands and similar central and northern European countries were not really conducive to high quality optical observing of the mid twentieth century. As a result, after the Second World War the UK and the Netherlands began to concentrate their attention on the new discipline of radio astronomy. But West Germany was unable to follow suit for a number of years until restrictions on all forms of radio research were lifted in 1950.

Subsequently West Germany developed its capabilities in radio astronomy, resulting in its decision in 1966 to build a 100 m diameter radio telescope at Effelsberg, near Bonn (see Section 23.1). Two years later the Max Planck Institute for Astronomy (MPIA) in Heidelberg began a site survey to find a suitable site in Europe on which to build a new optical observatory.(15) A number of possible sites were surveyed in Greece and Spain until one was chosen in 1970. This was at an altitude of 2,160 m (7,100 ft) on Calar Alto in the mountains of southern Spain. It had about 200 clear nights per year compared with about 50 in Germany.

Two years of negotiations between the West German and Spanish governments then followed covering the construction and operation of the new observatory.

[14] Kiepenheuer-Institut für Sonnenphysik: Optical Design in www.kis.uni-freiburg.de/index.php?id=165&L=1.

[15] The station at Haleakala was closed in 1991 and the instrument moved to Mount Wilson, California in the following year.

These resulted in an agreement that the Max Planck Institute would be responsible for the construction and operation of the telescopes and their buildings, whilst the Spanish government would build a 30-km access road, and provide power and water. In addition Spain would restrict the use of artificial lights within about 5 km (3 miles) of the observatory. In return Spanish astronomers would have access to the telescopes for a minimum of 10% of their observing time.

Construction of the Calar Alto Observatory was started in 1973 with the first permanent telescope, a 1.2 m aplanatic Ritchey-Chrétien reflector, being completed two years later by the Carl Zeiss company. By the end of that decade Carl Zeiss had also installed a 2.2 m Ritchey-Chrétien telescope with a 1° field. Its primary mirror, like that of the 1.2 m, was made of Zerodur which was produced by Schott of West Germany. At about the same time the Madrid Observatory also installed an independently operated 1.5 m reflector.

In 1980 the Hamburg Observatory's 80/120 cm Schmidt was moved from Germany to Calar Alto,[16] and in 1985 a new 3.5 m reflector was completed at the observatory. This 3.5 m was designed very much in the traditional mode with a classical f/3.5 monolithic primary mirror and horseshoe mount similar to that of the Palomar 5 m which had been completed just after the Second World War. The 3.5 m was also the last equatorially mounted telescope of its size to be built anywhere in the world. Some time later the Spanish authorities extended the light-pollution control region around the Calar Alto observatory to cover an area of about 3,800 square km.

No other large telescopes have been built at Calar Alto, as more recent German funding has been concentrated on building and equipping telescopes at the European Southern Observatory in Chile.

References

1. Jones, Anthony W., The Canary Islands – An Astronomer's Experiment, *Sky and Telescope*, September 1981, pp. 199–201.
2. Plaskett, H. H., Annual Address, *Monthly Notices of the Royal Astronomical Society*, 106, 1946, pp. 80–94.
3. Macdonald, Lee T., 'A Large Chunk of Glass': The 98-inch Mirror of the Isaac Newton Telescope, 1945–1959, *The Antiquarian Astronomer, Journal of the Society for the History of Astronomy*, Issue 6, January 2012, pp. 59–73.
4. Lancaster Brown, P., The 98-inch Isaac Newton Telescope, *Sky and Telescope*, December 1967, pp. 356–361.
5. Macdonald, Lee T., The Origins and Construction of the Isaac Newton Telescope, Herstmonceux, 1944–1967, *Journal of the British Astronomical Association*, 120, No. 2, 2010, pp. 73–86.

[16] The Schmidt was decommissioned in 2000.

6. Smith, David H., Approaching First Light on La Palma, *Sky and Telescope*, March 1984, pp. 214–217.
7. Boksenberg, A., The William Herschel Telescope, *Vistas in Astronomy*, 28, 1985, pp. 531–553.
8. Murdin, Paul, and Boksenberg, Alec, The William Herschel Telescope, *Astronomy Now*, July–September 1987, pp. 17–25.
9. Andersen, Torben, et al., Active Optics on the Nordic Optical Telescope, *ESO Conference and Workshop Proceedings No. 42*, ESO, Garching, April 1992, pp. 311–314.
10. Godoli, Giovanni, The Joint Organization for Solar Observations (JOSO) and The Large Earth-based Solar Telescope (LEST) Foundation, *Società Astronomica Italiana, Memorie*, 57, No. 4, 1986, pp. 631–654.
11. Brandt, P. N., and Wöhl, H., Solar Site-testing Campaign of JOSO on the Canary Islands in 1979, *Astronomy and Astrophysics*, 109, 1982, pp. 77–89.
12. Mattig, W., JOSO – The Protagonist for a Closer Cooperation in Europe – Some Historical Remarks, *Hvar Observatory Bulletin*, 28, No. 1, 2004, pp. 1–7.
13. Scharmer, Göran B., et al., The 1-meter Swedish Solar Telescope, *Proc. SPIE*, 4853, 2003, pp. 341–350.
14. Chaplin, William J., et al., BiSON Performance, *Solar Physics*, 168, 1996, pp. 1–18.
15. Robinson, Leif J., The German-Spanish Connection, *Sky and Telescope*, April 1980, pp. 279–286.

7

European Southern Observatory

7.1 La Silla

European Southern Observatory's Early Telescopes

After the Second World War the most immediate priority in western Europe was to feed and house its people aided by the American Marshall Plan. Once that problem had largely been solved these western European countries began to consider how to ensure that such a European conflict could not happen again. One way of doing this was to work together in joint ventures. On the scientific side the first such initiative was in the field of nuclear physics which eventually resulted in the establishment of CERN[1] in 1953.(1)

In the same year Walter Baade of the Mount Wilson and Palomar Observatories had been invited by Jan Oort to spend two months at the Leiden Observatory. During his visit Baade[2] suggested that it would be a good idea if European astronomers considered establishing a joint European observatory. At that time the largest optical telescopes were in the northern hemisphere so it would be best if the suggested new observatory were built in southerly latitudes. A southern observatory would also be beneficial as the Magellanic Clouds and the central region of the Milky Way were best observed from well south of the equator. Baade suggested that the observatory's main instruments should be a 120 inch (3 m) reflector similar to that of the Lick Observatory and a 48 inch (1.2 m) Schmidt like that on Mount Palomar (see Section 1.2).(2) Using these existing designs as a basis should enable the European versions to be built more quickly and cheaply than if

[1] CERN was an acronym of 'Conseil Européen pour la Recherche Nucléaire'.

[2] Walter Baade was born in Germany near the end of the nineteenth century but moved to the USA in 1931 where he remained until the year after his retirement in 1958. Although living in the USA for the best part of thirty years he still kept a close eye on astronomical developments in Europe both before and after the Second World War.

the telescopes had to be designed from scratch. Baade's idea was discussed shortly afterwards by a group of European astronomers who had gathered at Groningen, in the Netherlands, in June 1953 for a conference on galactic research. The meeting concluded that a meridian circle should be added to Baade's proposed instrument complement to undertake much-needed astrometric work in the southern hemisphere.

At the time South Africa was the envisaged location of the observatory in view of its known good observational conditions and the fact that a number of European countries already owned or used observational facilities there. Capital costs were estimated at about $2.5 million with running costs of about $100,000 per annum. In the following year a committee was established, which included representatives of Belgium, France, Great Britain, the Netherlands, Sweden, and West Germany, to examine the project's feasibility. They added a few smaller telescopes to the proposed inventory, resulting in increased capital costs of $3.5 million. Subsequently Great Britain withdrew from the project in favour of establishing an Anglo-Australian observatory in Australia.

Belgium, France, the Netherlands, Sweden and West Germany were still actively involved in planning the proposed European Southern Observatory (ESO) in 1959 when the Ford Foundation agreed to provide a grant towards its capital cost. The Ford grant of $1 million, against a capital cost then estimated at about $5 million, was on condition that at least four of the five interested countries signed the ESO Convention.(3) This grant was not only very useful funding in its own right, but it also demonstrated confidence in the project which helped to persuade the Dutch and French governments to participate. Subsequently, the ESO Convention was signed by the five Member States in October 1962. A few days later Otto Heckmann, director of the Hamburg-Bergedorf Observatory, was appointed acting director of ESO, initially for one year.[3,4] But in May 1963 his appointment was changed to that of full director with executive authority, a post he held until his retirement in 1970. In the meantime the ESO Convention had been ratified by four member states (the above five excluding Belgium) in 1963–1964, so allowing it to come into effect and allow the Ford grant money to be transferred to ESO. Then in 1967 Belgium and Denmark ratified the Convention and so joined ESO.

Site testing for the planned European Southern Observatory had been carried out in South Africa in the late 1950s and early 1960s. Unfortunately, not only were there no top quality sites found there, but also the political situation in South Africa was causing some anxiety in the early 1960s. However the American AURA

[3] Charles Fehrenbach had apparently been the first choice as director of ESO in view of his success in developing the Haute Provence Observatory in France, but he declined the invitation.

[4] To facilitate this appointment ESO's first headquarters were set up at Bergedorf so Heckmann could be director of both the Hamburg and ESO observatories.

(Association of Universities for Research in Astronomy) site surveys, which had been conducted essentially in parallel with those of ESO, had shown that there were excellent observing conditions in the Andes. As a result, in November 1962 AURA decided to locate their Inter-American Observatory at Cerro Tololo in the Chilean Andes (see Section 2.6). AURA also assisted ESO to review possible sites in the locality. As a result, eighteen months later the ESO Council decided to locate their European Southern Observatory 2,400 m (7,900 ft) above sea level at La Silla,[5] also in the Chilean Andes, just over 100 km from Cerro Tololo. ESO also sought to control the local environment around their observatory by purchasing 627 km² of the surrounding land.

The question of the location of ESO's headquarters was discussed in 1963 and 1964 by the ESO Committee[6] and ESO Council. Clearly it should be within reasonable travelling distance of La Silla, and should have foreign-language school facilities for employees' children. The medium-sized town of La Serena was about 100 miles (160 km) from La Silla and so fitted the first requirement, it also was home to AURA headquarters, but it had rather limited cultural activities for foreign staff. On the other hand, Santiago fitted the second requirement with a number of foreign schools, embassies, university facilities and cultural activities, but it was almost 400 miles (640 km) from La Silla. ESO were tending to favour Santiago when in August 1964 the Chilean government offered to donate 3.4 hectares (8.4 acres) of land next to the United Nations' South American Headquarters in the Vitacura area of Santiago. ESO accepted this generous offer and construction of their headquarters building was completed there in 1969.

The initial programme of telescope construction at La Silla (see Figure 7.1) was planned to include a 1.0 m (40 inch) telescope designed primarily for photoelectric photometry (later known as the Photometric Telescope), a 1.0 m telescope designed primarily for spectrographic work, and the Grand Prism Objectif (GPO) radial-velocity astrograph, which was a copy of the GPO installed at the Haute-Provence Observatory in France. Because of their different expertise and interests, the designs of the photometric and spectroscopic telescopes were delegated to Dutch and French astronomers, respectively.[7]

The 1.0 m photometric telescope, which operated from the ultraviolet to the infrared, was the first permanent telescope to be built on La Silla. Ben Hooghoudt, who had been responsible for the mechanical design of the Dwingeloo radio telescope in

[5] Amazingly this location was chosen without ESO having undertaken a thorough site survey to establish the observing conditions to be expected on La Silla. See Adriaan Blaauw, *ESO's Early History; The European Southern Observatory from Concept to Reality*, European Southern Observatory, 1991, p. 63, and *ESO Bulletin*, No. 1, November 1966, where A. B. Muller describes the first such survey undertaken in 1965.

[6] The ESO Committee was the forerunner of the ESO Council.

[7] At that time Jan Borgman and colleagues of the Kapteyn Laboratory at Groningen were working on photoelectric photometry, whilst Charles Fehrenbach and colleagues at Marseilles and the Haute Provence Observatory were undertaking spectroscopic work.

Figure 7.1. ESO's La Silla Observatory in the early 2000s. The main telescopes imaged wending one's way from the ESO 3.6 m at top left are the 1.2 m Leonhard Euler Telescope top centre, the dark coloured enclosure of the 3.6 m ESO New Technology Telescope, the white dome of the 1.0/1.6 m ESO Schmidt, the darker dome of the 2.2 m MPG/ESO telescope, three small white domes containing the 1.5 m Danish, the 1 m MarLy and the 1 m ESO telescopes, followed by the larger dome of the 1.5 m ESO telescope. (Courtesy ESO/C. Madsen.)

the Netherlands, was a consulting engineer overseeing its design and construction, whilst Jan Borgman of the Kapteyn Laboratory, also in the Netherlands, undertook the detailed design. The optics of the photometric telescope, which used Therman low expansion Schott glass for the primary mirror, were made by Jenoptik in Jena, East Germany.

It was expected when the main contracts were let in 1962 that the photometric and other telescopes would be built at a site in South Africa. But, as mentioned above, the decision was made in 1964 to build the European Southern Observatory on La Silla in Chile. Manufacture of the photometric telescope was completed in 1965 and in the following year it was mounted on La Silla in a temporary observatory building purchased from Astro Domes in the USA. The first photometric work was undertaken by Borgman and colleagues in December 1966 using a photometer borrowed from the Kapteyn Laboratory. Eighteen months later the photometric telescope was moved to its permanent observatory on La Silla.

As mentioned above, the ESO spectrographic telescope was expected to have a 1.0 m aperture. But French astronomers suggested in 1961 that it would be more cost-effective to purchase a duplicate of the 1.5 m spectrographic telescope then being designed for their Haute-Provence Observatory. This suggestion was accepted, but with the addition of an f/15 Cassegrain focus to the Haute Provence design. The contract to build the ESO telescope was given to the French firm REOSC who also had the contract to build the Haute-Provence instrument. A glass blank for the primary mirror was ordered from Sovirel Parra-Mantois in France and secondary mirror quartz blanks ordered from Corning.(4) The telescope was installed on La Silla in 1968 with the Coudé spectrograph, also built by REOSC, being operational in the following year. In the meantime a Cassegrain spectrograph had been borrowed from the Marseilles Observatory. It was replaced by a Boller and Chivens spectrograph in 1974.

The Marseilles Observatory, directed by Charles Fehrenbach, were hoping in the late 1950s to erect in South Africa a duplicate of the GPO already operating at the Haute Provence Observatory. The GPO design consisted of twin 40 cm photographic and visual refracting telescopes on a common mount, with the photographic telescope's objective lens being covered by an objective prism. This enabled it to measure stellar radial velocities. At about the same time it was suggested that the ownership of the new Marseilles GPO could be transferred to the ESO. The French delegation to an ESO meeting in July 1960 were strongly in favour of this suggestion as they hoped that it would help to persuade the French government to become members of ESO, particularly if the money already spent on the new GPO were considered as part of France's first financial contribution. In the following year the Marseilles Observatory moved their new GPO to Zeekoegat in South Africa as part of their radial velocity project where the instrument was also used as part of the ESO site testing programme.

The new Marseilles GPO completed its observational programme at the end of 1965. In the meantime it had been agreed that its ownership would be transferred to ESO. Consequently, the GPO's mechanical parts were shipped to La Silla and the optics returned to France for refurbishment. The completed instrument was installed on La Silla in 1968.

Initially the 'founding fathers' of ESO had anticipated incorporating a meridian circle in its original suite of instruments. But in 1962 this was changed to an astrolabe as, amongst other things, a modern astrolabe was not as susceptible to systematic errors as a meridian circle. It so happened at the time that the Dutch foundation ZWO had a spare astrolabe which they were prepared to sell to ESO at half price. Shortly afterwards the decision was made to base the ESO in Chile, and the University of Chile, who owned the observatory at Cerro Calán near Santiago, indicated that they would be willing to have this astrolabe located at their observatory where it could be linked to their modern quartz clocks. They were also willing to operate it. In the event the instrument was overhauled in France and installed at the Cerro Calán observatory at the end of 1965.

Walter Baade had suggested in 1953 that the proposed southern European observatory should have a 1.2/1.8 m (48″/72″) Schmidt like that on Mount Palomar (see Section 1.2). In the following year a 0.8/1.2 m (32″/48″) Schmidt had been installed at the Hamburg observatory.[8] But, unlike the large Palomar Schmidt,[9] the Hamburg Schmidt had an objective prism to produce spectra. European astronomers also wanted their proposed Schmidt to have an objective prism, which meant that they had to compromise on its size to make the telescope affordable. As a result, ESO decided in 1962 to build a 1.0/1.6 m which, although somewhat smaller than the large Palomar Schmidt, would produce the same plate scale of 67 arcsec/mm. Its main disadvantage compared with the large Palomar Schmidt was that its unvignetted image field was 5.4° × 5.4° compared with the latter's 6.5° × 6.5°, so requiring more images to cover the whole sky.[10]

Otto Heckmann, ESO's first director, had been the director of the Hamburg Observatory when the Hamburg Schmidt had been built. Naturally he took a very strong interest in the ESO Schmidt, and suggested that the firm of Heidenreich and Harbeck, which had been responsible for the mechanical parts of the Hamburg Schmidt, should be given a contract to manufacture the mechanical parts of the ESO Schmidt, as he was happy with their work. In the meantime, however, Walter Strewinski, who had led the mechanical design team at Heidenreich and Harbeck, had left the company to set up his own design bureau. So Heckmann suggested giving a contract to both Heidenreich and Harbeck and Strewinski to work together on the mechanical construction with Strewinski responsible for the design and Heidenreich and Harbeck for the manufacture.

But the ESO Instrumentation Committee wanted an in-house group to be responsible for developing all major instrumentation. As a compromise Strewinski and Hooghoudt agreed in 1961 to act as a pseudo in-house team responsible for the development of the ESO Schmidt. But there was little progress over the next two years or so, partly because Hooghoudt was also working on the 1 m photometric telescope and on a radio telescope project. So in 1965 Heckmann signed a contract with just Strewinski to undertake the design, although the ESO Council were

[8] Hamburg Observatory had been particularly interested in Schmidt telescopes ever since Bernhard Schmidt had built the first such instrument there in 1930. It was a 36/44 cm (14"/17") f/1.7 instrument.

[9] There was also a small, 46/66 cm (18"/26"), Schmidt at the Palomar Observatory.

[10] The saga of the ESO Schmidt outlined here is covered in more detail in Adriaan Blaauw, *ESO's Early History; The European Southern Observatory from Concept to Reality*, European Southern Observatory, 1991, pp. 187–200.

concerned that Strewinski's design bureau may become overloaded because they were also working on the ESO 3.6 m telescope. The manufacturing contract was signed with Heidenreich and Harbeck in 1967 even though Strewinski's drawings were nowhere near complete. In fact there were still drawings outstanding at the end of 1969.

In the meantime, ESO had in 1964 ordered the 1.62 m diameter spherical mirror made of low expansion Duran 50 borosilicate glass from Schott. In addition they ordered the 1 m diameter corrector plate made from ultraviolet-transparent Schott UK50 glass and the 4° objective prism made from UBK7 also from Schott. Zeiss-Oberkochen figured all three to shape. Both the spherical mirror and the corrector plate were ready for delivery in 1967. But Strewinski was late with his Schmidt mechanical drawings. Heidenreich and Harbeck, who were still responsible for its manufacture, were also late in making the Schmidt's mechanical parts. So it was not until the end of 1971 that the Schmidt actually arrived on La Silla and was installed in its dome which had been empty there for three years.

This was not the end of the story, however, as the new director, Adriaan Blaauw, had to send André Muller to La Silla in 1972 to sort out numerous problems that were being experienced with operating the Schmidt. This resulted in Muller, supported by a small team, making a number of major design improvements over the next two years or so. In fact, problems with the electronic control and mechanical drive systems were so bad that entirely new systems had to be installed. Improvements in optical collimation and mirror handling were also made.

It is easy to be wise after the event but the management of the Schmidt project clearly left much to be desired. Heckmann, because he was satisfied with the work of the company Heidenreich and Harbeck on the Hamburg Schmidt did not look elsewhere to get a competitive tender. But since the completion of the Hamburg Schmidt, Strewinski had left this organisation and set up his own firm, thus reducing the effectivity of Heidenreich and Harbeck. But simply giving a contract to both Heidenreich and Harbeck and Strewinski was not going to solve the problem as both organisations now had different interests, compared with when Strewinski was working for Heidenreich and Harbeck. Matters were not helped by contracting Strewinski to also work on ESO's 3.6 m telescope, as he could easily play off lack of progress on one against problems on the other. Although Strewinski was not working on his own, as he had set up his own bureau, he seems to have been so dominant in it that he was a single point failure.[11] In addition the earlier introduction of Hooghoudt into the team for a period of years did not help as he was

[11] According to Blaauw in his book *ESO's Early History; The European Southern Observatory from Concept to Reality* (European Southern Observatory, 1991, p. 195), Strewinski was a rather unstable character who was unable to delegate. Heckmann seems to have tried to turn a blind eye to this, possibly because he felt to blame for giving him the contract in the first place.

also working on two other telescope projects. If ever there was a case of how not to manage a technical project, this was it.

In addition to a Schmidt, Walter Baade had also suggested in 1953 that the proposed ESO observatory should have a 3 m (120 inch) telescope similar to that at the Lick Observatory. But it had since been found that the observer's prime focus cage on the Lick was really too small. As a result, in 1961 Fehrenbach and Heckmann suggested that the diameter of the planned European telescope should be increased to 3.5 m, which was accepted. In 1964 the ESO Convention was ratified and the Ford grant paid, allowing hardware orders to be placed. Subsequently ESO ordered a fused silica blank from Corning for the reflector's primary mirror. But when the blank was delivered in 1967 it was found that it had a larger useable diameter than expected of 3.6 m rather than 3.5 m. The mirror, after polishing by REOSC in France, was formally accepted by ESO in February 1972. In the meantime fused silica secondary mirrors blanks had been ordered for the Cassegrain and Coudé systems from the German firm of Heraaus-Schott. They were also figured by REOSC.

In the meantime, ESO had found that their main problem with the 3.6 m project was not with its primary mirror, as was usually the case with large reflectors, but with the telescope's mechanical parts. It had become clear over the period 1966–67 that ESO's chosen external design bureau, run by Strewinski, was under-resourced, creating continuing delays to their schedule of both the 3.6 m and the Schmidt. Various solutions were examined if Strewinski could not substantially increase his numbers of engineers and draughtsmen. These included limiting his design bureau to producing outline design drawings and finding another subcontractor to produce the manufacturing drawings. An alternative of getting a larger firm to work under Strewinski's supervision to undertake some of the outline drawings was essentially vetoed by Heckmann. So in 1969 ESO turned to a similar but much larger European technological organisation, CERN, for assistance.

Fortunately, at that time CERN was coming to the end of building their proton-proton collider, so they had staff available and consequently they agreed to help. This resulted in the creation of a joint ESO/CERN Technical Project (TP) Division, located at CERN's facilities near Geneva, to oversee the design and construction of the 3.6 m telescope. In the meantime Strewinski's design bureau had been asked to concentrate on completing the 1.2 m Schmidt. In 1971 the ESO/CERN team abandoned the original design of enclosure for the 3.6 m and started from scratch to produce a much smaller and more efficient enclosure at lower cost. By the following year work had progressed sufficiently with this 3.6 m project to allow ESO to place construction contracts for the enclosure and telescope structure. Finally the telescope saw first light in November 1976(5) to complete the ESO observatory's initial building programme. In spite of the radical change in enclosure design, the telescope was completed virtually on the schedule which the joint ESO/CERN team had outlined in February 1971.

A request to expand the ESO headquarters facilities in Santiago was rejected by the ESO Council in 1972. In a way this rejection acted as a catalyst for ESO to reconsider what facilities should exist in Santiago, what should exist in La Serena where ESO had a local office,[12] and what on La Silla. At the time Chile was becoming an unstable country politically and the large distance between Santiago and La Silla was affecting the efficiency of the organisation. Then on September 11 1973 the Chilean military overthrew the Allende regime in a bloody coup, raising the question as to whether the ESO should remain in Chile at all.

The ESO Member States and Council recognised that it would be difficult to remain in Chile but regarded it as their mission to support the science undertaken at the observatory. So they decided to remain in the country but to keep any political contacts to a minimum.(6) However the coup, and the planned arrival of a new Director General,[13] Lodewijk Woltjer[14] in 1975 brought the question of the location of the non-observational facilities to a head. The Santiago office had, through default, come to include a scientific centre which Woltjer argued should be located in Europe, as ESO was designed to support European astronomers. Woltjer wanted to go further than this and set up a small in-house scientific institution in Europe with both permanent and temporary staff. As a result, ESO would not just provide telescope facilities in Chile but would also have their own in-house scientists in Europe who would push, assist and challenge the in-house engineers to develop innovative facilities for current and future telescopes. Woltjer further suggested that the number of international staff in Chile should be reduced and replaced, as far as possible, by local staff. Finally he proposed that almost all ESO staff in Chile should be located at La Silla, now that an airstrip had been opened at the old La Silla base camp of Pelícano with flights to Santiago.

Woltjer's proposals, which were all put into effect, including that of creating a scientific-technical centre near Geneva, had the advantage in the short term of increasing the efficiency and security of ESO staff. But this was at the expense longer term of giving the impression in Chile that ESO did not care about the local community by withdrawing most of its staff there to the isolated area of La Silla. In the event most of the technical staff who remained in Chile had moved from Santiago to La Silla by the end of 1975. So in the following year the surplus space in the Vitacura building in Santiago was leased to the United Nations next door.

ESO's first administrative headquarters had been established in 1962 at the Hamburg-Bergedorf observatory. Then in 1970 the ESO/CERN Technical Project

[12] This was mainly to organise construction of the observatory.

[13] Heckmann's title was changed from ESO director to Director General in 1968. So this is what the director of ESO was now called.

[14] Chairman of the Department of Astronomy at Columbia University, New York. He was of Dutch nationality with a doctorate from Leiden University.

(TP) Division was set up near Geneva on Swiss territory, and for a time there was a suggestion that Switzerland may join ESO and that ESO's administrative headquarters could be moved to Geneva. At the time the West German government were concerned that relatively few international organisations had their headquarters in Germany, and they certainly didn't want to lose one of the few that was already there. So in 1973 the German delegate to the ESO Council offered to provide new facilities for ESO's headquarters. In the following year he suggested that this could be at Garching, near Munich, where there were already a number of German research organisations undertaking work in the physical sciences. The ESO Council accepted this offer in December 1975, which involved Germany providing land and a headquarters building free of charge for a period of 99 years. A temporary building was made available in 1976 until a permanent building had been built. This was completed four years later.

This idea of putting all of the ESO staff currently located in Europe under one roof at Garching did not go down too well with many people in Hamburg or in the TP division near Geneva. As a result a number of people left, significantly slowing down progress in instrument development and other technical activities.

National Telescopes on La Silla

In the early days La Silla was thought of as a European observatory for ESO telescopes. No one thought, at the time, that Member States would eventually ask if they could locate their national telescopes at the observatory. But over time La Silla's very favourable viewing conditions, together with ESO's local infrastructure, made the observatory an ideal location for some national telescopes, or even those of some technical institutes or universities.

The first request to place a national telescope on La Silla was made in the mid 1960s by Theodor Schmidt-Kaler, the director of the Bochum Observatory in West Germany, who wanted to install their new 60 cm telescope there. The idea was approved by the ESO Council in 1966, but it took three years to agree on the conditions under which the telescope would be located on La Silla. ESO astronomers were allocated 30% of its observing time, in return for ESO allowing the telescope to be located at their observatory. In the meantime the Bochum telescope had been installed on La Silla in September 1968, a year before the conditions of its location and use were formally approved.

There were two other potential national telescopes under discussion in 1966 for location on La Silla, one was a 1.5 m Danish telescope, that was still in the planning stage, and the other was Sweden's Uppsala Schmidt which had been installed at the Mount Stromlo Observatory, Australia, in 1957. At the time Denmark was not yet a member of ESO, but that changed in 1967. The Uppsala idea did not materialise.

In the event the next national telescope to be installed on La Silla was Copenhagen Observatory's 50 cm photoelectric telescope which was initially intended to be located there only as a temporary measure. As a result in February 1969 it was put in the building which had been used provisionally for the ESO 1 m photometric telescope. But, by the time this Copenhagen 50 cm had been installed on the mountain ESO had agreed to let it stay there on a more permanent basis.

ESO installed their own 50 cm photoelectric telescope on La Silla in 1971. It was built in Denmark as a copy of the Copenhagen telescope and was used, amongst other things, to develop control systems for the ESO 3.6 m telescope. In the same year Copenhagen's 50 cm was moved into its permanent dome which was identical to the one for the ESO 50 cm.

Since this period a series of national telescopes have been located on La Silla (see Table 7.1), including the Danish 1.5 m that was installed in 1978, and a 90 cm Dutch Telescope that was moved from South Africa and re-erected at La Silla in 1979. The Danish 1.5 m is still operational, whilst in 2006 the Dutch telescope was given by its owners, the Leiden Foundation, to Alain Maury[15] on condition that it was not used for financial gain. He then moved it to his observatory near San Pedro de Atacama where it was used to search for asteroids.

The Max Planck Institute for Astronomy (MPIA) in Heidelberg had ordered two 2.2 m telescopes from Carl Zeiss, one to be erected in Spain and the other in Namibia. Although the Namibia telescope had been built, it had not been located there for political reasons and had been in storage for some time.(9) Then in 1982 the Max-Planck-Gesellschaft (MPG, the parent organisation of the MPIA) decided to lend the telescope to ESO for 25 years provided that they installed it on La Silla at their cost. Observing time would be split 25%/75% MPG/ESO.

ESO were anxious that the 2.2 m telescope should be located on La Silla as quickly as possible. So they decided to purchase a commercially available Observa Dome from the US to house the telescope. Amazingly this was built on La Silla in just six weeks in late 1982. Erection of the telescope began in February of the following year and first light took place on June 22. Then in 1984 ESO began experiments in operating the 2.2 m telescope by remote control, first from close to the telescope, then from La Serena. Finally in 1987 ESO inaugurated the world's first routine system for remote control of telescopes between continents when they linked La Silla to ESO headquarters in Garching via an Intelsat communications satellite. Astronomers were then able to remotely control and receive images from the MPG/ESO 2.2 m. Shortly afterwards ESO fitted their first infrared imager, the 64 × 64 pixel InfraRed Array Camera (IRAC) to the 2.2 m.

[15] Alain Maury was a former astronomer at the European Southern Observatory who had set up an array of relatively small telescopes to get people interested in astronomy. He also hired out these small telescopes to interested amateurs.

Table 7.1. *Optical Telescopes Located on La Silla*

(a) ESO telescopes

Telescope	Year Construction Completed on La Silla	Comments
1 m photometric telescope	1966	Moved to permanent location on La Silla in 1968. Dedicated to DENIS (DEep Near Infrared Survey of the Southern Sky) project starting 1994. Later used by a collaboration between Bochum and UCN-Antofagasta universities.
1.5 m spectrographic telescope	1968	Design based on the 1.5 m Haute-Provence telescope. ESO telescope decommissioned 2002.
2 × 0.4 m GPO (Grand Prism Objectif)	1968	Twin 0.4 m photographic and visual refracting telescopes. As the Marseilles GPO it was initially located in South Africa in 1961. Decommissioned 1996 and replaced by the MarLy telescope (see below).
1.0/1.6 m Schmidt	1971	Original photographic camera decommissioned 1998. New Yale 160 megapixel QUEST camera installed 2009. Telescope now controlled remotely from Yale.
0.5 m photoelectric telescope	1971	A copy of the Danish 0.5 m telescope. Decommissioned in 1998. Later moved to Universidad Católica de Santiago.
3.6 m reflector	1976	COME-ON adaptive optics system installed 1990. HARPS (High Accuracy Radial velocity Planet Searcher) spectrograph installed 2003.
Coudé Auxiliary Telescope (CAT)	1981	Fed the 3.6 m telescope's Coudé echelle spectrometer through a light tunnel. In 1998 connected to ESO's 3.6 m via optical fibre.
3.6 m New Technology Telescope (NTT)	1989	For details, see text.

ESO's New Technology Telescope

In the late 1970s ESO looked into the possibility of building a very large telescope of about 16 m diameter. It was clear that it would be impossible to simply scale up existing designs to produce such a very large telescope. So they planned to build one of about 3.5 m aperture, their so-called New Technology Telescope (NTT), to try out all their new design concepts to check out their effectiveness. They took as their main targets minimum weight of primary mirror, minimum size of telescope building, and minimum number of focal positions.(10) In 1980 this resulted in ESO choosing, amongst other things, for their NTT:

- an altazimuth mount, as this required a smaller building than an equatorial and it was easier to meet pointing accuracy requirements
- an f/2.2 mirror, as this was fast enough to require a small dome diameter whilst being relatively easy to make
- a building that rotated with the telescope, similar to that of the MMT (Multiple Mirror Telescope) on Mount Hopkins, to minimise building size
- a Cassegrain focus, rather than a prime, Nasmyth and Coudé focus, as this required the smallest dome diameter

Shortly afterwards it was estimated that the 3.6 m NTT would cost less than one third that of the existing ESO 3.6 m, but ESO did not have the money at that time to build it. Then in 1982 Italy and Switzerland joined the organisation and provided the extra funding required.

By this time ESO had abandoned their Cassegrain design and decided to have two Nasmyth focus platforms instead. This provided a less complex telescope design, reducing its weight and simplifying its maintenance.(11) Ray Wilson, ESO's expert optical designer, had also worked out a scheme for significantly reducing the thickness of the primary mirror, and hence its weight, whilst retaining its focusing accuracy. Obviously the shape of a much thinner mirror would change as the angle of the primary mirror to the vertical changed. So Wilson proposed what he called an 'active optics' approach (as already outlined in Section 5.1) which involved the telescope monitoring its own optical quality and correcting it as required with an active support system for the primary mirror. Wilson thought that it was feasible to build an active optics system to correct a 20 cm thick primary mirror for the NTT. But as this design was untried Woltjer, the Director General, insisted that the telescope should be able to produce acceptable results even if the active system failed.(12) As a result a conservative thickness of 24 cm was chosen.

It had been suggested early on that the NTT should have a nickel-coated aluminium or aluminium alloy primary mirror in addition to its Zerodur glass ceramic primary. The NTT would be an ideal facility on which to compare the performance of both, having the proposed very large telescope in mind. Wilson argued

(b) Other telescopes

Telescope	Year Construction Completed on La Silla	Comments
60 cm Bochum telescope	1968	Now decommissioned.
50 cm Copenhagen observatory telescope	1969	Fully programmable photoelectric telescope. Now decommissioned.
Swiss 0.4 m telescope	1975	Geneva observatory telescope, decommissioned 1980. Replaced by Swiss T70 telescope.
1.5 m Danish telescope	1978	Completely overhauled 1993. Now operated remotely from Europe.
90 cm Dutch telescope	1979	Moved from South Africa. Re-equipped with a CCD 1991. Mothballed 1999. In 2006 donated to Alain Maury who moved it to his observatory near San Pedro de Atacama. Now used to search for asteroids.
Swiss T70 telescope	1980	Replaced Swiss 0.4 m telescope. Decommissioned 1998 following installation of 1.2 m Leonhard Euler Telescope on La Silla.
MPG/ESO 2.2 m telescope	1983	At the suggestion of the Max-Planck-Gesellschaft (MPG) their 2.2 m telescope was installed by ESO at La Silla as a shared facility of the MPG and ESO. A 67 megapixel Wide Field Imager was installed on the 2.2 m in 1999. Since October 2013 this telescope is an MPG-only facility.
Marseilles 0.36 m telescope	1989	Now decommissioned.
MarLy 1 m telescope	1996	Marseilles Lyon telescope. Decommissioned 2009 and moved to Mount Djaogari in Burkina Faso.
1.2 m Swiss Leonhard Euler Telescope	1998	Geneva observatory's telescope used with the CORALIE echelle spectrograph to search for large exoplanets.
0.6 m Italian Rapid Eye Mount (REM) telescope(7)	2003	Robotic optical and near infrared telescope. Used for rapid follow-up observations of gamma-ray bursts first detected by Swift and other spacecraft.
25 cm TAROT (Téléscope à Action Rapide pour les Objects Transitoires)(8)	2006	Robotic optical telescope. Very fast reaction time (pointing speed up to 80°/s) allowed follow-up observations of gamma-ray bursts detected by spacecraft.
Belgian/Swiss 0.6 m TRAPPIST (TRAnsiting Planets and PlanetesImals Small Telescope)	2010	Installed in building that housed Swiss T70 telescope. Operated remotely from Liège, Belgium to detect exoplanet transits and study comets.

that an aluminium mirror would be much cheaper than a Zerodur one and his proposed active optics system would correct for any warping of the metal. He also pointed out that the active optics system built for the Zerodur primary would require only minor adjustments to make it compatible with such a metallic mirror. (13) Unfortunately ESO were forced to reject the purchase of an aluminium primary mirror because the preparatory experimental work on samples had taken too long, and the Zerodur blank had been ordered much earlier.

A major cause of relatively poor images at observatories at that time was the heat produced by telescopes and their associated electronics. The NSO tackled this by actively cooling the observatory floor. In addition an air conditioning system maintained the telescope and observatory rooms at the nighttime ambient temperature.

Optical tests of the NTT's primary mirror at Carl Zeiss, its manufacturer, went extremely well in 1988.(14) But when the first optical tests were carried out of the whole telescope in January 1989 it was found that the curvature of the primary mirror had been built to the wrong figure.(15) It transpired that this had originally been caused by an error made in testing, rather like that of the Hubble Space Telescope (HST). However, Carl Zeiss had realised their mistake prior to the delivery of the NTT mirror and had attempted to correct it. Unfortunately, they compounded the problem by making another error in trying to make the correction. But salvation was at hand as, unlike the HST, the NTT had an active optics system which was able to correct the primary mirror's shape.

As a result at the astronomical first light in March 1989, the NTT (see Figure 7.2) produced the sharpest raw images yet produced by any ground-based optical telescope of 0.33 arcsec.[16,17] These images were immediately transmitted to ESO headquarters at Garching via a remote control facility that had first been used for the 2.2 m using the Intelsat spacecraft. At this time the NTT was the first large telescope anywhere to use a thin monolithic mirror with an active optics system. Almost unprecedentedly the telescope had also been completed within its financial budget.

La Silla Today

Over the last 25 years or so the complement of telescopes on La Silla has continually changed as indicated in Table 7.1. For example ESO withdrew from the MPG/ESO 2.2 m project in October 2013. But they continued to operate their own original 3.6 m reflector and 3.6 m NTT. In addition ESO supported the Swiss 1.2 m Leonhard Euler Telescope of the Geneva Observatory, the Italian 0.6 m Rapid

[16] This performance was achieved by operating the active optics system manually as the NTT software was not designed to make such corrections automatically. These were only possible when the VLT active optics software was installed on the NTT in 1996.

[17] By 1991 the NTT's raw resolution (i.e. without adaptive optics) had been improved to 0.25 arcsec.

Figure 7.2. ESO's 3.6 m New Technology Telescope in its rotating building on La Silla. (Courtesy ESO/C. Madsen.)

Eye Mount (REM) Telescope, the 25 cm French TAROT (Téléscope à Action Rapide pour les Objects Transitoires), and the Belgian/Swiss 0.6 m TRAPPIST (TRAnsiting Planets and Planetesimals Small Telescope).

Both the REM and TAROT telescopes were designed to observe the afterglow of gamma-ray bursts. Both telescopes quickly slewed automatically to the celestial coordinates of a gamma-ray burst after receiving location data from satellite observatories. REM produced images in the visible and near infrared,(16) whilst TAROT produced sub-arcsecond positions of the burst.(17) TRAPPIST was devoted to the detection of exoplanets and the study of comets.

7.2 Cerro Paranal

The VLT

ESO's Very Large Telescope (VLT) project of four 8 m telescopes and two 1.5 m telescopes was approved by ESO's eight member states in December 1987 with an estimated cost of 382 million Deutschmark (about $210 million) at 1986 rates. The fourth large telescope was expected to be completed in 1998.[18] In the month following programme approval Harry van der Laan took over from Lodewijk Woltjer as ESO Director General with the main task of transforming the paper design of the VLT,[19] which would be the world's largest ground-based optical telescope, into a working instrument. Of course, whilst directing this major project van der Laan also had to ensure that the existing telescopes on La Silla, including the almost complete NTT, provided the agreed services to the astronomical community.

The four 8 m VLT 'unit telescopes' or UTs, as they were called, were altazimuth-mounted Ritchey-Chrétien instruments. When the VLT programme was approved it was planned to have them mounted so they could operate either alone or together as an interferometer. The plan was to mount them individually in a line on a common rectangular service platform about 150 m long, and 5 to 10 m above the ground to minimise thermal turbulence. Two 1.5 m auxiliary telescopes were also proposed to extend the interferometer. (These were later changed to four 1.8 m auxiliary telescopes.)

When the programme was approved each of the unit telescopes was planned to be covered by an inflatable shelter, supported by a rigid structure, which could be completely collapsed when the telescopes were about to be used. This enabled the telescopes to operate in the open air, so minimising thermal problems. A louvre-type wind-shield was to be built along the long edge of the service platform to protect all four unit telescopes. This was to smooth the flow of air of the prevailing wind across their primary mirrors.

The f/1.8 primary mirrors were to be of a monolithic construction only about 15 to 20 cm thick. With such thin mirrors active optics were mandatory. As a baseline design it was assumed that these primary mirrors would be made of spun-cast Zerodur or fused silica, although steel or aluminium were considered as a possible backup. Each telescope was to have two Nasmyth foci, one for the visible and one for the infrared waveband, and a Coudé focus. A Cassegrain focus was excluded from the baseline design for cost reasons although such a system was under consideration. Cerro Paranal (2,640 m), 500 km north of La Silla, and Cerro Vizcachas, 5 km from La Silla were both being examined as possible sites for the VLT, with Paranal apparently having the better meteorological conditions. However, if Paranal were to be chosen it would entail the development of a new observatory from scratch some distance from La Silla.

[18] *Proposal for the Construction of the 16-M Very Large Telescope*, ESO, March 1987, pp. 322–324.

[19] The provisional design of the VLT was contained in *Proposal for the Construction of the 16-M Very Large Telescope*, ESO, March 1987 which was submitted to the ESO Council in the same month. The design is also summarised in the *ESO Annual Report, 1987*. The designs in the two documents are basically the same although some details vary.

In 1988, shortly after programme approval a Cassegrain focus was added to all four telescopes following a reconsideration of the scientific requirements placed on the VLT. This required the altitude axis to be raised.(18) It also complicated the design of the primary mirror cell and added a substantial amount of weight to the telescope structure.

In response to a request for quotations, bids for the four 8.2 m primary mirror blanks were received from Schott (who had made the blank for the NTT), who proposed making them of Zerodur, and from Corning (who had made the blank for the original 3.6 m) who proposed using fused silica. At that time no one had produced an 8 m size astronomical mirror blank before, and Schott proposed to use a relatively novel technique and spin cast it.[20] But they had already spun cast a 4 m mirror and their product was thought to be superior technically to Corning's. So ESO awarded Schott the contract in September 1988 at an agreed price of 55.5 million Deutschmark ($32 million) for four 17.5 cm thick blanks.[21] In the following year ESO awarded the French company REOSC the contract to grind and polish the four mirrors.[22] The idea of having metal primary mirrors as a possible backup solution for the VLT was eventually dropped.(19)

The design concept of the 150 m long common rectangular service platform was scrapped in 1989 when seeing measurements at La Silla indicated that there was no advantage in placing the telescopes so high above the ground. The louvre wind shield was also abandoned when studies showed that wind buffeting could be countered more economically using the telescopes' drives and active secondary mirrors.[23]

As mentioned previously, the active optics system on the NTT had produced excellent results following first light in 1989. This gave ESO confidence that their reliance on an active optics system for the VLT unit telescopes was not misplaced. But the system for the VLT would have to be appreciably more sophisticated than that for the NTT because the VLT's primary mirrors would be much more flexible. The initial active optics corrections on the NTT had been carried out manually, but in 1996 the VLT active optics software was installed on the NTT for the first time. It worked very well.

In the meantime, in December 1990 ESO had decided to build the VLT at Cerro Paranal in the middle of 725 sq km of land which had been donated to ESO by the government of Chile two years earlier,[24] on condition that ESO started construction

[20] Schott had been working on the technology since 1984, see *Sky and Telescope*, November 1988, p. 461.

[21] This was the maximum thickness considered possible with Zerodur, see R. N. Wilson, *Reflecting Telescope Optics II; Manufacture, Testing, Alignment, Modern Techniques*, Springer-Verlag, Corrected Second Printing, 2001, p. 298.

[22] *Sky and Telescope*, October 1989, p. 351.

[23] *Sky and Telescope*, January 1989, p. 8.

[24] *ESO Annual Report*, 1990, pp. 13–14.

within five years. Although this required building access roads, buildings and so on for an observatory at a considerable distance from La Silla, this location was clearly better astronomically. For example the median seeing at Paranal was 0.66 arcsec compared with 0.76 arcsec for La Silla, and Paranal had more than twice as many nights of exceptional seeing (better than 0.5 arcsec) as La Silla.(20)

Experiments had been carried out on La Silla using a half-scale inflatable enclosure in 1988. But when ESO chose Cerro Paranal it became clear that operating the telescopes there in the open air with these inflatable enclosures was no longer feasible. However experience gained with the NTT enclosure since its telescope's first light in 1989 had shown that a similar design would be acceptable. The main problem was that a rotating cylindrical carousel structure like that of the NTT, even when opened for observations, created an obstruction to the local airflow that the collapsed inflatable domes did not. So the relative location of the four domes now had to be carefully chosen to minimise the effect of one dome on the airflow of neighbouring telescopes. In addition the location of these large telescopes had also to be optimised from an interferometric point of view. Eventually in 1992 an acceptable arrangement was arrived at with the four large telescopes no longer in line. It was also decided to locate the beam combining system for the interferometer underground rather than above the surface as originally planned.

Meanwhile construction work had started at Cerro Paranal in September 1991. This began with blasting work and soil removal to reduce the height of the mountain by 28 m to create a platform to accommodate the four unit telescopes, the auxiliary telescopes[25] and interferometric passages linking them. Shortly after work started, a nitrate mine was opened only 21 km away from the planned observatory just outside the area owned by ESO. Fortunately, atmospheric pollution from the mine was found to be insignificant.[26]

There were also problems on the personnel front at ESO in the early 1990s as Chilean astronomers began to agitate for a guaranteed percentage of observing time on the various ESO telescopes then operational at La Silla as well as on the future VLT.[27] Spanish astronomers had been granted guaranteed observing time on the European Northern Observatory telescopes on the Canary Islands and Chilean astronomers had been granted similar access to the AURA telescopes at Cerro Tololo, so they appeared to have a good case. But ESO maintained that telescope time should be awarded on the basis of the scientific merit of proposed observing projects established by peer review, not on the nationality of the proposed

[25] In the following year, the number of auxiliary telescopes and delay lines was increased from two to three, see *ESO Annual Report*, 1992, pp. 10–11.

[26] The mine was actually closed a few years later.

[27] The background to this dispute is well covered in Claus Madsen, *The Jewel on the Mountaintop; The European Southern Observatory through Fifty Years*, ESO and Wiley-VCH, 2012, pp. 289–294.

observers. The Chilean staff also demanded wages and working conditions comparable to that of their European counterparts. At the start, the astronomers were on their own in airing their grievances but eventually the Chilean government backed their demands. For a time Chile threatened to pull out of the VLT project, but a compromise was reached in 1993 allowing the VLT project to continue. This resulted in Chilean astronomers being awarded 5% of observing time on the VLT and 10% on other ESO telescopes, and local personnel were allowed to join a trade union if they wished.[28]

Unfortunately as this personnel problem was reaching a solution the Andrés Santa Cruz family, who were descendents of the Chilean war hero Juan Latorre, sued the Chilean government and ESO in April 1993. This was because they claimed to be the rightful owners of Cerro Paranal.(21) This resulted in a number of court cases which went right up to the Chilean Supreme Court. In 1995 this Supreme Court instructed a local magistrate to visit the observatory site and, if necessary, force entry, which he did. The Chilean parliament also got involved and from time to time work was stopped on Cerro Paranal. Naturally this legal and political dispute unsettled ESO and they started to look for alternative sites for their VLT just in case, although the Chilean government assured ESO that they were on their side. In the end the government agreed to pay the Santa Cruz family for the site. The agreed amount eventually turned out to be $8.8 million, compared with the family's original claim for $130 million! At the end of 1995 ESO also agreed not to pursue a claim for damages from the Chilean government for the cost of work stoppages (estimated at about $5 million). And in the following year a new agreement between the government of Chile and ESO was approved by the Chilean parliament. It covered the working arrangements and remuneration for Chilean workers and the guaranteed observing time on ESO facilities for Chilean astronomers.

In the meantime Riccardo Giacconi, the highly respected X-ray astronomer and head of the Hubble Space Telescope Science Institute, had become Director General of ESO in January 1993. On taking up his duties he was concerned at what he found as, although the organisation had excellent engineers and scientists, it appeared to resemble a civil service agency rather than a research institution.(22) What was more, his predecessor had warned the ESO Council in the previous year that the VLT's final cost would be about 50 million Deutschmarks ($32 million) above budget.

As a result, Giacconi undertook a thorough review of the project and concluded in April 1993 that it was seriously late mainly due to a shortage of manpower. In addition the extra costs of including a Cassegrain focus had not been properly addressed, and the costs of interferometry had also been underestimated. All

[28] *Sky and Telescope*, December 1993, pp. 12–13.

in all he estimated that the project would now cost 573 million Deutschmarks ($345 million), compared with 430 million Deutschmarks ($260 million), which was the original estimate adjusted for inflation.[29] He also wanted to increase the number of staff from 193 to 239. The ESO Council did not reject Giacconi's estimates outright but requested an independent financial audit. In the meantime the Council decided to delay implementation of the interferometer system, the Coudé focus, and the associated adaptive optics.[30]

The effect of the Council decision on the VLT interferometer project was catastrophic as many people interpreted the decision as its death knell. A number of calls for tender, that had already been issued, had to be withdrawn. The French science delegate to Council resigned and a number of key people working on the interferometer left the organisation.(23) But Giacconi, hoping for better days, allowed the interferometer tunnel to be excavated and the general infrastructure to be prepared in readiness for a possible later go-ahead.

Meanwhile the first primary mirror blank for the VLT had been spun cast by Schott in early 1991, but that blank and two others had later cracked in the annealing oven.[31] An investigation showed that this had been caused by chemical contamination of the top millimetre of glass which had caused a differential thermal expansion of the top layer compared with the majority of the blank. Correcting for this, the next mirror blank was then successfully annealed. This was followed by 8 months of ceramisation to transform the glass into the glass ceramic Zerodur.(24) The blank was then delivered to REOSC in July 1993 for final grinding and polishing.

This first mirror blank suffered a serious problem at REOSC during the polishing sequence when a distorted polishing tool scratched the surface. Following an investigation, it was decided to polish out the scratches and accept a slightly different radius of curvature of the mirror, which was copied for the other mirror blanks. Polishing of this first mirror was then completed in November 1995 without further problems. But in spite of the delay the remainder of the project was running late so the mirror was put into storage and not delivered to Paranal until the end of 1997. There it was integrated into the first unit telescope in April of the following year.[32] Finally, the fourth and last primary mirror blank was delivered to REOSC in September 1996 and transported to Paranal in April 2000.

[29] I have to say that none of this surprises me having spent over twenty years of my career in the European space business. The problem seems to me to have arisen because the VLT project had not been subject to the normal rigours of what you would have expected of a space project, which start with systems engineering, regular cost to completions, and so on. The VLT also suffered from being the first large project (although in space terms it was a modest project) in an organisation unused to such.

[30] *ESO Annual Report*, 1993, p. 52.

[31] *Sky and Telescope*, May 1992, p. 491.

[32] Unfortunately, a problem with the magnetron in the mirror coating facility in Chile caused the coating of the first primary mirror to be delayed. So this integration of the primary mirror into the first unit telescope in April 1998 took place with an uncoated mirror.

The hyperbolic secondary mirror of the VLT was a complex unit that undertook focusing, coma correction, infrared chopping and fast tracking. These last two functions placed severe demands on the mass, moment of inertia and stiffness of the mirror. As a result ESO initially considered making it of beryllium or silicon carbide, rather than glass ceramic, owing to their much better mass and inertia properties.[33] But they had rejected beryllium as it was very expensive and machining it was a very hazardous operation. As a result ESO opted for silicon carbide.(25) But unfortunately the American company Carborundum Inc. that had been selected for provision of the silicon carbide substrate was unable to deliver it, so the contract had to be cancelled. As a result ESO decided in 1994 to have the secondary mirrors made of nickel-coated beryllium instead. Each had a diameter of 1.1 m and weighed just over 50 kg, including mount and cell.(26)

The first secondary mirror blank was produced satisfactorily by Brush-Wellman Inc, but another company responsible for light-weighting and machining it filed for bankruptcy in the middle of processing the blank! Happily ESO managed to have it transferred to another company for completion before it was transferred to REOSC for polishing. Fortunately, there were no such problems with the other secondary mirrors. Subsequently the first secondary mirror unit was delivered to Paranal in December 1997 in time for its integration into the first unit telescope in March the following year.

As mentioned previously, the active optics software for the VLT had been installed on the NTT in 1996, where it made automatic corrections for the first time. This test was highly successful. Then in April 1998 initial optical tests were carried out with the first 8.2 m VLT unit telescope even though its primary mirror was then uncoated. On just the third night it produced a resolution of 0.4 arcsec at 1.2 microns. The official first light took place in the following month after the primary mirror had been coated.

The inauguration of the Paranal Observatory took place in March 1999 when the four 8.2 m unit telescopes (see Figure 7.3) were given the names proposed by a 17-year-old schoolgirl, Jorssy Albanez Castilla, who had won the competition to name them. These were Antu (Sun), Kueyen (Moon), Melipal (Southern Cross), and Yepun (Venus) from the Mapuche language of a native Chilean people. The fourth and final unit telescope saw first light at Paranal in 2000.

ESO had produced a VLT instrumentation plan in 1990 which included ten instruments. Over the next eight years this plan saw a number of changes with some of these original instruments being cancelled and new ones proposed. In

[33] The choice of the material for the VLT secondary mirrors is covered in R. N. Wilson, *Reflecting Telescope Optics II; Manufacture, Testing, Alignment, Modern Techniques*, Springer-Verlag, Corrected Second Printing, 2001, pp. 237–240.

Figure 7.3. The four 8.2 m Unit Telescopes of ESO's VLT are seen in this image taken in 2005. The smaller building at the top right was built for the VLT Survey Telescope (VST). Domes of two of the 1.8 m Auxiliary Telescopes are seen near the centre of the image. (Courtesy ESO/G. Hüdepohl.)

the event the two instruments, ISAAC[34] and FORS 1, that had been envisaged to be installed first survived, with both of them being installed on the VLT unit telescope number 1, UT1 (Antu), in 1998 (see Table 7.2). ISAAC was placed at the telescope's Nasmyth B focus and FORS 1 at its Cassegrain focus.

The VLT instruments were highly complex and it was no great surprise that most of them over the years had had teething and other problems. For example several cryogenic mechanisms failed on ISAAC within the first year and in 2000 its mode selection mirror failed. In the following year progressive degradation was observed and operation of part of the instrument had to be suspended for five months because of problems caused by high-frequency oscillations of the telescope secondary mirror. FORS 1 had also experienced problems. For example it showed a significant sensitivity reduction in 1999, six months after its initial commissioning. Part of this was due to dust on the mirrors and part due to the inadvertent use of volatile materials in the cryostats which were contaminating the surface of the CCD. Then in 2001 a problem was found with the instrument's blue-UV sensitivity.

[34] This acronym and that of the other VLT instruments mentioned later in this Section are defined in Table 7.2.

Table 7.2. *VLT Instruments*

(a) First generation instruments

Instrument Name (including meaning of acronym)	Purpose	Waveband	Spectral Resolution (SR) and Image Resolution (IR)	Contractor/ Consortium	Comments
FORS (FOcal Reducer and low dispersion Spectrograph) **1 and 2**	Imaging in the visible and low resolution, multi-object spectroscopy.	330–1100 nm	SR 260–2600	Heidelberg, Munich and Göttingen observatories.	1998 September FORS 1 first light at Cassegrain focus of UT1. 1999 October FORS 2 first light on UT2. 2004 FORS 1 moved to UT2 and FORS 2 to UT1. 2009 FORS 1 decommissioned from UT2 to make way for X-shooter (see below), and FORS 2 rebuilt at UT1 with components from both instruments.
ISAAC (Infrared Spectrometer And Array Camera)	Infrared imager and spectrometer.	1–5 microns	SR 250–10,000	In-house	Planned to be first instrument to be fitted to a VLT. 1998 November First light at Nasmyth B focus of UT1. 2009 Moved from UTI to Nasmyth A focus of UT3. 2013 Removed from UT3 to allow installation of SPHERE (see below).

UVES (Ultraviolet and Visible Echelle Spectrograph)	Very high resolution echelle spectrometer	300–1,100 nm	SR up to 80,000 blue arm/ 110,000 red arm. IR ~0.3" in N band.	In-house	1999 September First light at Nasmyth B focus of UT2.
NACO which consisted of **NAOS** (Nasmyth Adaptive Optics System) and **CONICA** (COudé Near Infrared CAmera)	NAOS was an adaptive optics unit for CONICA, a high-resolution imaging camera.	1–5 microns	SR <1,500 IR 60 mas	NAOS built by the ONERA-INSU consortium. CONICA by the Max Planck Institute (Heidelberg and Garching), and Turin Observatory consortium.	CONICA was originally designed for a UT Coudé focus. It was then redesigned for fitting to a Nasmyth focus following postponement of Coudé system in 1993. 2001 NAOS-CONICA first light at Nasmyth B focus of UT4. 2007 laser guide star facility available. 2014 NACO moved to Nasmyth A focus of UT1 to make way for MUSE (see below).
VIMOS (Visible Multi-Object Spectrograph)	Multi-object spectrograph and imager	360–1,000 nm	SR 200–3,000	Consortium of 9 institutes led by the Laboratoire d'Astrophysique de Marseille.	Fast-track instrument. 2002 First light at Nasmyth B focus of UT3.
NIRMOS (Near InfraRed Multi-Object Spectrograph)	Near-IR Multi-Slit Spectrograph			Same consortium as VIMOS.	Fast-track instrument 1997 Contract let. 2002 Contract cancelled.[a]

(continued)

Table 7.2. *(continued)*

Instrument Name (including meaning of acronym)	Purpose	Waveband	Spectral Resolution (SR) and Image Resolution (IR)	Contractor/ Consortium	Comments
FLAMES (Fibre Large Array Multi-Element Spectrograph)	Multi-fibre echelle, integral field spectroscopy with wide field of view. Consisted of a fibre positioner (OzPoz), a medium-high resolution optical spectrograph (GIRAFFE), plus a link to the UVES spectrograph.	370–950 nm	SR 6,000–47,000 Corrected FoV 25 arcmin	Oz-Poz provided by the Anglo-Australian Observatory. GIRAFFE provided by the Paris observatory.	2002 First light at Nasmyth A focus of UT2. 2015 Discovers new kind of globular cluster.
SINFONI (Spectrograph for INtegral Field Observations in the Near Infrared)	Near-infrared integral field (3D) spectrograph fed by an Adaptive Optics module	1.1–2.4 microns	SR 1,500–4,000	SPIFFI spectrograph built by MPE-Garching. Adaptive Optics module built in-house.	2004 installed at Cassegrain focus of UT4. 2007 laser guide star facility available.
VISIR (VLT Imager and Spectrometer for the mid InfraRed	Diffraction limited imaging at high sensitivity in two mid-infrared atmospheric windows	N band 8–13 microns and Q band 16.5–24.5 microns.	SR 150–30,000 IR 0.3 arcsec in N band	CEA/DAPNIA/ SAP and NRFA/ ASTRON	2004 Installed at Cassegrain focus of UT3. 2008/9 Temporarily removed from Cassegrain focus of UT3 to make way for X-shooter (see below). 2014 Reinstalled at Cassegrain focus of UT3.

FUEGOS (Fibre Unit for European General Optical Spectroscopy)	Multi-fibre area spectrograph			Paris-Meudon Observatory	Contract signed 1995. Contract terminated 1997 due to cost and schedule problems.
CRIRES (CRyogenic high-resolution InfraRed Echelle Spectrograph)	Echelle slit spectroscopy	0.95–5.20 microns	SR 50,000–100,000 Image resolution can be improved by using the MACAO adaptive optics system.	In-house	2006 installed at Nasmyth A focus of UT1. Last of the first generation instruments to be installed. 2014 Removed from telescope for upgrade.

(b) Second generation instruments

Instrument Name (including meaning of acronym)	Purpose	Waveband	Spectral Resolution (SR) and Image Resolution (IR)	Contractor/ Consortium	Comments
HAWK-I (High Acuity Wide field K-band Imager)	Near-infrared wide-field imager.	0.85–2.5 microns	FoV 7.5 arcmin	In-house	2007 First light at Nasmyth A focus of UT4.
X-shooter	Spectrograph covering very wide range of wavelengths simultaneously. In the Rapid Response Mode (RRM) can observe transient phenomena.	0.3–2.45 microns	SR ~5,000–17,000		2008 First light at Cassegrain focus of UT3 temporarily replacing VISIR. 2009 Moved to Cassegrain focus of UT2.

(*continued*)

Table 7.2. *(continued)*

Instrument Name (including meaning of acronym)	Purpose	Waveband	Spectral Resolution (SR) and Image Resolution (IR)	Contractor/ Consortium	Comments
KMOS (K-band Multi Object Spectrograph)	Near-infrared multi-object integral field spectrograph.	0.8–2.5 microns	SR 1,800–4,000	Universities of Durham, Munich and Oxford, plus the Max-Planck Institute and the UK Astronomy Technology Centre.	2012 First light at Nasmyth B focus of UT1.
SPHERE (Spectro-Polarimetric High-contrast Exoplanet REsearch)	Discovery and study of extra-solar giant planets by direct imaging.	0.6–2.3 microns	SR 20–500 FoV ~1–10 arcsec	Consortium of 11 institutions led by the Institut de Planétologie et d'Astrophysique de Grenoble.	2014 Replaced ISAAC at Nasmyth A focus of UT3.
MUSE (Multi-Unit Spectroscopic Explorer)	Integral-field spectrograph.	0.46–0.93 microns	SR 2,000–4,000 FoV wide field 1 arcmin narrow field 7.5 arcsec.	AIP Potsdam CRAL Lyon ESO ETH Zurich IRAP Toulouse Leiden Observatory IAG Göttingen	2014 Replaced NACO at Nasmyth B focus of UT4.

(c) Interferometric instruments

Instrument name (including meaning of acronym)	Purpose	Waveband	Spectral resolution (SR) and image resolution (IR)	Contractor/ Consortium	Comments
MIDI (MID-Infrared interferometric instrument)	Interferometric instrument for photometry and spectroscopy.	8–13 microns.	SR 30 or 230. IR up to 15 mas at 10 microns.	Consortium of 13 institutions.	2002 First fringes using UT1 and 3.
AMBER (Astronomical Mulit-BEam combineR)	Combines the light of three telescopes for photometric and spectroscopic studies.	1.5–2.5 microns.	SR 30, or 1,500 or 12,000. IR 3 mas at 2 microns.	Consortium of 6 French, 1 German and 1 Italian research institutes.	2004 First fringes with three telescopes.

[a] *ESO Annual Report*, 2002, p. 110.

Figure 7.4. Domes of the four 1.8 m VLT Auxiliary Telescopes with the Unit Telescopes in the background. (Courtesy ESO.)

As far as the VLT interferometer was concerned, a contract had been signed in March 1998 with Fokker Space of the Netherlands for the provision of its delay lines. A few months later ESO also awarded a contract to AMOS[35] of Liège, Belgium to provide two 1.8 m auxiliary telescopes for the interferometer, with an option for a third. Eventually ESO purchased a fourth using special financial contributions from Belgium, Italy and Switzerland. When the interferometer was completed these auxiliary telescopes (see Figure 7.4) could be moved to any of 30 fixed stations along railway lines using a specially designed transporter. The system, which included the four unit telescopes, was designed to provide high sensitivity and milliarcsec resolution using groups of two or three telescopes with baselines of up to 202 m.

The basic interferometry system was checked out at 2.2 microns using two 40 cm test telescopes in March 2001, measuring the diameter of α Hydrae as 9.3 ± 0.2 mas. Later that year two 8.2 m unit telescopes[36] were connected 102 m apart to measure the diameter of α Eridani to be 1.92 ± 0.05 mas. The first auxiliary telescope arrived at Paranal Observatory in January 2004, and the first interference fringes were observed between this and the second auxiliary telescope in February 2005. The final two auxiliary telescopes became operational in 2006.

In the late 1980s, Gerard Rousset and colleagues had developed an early prototype of a VLT adaptive optics system. This system called COME-ON (see Section

[35] AMOS is an acronym of 'Advanced Mechanical and Optical Systems'.
[36] These two telescopes were UT1 (Antu) and UT3 (Melipal).

5.2), which operated in the infrared, was successfully tested on the Haute-Provence observatory's 1.5 m telescope in 1989. In the following year it was installed on ESO's 3.6 m telescope where it achieved a resolution of 0.17 arcsec at 2.2 microns. A second generation system called ADONIS was then developed that was much more user friendly. It was installed on ESO's 3.6 m telescope in 1994 where it produced a diffraction-limited resolution of 0.15 arcsec in the infrared.

ESO had planned to incorporate adaptive optics systems for the Coudé foci of each VLT unit telescope, mainly to support the VLT Interferometer (VLTI). But, as explained above, in 1993 the Council delayed work on the VLT interferometer system, the Coudé focus, and its associated adaptive optics. This resulted in CONICA, a VLT infrared camera, (see Table 7.2) being redesigned and located at the Nasmyth focus, and the contract for its adaptive optics system (called NAOS) being signed later than planned in 1997. Four years later the combined instrument NAOS-CONICA, or NACO, saw first light, producing almost diffraction-limited images of 0.07 arcsec on UT4 (Yepun) at about 2.2 microns (K-band). A little later a resolution of 0.04 arcsec was achieved at 1.2 microns.(27)

These adaptive optics systems required a computer to work out what corrections to apply to the secondary mirror. But in the 1990s a completely different sort of system was developed that required no such computer but used a curvature sensor linked to a bimorph deformable mirror instead. Over the period 2003–2005 a 60 actuator curvature sensor/bimorph system called MACAO[37] was installed in each of the four Coudé feeds to the VLTI. A similar system was installed in 2004 in the Cassegrain focus system of UT4 feeding the SINFONI spectrograph working in the near infrared. Two years later a similar unit was installed with the CRIRES high-resolution infrared échelle spectrograph.

Unfortunately there were parts of the sky devoid of sufficiently bright natural guide stars to allow an adaptive optics system to be used. But in this case a laser could be used to produce an artificial guide star by interacting with the Earth's upper atmosphere. Encouraging work on such a system had been carried out at the Lick Observatory in the mid 1990s (see Section 5.2). As a result ESO began work in 2003 to produce a laser guide star facility using a 4 watt continuous wave sodium laser in collaboration with MPE (Garching) and MPIA (Heidelberg).(28) The system, which was installed in 2006, became operational the following year on UT4 (see Figure 5.2 earlier) allowing operation with the NACO and SYMPHONI instruments.

The above systems were only able to produce a sharp image within a very small area, or isoplanatic patch, centred on the guide star. But in the 1990s a multi-conjugate system had been developed at various observatories which would

[37] MACAO is an acronym of 'Multi-Applications Curvature Adaptive Optics'.

produce sharp images over a much larger area. At about the same time ESO started to consider building a telescope appreciably larger than the VLT. If they eventually built such a machine it would be virtually essential to produce sharp images over an appreciably larger area than previously. So to provide such images they decided to build a multi-conjugate demonstrator called MAD[38] optimised for 2.2 microns, in collaboration with the University of Lisbon and the Observatories of Padua and Florence. It had three wavefront sensors and two deformable mirrors. MAD was first successfully tested in ESO's laboratory at Garching in 2005. It was then attached in March 2007 to the Nasmyth A visitor focus of UT3 (Melipal) where it used three natural guide stars. MAD produced images with resolutions of 0.10 arcsec over an area of about 20 arcsec diameter near the centre of the globular cluster Omega Centauri in seeing of about 0.7 arcsec. The nearest guide star was ~1 arcmin away from the centre of the field of view.

The development of these various adaptive optics systems has generally resulted in a complicated telescope design which is often difficult to operate and which has too much equipment in or near the focal plane. One way of simplifying this would be to make the secondary mirror itself deformable to produce an adaptive telescope. The first such deformable or adaptive secondary mirror had been fitted to the new MMT on Mount Hopkins in 2003. In the following year ESO started work on developing a deformable secondary mirror for the VLT. Eight years later the mirror was delivered to ESO's Garching laboratory for testing prior to its expected installation on UT4. This deformable secondary mirror had a Zerodur thin-shell mirror 112 cm in diameter and just 2 mm thick mounted 65 microns in front of its Zerodur light-weighted reference backplate which had holes for over one thousand actuators.(29) It was intended to operate this deformable secondary mirror using four laser guide stars.

In recent years interest in joining ESO has accelerated, helped considerably by the excellent performance of its telescopes. So Portugal joined in 2001, following by the UK in 2002, Finland (2004), Spain (2007), Czech Republic (2007), Austria (2009) and Poland (2015). In 2010 Brazil stated its intention of joining ESO but it has currently (as of early 2015) not ratified its accession.

VISTA

Towards the end of the 1990s a number of UK astronomers began to consider designs of a survey telescope as a follow on to the UK Schmidt (see Section 9.2) that was beginning to show its age. Initially a 2.5 m was considered, but in 1998 a design based on the 4.2 m William Hershel Telescope (WHT, see Section 6.1)

[38] MAD is an acronym of 'Multi-conjugate Adaptive optics Demonstrator'.

was proposed by a group of astronomers to the UK's PPARC.[39] Although PPARC accepted the basic idea of building a new large survey telescope, the design based on the WHT proved to be too expensive. So a more modest design was developed during the project's subsequent Phase A feasibility study.

At about the same time the UK had begun exploratory discussions with ESO about the possibility of the UK joining the organisation. Although not directly linked to these discussions, the UK also asked ESO if the UK's proposed new survey telescope, to be called VISTA (Visible and Infrared Survey Telescope for Astronomy), could be located in the Paranal area. ESO responded rapidly and positively to the latter enquiry resulting in an agreement in February 2000 to site VISTA on a 2520 m (8,270 ft) high peak, known as the NTT Peak[40], about 1.5 km from the VLT. It was also agreed that operationally VISTA would be considered to be part of the Paranal observatory. Then in December 2001 ESO and PPARC agreed that the UK would become a formal member of ESO, effective July 2002,[41] with the UK donating VISTA to cover part of its joining fee.[42]

Meanwhile ESO had also agreed to assist the UK in the design of VISTA. This was partly to ensure that the telescope was compatible with the Paranal operational system, but also to improve the telescope's design by using ESO's experience gained in designing the NTT and VLT telescopes. For example, VISTA had originally been expected to have a passive optics system, but following ESO advice this design was changed to one with a thin meniscus mirror and active optics.(30) It was also decided to use an extremely fast primary mirror of f/1.0 to reduce the size of the telescope. Unfortunately cost estimates showed that VISTA would exceed its agreed budget. Consequently it was decided to limit the telescope to infrared observations over the range 0.85 to 2.3 microns and eliminate the visible waveband requirement which would be covered by the 2.5 m VLT Survey Telescope (VST), see below.

The 4.1 m VISTA,(31) which had been conceived and developed by a consortium of eighteen UK universities, led by Queen Mary University of London, was a wide field (1.65°) infrared survey telescope. It had quasi-Ritchey-Chrétien optics. Its primary mirror, which was the most highly curved mirror made of its size and quality to date, was made of Schott glass and was polished by LZOS[43] in Russia.

[39] PPARC is an acronym of 'Particle Physics and Astronomy Research Council'.

[40] So called as at one stage it had been suggested that ESO's NTT could be located there.

[41] This decision followed the 1997 UK government's policy of increasing its investment in science and technology. This government of Tony Blair also signalled its political desire to increase collaboration with its European partners in a number of areas.

[42] The UK also agreed to pay about £50 million (~$75 million) as a one off 'joining fee', in addition to its £12 million ($18 million) annual subscription, see *Sky and Telescope*, October 2002, p. 26. But because VISTA would not be available for some time, the UK also agreed to provide survey data from the UK InfraRed Telescope (UKIRT, see Section 8.4) on Mauna Kea to ESO.

[43] LZOS is a Russian acronym of 'Lytkarino Optical Glass Factory'.

The telescope's 3-ton infrared camera called VIRCAM[44] was built by a UK consortium consisting of the Rutherford Appleton Laboratory, the UK Astronomy Technology Centre and Durham University. It consisted of an array of sixteen 2,048 × 2,048 pixel Raytheon CCDs (totalling 67 million pixels) which were cooled to 70 K. The CCDs were made of mercury cadmium telluride chips, rather than the usual silicon, to provide infrared coverage up to 2.5 microns. First light took place two years later than expected in 2008.

VLT Survey Telescope (VST)

An ESO working group to examine the future of La Silla concluded(32) in 1996 that 'ESO does not possess a truly competitive wide-field imaging telescope' as their 1.0/1.6 m Schmidt was somewhat dated.[45] At about the same time the director of the Naples Observatory, Massimo Capaccioli, received a grant from the Italian government to build a new 1.5 m telescope. But this was a one-off grant and the telescope would need an annual sum to enable it to operate efficiently and be updated from time to time. Capaccioli also hoped to build a larger telescope than this. He was, in particular, interested in building a medium-sized, wide-field, survey telescope at a high-quality site. Consequently he approached ESO to see if they were interested in providing a site for such an instrument and come to some agreement for operating it.

Shortly afterwards a UV-visible Wide Field Imager (WFI) with an eight 2,048 × 4,096 pixel (67 megapixel) CCD sensor(33) was fitted to the MPG/ESO 2.2 m on La Silla. This was to provide ESO with some wide-field coverage. But this system had only a 0.5° × 0.5° field of view, compared with the ESO Schmidt's 5.4° × 5.4° and Capaccioli's proposed telescope which would be significantly better than the WFI system. So ESO were interested in examining the Naples Observatory's proposal for what was to be called the VST Survey Telescope (VST).

Naples Observatory had been involved in the development of the WFI for the 2.2 m, so they had some experience in this field. Consequently ESO decided in 1998 to set up a joint venture with the Osservatorio Astronomico di Capodimonte of Naples in which Naples would build the VST along the lines proposed by Capaccioli and his team (but with some modifications), whilst ESO would supply its enclosure (see Figure 7.3 earlier), adjacent to the VLT, and would operate the telescope. It was anticipated that the VST would be completed by 2001.

The VST was a 2.6 m survey telescope operating in the 0.3 to 1.0 micron waveband with a 1° × 1° field of view. It had an active optics system and a 268 megapixel

[44] For 'Vista InfraRed CAMera'.

[45] It was to be retired as a traditional Schmidt using photographic plates two years later.

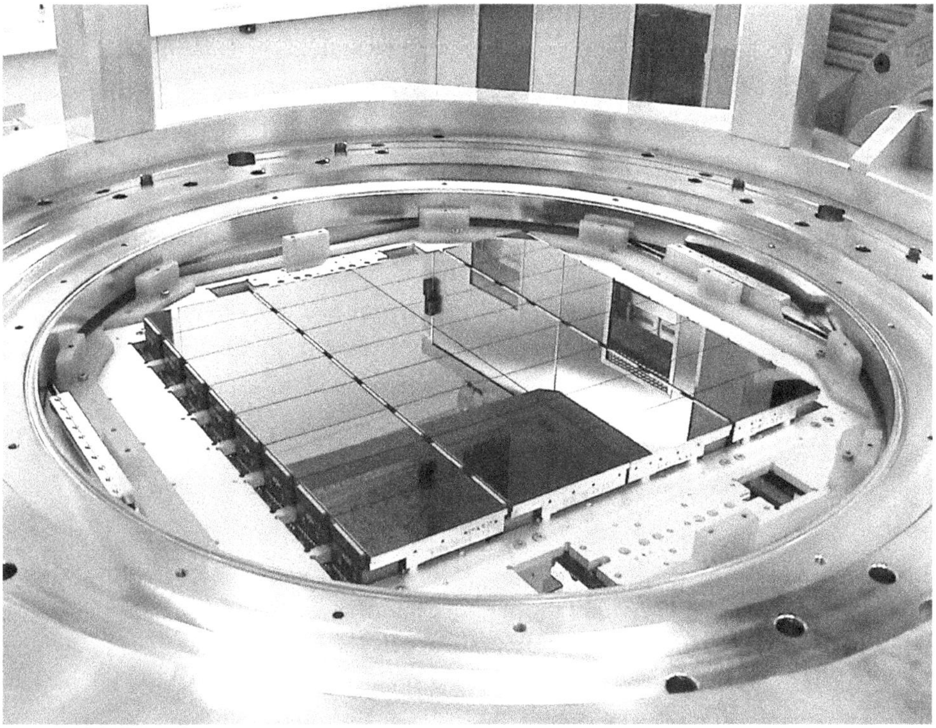

Figure 7.5. The 32 CCD detector array of VST's OmegaCAM. (Courtesy ESO.)

CCD camera, called OmegaCAM, which was the subject of a separate procurement from a consortium led by the Universitäts-Sternwarte München, the NOVA School (of the Netherlands) and the Osservatorio Astronomico di Padova (of Italy).[46] The large detector array for the camera of thirty-two 2,048 × 4,096 pixel CCDs (see Figure 7.5) was the responsibility of ESO who signed a contract for forty of these science-grade Marconi CCDs in February 2000.[47] Unfortunately, although the first detectors were delivered in the autumn of that year, the final delivery was not made until 2003 and the whole camera was not completed until 2006. But because of delays elsewhere it was then put into storage at ESO headquarters and not delivered to Paranal until 2008.

Carl Zeiss Jena, who won the contract for the VST telescope optics, subcontracted the provision of the primary and secondary mirrors to LZOS in Russia. The primary mirror was finished well ahead of schedule in early 2002 and sent by air,

[46] The camera was paid for governmental organisations in Germany, the Netherlands and Italy.

[47] The camera used 32 CCDs for the science array plus four auxiliary CCDs arranged around the edges of the field. Two of these four were to autoguide the telescope, and the other two for curvature-wavefront sensing and control of the active optics system, see *ESO Annual Report*, 2011, p. 34.

lorry and ship to Chile. But on opening its transport container in Chile it was found that the mirror was broken, apparently due to an accident on the ship. The secondary mirror also had some slight defects so it was returned to LZOS for repair.

The repaired secondary mirror arrived at La Paranal in 2006 followed by the new primary mirror in the following year. They where then put into storage awaiting the delivery of their mirror cells. Problems with the primary mirror cell resulted in it not being ready to be shipped to Chile until 2009, where yet another problem discovered. The transport container was found to be flooded with water; consequently the mirror cell was red with rust, which required it to be returned to Europe for a total overhaul. On completion it finally arrived at Paranal in the middle of the following year, enabling technical first light of the VST to finally take place in December 2010, about nine years late.

7.3 OWL and the E-ELT

In the late 1990s, as the VLT was approaching first light, Roberto Gilmozzi and others working at ESO began to think about building an even larger ground-based telescope. Gilmozzi, in particular, examined what size of optical telescope would be required to undertake spectroscopy on the faintest objects that would be observable with the Next Generation Space Telescope (now called the James Webb Space Telescope) then expected to be launched about 2009.[48] He concluded that this would require a telescope with a diameter of about 100 m which was a full order of magnitude larger than those currently available. Most of his colleagues dismissed building such a telescope as impossible. But Philippe Dierickx, the optical engineer in charge of the VLT mirrors, believed it could be done, even naming the proposed new 100 m diameter telescope OWL for 'OverWhelmingly Large'. (34) By the end of 1997 a small group of engineers and scientists at ESO had even produced an outline design and approximate scientific case for such a telescope.

So far the work had been undertaken by an informal group at ESO, but inevitably it eventually came to the attention of the Director General, Ricardo Giacconi. His initial reaction was to be concerned that the work may have taken the focus off getting the VLT operational. But in early 1998 Giacconi decided to back the nascent project, at least provisionally, and set up a small internal project team to examine its feasibility in more detail. Gilmozzi was appointed principle investigator and Dierickx project engineer.

At about the same time Arne Ardeberg, professor of astronomy at Lund Observatory in Sweden[49] had led a small team to produce the outline design of

[48] Current launch date (as of May 2015) is October 2018.
[49] And director of the La Silla Observatory from 1979 to 1984.

the largest optical telescope that they though feasible. This was the 50 m Swedish Extremely Large Telescope, whose name was changed to Euro50 in 2000 following the involvement of astronomers and engineers from institutes in Finland, Ireland, Spain and the UK. It was a Gregorian with a 50 m diameter f/0.85 segmented primary mirror and a 4 m deformable secondary. The cost estimate, including 15% contingency, was 722 million Euros or ~\$780 million (at 1999 rates).(35)[50]

At this stage the design of ESO's OWL consisted of segmented primary and secondary mirrors, integrated active optics and multi-conjugate adaptive optics.(36) The 100 m primary mirror was spherical and the 33.5 m secondary was flat, with spherical and field aberrations being corrected by a four mirror corrector. ESO thought that the only critical system needing development was the multi-conjugate adaptive optics, although they acknowledged that they would have to turn to the Keck designers for experience in designing and building segmented mirrors.[51] Costs were estimated at about 1,000 million Euros, including contingency, with a timescale of 11 to 12 years for completion after project approval. This outline plan assumed the demonstration of the principle of multi-conjugate adaptive optics on the VLT by 2003 (which was achieved in 2007).

Many outside observers thought that OWL was too large and risky. They also criticised ESO as, in their view, ESO had not fully involved the astronomical community to find out what sort of telescope was really required. In short, OWL was seen by many as an engineer's telescope for engineers rather than for astronomers. On their side ESO also had serious reservations about the Euro50 design, mainly its optical configuration which appeared to place excessive demands on the accuracy of alignment of its various elements and on the structural stiffness of the telescope.

So in the early years of the new millennium there were two radically different designs of European extremely large telescopes, the 100 m OWL proposed by ESO and the Euro50 proposed by another group independent of ESO. It was then that European Union (EU) became involved.

The European Community, the forerunner of the EU, had become engaged in funding technological Research and Development in a small way since the early 1980s.(37) Then in 2002 the EU introduced their sixth Framework Programme (FP6) which contained funds for developing new trans-European research infrastructures with limited funds for construction. The Euro50 team planned to apply for some of this funding from FP6. ESO, on the other hand, were more cautious about applying for funding as this may have come with unacceptable conditions.

[50] During 1999, the Euro exchange rate varied from about \$1.17 at the beginning of the year to about parity at the end.

[51] This was no longer the case after Spain joined ESO in 2006 because of their experience in building the 10.4 m Gran Telescopio Canarias (see Section 6.1) that also had a segmented mirror.

The EU's previous FP5 programme had provided funds for two European astronomical networks, OPTICON[52] for optical astronomy and RadioNet for radio astronomy. So in 2002 OPTICON arranged a conference in Turku, Finland hoping to resolve the differences between the Euro50 and OWL telescope programmes. This conference included astronomers from all over Europe as well as representatives of the OWL and Euro50 projects. It was followed shortly afterwards by a meeting in Bologna between OPTICON, ESO and Euro50 representatives at which it was agreed that both projects should continue for eighteen months. Subsequently a joint proposal would be submitted to the European Commission (EC) requesting funding for studies of the technologies required by both projects.

The joint proposal, which was submitted in 2004, asked the EC for 22 million Euros (about $27 million) out of a total project study budget of 42 million Euros ($52 million), of which ESO would contribute 8.5 million Euros ($10.5 million). In the event the EC decided to limit its contribution to 8.4 million Euros, which caused a number of institutions to pull out of the study. Nevertheless, the work continued with ESO hosting the project office. It was completed in 2009.

Meanwhile ESO had been concentrating in the early 2000s on completing the VLT's instrumentation programme, commissioning the VLTI and early work on the ALMA millimetre/submillimetre array (see Section 24.1). So, although the ESO Council were kept informed of the early work on OWL, there was no urgent need to start work at that time on another major project. However the involvement of the EU on helping to define a European extremely large telescope programme required a decision to be made of ESO's policy on such a programme. As a result in December 2004 Council decided that 'ESO's highest priority strategic goal must be the European retention of astronomical leadership and excellence into the era of Extremely Large Telescopes'.[53] (38) In addition they committed ESO to 'the construction of an Extremely Large Telescope on a competitive timescale', as they put it, seeking additional funds as required. Consequently ESO determined to complete their early conceptual work as soon as possible.

In the following year ESO/OPTICON published a study of the scientific case for the European Extremely Large Telescope (ELT), which was defined as a 50 to 100 metre telescope, thus including both Euro50 and OWL.(39) This scientific case had been developed at a series of meetings sponsored by OPTICON over the preceding few years. In the same year ESO published their Blue Book[54] outlining their baseline OWL design on completion of their OWL Conceptual Study.

A comprehensive review was then carried out of the results of this OWL conceptual study by an international panel that presented its findings in November

[52] OPTICON is an acronym of 'OPTical Infrared CO-ordination Network for astronomy'.

[53] *ESO Annual Report,* 2004, p. 10.

[54] *OWL Concept Design Report*, Phase A Design Review, OWL-TRE-ESO-0000-0001.

2005. They concluded that the design was basically feasible, but that there were still many risks involved. These were mainly connected with the segmentation of both the primary and secondary mirrors, the highly aspherical fourth mirror and the Laser Guide Star system. Consequently there was a significant probability that the telescope may not achieve its required scientific performance and there was a high likelihood of a significant schedule slip with an associated cost overrun. As a result the panel recommended that ESO considers building a smaller diameter, less complex ELT than OWL whilst, at the same time, speeding up the site selection and site procurement process.

The ESO Director General Catherine Cesarsky[55] was concerned that the work on OWL had, so far, not had enough involvement of the scientific community as it was vital that they buy in to the ELT design that was eventually chosen. So, following the international panel review, she set up five working groups involving both ESO members and people from the scientific community. These five working groups were to review the key themes of science, telescope and observatory design, instrumentation, adaptive optics and site selection. They were to examine the ELT design against existing capabilities, propose a list of key trade-offs and suggest an initial prioritisation of ESO ELT capabilities. The working groups started their work early in January 2006 and presented their draft reports by the end of February. These were integrated into a consolidated report to aid the ESO project team in designing the telescope. In the meantime, ESO had decided that the ELT should have a diameter of about 42 m. Then a conference was held at Marseille, near the end of the year, at which ESO and the working groups outlined the results of their work.

ESO presented two alternative optical designs at Marseille, both of which used a 42 m f/1 aspheric primary mirror with more than 900 hexagonal segments. Both designs also covered a 10 arcmin diameter field. One design was a Gregorian, similar to that of Euro50, with an adaptive concave 4.8 m secondary mirror (M2), whilst the other was a novel five-mirror system with an active 6 m convex secondary.(40) In the five-mirror design the 2.5 m quaternary mirror (M4) was deformable and the 2.7 m fifth mirror (M5) provided tip-tilt compensation. Naturally the strengths and weaknesses of both designs provoked a great deal of discussion prior to and at the conference itself. In the end it was decided to take the five-mirror design as the baseline, with the three-mirror design as backup. At this stage the cost of this European Extremely Large Telescope (E-ELT), as it was now called, was estimated to be about 850 million Euros (about $ 1,090 million) with a ten year design and construction timescale.(41)

[55] Catherine Cesarsky had taken over from Riccardo Giacconi as Director General in 1999.

Following this Marseille conference the ESO Council gave the go-ahead for the E-ELT three year Phase B study at an estimated cost of 57.2 million Euros ($73 million). During this study, which ended in 2010, a number of simulations were performed to examine the effect of telescope and instrument design parameters on the system's scientific performance. In addition, a number of contracts were let to industry for them to produce designs for key systems and subsystems, including particularly the telescope structure and dome. Several prototypes of key elements were also produced.

In April 2010, following a clear recommendation from the Site Selection Advisory Committee, the ESO Council decided to build the E-ELT on the 3,060 m (10,040 ft) high Cerro Armazones, about 20 km (12 miles) from Cerro Paranal. The Chilean government agreed to donate the land. Five months later the E-ELT Phase B study was completed with a highly satisfactory external panel review that unanimously concluded that the telescope design was mature enough for construction to begin.[56] Nevertheless, this panel observed that there were still some unresolved issues with the telescope's thermal emissivity and with the M2 and M4 mirrors, of which the project was already aware. The panel also felt that the planned construction period was unreasonably short, suggesting that it may take two or three years longer than ESO planned, and believed that the proposed financial contingency of 12% was too low.

It had been clear for a little while that the E-ELT was probably unaffordable particularly if, as suggested by the external review panel, the planned construction timescale was too short and the financial contingency was too low. Consequently ESO decided to examine how the risk and cost of the telescope could be reduced before starting the construction phase. The subsequent study resulted in the decision, taken by Council in June 2011, to reduce the size of the primary mirror from 42 to 39 m.

This reduction in mirror diameter impacted the science that could be undertaken by the E-ELT in a number of ways.(42) Clearly the image resolution would be reduced, so the planned imaging of Earth-sized planets could be undertaken to only 80% of the originally foreseen distance, thus reducing the size of space observable in this research by about 50%.[57] On the other hand observations of relatively dim galaxies and other objects could still be undertaken with the smaller telescope, but the exposure times would have to be extended by ~20–35%, with a consequent loss in the number of research tasks that could be undertaken in a given time. Unfortunately there were cases in which the observing time could not be usefully extended, for example those of non-recurring transient events like supernovae

[56] *Report of the Review Board*, E-ELT Phase B Final Design Review, September 21–24, 2010.
[57] *ESO Annual Report*, pp. 62–63.

Figure 7.6. An artist's impression of the 39 m diameter E-ELT in its enclosure. (Courtesy ESO/L. Calçada.)

or gamma-ray bursts which, by their very nature, were of limited duration. So the study of such objects would be somewhat compromised.

A number of changes in the telescope design ensued from making the primary mirror smaller. For example the mirror could be made faster (f/0.9 instead of f/1), so reducing the size and mass of the telescope structure, mount and enclosure even further. Reducing the size of the enclosure not only saved money but also reduced wind effects, improving the performance of the telescope. The secondary mirror could now also be made smaller. It was known that this mirror could be polished much more easily if its diameter was below 4.2 m, so the designers accepted that diameter as a constraint and evolved the telescope's optical design about it.

Technical first light for this 39 m telescope (see Figure 7.6) was expected to take place around the end of 2021, assuming a project start in January 2012.[58] Start of observatory observations was expected in the fall of 2022. The cost of the project was estimated at 1,083 million Euro (about $1,460 million) including a 100 million Euro ($135 million) contingency and 100 million Euro ring-fenced for instruments.[59]

[58] *E-ELT Construction Proposal*, ESO, 2011, pp. 13–15.
[59] This was expected to pay for four instruments.

In a little more detail; the telescope optical design consisted of a 39.3 m diameter f/0.9 primary mirror made of 798 hexagonal segments, each segment having a maximum corner-to-corner dimension of about 1.45 m. This primary had an 11 m central obstruction. The 4.2 m active thin meniscus convex secondary returned the beam through a hole in the 2.4 m M4 mirror to the 3.8 m mildly aspheric concave M3 located at the vertex of the primary. The M3 then reflected the beam to the flat adaptive M4 supported by about 6000 to 8000 actuators and the 2.6 × 2.1 m flat M5 that provided tip-tilt image correction and steered the beam towards the Nasmyth focus.

Provisional approval for the E-ELT was given by the ESO Council in June 2012, followed by project confirmation in December when the required two-thirds majority of Member States was reached. The first contract, that to the Italian firm AdOptica for design and construction of the 2.5 m deformable M4 mirror, was awarded in the summer of 2012. Two years later the task of flattening the top of Cerro Armazones was begun to provide an area for building the E-ELT. In the process the height of the mountain summit was reduced by about 18 m. Construction of the E-ELT was formally approved by the ESO Council in December 2014 with first light expected in late 2024.

References

1. Madsen, Claus, *The Jewel on the Mountaintop; The European Southern Observatory through Fifty Years*, ESO/Wiley-VCH, 2012, pp. 25–26.
2. Blaauw, Adriaan, *ESO's Early History; The European Southern Observatory from Concept to Reality*, European Southern Observatory, 1991, p. 4.
3. Ibid., pp. 10–12 and Appendix 2.
4. Westerlund, Bengt E., European Southern Observatory Dedicated in Chile, *Sky and Telescope*, June 1969, pp. 340–347.
5. Lausten, S., 'First Light' for the ESO 3.6-meter Telescope, *Sky and Telescope*, February 1977, pp. 97–103.
6. Madsen, Claus, *The Jewel on the Mountaintop; The European Southern Observatory through Fifty Years*, ESO/Wiley-VCH, 2012, p. 88.
7. Chincarini, G., et al., The Last Born at La Silla: REM, The Rapid Eye Mount, *ESO Messenger*, No. 113, September 2003, pp. 40–44.
8. Boër, M., et al., The Gamma-Ray Burst Hunt at La Silla; The TAROT-S Very Fast Moving Telescope, *ESO Messenger*, No. 113, September 2003, pp. 45–48.
9. Madsen, Claus, *The Jewel on the Mountaintop; The European Southern Observatory through Fifty Years*, ESO/Wiley-VCH, 2012, pp. 115–117.
10. Richter, W., New Technology Telescope, *The Messenger*, No. 21, September 1980, p. 19.
11. Wilson, R., Progress on the 3.5 m 'New Technology Telescope', *The Messenger*, No. 29, September 1982, pp. 24–27.
12. Wilson, R. N., The History and Development of the ESO Active Optics System, *The Messenger*, No. 113, September 2003, p. 5.

13. Wilson, R. N., *Reflecting Telescope Optics II; Manufacture, Testing, Alignment, Modern Techniques*, Springer-Verlag, Corrected Second Printing, 2001, pp. 200, and 232–233.
14. Wilson, R. N., et al., Active Optics: the NTT and the Future, *The Messenger*, No. 53, September 1988, pp. 1–7.
15. Wilson, R. N., The History and Development of the ESO Active Optics System, *The Messenger*, No. 113, September 2003, p. 5.
16. Chincarini, G., et al., The Last Born at La Silla: REM, The Rapid Eye Mount, *The Messenger*, No. 113, September 2003, pp. 40–44.
17. Boër, M., et al., The Gamma-Ray Burst Hunt at La Silla: The TAROT-S Very Fast Moving Telescope, *The Messenger*, No. 113, September 2003, pp. 45–48.
18. Madsen, Claus, *The Jewel on the Mountaintop; The European Southern Observatory through Fifty Years*, ESO and Wiley-VCH, 2012, p. 204.
19. Ibid., p. 205.
20. Ibid., pp. 183–184.
21. Ibid., pp. 295–304.
22. Ibid., pp. 211–212.
23. Ibid., pp. 247–248.
24. Mueller, R., et al., The 8.2 m Primary Mirrors of the VLT, *The Messenger*, No. 73, September 1993, p. 5.
25. Enard, D., Work Starts on the VLT M2 Units, *The Messenger*, No. 76, June 1994, pp. 2–4.
26. Stanghellini, S., and Michel, A., Performance of the First Two Beryllium Secondary Mirrors of the VLT, *The Messenger*, No. 94, December 1998, pp. 10–11.
27. Brandner, W., et al., NAOS+CONICA at YEPUN: First VLT Adaptive Optics System Sees First Light, *The Messenger*, No. 107, March 2002, pp. 1–6, and *Sky and Telescope*, March 2002, pp. 16–17.
28. Madsen, Claus, *The Jewel on the Mountaintop; The European Southern Observatory through Fifty Years*, ESO and Wiley-VCH, 2012, p. 240.
29. Arsenault, Robin, et al., Delivery of the Second Generation VLT Secondary Mirror (M2) Unit to ESO, *The Messenger*, No. 151, March 2013, pp. 14–19.
30. Madsen, Claus, *The Jewel on the Mountaintop; The European Southern Observatory through Fifty Years*, ESO and Wiley-VCH, 2012, p. 407.
31. Emerson, J. P., et al., VISTA, The Visible and Infrared Survey Telescope for Astronomy, *The Messenger*, No. 117, September 2004, pp. 27–32.
32. Andersen, J., Planning for La Silla in the VLT Era: What Came Out?, *The Messenger*, No. 83, March 1996, pp. 48–49.
33. Baade, D., The Wide Field Imager for the 2.2-m MPG/ESO Telescope: a Preview, *The Messenger*, No. 93, September 1998, pp. 13–15.
34. Madsen, Claus, *The Jewel on the Mountaintop; The European Southern Observatory through Fifty Years*, ESO and Wiley-VCH, 2012, p. 413.
35. Andersen, T., et al., The Proposed 50 m Swedish Extremely Large Telescope, Proceedings of the Bäckascog Workshop on Extremely Large Telescopes 1999, *ESO Conference and Workshop Proceedings No. 57*, pp. 72–82.
36. Gilmozzi, R., and Dierickx, P., OWL Concept study, *The Messenger*, No. 100, June 2000, pp. 1–10.
37. Madsen, Claus, *The Jewel on the Mountaintop; The European Southern Observatory through Fifty Years*, ESO and Wiley-VCH, 2012, p. 416.
38. Monnet, Guy, and Gilmozzi, Roberto, Status of the European ELT, *The Messenger*, No. 123, March 2006, pp. 4–5.

39. Hook, Isobel and Gilmore, Gerry (eds.), *The Science Case for the European Extremely Large Telescope; The Next Step in Mankind's Quest for the Universe*, ESO and OPTICON, 2005.
40. Gilmozzi, Roberto, and Spyromilio, Jason, The European Extremely Large Telescope (E-ELT), *The Messenger*, No. 127, March 2007, pp. 11–19.
41. Hook, Isobel, Monnet, Guy, and Cuby, Jean-Gabriel, Reports on the Conference 'Towards the European Extremely Large Telescope' held in Marseille, France, November 27 – December 1 2006, *The Messenger*, No. 127, March 2007, pp. 20–27.
42. McPherson, Alistair, et al., Recent Progress Towards the European Extremely Large Telescope (E-ELT), *The Messenger*, No. 148, June 2012, pp. 2–8.

8

Mauna Kea Observatory

8.1 Introduction

Hawaii was an important location for geophysical and astronomical observations during the International Geophysical Year (IGY) of 1957–58 because of its relative isolation in the middle of the Pacific Ocean. As a result, a number of such observations were planned but, unfortunately, there was not enough time nor money to establish all the facilities required. For example Walter Steiger of the University of Hawai'i (UH) had been hoping to establish a solar observatory in Hawaii, and in 1955 he had started site testing at the top of the 3,060 m (10,040 ft) high Haleakala, a virtually extinct volcano[1] on the island of Maui. These tests indicated that Haleakala would be an ideal location for a solar observatory, but because of financial and timescale limitations Steiger had to settle, instead, for establishing a sea-level observatory on the island of Oahu.[2]

Fred Whipple had also asked if the UH could build a satellite tracking station on Haleakala for the duration of the IGY. In this case he arranged for money to be provided in a rather round about way. He persuaded his old friend Kenneth Mees, a retired senior executive of Kodak and photographic scientist, to donate a number of Kodak plates to the UH so they could sell them and use the money to establish the tracking station. As a result the UH was able to acquire a plot of land on the summit not only for the satellite tracking station, which became operational in 1957, but, hopefully, to accommodate a future solar observatory there. At about the same time the United States Weather Bureau/National Bureau of Standards established an atmospheric research station at 11,120 ft (3,390 m) on the relatively stable northern slope of the active volcano Mauna Loa[3] on the main island of Hawaii.

[1] The last eruption had taken place around 400 years previously.

[2] Walter Steiger, *Origins of Astronomy in Hawai'i*, Institute for Astronomy, University of Hawai'i website, www.ifa.hawaii.edu/users/steiger/igy_period.htm.

[3] The last eruption on Mauna Loa had taken place in 1950. Subsequently, eruptions took place in 1975 (a very small one) and 1984.

Finally groundbreaking took place in February 1962 for the solar observatory on Haleakala. Dedicated two years later it was named the C.E. Kenneth Mees Solar Observatory in honour of the former Kodak senior executive.

Mitsuo Akiyama of the Hawaii Chamber of Commerce sent letters to many American and Japanese Universities in June 1963 suggesting that they consider the possibility of setting up an astronomical observatory on either of Hawaii's two large volcanoes, Mauna Kea[4] and Mauna Loa.(1) Gerard Kuiper, director of the University of Arizona's Lunar and Planetary Laboratory, was the only one who replied.(2) In fact he had already had his assistant Alika Herring undertake a testing programme on Haleakala, funded by NASA and the United States Department of Defense with a view to setting up a large telescope there for planetary observations. These tests indicated that Haleakala was not high enough above the normal inversion layer in the area. But Mauna Kea, on the other hand, should be suitable as at 4,200 m (13,800 ft) it was over one thousand metres higher than Haleakala.

Unfortunately, unlike Haleakala, there was no access road up Mauna Kea. But in January 1964 Kuiper persuaded John Burns, the Governor of Hawaii, to build a rough track up the mountain. Kuiper also agreed to carry out his investigations on Mauna Kea in collaboration with the UH. With these agreements in place Kuiper drew up a proposal to NASA to investigate Mauna Kea as a possible site for an optical observatory using a telescope in a 3.8 m (12.5 ft.) dome.

The rough track was completed up Mauna Kea by May 1964 and the dome and Herring's 32 cm (12½ inch) test telescope were installed shortly afterwards on the cinder cone Pu'u Poliahu at an altitude of 4,155 m (13,630 ft). The first round of tests that were concluded in October 1964 indicated that observing conditions on Mauna Kea were superb and even better than Haleakala. The air was very dry, so the site should be ideal for both visible and infrared observations. Kuiper then suggested extending these visual tests in collaboration with the UH's Institute of Geophysics (UH had no astronomy department at the time), using the University of Arizona's 70 cm (28 inch) photometric telescope that could be moved from their Catalina observatory. This would allow quantitative observations.

Unfortunately for Kuiper, NASA was becoming increasingly concerned that he was extending himself too much as he was already involved in a number of other research programmes. NASA's concern was known to the UH, so they backed off further involvement with Kuiper's plans for Mauna Kea. This resulted in him submitting a proposal to NASA in December 1964, independent of the UH, for the University of Arizona to build a 1.5 m (60 inch) reflector on Mauna Kea. Then, encouraged by NASA, the UH submitted a separate proposal two months later written by John Jefferies, Steiger's successor as head of a small team of solar astronomers, to build

[4] The Mauna Kea volcano was extinct.

a 2.1 m (84 inch) on the mountain.(3) Much to Kuiper's annoyance, the UH's proposal was accepted by NASA even though, as mentioned above, it had no astronomy department at the time.[5] NASA's chose the UH, rather than his, because the observatory was to be managed locally, and, in addition, it would provide a larger telescope. The contract was signed between NASA and the UH on July 1 1965, just over six months after the UH had made the decision to make a proposal.[6]

The UH had left the location of their proposed telescope open, because of internal Hawaii politics, suggesting that either Haleakala or Mauna Kea could be suitable. Following contract award they investigated atmospheric and meteorological conditions at both sites, concluding after six months that Mauna Kea had a clear advantage in seeing quality, freedom from cloud cover and atmospheric transmission in the infrared. So they recommended that the 2.1 m telescope should be built there, even though they recognised the problems caused by human beings operating at such high altitudes. NASA endorsed the decision in March 1966.

Although NASA was to pay for the telescope, they were not responsible for the provision of the dome, paved road, power supply nor local accommodation. These were the responsibility of Hawaii's local administration who were to build mid-level support buildings at an altitude of 2,835 m (9,300 ft) on the site of a ranger station at Hale Pohaku. These facilities enabled astronomers to get used to high altitudes and provided accommodation between observing runs.

The f/2.7 fused silica primary mirror blank produced by Cornings in 1967 for the UH's telescope turned out to be large enough to provide a clear aperture of 2.2 m (88 inches),(4) rather than the 2.1 m expected. However, severe weather delayed construction on the mountain so that the telescope was completed two years late in 1970.[7] At that time it was the eighth largest optical telescope in the world.

Optically the 2.2 m performed very well, but for the first few years the UH had problems with the telescope's control system. In those days computer control systems were relatively new for telescopes and the computer software was not really fit for purpose. In addition the computer emitted a large amount of heat. It took until 1976 for this problem to be definitively solved.

In the meantime the US Air Force had built a 60 cm (24 inch) to the south of the summit ridge in 1968 mainly for satellite tracking. Then two years later the Lowell Observatory built a 60 cm on the summit ridge mainly for planetary observations. This telescope has since been moved and the Frederick C. Gillett Gemini North telescope (see Section 3.3) has been built on its site.

[5] At that time the UH had just four solar astronomers, none of whom had experience of night-time observing. In their proposal, the UH promised to build up their astronomy programme. Subsequently, their Institute of Astronomy was formed with Jefferies as director.

[6] Incredible though it may seem to modern astronomers, this proposal was only 25 pages long.

[7] Twenty-two years later David Jewitt and Jane Luu discovered the first Trans-Neptunian Object, 1992 QB1, using this 2.2 m telescope.

In common with telescopes elsewhere, money spent on preventative maintenance for the UH 2.2 m has not been enough to avoid serious problems. In the case of the UH telescope, for example, the right ascension drive failed and concern was expressed in 2014 that the declination drive may also fail unless it is completely overhauled.[8] The dome hydraulic drive also sprang a leak and the windscreen was out of commission for some time. There were also problems in updating the control software in 2014 as the existing software ran on a 1998 vintage computer and the new software would not run on such a machine. Consequently updating the control software also involved purchasing a new computer. As if these problems were not enough, in June 2011 the UH 2.2 m and its associated weather station were struck by lightning which severely damaged both. It took over two months to repair the telescope.[9]

8.2 Canada-France-Hawaii (CFH) Telescope

French astronomers began to consider building a new large optical telescope in France in the late 1960s, but they were unable to find a good enough site there. Shortly afterwards, John Jefferies who was in France at the time, heard of the problem. As a result he suggested that they consider building their telescope on Mauna Kea. In the meantime the French had ordered the telescope's 3.6 m (142 inch) Cer-Vit primary mirror blank from Owens-Illinois, and in 1971 they decided to follow Jefferies' advice.(5) At about the same time Canadian astronomers had also been looking at potential sites for a large telescope and, on hearing of the French decision, suggested to them that they join forces and share the cost. This resulted in an agreement between Canada, France and the University of Hawai'i signed in 1974 in which it was agreed that a 3.6 m Canada-France-Hawaii (CFH) telescope would be built on Mauna Kea.(6) Canada and France would share the costs of building the telescope 50/50, whilst the University of Hawai'i would provide the site, access road and local facilities.[10] France was responsible for the telescope's mechanical parts, and Canada for the observatory building, telescope control system and for grinding and polishing the f/3.8 primary and the f/8 Cassegrain secondary mirrors. Canada, France and the University of Hawai'i agreed to share the operating costs 45/45/10. The University of Hawai'i would get 15% of the telescope time, with the remainder being shared equally between Canada and France. Unusually for modern telescopes the CFH would have a prime-focus cage, like that of the 5 m (200 inch) Palomar. The observing floor would be cooled to improve dome seeing.

[8] Institute for Astronomy, University of Hawai'i, Newsletter No. 52, 2014.
[9] Institute for Astronomy, University of Hawai'i, Newsletter No. 40, 2011.
[10] The 3.6-meter Reflector for Hawaii, *Sky and Telescope*, November 1973, p. 306.

Construction of this CFH telescope began in 1974 and was completed five years later. Then in 1996 an adaptive optics system was installed. The telescope, which had originally been built for photography, had an unusually wide field-of-view of 1° at prime focus. In 2003 the MegaCam imager was installed to cover most of this field-of-view with a 18,400 by 18,400 pixel CCD camera. The dome shutter system failed catastrophically in 2012, which was hardly surprising considering it was over 35 years old. In the event it was fully repaired by 2014.[11]

8.3 NASA InfraRed Telescope Facility (IRTF)

NASA decided in the early 1970s that they needed a large infrared telescope to support their Voyager planetary missions. A committee chaired by Jesse Greenstein of the California Institute of Technology (Caltech) recommended that it should have a diameter of about 3 metres.(7) James Westphal, also of Caltech, was contracted to investigate possible sites. He looked at a number in the United States, Mexico and Chile and concluded that Mauna Kea was the most suitable. Unfortunately, there was still bad feeling between the University of Arizona and the University of Hawai'i following NASA's decision to fund the latter's 2.2 m telescope. In addition the University of Arizona had a strong infrared group and wanted NASA's proposed telescope to be built on Mount Lemmon in Arizona, as they considered infrared viewing conditions there were just as good as on Mauna Kea. In the event NASA decided to locate their InfraRed Telescope Facility (IRTF) on Mauna Kea, and in 1974 awarded a contract to the University of Hawai'i for its design and construction, much to the chagrin of the Arizona group. A key requirement was to keep the costs down, so the use of new technology was minimised and existing designs were used wherever possible.(8)

NASA wished to be able to use their IRTF to observe infrared celestial objects in the daytime. But during the daytime only the very brightest stars, which would be used for guiding, could be seen. This created a problem as there could be a considerable angular distance between the nearest bright star and the object to be observed. Consequently NASA specified that the telescope should be able to be offset 15° from a tracking star with less than 2 arcsec error. The telescope also had an oscillating secondary mirror just 24 cm in diameter operating up to 40 Hz to enable the intensity of the infrared image to be calibrated.

A serious problem occurred in the programme when the IRTF's 3 m Cer-Vit primary mirror blank had to be written off in 1975. This followed the appearance of a crack in the mirror after cutting the Cassegrain hole in the optical shop of the Kitt Peak National Observatory.[12] Another mirror blank had to be purchased

[11] CFHT Annual Report 2014, p. 10.

[12] This mirror blank had originally been used in the Hubble Space Telescope programme.

and NASA took advantage of the resulting delay to re-examine the telescope's design. This resulted in a major redesign of the telescope's mounting system changing it from a fork to a yoke design. The latter was designed by a team of engineers from Kitt Peak, the Jet Propulsion Laboratory and the UH, and was made in Italy.

The NASA IRTF facility was completed in 1979, the same year as the CFH telescope.

8.4 United Kingdom InfraRed Telescope (UKIRT)

The UK's InfraRed Telescope (UKIRT) was the third telescope to be completed on Mauna Kea in 1979. UKIRT was the brainchild of Jim Ring of Imperial College, London and Gordon Carpenter of the Royal Observatory, Edinburgh who looked into the possibility of building a 4 m infrared telescope in the early 1970s.(9) But they were heavily constrained by lack of money. So they decided to use a very thin mirror, supported both axially and radially, together with the smallest possible dome. The very thin Cer-Vit f/2.5 primary mirror, which was figured by Grubb Parsons in the UK, weighed much less than a mirror of conventional thickness. This enabled a relatively lightweight supporting structure to be used, so saving money. The undersize convex secondary mirror (of 31.4 cm diameter) acted as an aperture stop, preventing thermal radiation from the mirror cell and tube from reaching the infrared detector.

UKIRT's design was derived from that of the 1.5 m Infrared Flux Collector (see Section 6.1, now called the Carlos Sanchez Telescope) built on Tenerife in the Canaries in 1971. UKIRT's telescope structure, which was designed and built by Dunford Hadfield's in the UK, was shipped to Hawaii at the end of 1977. When completed two years later the telescope, which could cover the range from 0.8 to 30 microns, was the largest in the world specifically designed for the operating in infrared. It had cost only \$5 million, which was very cheap for such a telescope. Although UKIRT worked exceptionally well optically, its tracking was not very good for the first few years.

The Joint Astronomy Centre was opened in Hilo on Hawaii in 1986 to act as a joint operations centre for both the UKIRT and the submillimetre James Clerk Maxwell Telescope (JCMT, see Section 21.2). Resident astronomers and support staff worked from there. A series of improvements was undertaken to UKIRT in the late 1990s to improve its imaging performance. This was followed by software improvements to create a highly efficient scheduling environment linking observations to the prevailing weather conditions.

For most of its early life the UKIRT was operated as a general-purpose observatory. But in 2005 a strategic review was carried out by an international panel chaired by Richard Ellis of Caltech. Its main recommendation was that the observatory should concentrate on large-scale, campaign-style astronomy in future. Consequently UKIRT was used to undertake the next generation near-infrared sky survey or UKIDSS (UK Infrared Deep Sky Survey); the successor to 2MASS (see Section 2.7) at three magnitudes deeper. Unfortunately four years later, following a re-evaluation of UK priorities in ground-based astronomical observatories, it was decided to undertake a phased withdrawal from UKIRT starting in 2011.

In October 2014, the ownership of the UKIRT was transferred from the UK to the University of Hawai'i. Since then the telescope has been operated by a new partnership of the University of Arizona and the Lockheed Martin Space Systems Company.

8.5 Subaru

In 1984 the Japanese Group of Optical and Infrared Astronomers (GOPIRA) began a concept study of the optical/infrared Japanese National Large Telescope (JNLT) which Japan hoped to build at one of the world's best astronomical sites.(10) As a result two years later a Memorandum of Understanding was exchanged between the Tokyo Astronomical Observatory and the University of Hawai'i about the construction of the JNLT at a candidate observatory site on Mauna Kea.(11) This site was subjected to a six month testing campaign in 1987.

It had been clear for a little while that the Tokyo Astronomical Observatory could not be expected to run the JNLT project as a broader Japanese user base was required for both technical and political reasons. So in 1988 the National Astronomical Observatory of Japan (NAOJ) was formed, representing many Japanese universities, to oversee this project.

The site offered by the University of Hawai'i was about 6 m lower and only 130 m from the Keck observatory. It was also to the leeward of the Keck when the trade winds were blowing. This naturally caused some concern to the NAOJ because of potential air turbulence caused by the Keck telescope.[13] To understand and resolve the issue the NAOJ team built a scale model of the mountain's summit and tested it in a wind tunnel. In addition, the design of the telescope's enclosure was determined using computer simulations and wind tunnel tests. The result was a near-cylindrical rotating enclosure 43 m high, air conditioned in the day, with fans to circulate the air at night.

Although the telescope was initially envisaged as being of 7.5 m diameter, when funding was approved in 1991 its size had been increased to 8 m. Shortly

[13] This was before the second Keck was approved in 1991.

afterwards the telescope's name was changed to Subaru (Japanese for the Pleiades) and the planned effective diameter of this Ritchey-Chrétien telescope was further increased to 8.2 m. Mitsubishi was appointed the main contractor.

The Subaru telescope's primary mirror was an f/1.8 monolithic thin meniscus type made using Corning's Ultra Low Expansion (ULE) fused quartz, just 20 cm thick, and supported by an active optics system of over 250 computer-controlled actuators. The mirror was made of hexagonal glass segments fused together into a single blank. The blank was then ground flat on both faces, and reheated and allowed to slump onto a convex mould of the appropriate shape. With its Suprime-Cam (SUbaru PRIME-focus CAMera) wide field camera the Subaru telescope became an f/2.0 system at prime focus with a field of view of about 30 arcmin. In addition it had a Cassegrain focus and two Nasmyth foci, one for the infrared via a small chopping mirror, and one for optical observations.

Construction of Subaru, which was planned as a low risk, very high quality telescope,(12) was started on Mauna Kea in 1992. Two years later the mirror blank was completed and driven over 200 miles (320 km) from the Corning plant in New York to an underground Contraves optics facility near Pittsburgh, Pennsylvania where it was ground and polished.(13) This facility was a converted limestone mine that had a 100 ft (30 m) deep vertical test shaft providing a vibration-free environment at a constant temperature.

The primary mirror was delivered to the summit of Mauna Kea in November 1998, followed in the following month by engineering first light for Subaru. The telescope (see Figure 8.1) was available for scientific observations in the following year. At that time the Subaru mirror was the largest monolithic telescope mirror in the world.

A 36 element adaptive optics system was attached to the Cassegrain focus in 2000 using a curvature wavefront sensor and bimorph deformable mirror. It produced nearly diffraction-limited resolution at K band (2.2 microns).(14) This was followed in 2006 by a 188 element wavefront sensor/bimorph system, with a laser guide star, located at the infrared Nasmyth focus. The system became available for general use with a natural guide star in 2008, but problems with the laser guide star system delayed its general use until 2011.

In the following year the Hyper Suprime-Cam (HSC) ultra-wide-field second-generation prime focus camera was installed with a new wide field corrector. This replaced the original Suprime-Cam wide field camera to increase the field of view at the prime focus from about 30 arcmin to an astonishing 1.5°. The HSC, which had a 870 megapixel detector, weighed 870 kg, compared with the 170 kg of its 84 megapixel predecessor.

Building large complex structures like modern large telescopes is not without personal risk in even the best managed construction environments. But the

Figure 8.1. The enclosure for the 8.2 Subaru Telescope on Mauna Kea. (Courtesy Denys via Wikimedia Commons.)

accidents that happened during the construction and operation of the Subaru telescope were especially traumatic, resulting in four fatalities. One worker was killed when his fork-lift truck toppled over on him in 1993, and in 1996 three workers died and 26 were taken to hospital when sparks from a welder ignited insulation, generating noxious smoke.[14]

8.6 The Kecks and Gemini North

Mauna Kea is also home to the two Keck telescopes and the Gemini North telescope (see Table 8.1). Their developments have already been described in some detail in Sections 4.2 and 3.3, respectively, so they will not be covered here. Mauna Kea is, in addition, home to a number of submillimetre radio telescopes which are covered in Section 21.2.

[14] Paul Lawrence died in 1993, and the three who died in 1996 were Marvin Arruda, Ricky Del Rosario and Warren Kaleo. They are remembered by a plaque outside the base of the telescope dome.

Table 8.1. *Mauna Kea Optical/Infrared Telescopes*

Telescope Name	Telescope Diameter	Year Construction Completed	Comments
University of Hawai‘i (UH) Hilo Hoku Ke’a Telescope	90 cm	2012	Replaced UH Hilo Telescope which had originally been built for the US Air Force in 1968.
Lowell Observatory telescope	60 cm	1970	Since removed to make way for the Gemini North telescope.
University of Hawai‘i 2.2 m telescope	2.2 m	1970	
Canada-France-Hawaii Telescope	3.6 m	1979	
NASA Infrared Telescope Facility (IRTF)	3.0 m	1979	
United Kingdom Infrared Telescope (UKIRT)	3.8 m	1979	Now owned by the University of Hawai‘i.
Keck I	10 m	1992	First scientific instrument installed 1993.
Keck II	10 m	1996	First light with adaptive optics 1999 and with laser guide star 2003.
Subaru	8.2 m	1999	
Gemini North Frederick C. Gillett telescope	8.1 m	1999	

8.7 Environmental and Other Concerns

The development of the Mauna Kea observatory (see Figure 8.2) had not been universally popular in Hawaii and elsewhere because of its effect on the local habitat. Various areas of the mountain also had a religious significance which must be respected. As in many similar situations elsewhere, it is difficult to determine what were the justifiable concerns of the local population and their supporters, and what are not, and how these should be balanced against the advantages of having a world-class observatory on the mountain.

John Burns, the governor of Hawaii from 1962 to 1974, had been very keen in the mid 1960s on the development of the summit area of Mauna Kea as a major international astronomical observatory complex.[15] The first major telescope to be

[15] Burns had a major impact on the development of Hawaii by attracting foreign tourism and investment. He supported the expansion of the University of Hawai‘i into a first-class university, attracting students from all over the world, and promoted Hawaii as a centre for oceanography. His support for an international astronomical observatory complex at the top of Mauna Kea fitted in very well with these other plans.

Figure 8.2. Aerial view of Mauna Kea showing the dome of the 3.6 m Canada-France-Hawaii telescope in the centre foreground with that of the 8.1 m Gemini North telescope to its left. The domes of the two 10 m Keck telescopes are on the right of centre in mid-distance and the dark coloured enclosure of the 8.2 m Subaru telescope is just to their left. (Courtesy NOAO/AURA/NSF.)

built there, the 2.2 m University of Hawai'i telescope, was approved by NASA in 1966. Then two years later the Hawaiian Board of Land and Natural Resources (BLNR), encouraged by Burns, leased all land on Mauna Kea above 12,000 ft (3,660 m) to the University of Hawai'i. This was to be the Mauna Kea Science Reserve,[16] with the observatory area in the middle.

The development of the summit area of Mauna Kea into an astronomical observatory complex created considerable concern amongst the local community. In particular hunters of wild pigs, sheep, goats, deer, and game birds were concerned that the development may adversely affect their sport, naturalists were worried about its impact on the mountain's unique ecosystem, and winter sport enthusiasts and hikers did not want their access to the mountain compromised. On the other hand the business community supported the planned observatory complex as they could see, as could governor Burns, that this would be good for business.

Mae Mull, the spokesperson of the Hawaiian Audubon Society (HAS), an organisation dedicated to researching and preserving the environment, was one of

[16] Mauna Kea Science Reserve Complex Development Plan, February 1983, p. 2.

the most vociferous objector. Importantly the HAS had the support of the acting governor, George Ariyoshi. Mull called for a halt to building any more than the six existing and approved telescopes on the mountain. Consequently the University of Hawai'i was asked to prepare a master plan for the mountain taking account of the views of all valid objectors in addition to those of the scientific community.

In response the university produced a conceptual plan for Mauna Kea in 1975. This was followed by the Mauna Kea Plan which was adopted by the BLNR in 1977. But Mull and others felt that this did not adequately take their concerns into account. She was particularly concerned about the impact of the observatory on Hawaiian arthropods, plants and birds; particularly the palila bird which was on the list of endangered species as it was the last remaining finch-billed Hawaiian honeycreeper.(15) It fed only on the green seeds of the māmane tree and the fruit of the naio that grew in the māmane forest on Mauna Kea.[17]

Taking account of these concerns the University of Hawai'i published the Mauna Kea Science Reserve Complex Development Plan in 1983. It proposed limiting the number of telescopes to 11 large and 2 small telescopes (including submillimetre and millimetre-wave instruments) on Mauna Kea by the year 2000.

Since the publication of this 1983 Development Plan there have been a number of concerns expressed about its implementation as the astronomy complex grew. For example, although the plan called for protecting the wēkiu bug, which has an antifreeze in its blood,(16) almost the whole population on the mountain had apparently been wiped out. This was partly attributed to the massive amounts of dust created during the construction of new telescopes and to traffic on the road to the summit. Many Hawaiians were also critical of the deal done by the University of Hawai'i to attract foreign telescopes to the mountain that charged only nominal rents. In addition there were also some unsubstantiated rumours that human bones had been dug up during construction and quietly removed. This was a critical issue bearing in mind the sacredness of the mountain to many Hawaiians.

With these concerns in mind the Hawaiian senate commissioned an independent audit of the University of Hawai'i's management of the science reserve. The audit report, which was published in 1998, was highly critical of the university. (17) In response the university produced a new Master Plan which, after a number of drafts, was finally accepted in 2000 by Ahahui Ku Mauna (Group for the Mountain), an organisation set up dedicated to the cultural and spiritual protection of Mauna Kea who had been particularly critical of the university's performance since 1983.

One problem that had to be resolved in producing this new Master Plan was how to count the number of telescopes on the mountain. The university counted the

[17] Mauna Kea Science Reserve Master Plan, 2000, p. IV – 6.

two Keck telescopes and their six 1.8 m outriggers (see Section 4.2) as one telescope, and the eight 6 m diameter dishes of the Smithsonian Submillimeter Array (see Section 21.2) as one telescope, the latter making it the last telescope of those 13 agreed in the 1983 plan. In the event the 2000 plan used the university interpretation. But the 2000 plan called for only three telescopes to be built on new sites at the summit by 2020, although new telescopes could be built to replace existing telescope. Any proposed structure or construction at the summit would now have to meet the approval of the newly appointed Mauna Kea Management Board on which Ahahui Ko Mauna would have an advisory role.

The 2000 plan also designated 525 acres (212 hectares) of the 11,288 acre (4,570 hectare) science reserve as an astronomy precinct, thus limiting the area for future astronomical development. This astronomy precinct would be subject to architectural, environmental and cultural controls, and all currently undeveloped summit pu'u (cinder cones) would remain undeveloped.

The proposed Next Generation Large Telescope of at least 25 m diameter was pencilled into the 2000 plan as one of the three new telescopes. This became the Thirty Meter Telescope (TMT) whose board of directors selected Mauna Kea as their preferred location in 2009. But, before construction could begin, the TMT Observatory Corporation needed to obtain a Conservation District Use Permit from the Hawaiian BLNR. The Corporation received a provisional permit in February 2011, conditional on the successful conclusion of legal proceedings which were under way at the time. This provisional permit was upgraded to a full permit in April 2013, but the people contesting the legal case filed an appeal.

On October 7 2014, the dedication and groundbreaking ceremony took place for the TMT on Mauna Kea. The ceremony had been delayed for several hours by a peaceful protest by native Hawaiians who opposed the building of additional telescopes on the mountain. In the following year protestors continued their opposition to the telescope, resulting in a temporary halting in construction. It remains to be seen how some sort of compromise can be reached.

References

1. Krisciunas, Kevin, *Astronomical Centers of the World*, Cambridge University Press, 1988.
2. Parker, Barry, *Stairway to the Stars; The Story of the World's Largest Observatory*, Perseus Publishing, 1994, p. 25.
3. Jefferies, John T., *Dawn of a Brilliant Opportunity*, University of Hawai'i Institute for Astronomy, www.ifa.hawaii.edu/users/jefferies/Dawn_of_a_Brilliant_Opportunity.htm.
4. Jefferies, John T., and Sinton, William M., Progress at Mauna Kea Observatory, *Sky and Telescope*, September 1968, pp. 140–145.

5. Morrison, David, and Jefferies, John T., Hawaii's Mauna Kea Observatory Today, *Sky and Telescope*, December 1972, pp. 361–365.
6. Locke, J. L., Report on Construction of the Canada-France-Hawaii Telescope, *Journal of the Royal Astronomical Society of Canada*, 71, 1977, pp. 9–20.
7. Parker, Barry, *Stairway to the Stars; The Story of the World's Largest Observatory*, Perseus Publishing, 1994, p. 46.
8. Smith, Gerald M., Progress on NASA's 3-meter Infrared Telescope, *Sky and Telescope*, July 1978, pp. 25–27.
9. Humphries, Colin M., The United Kingdom's Giant Infrared Reflector, *Sky and Telescope*, July 1978, pp. 22–24.
10. Kodaira, Keiichi, Japanese National Large Telescope (JNLT) Project, *Proceedings of a ESO Conference on Very Large Telescopes and their Instrumentation*, Garching March 1988, European Southern Observatory, 1988, Ulrich, Marie-Helene, (ed.), pp. 47–57.
11. Kodaira, Keiichi, International Symposium on the Japanese National Large Telescope and Related Engineering Developments, Tokyo, Japan, November/December 1988, in *Astrophysics and Space Science*, 160, No. 1–2, October 1989, pp. 137–144.
12. Wilson, R. N., *Reflecting Telescope Optics II; Manufacture, Testing, Alignment, Modern Techniques*, Springer-Verlag, Corrected Second Printing, 2001, p. 213.
13. Sinnott, Roger W., Subaru Mirror on the Move, *Sky and Telescope*, November 1994, pp. 38–39.
14. Takami, Hideki, et al., Performance of Subaru Cassegrain Adaptive Optics System, *Publications of the Astronomical Society of Japan*, 56, 2004, pp. 225–234.
15. Gordon, M. A., *Recollections of 'Tucson Operations', The Millimeter-Wave Observatory of the National Radio Astronomy Observatory*, Astrophysics and Space Science Library, Vol. 323, Springer, 2005, p. 139.
16. Kay, E. Alison, (ed.), *A Natural History of the Hawaiian Islands: Selected Readings II*, University of Hawai'i Press, 1994, p. 357.
17. Tytell, David, Sharing Mauna Kea; Understanding the Deep-seated Conflict Between Two Passionate Groups Who Equally Cherish the Same Mountaintop, *Sky and Telescope*, August 2001, pp. 40–48.

9

Australian Optical Observatories

9.1 Mount Stromlo, the Early Years

Unlike many optical observatories discussed in this book the foundation of the Mount Stromlo observatory in Australia goes back to before the Second World War. In fact the idea of building a solar observatory in Australia goes back to before the First World War when Walter Duffield, an Australian working at the University of Manchester in the UK, pointed out its potential advantages. Not only would it fill the coverage gap in observing the Sun from the western Pacific, but it would also offer excellent observing conditions during the northern winter when the northern solar observatories were often compromised by poor weather and limited daylight.

In 1909 James Oddie, a wealthy philanthropist, donated a 9 inch (22 cm) Grubb refractor to the Commonwealth government of Australia to be the basis of a new national observatory, and the trustees of Lord Farnham's estate donated a 6 inch (15 cm) refractor to the observatory 'if and when it was founded'.(1) A board was then appointed by the government to find a suitable site near Canberra, the capital of the new Commonwealth of Australia.[1] Two years later the board chose Mount Stromlo, at an altitude of 2,530 ft (770 m), where the 9 inch was installed to undertake site testing. But the First World War was to intervene before a firm decision could be taken to build the observatory. In the event, building of the Commonwealth Solar Observatory was finally begun at Mount Stromlo in the mid 1920s.

Although the main work of the observatory was observing the Sun, it also undertook other astronomical observations. For example, in 1924 John Reynolds, a British engineer and amateur astronomer donated a 30 inch (75 cm) reflecting telescope with equatorial mounting for photographing nebulae of the southern sky, and William Rimmer used the Oddie refractor to measure the parallaxes of the brighter

[1] The decision to locate the new observatory near Canberra was to a large extent political as it avoided interstate rivalry and, in its own small way, showed that this small country town was now the national capital, see Raymond Haynes, et al., *Explorers of the Southern Sky; A History of Australian Astronomy*, Cambridge University Press, 1996, pp. 154–155.

stars. But the most important work of this early period was undertaken by Clabon Allen who used a newly completed vertical solar telescope with an 18 inch (45 cm) coelostat and 12 inch (30 cm) objective to produce a comprehensive photometric atlas of the solar spectrum.

After the Second World War the Reynolds telescope was used by Gerard de Vaucouleurs, after some modifications, to undertake a photographic survey of relatively bright southern galaxies. Then over the period 1969–1972 John Hart made further substantial modifications to the instrument to improve its use for spectroscopic observations.(2)

In the meantime, Richard v. d. R. Woolley, who was later to become the UK Astronomer Royal, had been appointed director of the Commonwealth Solar Observatory in 1939. Woolley wanted to change the work of the observatory from solar to stellar research, and in 1945 he purchased the problem-prone Great Melbourne Telescope for its scrap value of £500. He had its speculum mirror changed to aluminised Pyrex, reducing the focal length in the process[2], and had an electric drive added. This heavily modified telescope began operation in 1958.

Woolley had also hoped to acquire a 74 inch (1.9 m) reflector, similar to that planned for the Radcliffe Observatory in Pretoria, but delays due to the Second World War meant that the Mount Stromlo 74 inch was not ordered until 1948. Although the telescope was exhibited in London at the 1951 Festival of Britain Exhibition, its building and dome were delayed so that it was not inaugurated on the mountain until four years later. Unfortunately it was then found that the primary mirror suffered from astigmatism. Consequently the mirror was returned to its supplier, Grubb Parsons, for refiguring when a spare mirror became available in 1959.[3] The original mirror was reinstalled two years later.

The Commonwealth Solar Observatory had been renamed the Commonwealth Observatory in 1945. Seven years later a bushfire destroyed its workshop and the records of the Great Melbourne Telescope. Fortunately, the remainder of the observatory was largely unharmed. Then in 1957 the observatory was renamed the Mount Stromlo Observatory when control was transferred from the Department of the Interior to the new Australian National University (ANU) in Canberra. In the same year Bart Bok of Harvard was appointed director of the observatory and professor of astronomy at the ANU, following his failure to be appointed director of the Harvard College Observatory.[4] At about the same time Bok persuaded

[2] To reduce tube flexure and vibration.

[3] This was the repaired 74 inch mirror produced for the Helwan Observatory in Egypt that had been cracked during figuring.

[4] The exact reason for his non-appointment is unclear, but his case was not helped by his support of Harlow Shapley, the incumbent director, during the latter's run-in with the House Committee on Un-American Activities, see David H. Levy, *The Man Who Sold the Milky Way; A Biography of Bart Bok*, The University of Arizona Press, 1993, pp. 74–82.

Theodore Dunham, who had built a Coudé spectrograph for the Mount Wilson Observatory, to design a similar instrument for the Stromlo 74 inch.

Meanwhile in 1953 a 26 inch (65 cm) photographic refractor had been moved from the Yale-Columbia southern station in South Africa to Mount Stromlo because of the better observing conditions in Australia. Ten years later the telescope was given to the Australian National University when the Yale-Columbia southern station was moved to Argentina. In addition, the Swedish University of Uppsala had, at Woolley's invitation, located its new 20/26 inch (50/66 cm) Schmidt telescope on Mount Stromlo in 1957.[5]

9.2 Siding Spring

Frequent cloud cover[6] and light pollution from the expanding city of Canberra was becoming a serious problem at the Mount Stromlo Observatory after the Second World War. This led Bok to initiate the search for a new observing site or sites to satisfy the observatory's short- and long-term needs. The short-term requirement was to find a field station to locate its best telescopes, whilst also keeping an observatory at Mount Stromlo linked to the ANU in Canberra. This field station was required to have a clear weather advantage over the existing site and be close enough to it to allow frequent visits and easy transportation of equipment.(3) In the longer term Bok wanted to find the best observing site in Australia on which to build telescopes of a similar size to the largest telescopes on mainland USA.

Arthur Hogg, a senior astronomer at Mount Stromlo, was put in charge of the site testing programme by Bok. Five possible locations for the field station were found in New South Wales.[7] One of these, Mount Bingar, was chosen as a reference site and a 26 inch (66 cm) reflector installed there in 1959 to undertake site testing. By 1962 the short list for the field station had been reduced to Mount Bingar and Siding Spring. The ANU Vice-Chancellor, Leonard Huxley, then called a meeting to discuss the choice of site. Bok favoured Mount Bingar, as the weather there was less cloudy than Siding Spring in the summer when the Magellanic Clouds were best observed, whilst Hogg and other senior astronomers favoured Siding Spring, as it had better seeing. Following his review of the advantages and disadvantages of both sites, Huxley announced in May 1962 that Siding Spring (see Figure 9.1) at an altitude of 1,135 m (3,720 ft) near Coonabarabran, about 350 miles (560 km) north of Mount Stromlo, would become the Observatory's permanent field station. The search for the best optical site in Australia still continued, however, as at

[5] The Uppsala Schmidt was later moved to Siding Spring to escape the increasing light pollution of Canberra.

[6] The initial site survey had proved to be unduly optimistic.

[7] Which surrounds the Australian Capital Territory in which Canberra is located.

Figure 9.1. A modern image of the observatory at Siding Spring with the Anglo-Australian Telescope in the left foreground. (Courtesy Ssopete via Wikimedia Commons.)

that time the UK and Australia was considering building a 150 inch (3.8 m) joint Anglo-Australian telescope (see below).

By 1964 a 40 inch (1.0 m) and a 16 inch (40 cm) Boller and Chivens reflector had been installed at Siding Spring to be used mostly for photometry and photography. The Mount Stromlo Observatory and its Siding Spring field station were subsequently renamed the Mount Stromlo and Siding Spring Observatory.

One of the reasons behind Bok's appointment as director had been his knowledge of both optical and radio astronomy. Consequently since his appointment he had tried to improve the relationship between the radio astronomers of the CSIRO[8] Radiophysics Laboratory (see Chapter 16) and the optical astronomers of his observatory who tended to view each other with suspicion. In fact encouraged by Bok some of his students then spent as much time working at CSIRO's Radiophysics Laboratory as they did at their own optical observatory. Bok, an excellent communicator, also introduced monthly visitors' nights which he invariably supervised himself. He also became friends with Australia's prime minister, Robert Menzies, who was well-disposed towards scientific research and who became a regular visitor to Mount Stromlo. All this helped to educate people and keep the observatory in the public eye, showing legislators and the general public alike where Australian tax dollars were being spent.

[8] CSIRO is an acronym of Commonwealth Scientific and Industrial Research Organisation.

Olin Eggen, who took over from Bok in 1966, was a very different personality with very different priorities.(4) Eggen put low priority on public relations and discontinued visitors' nights. Bok had concentrated on observational astronomy, even insisting in his early years as director that theoretical astronomers spend some time at the telescope, whilst Eggen wanted to build up a parallel theoretical group. Bok had concentrated on observations of the Milky Way and the Magellanic Clouds, arguing that the observatory should take full advantage of its position in the southern hemisphere. Eggen, on the other hand, wanted to concentrate on stellar research, leaving the large telescopes in America to use their large light gathering power to work on distant galaxies. In the event, however, whilst Eggen continued his stellar evolution research many astronomers at Mount Stromlo concentrated on trying to understand the structure and dynamics of normal galaxies.

Anglo-Australian Observatory

(a) Anglo-Australian Telescope

Richard Woolley had suggested as long ago as 1953, even before the 74 inch was completed, that a very large telescope should be built on Mount Stromlo. His idea was for a 200 inch (5 m) to be built as a collaborative British Commonwealth project[9] between Australia, Canada and the UK.(5) He did not get very far at the time, and any Canadian interest subsequently disappeared. But around the time of Bok's appointment in 1957 things began to change in Australia when their prime minister, Robert Menzies, led a major initiative to improve the position of science in his country.(6) This was one of the reasons why Bok had cultivated Menzies' friendship, hoping to persuade him to back the proposed new telescope project.

Woolley, who was by now Astronomer Royal in London, tried to persuade his colleagues that the UK should join Australia to build the new large British Commonwealth telescope. However, at that time the UK already had access to the southern sky with telescopes in South Africa and were already considering whether to join other European countries (see Chapter 7) in building a European Southern Observatory. Alternatively, as a number of astronomers suggested, the UK could collaborate with American and Australian astronomers in constructing a large telescope somewhere else in the southern hemisphere.

The UK Royal Society then examined the various alternatives for the development of UK optical astronomy in the southern hemisphere and concluded in 1959 that there was both a short-term and long-term solution. In the short-term

[9] Sometimes in the literature this is referred to as a 'Commonwealth project'. But this is ambiguous as there is both a British Commonwealth and a Commonwealth of Australia. As a result I have used the term 'British Commonwealth project' to indicate that it was not just an Australian project.

the Royal Society recommended that facilities at the Radcliffe Observatory in South Africa should be developed. But in the longer term they supported a British Commonwealth Observatory under joint management with Australia. Consequently Cyril Hinshelwood, President of the Royal Society, made a formal approach to the Australian Academy of Science (AAS) to discuss the possibility of such a joint project. But the AAS had other calls on its funding at the time and were unable to make a commitment. This caused Bok to become increasingly frustrated at what he saw was an unnecessary delay, especially when his organisation had already found a suitable site for the proposed large telescope at Siding Spring. Then in October 1962, the UK removed one of the obstacles from the proposed joint Anglo-Australian project when it became clear that they would not become a member of the European Southern Observatory.[10]

Woolley in the UK and Bok in Australia continued to push for a joint Anglo-Australian telescope. Then towards the end of 1962 there were the first signs that senior people both inside and outside of the Australian government were beginning to be won round. Consequently in May 1963 Woolley suggested to Bok that they set up a joint working party of astronomers to consider its possible design. The working party's meeting which took place at Herstmonceux, UK a few months later agreed that the telescope should be a 150 inch (3.8 m) with an equatorial mounting. In the following year another Anglo-Australian meeting concluded that Siding Spring was probably the best available site in Australia, although testing of two other sites, at Mount Serle and Mount Singleton both outside New South Wales, continued for a time.

In June 1965, the AAS and the Royal Society agreed to make a joint submission to both governments asking for funding on a 50/50 basis. But the Australian government was also expecting an offer of a joint project with the United States at the time, and Menzies was somewhat cautious about a joint project with the UK as he wondered whether the UK would carry the project through to completion. In parallel Woolley, frustrated with the lack of progress of the Anglo-Australian project, began talks in June 1966 with Horace Babcock, the director of the Mount Wilson and Palomar Observatories in California, about the possibility of an Anglo-American optical telescope project. In the meantime Bart Bok, who was also frustrated, had left his job at Mount Stromlo in March 1966 to return to the United States.

Then, much to most people's surprise, on April 14 1967 the British government finally received Australia's formal agreement to open substantive negotiations on the Anglo-Australian Telescope project. This was as equal partners in both building

[10] The UK had been somewhat lukewarm about joining this observatory organisation for some time. But it was not until the ESO Convention was open for signature in October 1962 that it became absolutely clear that they would not be joining (although they changed their minds forty years later, see Section 7.2).

and operating the telescope. There were a number of conditions attached, including that it should be located at Siding Spring, and that its design should be based on the design of the 4 m Kitt Peak telescope (see Section 2.3) to limit the possibilities of timescale and cost overruns. These points, and others, were generally accepted by the British. But having got that far, after a number of years delay, the drafting of the formal agreement still took another two years. Then an act of parliament had to be passed in Australia because the telescope was to be located there. As a consequence the royal assent was not received until September 1970 and the Anglo-Australian Telescope Agreement Act, 1970 did not come into force until February 1971.

In the meantime the AAS and Royal Society had decided in 1967 to set up a Joint Policy Committee (JPC), later to become the Anglo-Australian Telescope Board (AATB), without waiting for formal project approval. The JPC was to be responsible for planning and starting the construction phase, finalising specifications, and letting construction contracts. Subsequently the order for the monolithic Cer-Vit primary mirror blank was placed with Owens-Illinois in November 1967[11] and that for the mount was placed in August 1970, one month before the royal assent.

The next three years saw a heated argument between members of the AATB about how the project should be managed. In particular should the Anglo-Australian Telescope team be responsible to the ANU's management of the Siding Spring Observatory, where it was to be built, or should it be independent? Contrary to Eggen's wishes the Australian government decided to accept the majority decision of the AATB that the project should be independent of the ANU. Eggen responded to this decision by resigning his membership of the AATB in August 1973.

Meanwhile the Cer-Vit primary mirror blank had arrived at the end of 1969 at Grubb Parsons in the UK where it was to be ground and polished in two stages. In stage I it was to be polished such that 95% of the incident light would be within a circle of 0.7 arcsec diameter. Then if stage I was successful there was an option to go on to stage II where the 0.7 arcsec specification was reduced to 0.5 arcsec.[12] In the event the stage II specification was achieved.

The resulting Anglo-Australian Telescope (AAT) was one of the first telescopes to take full advantage of the development of computer control technology. Its computer slewed the telescope to the objects to be studied, automatically taking account of atmospheric refraction, telescope misalignment and flexure.(7) The

[11] As explained in Section 2.6, this order was a joint order with the French National Observatory and AURA who ordered similar-sized blanks.

[12] The complete specification was for stage I (in arcsec) 0.4, 0.7 and 1.0, for 80%, 95% and 99% of light, respectively, and for stage II 0.3, 0.5, 0.8. The achieved values were 0.30, 0.48 and 0.65, see S. C. B. Gascoigne, K. M. Proust and M. O. Robbins, *The Creation of the Anglo-Australian Observatory*, Cambridge University Press, 1990, p. 93.

Figure 9.2. The 3.9 m Anglo-Australian Telescope in 1980. (© Australian Astronomical Observatory/David Malin Images.)

telescope (see Figure 9.2) had a horseshoe mount, manufactured by the Mitsubishi Electric Corporation of Japan, that was similar to that of the 4 m telescope on Kitt Peak. It was fixed on the top of a 55 ft (17 m) high hollow concrete cylinder, 33 ft (10 m) in diameter that was completely isolated from the surrounding observatory building to reduce telescope vibration. The 16 ton, 2 ft (60 cm) thick mirror was supported from the back by 36 pads arranged in two circles and from the sides by 24 mechanical counterweights acting through levers. The telescope had a triplet corrector lens which covered a 60 arcmin field, and a doublet that produced better definition but with a smaller coverage of about 25 arcmin.

Ben Gascoigne, the commissioning astronomer, took the first photograph with the 153 inch (3.9 m) AAT in April 1974. The primary mirror was aluminised just over six months later. The telescope had cost a total of $A16 million, which, after taking account of inflation, was slightly less than the original estimate.

The AAT was used for both imaging and spectroscopy. In the late 1970s Peter Gray developed a radically new spectrograph known as FOCAP (Fibre Optic Coupled Aperture Plate) for the AAT to produce spectra of a number of objects simultaneously. It consisted of 25 fibres each 2 metres long attached to holes drilled in an aperture plate located at the telescope's f/8 Cassegrain focus.[13] The position of these holes matched that of the sources to be observed whilst the aperture plate cut out most of the background skylight. The system had first light in 1981.

Aperture plate systems were much more efficient in telescope time than conventional spectrograph systems. But every time a new object field was to be observed a new aperture plate had to be drilled. As a result, in spite of their advantages, these systems were inflexible and the time to produce the aperture plates was significant. Consequently a new multi-object spectroscopy system, called Autofib, was developed for the AAT in which a single computer-controlled robotic positioner, mounted on an *x-y* carriage, picked up each of 64 fibre-optic probes sequentially. It then placed them on a flat steel plate in the Cassegrain focal plane where they were held in position by a small magnet on each probe. A small prism at the end of each fibre reflected light from the telescope through 90° to enter the fibre. Astrofib was first used on the AAT in 1987.

Astronomers using the AAT wanted to be able to determine the redshift of hundreds of thousands of faint galaxies in a major galaxy survey. Consequently they wanted to be able to measure the redshift of as many of these faint galaxies as possible in a given time. To do this a new system called 2dF (two-degree field) was designed that allowed the spectra of 400 objects to be obtained simultaneously over a wide field of view by using robotically positioned optical fibres.(8) But 2dF was more than just a multi-object spectroscopy system as it had a four-lens aberration corrector to provide an aberration-corrected two-degree diameter field at prime focus, and an atmospheric dispersion compensator, as well as an automatic fibre positioner. To increase the throughput rate there were two stainless-steel plates for fibre placements with 400 fibres on each plate. Whilst one plate was being used in the focal plane, the other one was being populated with fibres. A tumbler mechanism then changed plates when the observation of the first plate was completed. This system was used to produce the 2dF Galaxy Redshift Survey completed in 2002 that measured the position of over 220,000 galaxies.

(b) UK Schmidt

The UK Science Research Council (SRC) set up a committee in December 1967, following the decision to build the AAT, to examine the facilities for British astronomers in the southern hemisphere. This resulted in a reorganisation of the

[13] The basic idea of using an optical fibre to transmit starlight from the focal plane of a telescope to a spectrograph had been proposed by Narinder Kapany as long ago as the 1950s. But it was not until low-loss optical fibres became available in the early 1970s that such a system became an attractive possibility in astronomy, see John Hearnshaw, *Astronomical Spectrographs and their History*, Cambridge University Press, 2009, pp. 185–186.

telescopes in South Africa including the movement of the Radcliffe 74 inch (1.9 m) to a new site at Sutherland. The committee also observed that work with the 48/72 inch (1.2/1.8 m) Schmidt on Mount Palomar (see Section 1.2) was vital to the success of the Palomar 200 inch (5 m). So they proposed that a similar large Schmidt should be located near the AAT at Siding Spring, if possible within the next five years, to perform a similar function in Australia.[14] This would enable the Palomar Schmidt photographic sky survey to be extended to the south celestial pole. The committee also suggested, as a second priority, that an optical telescope with an aperture of about 60 inches (1.5 m) should be built of an advanced design.

Unfortunately for UK optical astronomers this recommendation to build a large Schmidt telescope coincided with a request from Jodrell Bank to provide funds to build a new radio telescope there, the Mark V. However in October 1969, without waiting for a decision on the Schmidt, the SRC obtained the agreement of the ANU to build and operate such a Schmidt telescope at Siding Spring. In addition Horace Babcock, director of the Hale Observatories, agreed to provide a complete set of drawings of the Palomar Schmidt. In return the SRC agreed to provide access for Palomar observers to the proposed Siding Spring Schmidt for 12 to 15 nights per year. UK government approval was given for the project in early 1971.

Vincent Reddish of the Royal Observatory, Edinburgh was given the task to manage the project from the design to the operational phase.(9) It was important that this new Schmidt be available at the same time as the AAT, and money was tight, so Reddish was asked to make only the absolute minimum of changes to the Palomar design. Reddish could have produced such a telescope but he would have been subjected to the criticism that he could have produced a far better design using materials that were not available when the Palomar Schmidt had been built over twenty-five years earlier. Consequently, although he produced a design based on the Palomar instrument, he introduced many modifications. The telescope's drive and anti-backlash systems, the plateholder and focussing systems, and the support system for the Cer-Vit mirror were all redesigned. The telescope was made more rigid and a thin objective prism was mounted at its entrance aperture to provide slitless spectroscopy, whilst the visual guide telescope was replaced with a fully automatic acquisition and guidance system.

The 1.2/1.8 m Schmidt (see Figure 9.3) was built by Grubb Parsons in the UK in record time and delivered to Siding Spring in May 1973. The dome was built by the Boller and Chivens Division of Perkin-Elmer. The first plate was taken in July 1973 and the telescope was formally opened in the following month when it became known as the UK Schmidt. It entered service on September 3 1973, a year

[14] The origin and development of this UK Schmidt parallels, to some extent, that of the ESO Schmidt that was completed on La Silla in the early 1970s (see Section 7.1).

Figure 9.3. The 1.2/1.8 m UK Schmidt in 1977. (© Australian Astronomical Observatory/David Malin Images.)

ahead of the AAT. Initially the UK Schmidt facility was treated as an outstation of the Royal Observatory, Edinburgh. But in 1988 it officially became part of the Anglo-Australian Observatory.

In the meantime contact had been made between SRC and ESO to discuss the possibility of undertaking a joint survey programme with their Schmidt telescopes to extend the Palomar Schmidt survey (see Section 1.2) down to the south celestial pole.[15](10) Subsequently a conference on 'The role of Schmidt telescopes in Astronomy' took place in Hamburg in early 1972 sponsored by ESO, SRC and the Hamburg Observatory. This was followed by a working session which included representatives of ESO, SRC and people who had been involved in the Palomar survey. Finally in January 1974, ESO and the SRC signed an agreement to carry out a joint survey of the southern sky. The first element would be the Quick Blue Survey undertaken by ESO using course-grain, unsensitised IIa-O plates. It was completed in 1978.

[15] Because of its location in California the Palomar survey had stopped at –30° declination.

Reddish wanted to use on the UK Schmidt the Kodak fine grain, high contrast, high resolution IIIaJ plates which had been put on the market in the mid 1960s. The only problem was that they were very slow and it was not clear whether the hypersensitisation techniques then in use would produce the speed increase required. In the event the Schmidt design team[16] developed a new hypersensitisation system for both the blue sensitive IIIaJ plates and the later red sensitive IIIaF emulsions.

The first survey plates produced using the UK Schmidt were blue sensitive exposures using hypersensitised IIIaJ plates from a latitude of – 17° to the south celestial pole.(11) ESO then followed their Quick Blue Survey by taking red sensitive exposures using hypersensitised IIIaF plates over the same part of the celestial sphere as the UK Schmidt. When the UK Schmidt team had completed their blue survey from – 17° southwards, they undertook an equatorial survey from 0° to – 15° using both IIIaJ and IIIaF plates in support of the Hubble Space Telescope programme.[17]

The UK Schmidt originally had a corrector plate of Schott type BK7 glass optimised for operation at 420 nm. But this was replaced in 1977 by an achromatic correcting plate designed by C. G. Wynne to enable it to cover the photographic range from about 330 to 1200 nm.(12) This corrector plate was so successful that similar achromatic correctors were later made by Grubb Parsons for the ESO 1.0/ 1.6 m Schmidt and for the Palomar 1.2/1.8 m Schmidt.

In 1986 Fred Watson developed a fibre-fed multi-object spectroscopy system called FLAIR (Fibre-Linked Array Image Reformatter) for the UK Schmidt. His aperture plate was simply a photographic positive glass plate, held in the normal Schmidt plate holder, with fibres glued to the emulsion side. This system produced a good registration between the fibres and the objects to be studied. As the Schmidt covered a much larger field of view than a normal reflecting telescope, the 39 fibres were each much thinner than normal with a 40 micron width covering 2.7 arcsec. FLAIR was replaced on the UK Schmidt in 1988 by PANACHE[18] which had an improved fibre positioning system and an improved CCD detector in the spectrograph. Four years later that was replaced by FLAIR II with 90 fibres and a much-improved spectrograph.

All three of these fibre-fed multi-object spectroscopy systems used manual positioning of the fibres. But in 2001 this was replaced by 6dF (six-degree field) which had a robotic fibre positioner. Four years later it completed its first main task, the 6dF Galaxy Survey, where it measured the redshifts of over 120,000 galaxies. This survey was followed by RAVE (RAdial Velocity Experiment) which measured the radial velocities of over 500,000 stars in the Milky Way.

[16] Called the 'UK Schmidt Telescope Unit'.
[17] *Sky and Telescope*, December 1982, p. 543.
[18] For 'PANoramic Area Coverage with Higher Efficiency'.

(c) The Anglo-Australian Observatory becomes the Australian Astronomical Observatory

The United Kingdom joined the European Southern Observatory in 2002 and consequently announced in 2005 its intention to withdraw from the Anglo-Australian Telescope Agreement with effect from July 1 2010. In the interim the UK had gradually reduced its funding of the Anglo-Australian Observatory so that by 2010 the observatory had become 100% financed by Australia. Renamed the Australian Astronomical Observatory, with an unchanged abbreviation (AAO), it continued to operate both the Anglo-Australian Telescope and the UK Schmidt.

Advanced Technology Telescope

The Anglo-Australian Telescope had been heavily oversubscribed when it came into operation in the mid 1970s, so that by 1977 it had become clear that another large telescope was required at Siding Spring observatory to meet demand. As a result Don Mathewson, the acting director of the Mount Stromlo and Siding Spring Observatories, wrote to Vincent Reddish in early 1978 suggesting that the UK SRC and the ANU collaborate in building a 2.5 m telescope with a thin mirror, altazimuth amount and rotating building (like that of the MMT then being built on Mount Hopkins, see Chapter 10). Unfortunately Reddish thought that the SRC could not fund the telescope at that time.(13)

Shortly afterwards Mathewson chaired an Australian working party at Mount Stromlo Observatory on New Generation Telescopes. The working party supported the idea of a 3 m telescope, named the Universities Telescope, to be funded by all Australian universities. Its primary mirror would be very thin with an aspect ratio of 20:1. Like the previously proposed 2.5 m telescope, the Universities Telescope would have an altazimuth mount and rotating building and be built at the Siding Spring observatory. Mathewson estimated that it would cost about $3 million, compared with $12 million for a similar-sized telescope of conventional construction.(14)

In spite of the enthusiastic support of all Australian universities, the 3 m project collapsed in May 1979 due to lack of funding. Subsequently Mathewson became aware that Owens-Illinois had a spare 2.5 m Cer-Vit mirror blank available and, although it was somewhat smaller than the desired 3.0 m and was badly chipped, he persuaded the ANU to purchase it as the basis for a new Australian telescope which Mathewson estimated would cost $1 million. Mathewson then had the blank cut in half horizontally to produce two 9 inch (23 cm) thick blanks, one of which he kept for the new Australian telescope. The other half was sold in the following year to the McGraw-Hill observatory for almost the same amount of money ($200,000) that the ANU had paid for the complete blank. (This half was later used in the Hiltner Telescope on Kitt Peak, see Section 2.4.)

Fortunately, the University of Wyoming (UW) had had a 2.3 m telescope built on Jelm mountain. Whilst, once completed, the Australian mirror blank was of virtually the same diameter as the UW's blank. Consequently the ANU gave contracts to the manufacturers of the UW's mirror housing (L & F Industries) and support system (Boller and Chivens) for them to produce duplicate items for the new Australian telescope. Norman Cole of Tucson, Arizona was contracted to carry out the grinding and initial polishing of the primary mirror, which he did in his former garage that he had converted into an optical laboratory. Subsequently the final polishing and figuring were undertaken by the University of Arizona's Optical Sciences Center.

Mathewson's estimate for building the 2.5 m telescope had risen from $1 million in May 1979 to $1.4 million in November of that year. By the end of construction in 1984 the total cost had increased to $3.2 million. Amazingly, the project's budget had never been formally approved by the ANU. Instead the money had been found on an ad-hoc basis year by year.

When completed the fast (f/2.05) mirror of this new Australian Advanced Technology Telescope (ATT), as it was called, was relatively thin with an aspect ratio of 13:1. Its support system consisted of an air bed underneath the mirror and a mercury collar around it. The telescope had two different top-end structures to accommodate its two different secondary mirrors, one for the Cassegrain and one for the Nasmyth configuration. Switching between these two different configurations took just 20 minutes using a built-in crane, whilst it took only a few seconds to switch the light beam between the two different Nasmyth foci.

The ATT was finally inaugurated by Bob Hawke, the prime minister of Australia, in May 1984. Subsequently a Dual-Beam Spectrograph (DBS) that had been designed and built at the Mount Stromlo observatory workshop was installed at one of the two Nasmyth foci whilst a combined imager and low-resolution spectrograph was installed at the other Nasmyth focus. Then in 1993 the Cryogenic Array Spectrometer-Imager (CASPIR) came into operation at the Cassegrain focus.

9.3 Bushfires on Mount Stromlo

As mentioned previously a major bushfire had damaged the Stromlo Observatory in 1952, but fortunately all the major facilities had survived, although the records of the Great Melbourne Telescope had been destroyed along with some expensive machine tools.

Unfortunately, the January 2003 bushfire at Stromlo Observatory was a much more serious affair.(15) All the major domes and telescopes, except for an old 6 inch (15 cm) refractor, were destroyed. The workshops, which contained the almost-completed NIFS (Near-Infrared Integral-Field Spectrograph) for the Gemini North telescope in Hawai'i, were also destroyed. But, fortunately, two office buildings survived which contained all Mount Stromlo's computerised archives, and Siding Spring was not affected. A new NIFS was subsequently built by the Canberra-based aerospace company, Auspace Ltd, to speed up delivery, and was shipped to Hawai'i in 2005. It included a state-of-the-art detector that had missed the fire as it had been held up in Sydney customs at the time.

Penny Sackett, who had been appointed director of the Mount Stromlo Observatory in 2002, was faced with an enormous problem after the fire as all Stromlo's research telescopes had been destroyed. So there was now a question as to whether the Stromlo Observatory still had a rôle, and whether it should be rebuilt, as all the most modern optical telescopes had been built at Siding Spring in recent years as it suffered much less from light pollution. Although the Federal government provided \$7.3 million to help to rebuild the Stromlo workshops, Sackett's problem was not helped by the fact that the insurers argued long and hard over their financial liability.

In the event it was decided to build an Advanced Instrumentation Technology Centre (AITC) at Mount Stromlo with a large integration and assembly hall attached to new design workshops and laboratories.(16) New telescope instruments would be built there whilst any new large Australian optical telescopes would be built at Siding Spring. Australia's involvement in any international telescope elsewhere in the world would be managed from Mount Stromlo.

The Stromlo team then designed the wide field 1.35 m diameter SkyMapper survey telescope that was built at Siding Spring to replace the 1.27 m Great Melbourne Telescope. SkyMapper was a fully automated telescope system with a 268 megapixel CCD detector to undertake what was called the Stromlo Southern Sky Survey. The telescope sent the survey images over a fibre optic link to a super-computer at the ANU in Canberra, where they were calibrated, reduced and sent to astronomers for analysis. SkyMapper saw first light in 2008.

In the meantime the Australian Gemini Project Office, which managed the Australian involvement in the multinational Gemini project (see Section 3.3), had been set up at Mount Stromlo in 1998. Subsequently the Mount Stromlo Observatory built the near-infrared Gemini South Adaptive Optics Imager (GSAOI) that recorded diffraction-limited images over an 85 × 85 arcsec field of view. It was shipped to Chile in 2006 where it was to work with Gemini South's multi-conjugate adaptive optics system. The Mount Stromlo Observatory also designed and build the Wide Field Spectrograph (WiFeS) which replaced the DBS on the Advanced Technology Telescope at Siding Spring in 2009. It was an integral field spectrograph able to produce 950 spectra simultaneously.

Probably Sackett's most important contribution to Australian astronomy was getting the Mount Stromlo Observatory involved in the American-led Giant Magellan

Telescope (GMT) project in 2006 (see Section 12.2) just before she left her post as director at Mount Stromlo. The GMT was designed to have six 8.4 m diameter mirrors which surrounded a single mirror of the same size. Sackett's successor, Harvey Butcher, extended this GMT cooperation by transforming the Australian role to that of a full member of the GMT consortium. Then in 2010 the Mount Stromlo team won a contract, partly funded by the Australian government, to undertake the design of the near-infrared GMT Integral Field Spectrometer (GMTIFS)(17) using their experience with the NIFS and GSAOI instruments.

9.4 Optical Interferometers

The British astronomers Robert Hanbury Brown and Richard Twiss outlined the design of an optical intensity interferometer in 1954. This was based on a similar radio-based system that Hanbury Brown had invented some years before at Jodrell Bank Observatory in the UK.(18) The optical intensity interferometer used two independently controlled optical telescopes, and had the big advantage over the Michelson interferometer in that only the intensity needed to be detected, not the relative phase of the signals. As a result, unlike a Michelson interferometer, the intensity interferometer was not effected by atmospheric turbulence, although it was only practical for the brightest stars. Hanbury Brown produced a proof-of-concept instrument at Jodrell Bank in 1955 using two modified searchlights. But unfortunately the weather conditions in England are not really conducive to optical astronomy so the optical observations proved rather tedious to undertake. But by the following year Hanbury Brown was able to announce that the had successfully used this intensity interferometer to measure the diameter of Sirius.

With this successful experiment under his belt Hanbury Brown managed to persuade the Department of Scientific and Industrial Research in the UK and the University of Sydney[19] to share the costs of an operational intensity interferometer which would be located in Australia. He then moved to Australia in 1962 where he supervised the construction of this operational system which used two 22 ft (6.7 m) parabolic mirrors on a 618 ft (188 m) diameter railway track. It was built at Narrabri Observatory, New South Wales about 325 miles (525 km) north of Sydney. At first sight the idea of building two 22 ft diameter mirrors appears daunting, but their surface accuracy could be very modest as their primary function was to focus starlight onto a photocathode, not to produce an image.(19) Both mirrors were composed of about 250 sixteen inch (40 cm) aluminium-coated, hexagonal glass mirrors each mounted on an adjustable three-point suspension for focusing. This intensity interferometer system, which was operated by members of

[19] Via Harry Messel, the professor of physics at the University of Sydney.

the University of Sydney from 1963 to 1972, produced the first accurate measurements of the diameters of 32 mostly main sequence stars.

Unfortunately this system was only usable with stars brighter than 2.5 magnitude. To significantly improve the design would require much larger mirrors and be very expensive. But, in the meantime, there had been significant advances in the field of active optics and in understanding atmospheric scintillation. Consequently Hanbury Brown, who was now at the University of Sydney, and his colleague John Davis decided that a modern Michelson interferometer would be able to give the improved sensitivity and resolving power that they required. So in 1975 Hanbury Brown submitted a proposal to the Australian Department of Science to build a Michelson interferometer as a follow-on instrument to the intensity interferometer. Following approval, a prototype system was designed by John Davis and Bill Tango[20] with most of the features of the final instrument but with a single relatively short baseline of 11.4 m to minimise complexity and cost. This was followed over ten years later by the full-sized instrument, called the Sydney University Stellar Interferometer or SUSI, which was designed to undertake very high angular resolution studies of single stars and close binary systems.(20)

SUSI was designed to operate with baselines ranging from 5 to 640 m, producing resolutions of from 20 to 0.07 mas (milliarcsec) at a wavelength of 450 nm. Only two apertures were used at a time as SUSI was not intended to produce images, unlike the multi-aperture systems of COAST[21] (see Section 14.3) and NPOI[22] (see Section 14.6) which produced images. The SUSI mirrors were limited to 20 cm in size, with a clear aperture of 14 cm, to reduce problems with atmospheric turbulence. Construction started in 1987 and the instrument was opened in 1991 with the 'blue' system operating in the wavelength range 400 to 540 nm. It was then modified to also allow operation with the red system in the range 500 to 950 nm in 2002.

References

1. Haynes, Raymond, et al., *Explorers of the Southern Sky; A History of Australian Astronomy*, Cambridge University Press, 1996, p. 156.
2. Frame, Tom, and Faulkner, Don, *Stromlo, An Australian Observatory*, Allen and Unwin, 2003, p. 182.
3. Ibid., pp. 140–143.
4. Haynes, Raymond, et al., *Explorers of the Southern Sky; A History of Australian Astronomy*, Cambridge University Press, 1996, p. 179.

[20] Bill Tango was specifically recruited by Hanbury Brown because of his previous experience in building a Michelson interferometer for the Royal Observatory, Edinburgh at its Monte Porzio Catone observing station near Rome.

[21] Cambridge Optical Aperture Synthesis Telescope at Cambridge, UK.

[22] Navy Prototype Optical Interferometer near Flagstaff, Arizona.

5. Gascoigne, S. C. B., Proust, K. M., and Robbins, M. O., *The Creation of the Anglo-Australian Observatory*, Cambridge University Press, 1990, p. 29.
6. Frame, Tom, and Faulkner, Don, *Stromlo, An Australian Observatory*, Allen and Unwin, 2003, p. 155.
7. Wampler, E. Joseph, The 153-inch Anglo-Australian Telescope, *Sky and Telescope*, October 1975, pp. 224–228.
8. Hearnshaw, John, *Astronomical Spectrographs and their History*, Cambridge University Press, 2009, pp. 203–204.
9. Gascoigne, S. C. B., Proust, K. M., and Robbins, M. O., *The Creation of the Anglo-Australian Observatory*, Cambridge University Press, 1990, p. 193.
10. West, Richard M., The Southern Sky Surveys – a Review of the ESO Sky Survey Project, *ESO Bulletin*, No. 10, May 1974, pp. 25–40.
11. Cannon, Russell, Schmidt Views of the Southern Sky, *Sky and Telescope*, April 1984, pp. 301–303.
12. Wynne, C. G., The Optics of the Achromatized UK Schmidt Telescope, *Quarterly Journal of the Royal Astronomical Society*, 22, 1981, pp. 146–153.
13. Mathewson, Don, et al., The Australian National University's 2.3 m New Generation Telescope at Siding Spring Observatory, *Journal of Astronomical History and Heritage*, 16, No. 1, 2013, pp. 2–28.
14. Frame, Tom, and Faulkner, Don, *Stromlo, An Australian Observatory*, Allen and Unwin, 2003, p. 192.
15. Beatty, J. Kelly, Australian Scopes Lost to Wildfire, *Sky and Telescope*, April 2003, pp. 18–19, and Tytell, David, Mount Stromlo: Out of the Ashes, *Sky and Telescope*, August 2003, p. 26.
16. Bhathal, Ragbir, Sutherland, Ralph, and Butcher, Harvey, *Mt Stromlo Observatory, From Bush Observatory to the Nobel Prize*, CSIRO publishing, 2013.
17. McGregor, Peter J., et al., GMT Integral-Field Spectrograph (GMTIFS) conceptual design, *Proc. SPIE*, 8446–053, 2012.
18. Hanbury Brown, R., *Boffin; A Personal Story of the Early Days of Radar, Radio Astronomy and Quantum Optics*, Adam Hilger, 1991.
19. Hanbury Brown, R., The Stellar Interferometer at Narrabri Observatory, *Sky and Telescope*, August 1964, pp. 64–69.
20. Davis, J., et al., The Sydney University Stellar Interferometer – I. The Instrument, *Monthly Notices of the Royal Astronomical Society*, 303, 1999, pp. 773–782.

10

Mount Hopkins' Whipple Observatory and the MMT

In the mid 1960s the Smithsonian Astrophysical Observatory (SAO) decided to move their satellite tracking station from White Sands to a better observing site at higher altitude and to make this new site the location for an astronomical observatory.(1) Fred Whipple, the director of the SAO, began to investigate possible locations and settled on the Tucson area in southern Arizona as the most suitable. There he considered three possible sites, namely Kitt Peak, Mount Lemmon, and Mount Hopkins. He rejected Kitt Peak because of its relatively low altitude and potential light pollution. Mount Lemmon was the highest peak, and so should be the best for infrared observations. But Mount Lemmon was only 16 miles (25 km) from Tucson and was already suffering from light pollution. So Whipple chose Mount Hopkins in the Coronado National Forest, about 35 miles (55 km) from Tucson, as the site for the new SAO observatory.

The SAO began to build their new observatory in 1966 on a 7,600 to 7,800 ft (2,320 to 2,380 m) high ridge on Mount Hopkins, leaving the 8,590 ft (2,620 m) summit for the construction of a large optical telescope later.(2) The new observatory initially included a laser and f/1.0 Baker-Nunn camera for satellite range-finding and tracking, and a 10 m diameter optical reflector for gamma-ray astronomy. Then in 1969 the SAO built a 60 inch (1.5 m) telescope on the ridge, to be used for photoelectric spectrophotometry. It was named after Carlton W. Tillinghast, a Smithsonian administrator who died in 1969 at the age of 36.[1]

Whilst the 60 inch was being constructed, Fred Whipple and colleagues investigated possible designs for the SAO's projected large optical telescope. At first they considered building a telescope with a fixed, large, spherical segmented primary mirror, similar to one proposed by Aden Meinel when he had been at Yerkes in 1953 as the optical equivalent of the Arecibo radio dish.(3) But they rejected this

[1] *Sky and Telescope*, November 1970, p. 272.

as the reflecting area would continuously change during an observation making the interpretation of infrared observations difficult.

At about the same time Frank Low of the University of Arizona's Lunar and Planetary Laboratory (LPL) had been developing observational techniques to detect faint objects in the infrared with a 1.5 m telescope. He then began to consider how to detect even fainter infrared objects with a larger telescope. The problem was that the thick conventional mirrors of large traditional optical telescopes of the time emitted too much infrared noise of their own. In addition their secondary mirrors were generally large and could not be used to chop the infrared beam on the sky and so enable the general sky background to be eliminated. So Low envisaged using a number of 1.5 m infrared telescopes with thinner primary mirrors combining their light at a common focus. In 1967 he presented this idea to Aden Meinel, the chair of University of Arizona's astronomy department.

It so happened that Meinel at that time had been considering how to build a space telescope for the military with an effective diameter larger than the diameter of its launch vehicle envelope. He proposed that, instead of using one large primary mirror, a number of smaller mirrors could be packaged in the launch vehicle envelope and deployed after launch to provide the effective diameter of the larger mirror.(4) Consequently when Low approached him, Meinel had already been considering the design of a telescope with a number of small primary mirrors, albeit for space applications.

Interestingly, although Low and Meinel did not realise it at the time, the basic idea of a multiple mirror telescope had originally been published in 1930 by the Irish scientist Edward Synge. Twenty years later Yrjö Väisälä, the director of the Turku Observatory in Finland, had actually built a model of a multiple mirror telescope with six 32 cm diameter mirrors arranged in a ring and with a seventh mirror in the centre. But he did not build an operating instrument.(5)

Following Low's approach, Meinel started to design a ground-based infrared telescope based on Low's ideas. Meinel was aware that the US military had a number of 1.8 m (72 inch) diameter fused quartz mirror blanks surplus to requirements. These were lightweight, which was an advantage for infrared observations, as they had originally been purchased as part of a space programme. Consequently Meinel proposed that the new telescope should consist of an hexagonal array of six of these 1.8 m mirrors to produce the light-gathering power of a 4.5 m (177 inch) diameter telescope at a fraction of the cost of a conventional instrument. One possible configuration would have the six mirrors focusing directly at a common prime focus. But this would require difficult and expensive off-axis optics, and the infrared image would be contaminated by radiation from the voids between the primary mirrors. The other design configuration, which Meinel adopted in his so-called Project Colt (because of its six-shooter concept), was to have a secondary mirror for each primary which then reflected the light to a common focus. This also had the advantage of reducing the telescope's length.[2] Unfortunately the University of Arizona foresaw problems in raising funding to build the telescope.

Low and Meinel's work on a multi-mirror infrared telescope was being undertaken at the University of Arizona whilst Whipple's team at the SAO were considering their various design options for a large optical telescope to be built at their observatory on Mount Hopkins. Whipple had some knowledge of the University of Arizona's design as Gerard Kuiper, the director of the University of Arizona's LPL, had kept him informed from time to time. Then in February 1970, Meinel told Whipple that the US military had a number of surplus 1.8 m diameter fused quartz mirror blanks available. The SAO were more likely to get funding than the University of Arizona, so maybe Whipple could develop Meinel's design and acquire the surplus military blanks.

Consequently in November 1970, the SAO submitted a proposal to Congress for 'A Large Astronomical Telescope at Low Cost' with a telescope based on Meinel's design. Eight months later Congress approved the first instalment of 50% of the expected $3 million cost of the programme. Then in late 1971 SAO signed a Memorandum of Understanding with the University of Arizona to build and operate the telescope called the Multiple Mirror Telescope or MMT for the benefit of astronomers from both organisations. Under this agreement the SAO would provide the mount, optics support structure (the open telescope 'tube') and building, whilst the University of Arizona would provide the passive and active optics and supporting cells. The SAO would have overall programme management responsibility and would provide the programme manager for the development and construction of the telescope.

The MMT design had an unvignetted field of view of only about 50 arcsec, with a significant loss of focus a short distance from the optic axis. But that was not a major concern to Low and Kuiper as they wanted to use the telescope for infrared photometry or spectroscopy of individual faint objects like quasars. Ray Weymann, the recently appointed director of the University of Arizona's Steward Observatory, also thought that the telescope would be just as useful for optical spectroscopy and small-field optical astronomy in general.

Three possible locations were considered for the MMT, namely Kitt Peak, Mount Lemmon, where the University of Arizona's new observatory was being built, and Mount Hopkins, where the SAO's observatory was located. Kitt Peak was dismissed early on as its altitude was considered to be too low for infrared observations. Kuiper and Low, of the University of Arizona, naturally wanted to

[2] Blanco et al., *The Optics Support Structure of the MMT*, accessed via www.mmto.org/MMTpapers/pdfs/tr/tr21.pdf.

Figure 10.1. The MMT with its array of six 1.8 m diameter telescopes. (Courtesy Howard Lester, MMT Observatory.)

build the telescope on Mount Lemmon. They argued that the mountain was developed and accessible, and should be better for infrared observations than Mount Hopkins as it was slightly (570 ft or 170 m) higher. But the SAO wanted to build the telescope on the top of Mount Hopkins, near their observatory which was about 900 ft (275 m) lower down the mountain. Unfortunately the top of Mount Hopkins was rather sharply peaked, limiting the amount of room available.

Interestingly Meinel and Weymann, although working for the University of Arizona, both favoured Mount Hopkins. As Weymann explained it,(6) he was concerned about the effect of light pollution at Mount Lemmon on optical astronomy. He also thought that it would be easier to get funding for the SAO for a telescope on Mount Hopkins than for a telescope at the University of Arizona's observatory on Mount Lemmon. In the event in 1973, after a great deal of debate that almost destroyed the project, Mount Hopkins was chosen as the site for the MMT.

A number of design features of the MMT (see Figure 10.1) had already been established by the start of the joint project at the end of 1971. In particular it had been decided that the MMT would consist of an hexagonal array of six 1.8 m diameter telescopes, as proposed by Meinel, each of which had an f/2.7 primary mirror. These short focal length mirrors had been selected as it was thought that the cost savings achieved with the shorter telescope structure would more than offset the extra costs of shaping the more highly curved mirrors. The six telescopes each had two secondary mirrors, one hyperbolic for optical and one bi-convex for infrared observations. They were in a Cassegrain configuration to produce a final f/31.6 (optical) and f/33.6 (infrared) system for each telescope. The infrared secondaries were smaller and thinner than the optical ones to avoid infrared radiation coming from the edges of the primary mirrors and to provide low-inertia infrared secondaries. This was to facilitate them being mechanically oscillated to compare the infrared intensity of cosmic sources with that of the background sky. The MMT also had an active optics system to keep each of the six independent telescopes co-aligned. Finally an altazimuth mount had been chosen as not only did it produce a more compact telescope, and hence a smaller observatory building, but it also simplified the design of the optics support structure. It was thought that the extra costs of eliminating the subsequent field rotation would be relatively small.

The ex-military lightweight primary mirror blanks, which were of an egg-crate construction, had to be returned to Corning Glass Works to have their front surface slumped to an approximation of their final f/2.7 shape. Final figuring and polishing were done in the optical shop of the University of Arizona's Optical Sciences Center. The design goal was to have 90% of the light energy within 0.5 arcsec for each of the six mirrors. In the event five of the mirrors had values between 0.54 and 0.64 and one had a value of 0.86.

As mentioned above, the amount of room at the top of Mount Hopkins was very limited, but the use of an altazimuth design meant that the telescope enclosure could be made much smaller than one enclosing an equatorial mount. In addition, Thomas Hoffmann and Nathaniel Carleton of the SAO suggested that the telescope enclosure could be brought much closer to the telescope making it even smaller, whilst the swing of the telescope could be accommodated by having its barn-like enclosure rotate with the telescope. This revolutionary idea naturally encountered much opposition, but, in the end, it turned out to be one of the most important ideas implemented in the design of the MMT.(7) Co-rotating enclosures have since been incorporated into the design of the European Southern Observatory's New Technology Telescope (first light 1989), the ARC telescope at Apache Point (1994), the Japanese Subaru telescope (1998) and the Large Binocular Telescope (2005).

To have six 1.8 m telescopes producing the same resolution as one larger telescope it was essential to not only superimpose all six images, but the images must also be in phase. If this was achieved, any two of the 1.8 m telescopes, which had a maximum edge-to-edge distance of 6.9 m, could be used as a Michelson or speckle interferometer to produce high resolution, and all six telescopes could be used to provide better resolution in two dimensions. However, in order to avoid trying to resolve too many issues at once, it was decided early on to postpone the required phasing to a later stage, as high resolution was not the major driving force for the telescope design.[3]

Shortly after project go-ahead Michael Reid and Gregory Sanger, both of the Steward Observatory, produced a design of the active optics system in which the secondary mirrors each had two tilt and one focus motion driven by an optical error-sensing feedback system. If a natural guide star were to be used with such a system it would have to have been reasonably bright, and the chances of finding such a star within the limited field of view of the MMT was rather small. So an internal laser alignment system was chosen. Consequently a 75 cm (30 inch) diameter, wide field Ritchey-Chrétien telescope was added to the MMT in the centre of the hexagonal array of six telescopes to collimate the laser beams and to serve as a guide telescope. The overall design may sound rather simple but it involved adding more than 100 optical elements to the design, all of which had to be aligned and kept in collimation.

Two targets were set for the performance of the laser alignment system. The first was to maintain the coalignment of the centres of the six images to within 0.5 arcsec diameter, the second was to achieve a coalignment of 0.1 arcsec. The first goal was met by late 1979, but it required an enormous amount of work adjusting all the optical components. In the following year Beckers, who was then the director of the MMT Observatory (MMTO) decided to remove the laser system and rely on the structural integrity of the optical support structure instead.

The primary mirrors had been installed in the MMT and first light achieved with two of the six telescopes in May 1978. In the following year trial astronomical observing was begun. At that time the MMT with its unique design of six telescopes on a common mount was the first large modern optical telescope in the West to use an altazimuth mount, the first large telescope to use a form of active optics, and the first to use a building that co-rotated with the telescope. The net result of all these innovations was that the MMT had been built at about 40% of the capital cost of a normal reflector of the equivalent diameter of 4.5 m from an image intensity point of view. On the other hand, it took quite a number of years to get the telescope to work correctly.

Funding was a continuing problem with the MMT for a number of years. Although \$1.5 million had been approved in 1971 as 50% of the expected project costs, the estimated costs of the project had increased to almost \$7 million

[3] The telescope's primary purpose was optical and infrared astronomy of faint and low-contrast objects.

by 1975, a real increase (after taking account of inflation) of about 75%. Unfortunately, according to Weymann,(8) Meinel had given the NSF the impres sion early on that the Department of Defense would cover the majority of the University of Arizona's costs, which turned out not to be the case. Consequently the University of Arizona was obliged to turn to their State Legislature for financial support.

In November 1981, the Smithsonian Institution decided on Fred Whipple's 75th birthday to call their observatory on Mount Hopkins the Fred Lawrence Whipple Observatory.

Six years later it was decided to radically modify the MMT as it was clear by then that it was possible to build very large lightweight monolithic mirrors, which had been very problematic when the MMT had been originally designed. It was found to be possible to replace the six telescopes of the MMT by a single f/ 1.25, 6.5 m diameter lightweight monolithic mirror in the same observatory building. This new telescope would have 2.5 times[4] the light-gathering power of the MMT and, with suitable design, its angular field of view could be increased by a factor of 15. In the event its 6.5 m primary mirror blank of borosilicate glass was spun-cast by the University of Arizona's Mirror Laboratory in 1992.(9)

This new MMT (see Figure 10.2), with a total conversion cost of about $20 million, saw first light in the year 2000. Its instruments covered the wavelength range from 300 nm in the near ultraviolet to 25 microns in the mid-infrared. Three years later a deformable secondary mirror was fitted to the telescope made by the Steward Observatory and Italy's Osservatorio Astrofisico di Arcetri in Florence. It consisted of a 1.6 mm thick reflective shell controlled by 336 computer-controlled actuators. A larger version of this deformable secondary was fitted on the Large Binocular Telescope in 2010.

The modified MMT actually had three secondary mirrors, namely an f/9 tip- tilt, an f/5 tip-tilt and an f/15 deformable mirror for adaptive corrections with either natural or laser guide stars. These three interchangeable secondaries supported four Cassegrain configurations, namely an f/9 with a field of View (FoV) of 13 arcmin to enable instruments made for the original MMT to be used, an f/5 with a corrected FoV of 60 arcmin for wide-field spectroscopy, an f/5 with a corrected FoV of 30 arcmin for imaging, and the fully adaptive f/15 with a maximum FoV of 5 arcmin. The name MMT was retained for the telescope, even though the acronym doesn't now stand for anything.

In 1996 a 1.3 m (50 inch) reflector had also been built on Mount Hopkins specifically optimised for the infrared. It was used for a 2 micron All Sky Survey, led by

[4] This is because, not only had the collecting area been increased, but the new design had eliminated two reflections.

Figure 10.2. The modified MMT with its one 6.5 m diameter telescope. (Courtesy Howard Lester, MMT Observatory.)

the University of Massachusetts. This 2MASS telescope, as it was called, covered the northern part of the sky, whilst an identical telescope, located on Cerro Tololo covered the southern part (see Section 2.7).(10) Both telescopes housed three infrared arrays cooled down to 77K by liquid nitrogen, enabling the sky to be surveyed in the J (1.25 microns), H (1.65 microns) and K_s (2.16 microns) photometric bands.(11) All the operations were automatic. The 2MASS survey was completed in 2001 and the resulting All Sky Catalog was published two years later. For a time the Mount Hopkins 2MASS telescope was used for the PAIRITEL[5] project as a fully automatic infrared imaging system starting in 2004. It was designed to respond automatically to new gamma-ray burst transients, which it first achieved in November 2005, and was used to study supernovae and other transient sources. The telescope has since been decommissioned. The Whipple Observatory has also been host to the IOTA Michelson interferometer (see Section 14.4).

[5] PAIRITEL is an acronym of 'Peters Automated IR Imaging TELescope'.

References

1. Whipple, Fred L., Mount Hopkins and the MMT, in Weekes, Trevor C., (ed.), *The MMT and the Future of Ground-Based Astronomy, Smithsonian Astrophysical Observatory Special Report No. 385*, 1979, pp. 3–7.
2. Burke, John J., and Kirchhoff, Werner, Mount Hopkins Observatory, *Sky and Telescope*, November 1968, pp. 284–286.
3. McCray, W. Patrick, *Giant Telescopes: Astronomical Ambition and the Promise of Technology*, Harvard University Press, 2004, pp. 51–53, and 65.
4. Meinel, A. B., Multiple Mirror Telescopes of the Future, in Weekes, Trevor C., (ed.), *The MMT and the Future of Ground-Based Astronomy, Smithsonian Astrophysical Observatory Special Report No. 385*, 1979, p. 9.
5. Jacchia, Luigi, Forefathers of the MMT, *Sky and Telescope*, February 1978, pp. 100–102.
6. Beckers, Jacques M., and Ulich, Bobby L., The Multiple Mirror Telescope, in Burbidge, G., and Hewitt, A., (eds.), *Annual Reviews Monograph; Telescopes for the 1980s*, Annual Reviews Inc., 1981, p. 75.
7. Wilson, R. N., *Reflecting Telescope Optics II; Manufacture, Testing, Alignment, Modern Techniques*, Springer-Verlag, Corrected Second Printing, 2001, p. 187.
8. Beckers, Jacques M., and Ulich, Bobby L., The Multiple Mirror Telescope, in Burbidge, G., and Hewitt, A., (eds.), *Annual Reviews Monograph; Telescopes for the 1980s,* Annual Reviews Inc., 1981, p. 75.
9. Robinson, Leif J., Spinning a Giant Success, *Sky and Telescope*, July 1992, pp. 26–29 and 31.
10. Skrutskie, Michael, 2MASS: Unveiling the Infrared Universe, *Sky and Telescope*, July 2001, pp. 34–42.
11. Skrutskie, M. F., et al., The Two Micron All Sky Survey (2MASS), *Astronomical Journal*, 131, 2006, pp. 1163–1183.

11
Apache Point Observatory

11.1 ARC 3.5 m Telescope

The University of Washington (UW) decided in 1965 to expand its astronomy department and hired Paul Hodge and George Wallerstein from the University of California at Berkeley to increase the number of astronomers in the department to three!(1) Not surprisingly, the facilities available at UW were quite inadequate for modern observational research so Hodge and Wallerstein began to consider how to improve them by building a large modern observatory. There were two major issues to resolve before they could go any further. It was obvious that the UW astronomy department was too small to afford a large astronomical telescope, and there were not enough astronomers to properly utilise such an instrument even if they could raise the funding. In addition there were no suitable locations in Washington State to make optimum use of a large modern telescope, so a site survey was required to find a suitable site elsewhere. The UW Regents authorised construction of a large telescope providing funding could be found, but Wallerstein and Hodge were unable to organise a suitable consortium to finance the project.[1] As a result the UW built a small observatory on Manastash Ridge in Washington State as an interim measure at which they installed a 30 inch (80 cm) in 1972.

In the meantime the astronomy department at UW had continued to grow and rely heavily on the national observatory telescopes on Kitt Peak and Cerro Tololo for large modern telescopes. But, unfortunately, these were beginning to become heavily oversubscribed. Clearly the UW needed their own large telescope, even if it had to be with one or two other organisations in order to raise the finance.

When Alex Kane of Ashland, Oregon died in 1975 he left an estate worth $250,000 to build a large telescope. He initially intended to leave the money to

[1] Wallerstein and Hodge spoke to JPL and the University of Wisconsin at Madison and also tried to raise private funds, but without much success. Unfortunately at the time National Science Foundation funding was already committed to the Kitt Peak National Observatory and other facilities.

Oregon State University but they were not interested in building a new large telescope, so he left it to the UW instead. Unfortunately the money was not sufficient, on its own, to pay for the facility that the UW had in mind. At first they discussed the possibility of building a joint telescope on Mauna Kea with Stanford University or the University of Vienna, but these discussions came to nothing.

A few years later Bruce Balick, who had joined the UW in 1975, began a new drive to interest other research organisations in working with the UW on an advanced technology telescope. He suggested using ideas from the Kitt Peak Advanced Development Program and from radio astronomy. His new initiative resulted in positive interest being shown by Howard University (HU), New Mexico State University (NMSU) and Washington State University (WSU) in working on a joint project with the UW. They then agreed to work together on the outline design of a possible telescope.

The joint university design team proposed that the new telescope should have a lightweight 2 m diameter primary mirror, just 6 cm thick, with active optics on an altazimuth mount in a low-cost enclosure. The design also included a remote observing capability and rapid instrument changes to optimise the use of telescope time. A suitable site, later called Apache Point, was quickly identified at an altitude of 2,800 m (9,200 ft) near the Sacramento Peak campus of the National Solar Observatory, New Mexico in a region protected for astronomical use.[2] The seeing was excellent. In 1981 the four universities formed a provisional consortium with expected financial contributions of UW 40%, HU 30%, NMSU 16% or possibly 25%, and WSU 5%. The project cost was estimated to be about $3.6 million, including a 15% contingency.

Unfortunately, Howard University had to pull out of the project in early 1982 which put the project's future in jeopardy. However shortly afterwards the project was back on track when Princeton University (PU) and the University of Chicago (UC) joined the consortium. But Princeton really wanted to build a larger telescope than the one that the consortium were then investigating. Subsequently the consortium team looked at the possibility of meeting Princeton's requirements and by 1983 the size of the telescope had been increased to 3.5 m.

The five university consortium was then legally established in January 1984 as the Astrophysical Research Consortium (ARC). Allocated telescope time, based on members' contributions, was agreed to be 31.25% for both the UW and UC, 15.625%

[2] The New Mexico State University had considered building an optical telescope in the Sacramento Peak area some years before, and so had extensive meteorological data available going back to the 1950s. (More recent surveys have shown that night-time observations can be made about 65% of the time, with about 25% of the total time providing photometric conditions. The median site seeing is about 0.9 arcsec, with the best conditions producing 0.5 arcsec. See Jon Holtzman, et al., Apache Point Observatory: Facilities, Operations and Partnerships, March 15 2009, produced as *Position Paper No. 23, for Astro2010: The Astronomy and Astrophysics Decadal Survey*.)

for both the NMSU and PU, and 6.25% for WSU. Unfortunately by this time the total of members' promised financial contributions had fallen far short of the money required to build the observatory which by early 1984 had risen to $10 million.

It had been clear for some time that the ARC would have to find substantial financial contributions from external bodies. In particular they had hoped that the National Science Foundation (NSF) could be persuaded to provide a large financial contribution now that the project was supported by five American universities. Consequently in September 1983, consortium members sent a letter to the NSF telling them that they had decided to create a consortium and that they planned to ask the NSF for a grant of about $3.75 million. By May of the following year, when the grant request was actually submitted, the sum asked for had risen to $5.565 million!

Discussions between the ARC and the NSF continued through 1984, 1985 and 1986 whilst the project team was, in parallel, designing the telescope with a minimum of funding. In February 1986 the NSF suggested a grant of $3.3 million, with more if the ARC partners could increase their total contributions by $750 K, which they agreed to do. In the event the NSF agreed in July 1986 to grant the consortium a sum of $ 3.74 m towards the telescope's costs.

One of the main reasons that the NSF agreed to provide financial support was that the project fitted the criteria outlined in the National Research Council's Field Report of 1982/83.(2) In particular, in pages 146–148 of that report the Field committee recommended the construction of optical/infrared telescopes in the 2 to 5 metre class. It also encouraged federal assistance for those projects that received significant nonfederal funding for construction and operation from private and state funds. The ARC project met all these conditions.

As approved the ARC telescope was expected to be a general-purpose instrument that could be used to cover the very broad interest range of the member universities. Astronomers also hoped that it would be easy to reschedule observations and that the telescope would be able to be operated remotely. This was to enable telescope users to match observations to seeing conditions and to respond quickly to transient events like supernovae. Remote operations, without the need for travel, would also improve the telescope's availability to students.

In early 1983 Roger Angel of the University of Arizona (UA) Mirror Laboratory told the UW that the UA planned to make a 3.5 m diameter lightweight mirror using a new rotating furnace, and they wanted to check the quality of the mirror by using it in a real telescope. The NSF was already funding the laboratory so the UA could make the mirror available free of charge to whichever observatory had the best plans for using it. The mirror was to be supplied on a best efforts basis, provided the user agreed to carry out thermal and optical tests once it was installed in the telescope. This the ARC agreed to do, so the UA agreed to supply the mirror. However one problem of using an experimental mirror in the ARC telescope was timescale. Initially, the Mirror Lab expected to deliver the 3.5 primary mirror in August 1985 to enable telescope first light to take place in 1987. Then in 1985 the Mirror Lab estimated that it would cast the ARC 3.5 m mirror blank in September 1986, whilst in August 1986 the casting was expected to take place very soon. In the event casting only took place in April 1988, over two and a half years later than originally expected. After cooling and cleaning the mirror blank it was delivered in August 1988 to Norman Cole's Arizona Technologies Inc. for polishing, with the ARC then estimating first light of their telescope in early 1990.

The initial delay in providing the mirror blank was partly caused by the NSF not granting the Mirror Lab its full funding request. Eventually the ARC was asked to pay the Mirror Lab for materials and labour. In addition the ARC had had to increase member contributions by $1.5 million above their initial $3.95 million[3] owing to the initial shortfall in its NSF grant.

As far as the focal length of the mirror was concerned, the smaller the focal length the smaller the observatory needed to protect it. As a result some astronomers favoured an f/1.5 mirror, but this was considered too risky and time-consuming to produce. As a compromise the ARC settled on an f/1.75 mirror. Two interchangeable secondaries were designed, one to provide an f/10 telescope and the other an optional f/35 that could be used for infrared observations.

Norman Cole had considerable difficulty in producing a mirror of the required f/1.75 shape so that he had only finished rough grinding it by early 1990. By this time it had been expected that the telescope would have seen first light. The ARC had actually chosen Cole's small organisation to undertake the job, even though they knew that may be risky, because alternative organisations were very expensive. The UA's Mirror Lab had wanted to do the job but, at the time the contract was signed, they did not have the required facilities. In the meantime, however, Roger Angel had developed new facilities at the UA and offered to finish the 3.5 m mirror if the ARC would pay for the development costs of his testing system. In February 1992 the ARC accepted this proposal, so the mirror was moved back to the Mirror Lab. After so many delays, the subsequent steps occurred rather quickly. Mirror polishing was completed in August 1992 and aluminisation was completed at Kitt Peak on September 15. The mirror was delivered to Apache Peak three days later.

In the meantime the telescope structure had been completed by the end of 1987. Its barn-like housing, with a design based on that of the MMT (see Chapter 10), was completed in parallel. As a result at the end of 1987 the ARC had a telescope and no mirror, but the University of Calgary had a 1.8 m mirror and no telescope. So the ARC arranged to borrow the Calgary mirror, in return paying for it to be polished and aluminised. This mirror was installed in the ARC telescope in early 1991

[3] $3.2 million plus $750,000.

to enable the system to be tested with temporary secondary and tertiary mirrors that provided the same image scale as the final 3.5 m telescope. In the following year the ARC was able to carry out full remote operations from the campuses of its member universities with a system using a CCD camera, a guide camera and an infrared imager. The images obtained confirmed the good seeing of the site.

Amazingly even though the primary mirror had finally been delivered in September 1992 after many delays, the telescope was still not complete as neither the secondary nor tertiary mirrors had been delivered. The secondary, cast by the Hextek Corporation (a technology spin-off of the UA's Steward Laboratory), had only been delivered to the UA's Optical Sciences Center in February 1992 where it was completed in January 1994. Its delivery finally enabled the three mirror ARC telescope to achieve first light in April.

Although earlier tests with just the primary mirror had produced excellent results, those with all three mirrors now in place were disappointing. This was eventually traced to incorrectly figured zones on the secondary. Consequently it was sent to Lick Observatory for measurements and then to Kodak for ion polishing. Unfortunately the secondary was too thin to be completely corrected, so the final images, although much better than previously, were still below initial expectations.

The ARC telescope when completed provided rapid access to a variety of instruments permanently located at different telescope ports and accessed by the movable tertiary mirror. Typically, different instruments could be accessed in less than 5 minutes, if they were already located at one of the telescope ports, or 15 minutes if they had to be moved there. Almost all of the observing was undertaken remotely, although decisions about the opening and closing of the dome were taken on-site.

Since the ARC's foundation in 1984, the WSU has withdrawn from the consortium while the University of Colorado and the University of Virginia have joined. New members were required to contribute a share of the capital costs and a share of the running expenses. In a number of cases members have contributed new instruments in lieu of paying capital costs, which has helped to keep ARC instrument capabilities up to date. In addition the Institute for Advanced Study at Princeton[4] and Johns Hopkins University joined the ARC to participate in the Sloan Digital Sky Survey (see below) and not the 3.5 m.(3, 4)

11.2 Sloan Digital Sky Survey (SDSS)

The idea behind what became known as the Sloan Digital Sky Survey first surfaced in February 1987 at a meeting to discuss the potential use of a spare experimental

[4] The Institute for Advanced Study is not part of Princeton University.

Kitt Peak mirror.(5) At this meeting Jim Gunn of Princeton University suggested that a 2° field-of-view telescope could be built with it to undertake a wide-area digital survey of galaxies using state-of-the-art CCDs and the rapidly improving computational power of the time.(6) The telescope would have a camera for imaging and a fibre-fed spectrograph that could record the spectra of hundreds of galaxies at once. This would enable their red-shifts, and hence their distances, to be determined, amongst other things. He also suggested that the telescope should operate in a drift scanning mode in which it was fixed and the rotation of the earth was used to record small strips of the sky. Using such a system over a period of about five years should enable a three-dimensional, well calibrated map of a large slice of the distant sky to be produced. On further consideration Gunn thought that a 2.5 m mirror, which would be smaller than the spare Kitt Peak mirror, should be large enough to undertake the survey, bearing in mind the high efficiency of the CCDs then available. In spite of the initial enthusiasm of other astronomers his idea was not taken any further at that time.

Not knowing of Gunn's proposal, in April 1988 Richard Kron of the University of Chicago also suggested undertaking a galaxy survey using a telescope camera with a matrix of CCDs. His proposed telescope would also include a fibre-fed spectrograph to take large numbers of spectra, and hence redshifts, simultaneously.

It so happened that Gunn, Kron and Don York, also of the University of Chicago and the director of the ARC observatory, regularly met to discuss progress with the ARC's 3.5 m telescope. On one such occasion they also began to discuss a possible galaxy survey programme using a new 2.5 m telescope. They subsequently contacted cosmologists from other institutions to see if they were interested in joining such a programme. Then later that year Gunn, Kron and York and a group of interested researchers from other organisations sketched out the design of a possible telescope which they suggested could be located at either Kitt Peak or Apache Point. The biggest technical problem that they foresaw was handling the enormous quantity of data that the survey programme would produce.

At about this time the University of Chicago's request for funding a larger telescope had been rejected by the NSF. So Don Lamb, the head of the university's astronomy department, agreed to consider the 2.5 m telescope survey programme as a possible replacement. He contacted John Peoples, director of the nearby Fermi National Accelerator Laboratory to see if they would be interested in joining the project to solve the data processing problem, as they were used to handling large quantities of data. As it happened Fermilab had just been notified that they would not host the Superconducting Super Collider so they were looking for new outlets for some of their staff. It seemed as this galaxy survey programme. although small in Fermilab's terms, could be a useful addition to their normal work.

In parallel Jeremiah (Jerry) Ostriker, the head of Princeton University's astronomy department, contacted John Bahcall of the nearby Institute for Advanced Study to see if they were interested in joining the galaxy survey programme. Bahcall was keen to participate and promised a small amount of money in exchange for access to the data. So it was that in April 1990 the University of Chicago and Princeton University were joined by the Institute for Advanced Study (IAS) as initial members of the new programme. All three agreed to donate $350,000 each as a sign of their interest. By the end of the year the IAS had joined the other two institutions in becoming a formal member of the ARC. At the same time Fermilab had also decided to join the programme.

Meanwhile in 1989 Don York had made the first real estimate of programme costs, excluding those of university salaries, to be about $12 million, including about $600 k for the mirror. About a year later the estimate had increased to $14 million, with about $2 million for the CCDs and $1 million for the mirror. But by late 1991 the total had almost doubled to $25 million.

In the meantime the ARC had, in 1988, begun to consider the possibility of telescopes other than their 3.5 m being built on Apache Point. Then in December 1990, they formed a subcommittee to specifically investigate building the 2.5 m survey telescope there. But before the project could become a reality the money to build it had to be found as the interested institutions were nowhere near raising the $25 million required.

Representatives of Princeton University and the University of Chicago, as members of the ARC, persuaded the Alfred P. Sloan Foundation in 1991 to provide a grant of $8 million towards the programme costs. Consequently the ARC, who had agreed to adopt this galaxy survey programme, agreed to call it the Sloan Digital Sky Survey (SDSS) in recognition of the Foundation's substantial financial contribution.(7) By this time Fermilab had also agreed to provide $5 million. But there was still some way to go before the required funding was covered, particularly as the Sloan Foundation had received an assurance from the universities that they, the universities, would fund any cost overruns.

At about the same time Johns Hopkins University had decided to pull out of the 8 m Magellan telescope project (see Section 12.1) at the Las Campanas Observatory as it was becoming too expensive. But this left them without access to a large telescope. Hearing of this, Ostriker suggested to Johns Hopkins that they could join the Sloan telescope project for substantially less cost than their subscription to the Magellan project, and they could probably build the Sloan spectrograph as well using their extensive experience in spectroscopy. So not only would they get access to data from a large telescope more economically than with Magellan, but they would also be able to build a substantial piece of hardware which was not on offer with the Magellan programme. Johns Hopkins were delighted with this suggestion and in June 1992 joined the ARC Sloan consortium. The only problem was that it had already been agreed that the University of Washington would build the Sloan telescope, its enclosure and spectrograph based on its experience of building the ARC's 3.5 m telescope. But Johns Hopkins would not join the Sloan project unless they made the spectrograph and the project needed as many members as possible to fund the work. So the spectrograph went to Johns Hopkins.

As a result by mid 1992 the Sloan survey team was composed of four astronomical academic partners all of whom were members of the ARC, a national laboratory (Fermilab) used to handling large amounts of data, and an agency (the Sloan Foundation) which had agreed to make a substantial financial contribution. With this breadth of support the Sloan project also had reasonable prospects of getting a grant from the NSF.

It was agreed that the SDSS telescope would be dedicated to the galaxy survey so, unlike most professional telescopes, it was not available for other observational programmes. The telescope was an altazimuth-mounted, 2.5 m diameter f/5 modified Ritchey-Chrétien design with a Gascoigne astigmatic corrector. It also had a pair of interchangeable highly aspheric correctors near the focal plane, one for imaging and one for spectroscopy.(8) Its 120 megapixel camera, mounted at the Cassegrain focus, consisted of thirty cryogenically cooled 2048 × 2048 pixel CCDs to produce images in five different wavebands centred at wavelengths from 354 to 925 nm.(9) In addition a pair of fibre-fed spectrographs, permanently mounted on the image rotator would obtain 640 spectra simultaneously over a similar range of wavelengths. The telescope, which was to operate in a drift-scanning mode, would produce excellent image quality and small geometric distortions over a 3° diameter focal plane, and very small lateral and longitudinal colour errors in the spectroscopic mode.

The main problem with building relatively large telescopes at that time was the production of the primary mirror. At 2.5 m diameter it was then too small to interest the University of Arizona's Mirror Lab, so Hextek, who had produced the secondary mirror blank for the ARC's 3.5 m telescope, was contracted to supply it. Hextek started their work straight away by melting and cooling a large amount of glass in a mould. But, unfortunately, when the mould was opened in July 1992 it was found that the faceplate of the mirror blank's sandwich construction had three large cracks in it. Happily this was corrected by reheating the faceplate just enough to remove the cracks which did not reappear on recooling. But that was not the end of problems as an accident during the blank's subsequent grinding process caused a cylindrical crack to appear in the faceplate and its supporting honeycomb. Fortunately, the crack was not very far from the mirror's central hole, so enlarging the hole a little solved the problem with the loss of only a small amount of reflecting surface.

Figure 11.1. The Sloan Foundation's 2.5 m telescope at Apache Point. (Courtesy Sloan Digital Sky Survey.)

As in the case of the 3.5 m's secondary mirror, the 2.5 m primary was then sent to the UA's Optical Sciences Center for polishing. There it met another problem as on completion the mirror was found to be astigmatic. This was corrected by actuators fitted to the back of the mirror when it was installed in the telescope. Timescale problems associated with the mirror were matched elsewhere in the project, particularly in writing the software. Consequently the SDSS telescope (see Figure 11.1) did not see first light until May 1998, which was four years late. In addition it cost over $70 million compared with the initial estimate less than ten years earlier of $12 million, or about $16 million after correcting for inflation. In the end the Sloan Foundation gave the programme $20 million, rather than the initial cap of $8 million, to keep the show on the road.

But why was the Sloan project so late and so over budget? Ann Finkbeiner in her book *A Grand and Bold Thing*[5] gives a clear description of the management problems experienced and the solutions adopted.

[5] See Ann Finkbeiner, *A Grand and Bold Thing: An Extraordinary New Map of the Universe Ushering in a New Era of Discovery*, Free Press, 2010, mainly Chapter 5.

As in the case of a number of other telescope programmes mentioned in this present book, one of the main problems with the Sloan project was the lack of proper project management in virtually all aspects of the programme. Work breakdown structures and critical path analysis may be boring, but at least they help the project manager to keep his/her finger on the pulse and direct attention to the critical areas of the project. But there is no point in having a project manager who is just a glorified administrator. Successful project managers need to understand the technicalities behind the project well enough to be able to argue with the specialists whose natural tendency is to aim for perfection, no matter how much it costs and how long it takes, whilst other people are waiting for their end product to be delivered. Perfection is, in my experience, the enemy of the good.[6] The Sloan team seemed to be running in various directions like headless chickens for a number of years.

The first project manager had been appointed in 1994 as a condition of the NSF funding the project. The person appointed was a professional manager called Tom Dombeck, but he was given no management or budget authority. Not surprisingly he resigned two years later – in fact the only surprise is that he stayed as long as he did as, without control of the leavers of power, he could do very little. At least Dombeck had had some management experience when he was appointed. But his replacement, an astronomer from the US Naval Observatory called Jeff Pier, had had none. So, not surprisingly he had no more success than his predecessor as he, like Dombeck, was also given no management or budget authority.

By early 1997 things were getting desperate as not only was the project getting later and later it was haemorrhaging money, and no one seemed to know when that was likely to stop. So the ARC board appointed a committee of enquiry consisting of three external experienced people to review the project and recommend a way forward. Following completion of the review the ARC appointed one of the reviewers, Jim Crocker, who was then managing the European VLT, to be the new project manager.[7] Crocker quickly established a more formal approach, forcing team members to define what were the acceptable rather than the ideal requirements for various parts of the system and getting team members to sign off on

[6] I am loath to bring my personal experiences of running spacecraft and other projects in here. But in so many programmes that I have covered in this book, major problems have been experienced due to a lack of proper project planning and control. Yes, such an approach costs money, but it generally saves money and timescale in the long run. The main problem with such a systematic approach, however, is that if it is done properly the estimated cost appears too high and frightens off potential financial sponsors. Consequently, the project may never get off the ground. To avoid that happening, projects are then forced to make highly optimistic cuts to the estimated cost, only to see the cost escalate later.

[7] Crocker had previously been responsible for managing the hardware that had successfully corrected the Hubble Space Telescope's astigmatism. So his management experience was ideal.

them. His approach worked wonders so that when he left in 1999[8] first light had been achieved and the survey phase was well on the way to starting, which it did, in the following year.

Unfortunately early operation of the telescope indicated that it had been installed with a slight tilt, which caused problems with the scanning software. That problem was easily solved but a much more threatening one was the discovery in October 1999 that there was a crack near the centre of the secondary mirror. In this case the Mirror Lab cut out the centre of the secondary and capped the hole. On investigation it turned out that the problem had not been caused by the mirror itself but by a detailed fault in the telescope's mechanical design which had resulted in a small metal rod fouling and eventually cracking the mirror.

The initial five year survey programme, called SDSS-I, observed about a fifth of the sky around the North Galactic Polar Cap down to a limiting magnitude of between about 23 and 25, depending on colour. When completed in 2005 the telescope had imaged 215 million individual objects, obtained the spectra of 675,000 galaxies, 90,000 quasars and 185,000 stars, and had found over 100,000 asteroids. The measured redshifts of distant objects enabled a three-dimensional map of the universe to be constructed over a volume about two orders of magnitude larger than previous surveys.

This first survey programme was followed from 2005 to 2008 by SDSS-II which consisted of three parts: the Sloan Legacy Survey which completed the original survey of the North Galactic Cap, SEGUE (Sloan Extension for Galactic Understanding and Exploration) which produced a detailed three-dimensional map of the Milky Way, and the Sloan Supernova Survey (SSS) which undertook repeated imaging of 300 square degrees of the sky to discover and measure supernovae, amongst other things, to characterise the expansion of the universe.(10)

In the event SEGUE produced a spectroscopic survey of about 240,000 faint Milky Way stars of a wide variety of spectral types. These enabled a study to be undertaken into the kinematics and populations of stars in the Milky Way and its halo.(11) In addition over the course of three 3-month campaigns the SSS discovered about 500 Type Ia supernovae, which could be used to determine the rate of expansion of the universe over the last 4 billion years. SDSS-II was followed in 2008 by a series of four new surveys in the SDSS-III programme which finished in 2014. This involved upgrading the spectrograph with more fibres, new detectors and gratings to effectively double the telescope's output.

A number of new institutions joined the ARC over time, or worked with it, to enable them to participate in all or part of the SDSS programme. So that by 2005 the Advisory Council responsible for the SDSS included members from

[8] To manage the design team for the next space telescope.

25 different institutions in America, Europe and Asia. Additional institutions later joined specifically to participate in the SDSS-III programme.

11.3 ARC 0.5 m Photometric Telescope

A 0.6 m (24 inch) telescope had been located at Apache Point in 1995 to produce calibrations of standard stars to be used in the SDSS programme. Unfortunately this monitor telescope's performance was very poor(12) and it was replaced in 1998 by a 0.5 m (20 inch) which was then being used at Johns Hopkins University at Baltimore for public outreach. This telescope, which had been donated to Johns Hopkins by Michael Bloomberg, a university trustee, was moved with his approval from Baltimore to Apache Point. There it was used during the SDSS-I and -II surveys to continuously monitor the brightness of the sky and the transparency of the Earth's atmosphere above the observatory, and to provide photometric calibration for the 2.5 m telescope. After SDSS-II this telescope, renamed ARCSAT (ARC Small Aperture Telescope), was used as a general community resource. It could make observations remotely, although it required onsite assistance for opening, closing and weather monitoring.

11.4 NMSU One-Meter Telescope

A One-Meter Telescope had been built on Apache Point in the early 1990s owned and operated by New Mexico State University (NMSU). This Ritchey-Chrétien altazimuth telescope, which was completed in 1994, was optimised for observing extended objects such as nebulae, galaxies, star clusters and clusters of galaxies. Automated data reduction was also undertaken of objects like eclipsing binaries and cataclysmic variables. It could be operated at Apache Point or from the NMSU campus. The NSF provided funding for improvements to the telescope on condition that non-NMSU astronomers could use it for research.

References

1. Peterson, Jim, and Mackie, Glen, A Brief History of the Astrophysical Research Consortium and the Apache Point Observatory, *Journal of Astronomical History and Heritage*, 9, No. 1, June 2006, pp. 109–128.
2. Field, George B., et al., *Astronomy and Astrophysics for the 1980's*, Vol. 1 (1982) and Vol. 2 (1983), National Research Council, National Academy Press, Washington, DC.
3. Holtzman, Jon, et al., Apache Point Observatory: Facilities, Operations and Partnerships, *Position Paper No. 23, for Astro2010: The Astronomy and Astrophysics Decadal Survey*, March 15 2009, p. 8.
4. Peterson, Jim, and Mackie, Glen, A Brief History of the Astrophysical Research Consortium and the Apache Point Observatory, *Journal of Astronomical History and Heritage*, 9, No. 1, June 2006, p. 122.

5. Finkbeiner, Ann, *A Grand and Bold Thing; An Extraordinary New Map of the Universe Ushering in a New Era of Discovery*, Free Press, 2010, p. 23.
6. Gunn, James E., et al., The 2.5 m Telescope of the Sloan Digital Sky Survey, *Astronomical Journal*, 131, 2006, pp. 2332–2359.
7. Margon, Bruce, *Bulletin of the American Astronomical Society*, 25, 1993, p. 678.
8. Gunn, James E., et al., The 2.5 m Telescope of the Sloan Digital Sky Survey, *Astronomical Journal*, 131, 2006, pp. 2332–2359.
9. Knapp, Gillian R., Minding the Heavens: The Sloan Digital Sky Survey, *Sky and Telescope*, August 1997, pp. 40–48.
10. Frieman, Joshua A., The Sloan Digital Sky Survey-II Supernova Survey: Technical Summary, *Astronomical Journal*, 135, 2008, pp. 338–347.
11. Yanny, Brian, et al., SEGUE: A Spectroscopic Survey of 240,000 stars with g=14–20, *Astronomical Journal*, 137, 2009, pp. 4377–4399.
12. Finkbeiner, Ann, *A Grand and Bold Thing; An Extraordinary New Map of the Universe Ushering in a New Era of Discovery*, Free Press, 2010, pp 86–89.

12

Carnegie Southern Observatory (Las Campanas)

An Advisory Committee on Astronomy had been set up by the Carnegie Institution of Washington in 1902 to report on the likely advances to be made in astronomy and how they could be supported.(1) Chaired by Edward C. Pickering, the director of the Harvard College Observatory, the committee recommended that an observatory should be set up in the southern hemisphere as there were, at that time, ten times as many observatories in the northern hemisphere as in the southern.(2) Although William Hussey of the Lick Observatory was given a small grant to undertake a site survey, nothing came of the idea of a southern observatory at that time.

The idea of a large new observatory in the southern hemisphere was resurrected by George Ellery Hale in the 1920s when he suggested that it should include a 100 inch (2.5 m) or 60 inch (1.5 m) telescope. He even suggested that if a new 60 inch could not be afforded then the current 60 inch at the Carnegie Institution's Mount Wilson Observatory should be relocated there. Walter Adams continued to push for a Carnegie southern observatory after he became director of the Mount Wilson Observatory in 1923. In 1931 Henry Norris Russell proposed that the southern observatory should have a 150 inch (3.8 m) telescope whilst three years later Adams suggested that it should be a 120 inch (3 m). But the collapse of the United States stock market in 1929, and the ensuing depression, meant that such proposals had to be put on ice.

The idea of a Carnegie southern observatory was reintroduced in 1962 when Merle Tuve, a director at the Carnegie Institution, asked Ira Bowen, Horace Babcock and Robert Leighton to outline long-range plans for the Institution's Mount Wilson Observatory. In response they produced a list of six major projects, one of which was the southern observatory. Separately Bowen suggested to Caryl Haskins, the president of the Carnegie Institution, that this observatory should include a 140 or 150 inch in order to compete with the telescopes planned in the southern hemisphere for the European Southern Observatory, the Anglo-Australian Observatory and AURA's Cerro Tololo Inter-American Observatory. Then in 1963 the Carnegie

Trustees approved funds for a site survey. It was, at that time, envisaged that the observatory would include a copy of the 200 inch (5.1 m) Hale reflector (see Chapter 1), a 48 inch (1.2 m) Schmidt, plus a few smaller instruments.

Horace Babcock supervised the site testing in New Zealand, Australia, and Chile, starting in 1963, for this proposed Carnegie Southern Observatory (CARSO). Early on he concluded that it should be located in the Chilean Andes as New Zealand proved to be too cloudy and the sites in Chile had better seeing than those in Australia. Eventually the decision was made in 1968 to set up the observatory at an altitude of between 2,290 and 2,520 m (7,500 to 8,250 ft) on the Las Campanas mountain ridge(3) in Chile, about 90 miles (150 km) north of La Serena where the observatory's headquarters would be located. The proposed observatory site on Las Campanas and the surrounding area was purchased from the government of Chile in 1968. Construction was started in the following year on the first telescope, a 1.0 m (40 inch) Boller and Chivens reflector to conduct optical research on celestial X-ray sources.[1] It was dedicated in 1971 and named after the former Carnegie astronomer, Henrietta Swope, who had helped to finance it. In the same year a 24 inch (60 cm) reflector was installed by the University of Toronto at the Las Campanas Observatory to carry out photoelectric photometry and low-dispersion image-tube spectrography.[2]

Returning to the situation in 1963, Babcock had started discussions with AURA about CARSO's plans for a new observatory in Chile as, at that time, AURA were starting to build their Cerro Tololo Inter-American Observatory also in Chile.(4) In the following year the Carnegie Institution approached the Ford Foundation to see if they were prepared to help pay for a 5 m telescope at Las Campanas.[3] This apparently straightforward request involved a number of complex issues including the question of how would astronomers who were not affiliated with Carnegie gain access to their 5 m southern telescope if the Ford Foundation helped to pay for it.[4] The discussions about a possible Ford contribution of around $20 million continued until 1966 when McGeorge Bundy became the Ford Foundation's new president. Shortly after taking office Bundy consulted Leo Goldberg and other senior American astronomers about the proposed Carnegie telescope. As a result of their reaction he decided to put the suggested grant on hold. Goldberg, in particular, had been surprised to learn of the proposed telescope only shortly before the Bundy enquiry, and a number of astronomers had nor heard about it at all.

[1] *Sky and Telescope*, December 1969, p. 389.

[2] *Sky and Telescope*, March 1971, p. 145.

[3] The complex saga of the Ford Foundation grant is discussed in Frank K. Edmondson, *AURA and its US National Observatories*, Cambridge University Press, 1997, pp. 178–195 and p. 219.

[4] Access to the Mount Wilson and Palomar observatories, and particularly to the 5 m on Mount Palomar, was generally restricted to astronomers from affiliated organisations. But this would not be acceptable if Ford were to help to pay for this new 5 m.

Consequently they were taken by surprise at what was proposed and didn't like the sound of it. It seemed wrong to them that the same restricted list of astronomers who had access to the largest telescope in the northern hemisphere (the 5 m on Mount Palomar) should have access to the largest in the southern hemisphere also. Consequently there was a great deal of ill-feeling against the Carnegie proposal. Eventually Bundy decided, instead, to provide a $5 million grant towards AURA's 4 m on Cerro Tololo (see Section 2.6), as all American astronomers would have access to that.

12.1 Irénée du Pont Telescope

The Carnegie Institution, having being unable to raise additional funding for their 5 m, decided to build a 2.5 m (100 inch) instead, with possibly a larger telescope later. Finance of this optical/infrared 2.5 m, which had an exceptionally wide field of view of 2.1° diameter, was provided by Mr and Mrs Crawford Greenewalt supplemented by the Carnegie Institution.(5) Ira S. Bowen and Arthur H. Vaughan Jr, undertook its optical design (6,7) correcting astigmatism with a 74 cm (29 inch) diameter Gascoigne correcting lens, whilst bending the photographic plates slightly to eliminate the need for a field flattener. The telescope had an f/3.0 primary mirror of fused silica and a 95 cm (37 inch) secondary of ultra-low-expansion quartz, both supplied by Corning, whilst the floor of the observatory had a chilling system to improve internal seeing. The telescope was completed in 1976, and named the Irénée du Pont telescope after Mrs Greenewalt's father. Owned by the Carnegie Institution, Las Campanas Observatory had in 1969 become part of the Hale Observatories which included Mount Wilson Observatory and Palomar Observatory in California. Then about ten years later the Hale Observatories were dissolved and the Carnegie Institution created the Mount Wilson and Las Campanas Observatories instead.

This new organisation did not last very long as in 1985 the Carnegie Institution withdrew its financial support from the Mount Wilson Observatory and decided to concentrate their financial resources in astronomy on the Las Campanas Observatory. Then in the following year the Carnegie Institution, plus the University of Arizona and the Johns Hopkins University, agreed to build an 8 m telescope at the Las Campanas Observatory.(8) Johns Hopkins were forced to withdraw from this so-called Magellan Project in 1991 for financial reasons, but the other two partners decided to proceed with a 6.5 m, *f*/1.25 telescope instead. This telescope, which came to be called the Walter Baade Telescope, had a spun cast, borosilicate glass primary mirror produced in 1994 by the University of Arizona's Steward Observatory's Mirror Laboratory. In the meantime construction of the telescope had started in 1993 on an isolated peak called Cerro

Manqui at the Las Campanas Observatory, about 100 m (330 ft) higher than the du Pont Telescope. The primary mirror arrived at the observatory in 1999, where it was coated with aluminium. First light for the telescope occurred in September 2000.

Harvard University had, in the meantime, joined the Magellan Project in 1995, and the University of Michigan and the Massachusetts Institute of Technology joined in the following year. This enabled the construction of a twin telescope at Las Campanas about 60 m from the first. First light occurred for this second telescope, now named after Landon T. Clay, a Harvard graduate and Boston philanthropist, two years after that of the first.

12.2 Giant Magellan Telescope

The University of Arizona had been involved with the Smithsonian Astrophysical Observatory in building the Multiple Mirror Telescope (MMT) in the 1970s which consisted of six 1.8 m telescopes on a common mount (see Chapter 10). But problems with phasing the six telescopes resulted in them being replaced by a single 6.5 m instrument in the 1990s. The mirror for this new telescope was produced by Roger Angel and the University of Arizona's Steward Observatory's Mirror Lab (see Section 4.1) using their spinning furnace in 1992. Since then the Mirror Lab had produced the 6.5 m mirrors for the two Magellan telescopes and the two 8.4 m blanks for the Large Binocular Telescope (LBT).

No sooner had the two Magellan telescopes been built than Roger Angel started to plan for an even bigger telescope. By 2003 he and his team had produced an outline design of a Gregorian reflector with a segmented primary mirror.(9) But instead of a large number of modest sized hexagonal mirror segments as in the Keck Telescope (see Section 4.2) this segmented mirror consisted of six identical 8.4 m off-axis circular mirrors surrounding a central 8.4 m circular mirror (see Figure 12.1).[5] Naturally Angel expected these seven 8.4 m mirrors[6], of what came to be called the Giant Magellan Telescope (GMT), to be produced by the University of Arizona's Mirror Lab using their spinning furnace, following on from the production of the two 8.4 m mirrors for the LBT.

The selected aplanatic Gregorian design of the GMT had two key features. Firstly the secondary mirror could be optically conjugate to a position about 160 m above the ground, enabling excellent ground layer adaptive optics correction over

[5] Initially, the primary mirror had consisted of a closely packed array with a central hexagonal segment and six partial hexagons. But this was changed to the present design with circular segments because it maximised the angular resolution and simplified the manufacture. (See *Manufacture of 8.4 m off-axis segments: a 1/5 scale demonstration*, H. H. Martin, et al., GMT 5494–8 accessed via GMT website.)

[6] Plus a spare off-axis mirror to enable the telescope to continue operation when each of the six off-axis mirrors is taken off the telescope, one at a time, for recoating.

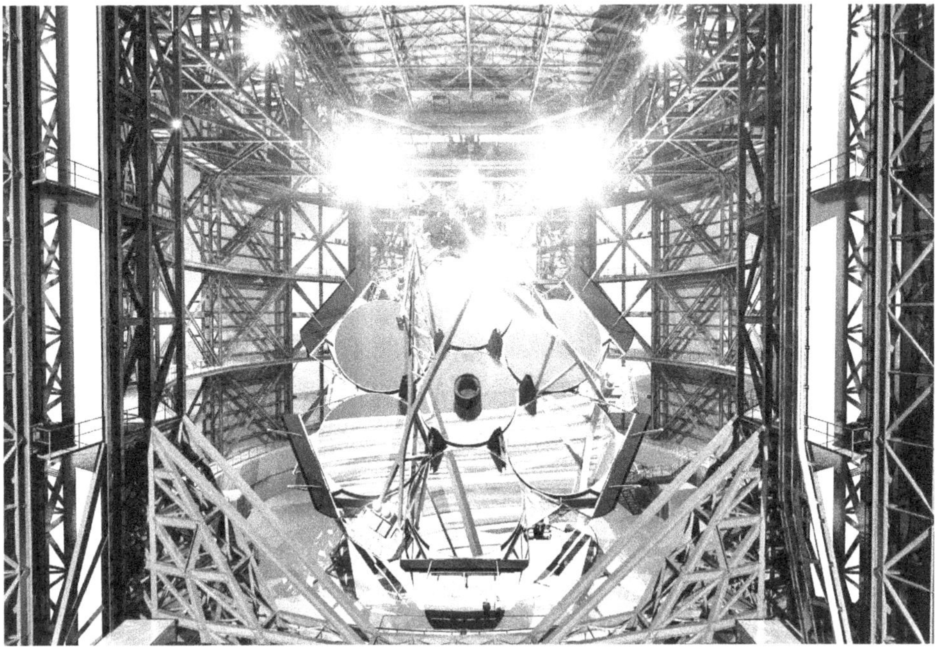

Figure 12.1. An artist's impression of the Giant Magellan Telescope showing its primary mirror array of seven 8.4 m mirrors. (Courtesy Giant Magellan Telescope – GMTO Corporation.)

a large field of view.[7] And secondly the design allowed the optimisation of an adaptive optics system with an artificial star at prime focus. The biggest challenge of the proposed design was to produce, test and align the six highly aspheric off-axis optics systems. In addition the radius of curvature of each of the individual 8.4 m mirrors had to be closely matched to high accuracy.

The initial GMT consortium consisted of the one that had built the two Magellan telescopes plus the University of Texas at Austin, Texas A & M University, and the Smithsonian Astrophysical Observatory. Later the Australian National University joined the Magellan project, followed by Astronomy Australia Ltd., the Korean Astronomy and Space Science Institute, the University of Chicago and the most recent new member the University of São Paulo in 2014. In the meantime the University of Michigan and MIT of the Magellan consortium had pulled out.

The maximum outer diameter of the telescope's primary mirror was[8] 25.4 m which produced a collecting area equivalent to a filled aperture of 21.9 m diameter. The focal length of the primary mirror was 18.0 m giving a focal ratio of f/0.7 for

[7] *Giant Magellan Telescope Project Overview*, June 2 2005, p. 3, accessed via www.gmto.org.

[8] The subsequent text outlines the design of the GMT as at 2012. Whether this will still be the design when the telescope is built remains to be seen. Nevertheless, it gives a good idea of how the design of a modern very large telescope has integrated and modernised many of the design concepts described earlier in this book.

the primary mirror assembly, whilst the focal length of the whole optical system was 207.6 m giving a focal ratio of f/8.2. The Gregorian focus was 5.5 m below the vertex of the primary mirror and the telescope's unvignetted field of view was 20 arcmin. The telescope was designed to operate in both the visible and infrared wavebands between 320 nm and 25 microns.

The design of the secondary mirror, which was adaptive, consisted of seven 2 mm thick Zerodur segments, which were deformable, conjugated 1:1 with the seven primary mirrors. This arrangement greatly simplified the alignment of the telescope optics, allowing slight errors in the alignment of the primary mirror segments to be corrected by the less massive secondary mirror segments. The design of the individual segments of the secondary mirror was a natural development of that of the MMT and LBT adaptive secondary mirrors (see Section 5.2). The seven secondary mirror segments of the GMT were each supported by 672 electromagnetic actuators attached to a rigid reference plate, with magnets glued to the face sheet.

There were three basic modes of adaptive optics foreseen for the GMT namely Natural Guide Star Adaptive Optics (NGSAO), Laser Tomography Adaptive Optics (LTAO), and Ground Layer Adaptive Optics (GLAO). The NGSAO used a single guide star and wavefront sensor to provide diffraction-limited imaging over a field of view limited by atmospheric isoplanatism centred on the guide star. It was designed to operate over wavelengths from 0.9 to 25 microns, The LTAO system used a ~1 arcmin diameter constellation of six laser guide stars, plus one natural guide star,[9] to provide diffraction-limited imaging over wavelengths from 0.9 to 25 microns. The GLAO system was designed to operate down to about 0.8 microns over a field of view of up to 10 arcmin, taking advantage of the close correspondence between the height of the turbulent boundary layer and the conjugate height of the Gregorian adaptive secondary mirror. The system used natural guide stars for the wider fields of view and laser guide stars for more limited fields of view when natural guide stars were not available.

The GMT also had a wide-field corrector and atmospheric dispersion compensator (ADC) to cover the full 20 arcmin field of view for multi-object spectroscopy in the visible. The corrector/ADC would be removed for infrared/adaptive optics observations. In addition a second secondary mirror, called the Fast-steering Secondary Mirror (FSM), was designed in addition to the Adaptive Secondary Mirror (ASM). The FSM was to be used during telescope commissioning and when the ASM was taken off the telescope for servicing.

There were no Nasmyth platforms, but there was a 9 m diameter rotating platform, called the Gregorian Instrument Rotator (GIR), below the primary mirror to

[9] To measure tip-tilt and other corrections.

carry scientific instruments.(10) It rotated about the telescope optical axis to compensate for field rotation caused by the telescope's altazimuth tracking. Scientific instruments could be mounted at the Direct Gregorian (DG) focus inside the GIR and at the so-called Folded Ports on the top surface. A deployable steerable mirror, called the tertiary mirror, directed the telescope beam to instruments at the Folded Ports. Four bays were provided in the GIR to take instruments each weighing up to 11.5 tons. The top surface of the GIR, which was 2 m below the bottom of the primary mirror cells, was used for both scientific instruments, with a maximum weight of 6.5 tons each, and adaptive optics units. A Gravity-Invariant Station (GIS) was also provided on the top surface of the azimuth disc to take a single instrument weighing up to 20 tons.(11)

A number of potential sites were tested at the Las Campanas Observatory over the period from 2005 to 2008.(12) They ranged in height from the Saddle near the duPont Telescope at 2308 m to Las Campanas Peak at 2551 m. The latter was eventually chosen, and site levelling was begun in March 2012.

The pacing item on the GMT was expected to be the production of the seven 8.4 m mirrors by the University of Arizona's Mirror Lab. So the first 8.4 m honeycomb mirror blank was successfully cast in July 2005 long before the project had enough funds to complete the telescope.[10] The second mirror blank was cast in January 2012 and the third in August 2013. It was expected that each mirror would take a further 3.5 years to polish and complete once the production pipeline was well established. Polishing of the first mirror blank, which took seven years, was completed in October 2012.(13)

The instrument development programme was essentially started in 2010 when funding was provided for six teams (out of nine) to produce conceptual designs of their instruments listed in Table 12.1 below. Although MANIFEST, which is included in the table, was not really an instrument, it was also funded to demonstrate the feasibility of the proposed fibre positioning scheme, and to ensure that its three host instruments (G-C-LEF, GMACS, NIRMOS) would be designed to accommodate its feed. On conclusion of the studies, in 2012 the GMTO project selected the first four instruments in the table for further study as potential first generation instruments. If finally selected for development, they were expected to be available for the telescope sequentially in the order shown in the table (from top to bottom).

In June 2015, the Giant Magellan Telescope Organization announced that its partner organisations had committed over $500 million to allow construction of the telescope to begin. It was expected, at that time, that the telescope would be ready with three of its primary mirrors attached in 2021, and with all seven primary mirrors about three years later.

[10] Even by 2012 firm commitments had been made for only 45% of the funds required to build the telescope.

Table 12.1. *Candidate First Generation Instruments for the GMT*

Instrument	Function	Wavelength Range (microns)	Spectral Resolving Power	Field of View	Location
G-CLEF	Optical High Resolution Spectrometer	0.35–0.95	25K – 120K	Fibres with diameters of 0.7 and 1.2 arcsec	Gravity Invariant Station
GMTIFS	NIR AO-fed IFU/Imager	0.9–2.5	5K, 10K	10/400 $arcsec^2$	Folded Port
GMACS	Optical Multi-Object Spectrometer	0.36–1.0	1K – 5K	40–80 $arcmin^2$	Direct Gregorian
GMTNIRS	Near-IR AO-fed High Resolution Spectrometer	1.2–5.0	50K – 100K	Single Object	Folded Port
NIRMOS	NearIR Multi-Object Spectrometer	1.0–2.5	3K – 5K	42 $arcmin^2$	
TIGER	Mid-IR AO-fed Imager and Spectrometer	1.5–14	300	0.25 $arcmin^2$	
MANIFEST	Facility Robotic Fiber Feed	0.36–1.1		300 $arcmin^2$	Direct Gregorian

References

1. Sandage, Allan, *Centennial History of the Carnegie Institution of Washington, Volume 1 The Mount Wilson Observatory*, Cambridge University Press, 2004, pp. 31–42.
2. Edmondson, Frank K., *AURA and its US National Observatories*, Cambridge University Press, 1997, pp. 215–223.
3. Irwin, John B., Chile's Mountain Observatories Revisited, *Sky and Telescope*, January 1974, pp. 10–16.
4. McCray, W. Patrick, *Giant Telescopes: Astronomical Ambition and the Promise of Technology*, Harvard University Press, 2004, pp. 76–79.
5. Babcock, Horace W., First Tests of the Irénée du Pont Telescope, *Sky and Telescope*, August 1977, pp. 90–94.
6. Bowen, I. S., and Vaughan, A. H. Jr., The Optical Design of the 40-inch Telescope and of the Irenee DuPont telescope at Las Campanas Observatory, Chile, *Applied Optics*, 12, 1973, pp. 1430–1434.
7. Babcock, Horace, A Biographical Memoir of Ira Sprague Bowen 1898–1973, *Biographical Memoirs of the National Academy of Sciences*, 53, 1982, p. 106.
8. Robinson, Leif J., Monster Telescopes for the 1990's, *Sky and Telescope*, May 1987, pp. 495–496.
9. Zirker, J. B., *An Acre of Glass; A History and Forecast of the Telescope*, The Johns Hopkins University Press, 2005, pp. 269–271.
10. Johns, Matt, et al., *Giant Magellan Telescope – Overview, Proc. SPIE*, 8444–052, 2012.
11. Jacoby, George H., et al., *The Instrument Development and Selection Process for the Giant Magellan Telescope, Proc. SPIE*, 8446–050, 2012.
12. Phillips, Mark and Thomas-Osip, Joanna, *GMT Site Evaluation at LCO*, Presentation, Pasadena, April 2009 (accessed via GMT website).
13. Zimmerman, Robert, How to Build a Giant Telescope, *Sky and Telescope*, March 2014, pp. 24–28.

13

Mount Graham International Optical Observatory

Astronomers at the University of Arizona started planning in the early 1980s for an international observatory on the 10,700 ft (3,250 m) Mount Graham, the highest peak in the Pinaleno range, about 75 miles (120 km) from Tucson, Arizona. Mount Graham was chosen because of its low light pollution, low atmospheric water vapour, ease of access, and excellent seeing which was only a little worse than that on Mauna Kea.[1] The planned telescopes included the 1.8 m Vatican Advanced Technology Telescope (VATT) of the Vatican Observatory, a submillimetre telescope (later called the Heinrich Hertz Submillimeter Telescope – see Section 23.6), and an optical-infrared binocular telescope, then called the Columbus Project,[2] which consisted of two 8 m primary mirrors on a common mount.

For a time Mount Graham was also considered as a possible site for the National New Technology Telescope (NNTT, see Section 3.2). This was strongly advocated by the University of Arizona as they were hoping to supply its mirror. But Mauna Kea was eventually chosen in 1987 as its location after three years of site surveys, with Mount Graham as a backup.

13.1 Vatican Advanced Technology Telescope (VATT)

In the meantime the University of Arizona's Mirror Laboratory had spun-cast the 1.8 m, f/1.0 primary mirror blank for the VATT in 1985.[3] In the same year construction work was also expected to begin on the VATT's new observatory building

[1] Median seeing on Mount Graham was measured as 0.55 to 0.63 arcsec, depending on the exact location on the peak, compared with 0.45 arcsec for Mauna Kea, see *Site Testing – The Early Days* on mgio.arizona.edu/mount-graham-site-testing-era.

[2] The Columbus project was so called as it was hoped to have the telescope completed in 1992, the 500th anniversary of Columbus voyage in which he discovered the New World.

[3] This very fast (f/1.0) mirror was then polished by the Mirror Lab using their new stressed-lap polishing technique, see Section 4.1.

on Mount Graham, but environmental concerns over the mountain's red squirrels and its other sensitive habitats led to a series of delays.

The prospect of being able to build an astronomical observatory on Mount Graham received a major setback in June 1987 when the Mount Graham red squirrel was declared an endangered species.[4] But it appeared as though the environmental impasse had been broken in October 1988 when university-sponsored legislation[5] was passed by the United States Congress to allow the three planned telescopes to be built on the mountain without Forest Service approval. This legislation allowed the university to build four more telescopes, if the first three were found to have had a minimal impact on the red squirrel.[6] But in March 1990, a Federal judge ordered a temporary halt to the observatory construction. Although this temporary injunction was overturned by the US Court of Appeals two months later, other potential legal problems had also arisen. The Vatican Observatory even began looking for an alternative site to locate their VATT telescope. But eventually the legal problems associated with building the VATT on Mount Graham were solved, and the telescope, which was designed to operate in the visible and infrared, was dedicated in September 1993. It achieved first light in the following year with the secondary mirror in place. Two major donors had subscribed to its costs, Fred A. Lennon and Thomas J. Bannan. As a result the telescope was named the Alice P. Lennon Telescope, after Fred Lennon's wife, and its associated astrophysics facility was named after Thomas Bannan.

13.2 Columbus Project or Large Binocular Telescope (LBT)

The telescope of the Columbus Project was to have two 8 m primary mirrors (see Figure 13.1) with their centres separated by 14 m on a common mount. Such a design would not only have the light-gathering power of a 11.3 m telescope, but it could also be used as an interferometer with the resolving power of a 23 m. The conceptual design was initially produced by a joint University of Arizona and Ohio State University team, later augmented by the University of Chicago and Italy's Arcetri Astrophysical Observatory. The estimated cost of about $50 million(1) was to be divided equally between the four members of the consortium.(2)

[4] The complex saga of the Mount Graham red squirrel and the astronomical observatory are covered in Conrad A. Istock and Robert S. Hoffmann (eds.), *Storm over a Mountain Island; Conservation Biology and the Mt. Graham Affair*, The University of Arizona Press, 1995, and L. Sage, A Brief History of the Controversy Surrounding the Mount Graham International Observatory, in André Heck (ed.), *Organizations and Strategies in Astronomy, Volume 4*, Kluwer Academic Publishers, 2003, pp. 75–91.

[5] The Mount Graham observatory was added as a rider to the Arizona-Idaho Conservation Act (1988) that was already being considered by Congress.

[6] *Sky and Telescope*, February 1989, p. 130.

Figure 13.1. Large Binocular Telescope showing its two 8.4 m diameter primary mirrors. (Courtesy Enrico Sacchetti and LBTO.)

The feasibility study for the Columbus telescope was completed in 1989, then on September 7 1991 Ohio State University (OSU) pulled out of the project.[7] Amazingly this was the very day that representatives of the OSU were due to attend the official signing signifying their continued support to the project. This withdrawal resulted in the OSU's dean of the college of mathematical and physical sciences resigning his administrative position in protest, as he had only been told of the OSU's withdrawal on the morning of the planned signing ceremony. Likewise Eugene Capriotti, chairman of the OSU's department of astronomy, who had also learnt of the decision on the morning of the signing, stepped down as he would now no longer be involved in the project. This was particularly hard on him as he had been key in getting the joint project off the ground in the first place.

The Research Corporation stepped in to take the place of the OSU in the Columbus telescope consortium in 1992. Then in the same year the four partners decided to proceed with the construction phase of the project, even though the funds available were only sufficient to install one primary mirror. Two years later a contract was signed for the detailed mechanical design of the telescope with a consortium formed by the Italian firms of ADS International of Lecco and European Industrial Engineering of Mestre.

[7] *Sky and Telescope*, April 1992, p. 370.

Meanwhile litigation was still ongoing to try to stop the construction of this Columbus binocular telescope, now called the Large Binocular Telescope (LBT) on Mount Graham. As a result the construction was not started until July 1996, when a Federal judge rejected the last case against it.[8] Shortly afterwards, the first of the two 8.4 m, f/1.14 blanks for the LBT was spun by the University of Arizona's Mirror Laboratory. Unfortunately it transpired that several square metres of its surface were too thin owing to leakage of glass. So two more tons of glass were placed onto the 17-ton blank in April 1997 where it was melted to make the necessary correction. By the summer the telescope's foundation and ring wall had been completed.

In 1997 the LBT Beteiligungsgesellschaft (LBTB), which was a consortium of German institutions led by the Max-Planck-Institut für Astronomie (Heidelberg),[9] joined the project and the Ohio State University returned to the fold providing enough funds to finish the two mirror project. Then in the following year a contract was awarded to Ansaldo Energia of Milan for the fabrication and erection of the main telescope structure in Italy. This was to check that everything was in order before the structure was dismantled and shipped to the USA which was done in 2002. In the meantime the second mirror blank had been produced in the University of Arizona's Mirror Lab with additional tension on the Inconel restraining bands to avoid glass leakage.

The $120 million LBT saw first light in October 2005 with just one mirror installed, and with both mirrors installed just over two years later. By then the consortium required to finance, build and operate the LBT had increased to include numerous institutes in the United States, Italy and Germany.[10] First binocular light was achieved in 2008. Then two years later first fringes were observed using the Large Binocular Telescope Interferometer (LBTI) which had been built by the University of Arizona's Steward Observatory and NASA's Jet Propulsion Laboratory (JPL).

The LBT's adaptive optics system design used a deformable secondary mirror based on the design of the smaller deformable secondary installed on the new MMT in 2003 (see Section 5.2). In the case of the LBT each of its two 91 cm diameter deformable secondary mirrors had a 1.6 mm thick Zerodur reflective shell controlled by 672 computer-controlled actuators. The first system, which was installed on the LBT in 2010, produced a resolution of 0.040 arcsec in H band

8 *Sky and Telescope*, October 1996, p. 15.

9 The other German institutions were the Max-Planck-Institut für Extraterrestrische Physik (Munich), the Max-Planck-Institut für Radioatronomie (Bonn), the Astrophysikalisches Institut Potsdam, and Landessternwarte Heidelberg (Centre for Astronomy, Heidelberg University).

10 The funding split between the partners was: Arizona (25%), consisting of the University of Arizona (Tucson), Arizona State University (Tempe), and Northern Arizona University (Flagstaff).
Instituto Nazionale di Astrofisica (INAF) in Italy (25%), which represented all Italian observatories and astronomical research institutes. Initially, the Arcetri Astrophysical Observatory in Florence had led a consortium of Italian astronomical observatories, a role later undertaken by INAF.
LBT Beteiligungsgesellschaft (LBTB) in Germany (25%).
Research Corporation (12.5%), which had also facilitated the involvement of the University of Notre Dame, the University of Minnesota and the University of Virginia.
Ohio State University (12.5%).

(1.65 microns) with a natural guide star system.(3) The second deformable secondary was delivered to the mountain in 2012.

The LBT scientific instruments were of basically two types, those that were located on one of the two telescopes and those which took light from both telescopes. Both of the telescopes was to be equipped with three single beam instruments, namely a large binocular camera (LBC), an optical spectrograph (Multi-Object Double Spectrograph or MODS) and a near-infrared interferometer (LUCI, previously called LUCIFER[11]). In addition the beams of the two telescopes were combined in LINC-NIRVANA[12] to provide interferometric imaging in the near infrared, and in the LBTI, an optical to near-infrared interferometer.

The LBCs were a pair of wide field (23 × 25 arcmin) mosaic CCD cameras built by Rome observatory and provided by the Italian Instituto Nazionale di Astrofisica (INAF).[13] The two cameras, which operated in different wavebands, were mounted at the prime focus of each of the two telescopes. One of the cameras, LBC Blue, was optimised for observations from about 350 to 650 nm, whilst LBC Red covered the range from about 550 nm to 1 micron. The first LBC (LBC Blue) was, in 2005, the first instrument to be installed on the LBT, whilst LBC Red was delivered in November 2007 to allow first binocular light in the following year.

MODS 1 and 2 were a pair of dual-beam blue-red optimised long-slit spectrographs plus imagers operating over wavelengths from 320 nm to 1.1 microns. There was one MODS for each of the direct f/15 Gregorian foci. They incorporated multiple slit masks for multi-object spectroscopy over a 6 arcmin field of view giving spectral resolutions up to 2300. The two MODS instruments, which were capable of running in single mirror or binocular mode, were provided by Ohio State University. MODS 1 was successfully installed on the LBT in 2010 but MODS 2 was not installed until the spring of 2014. This was some four years later than expected.

LUCI were a pair of near-infrared imager and spectroscopic instruments operating from 0.9 to 2.5 microns at the bent Gregorian foci. Both LUCI imagers had two cameras and could observe in both seeing and diffraction limited modes with adaptive optics. The pair of LUCI instruments were provided by a German consortium led by Landessternwarte Heidelberg (Centre for Astronomy, Heidelberg University). First light for LUCI 1 was achieved in 2008, with regular observing starting at the end of the following year. But LUCI 2 did not arrive until 2013, being like MOD 2 a number of years behind schedule. Both the LUCI and MOD suppliers found it difficult enough to provide one instrument each, let alone two.

[11] LUCIFER is an acronym of '*L*BT Near Infrared Spectroscopic *U*tility with *C*amera and *I*ntegral *F*ield Unit for *E*xtragalactic *R*esearch'.

[12] LINC is an acronym of '*L*BT *IN*terferometric *C*amera', and NIRVANA is an acronym of '*N*ear-*IR*/*V*isible *A*daptive i*N*terferometer for *A*stronomy'.

[13] As mentioned in an earlier footnote INAF represented all Italian observatories and astronomical research institutes.

LINC-NIRVANA, which was built by LBTB in Germany, in conjunction with INAF in Italy, saw first light in 2013. The LBT's adaptive secondary mirror provided correction of ground layer turbulence, whilst LINC-NIRVANA's multi-conjugate adaptive optics (MCAO) system provided correction of higher level turbulence. As mentioned above LBTI, which was funded by NASA, saw first light in 2010.

Although the weather is usually good for observing on Mount Graham, in January 2010 a snow storm delivered 1.3 m of snow, and in the following year an exceptionally low temperature of – 27.5°C was recorded. Two wild fires, one in April 1996 due to human activity and the other in July 2004 due to lightning strikes, also threatened the LBT. Fortunately, even though the fires came to within 200 m and 40 m, respectively, of the LBT, neither came close enough to create any damage. A large number of conifers surrounding the observatory had previously died due to drought and insect infestations which had exacerbated the fire problem on the mountain. It is often forgotten that mountain observatories, even in the most advantageous places astronomically, have to withstand the elements, normally in the form of high winds, low temperatures or snow. But whilst the highest observatories were often above the tree line, those that were not were also often threatened by wild fires.

The LBT project had been very loosely organised with no strong central management control.(4) As a result the project partners had each been in charge of building their own scientific instruments with little input from the project as a whole. Consequently the LBT project had to accept whatever delivery schedule each of the partners were prepared to sign up to. This arrangement was highly unsatisfactory so in 2013 the LBT partnership decided to appoint Christian Veillet as project director with the appropriate authority to manage the project. This will hopefully improve the scientific productivity of the LBT which has been rather poor since first light.[14]

References

1. Robinson, Leif J., Monster Telescopes for the 1990's, *Sky and Telescope*, May 1987, pp. 495–496.
2. Strittmatter, P. A., *Columbus Project Overview*, in Ulrich, Marie-Helene, (ed.), *Proceedings of a ESO Conference on Very Large Telescopes and Their Instrumentation, held in Garching, March 21–24, 1988*, European Southern Observatory, 1988, pp. 29–46.
3. Esposito, S., et al., Large Binocular Telescope Adaptive Optics System: New Achievements and Perspectives in Adaptive Optics, *Proc. SPIE* 8149, September 2011.
4. Zimmerman, Robert, Flawed Giants, *Sky and Telescope*, January 2015, pp. 60–67.

[14] Christian Veillet, *Current Status of LBTO*, abell.as.arizona.edu/~lbtsci/UM2014/Presentations/S1/2014UMstatus.pdf, and Alexandra Witze, TeethingTroubles at Huge Telescope, *Nature News*, July 12 2013.

14

Modern Optical Interferometers

14.1 Mount Wilson

Michelson had designed a stellar interferometer for the Mount Wilson Observatory in 1920 that consisted of a 6 m (20 ft) long steel girder and two 15 cm (6 inch) movable mirrors which he mounted across the open end of the 2.5 m (100 inch) telescope.(1) This enabled Michelson and Pease to measure the diameter of a number of red giants. However, little more interferometric work was done on Mount Wilson until 1979 when Michael Shao and David Staelin of the Massachusetts Institute of Technology designed a two telescope, white light, 1.5 m (5 ft) baseline interferometer. It was the first interferometer to compensate for atmospheric turbulence by making continuous phase and amplitude measurements with a 4 ms integration time. The Mark I was succeeded by the 3.1 m Mark II interferometer which was used as a technology testbed for astrometric measurements from 1982 to 1984. It incorporated a high speed, ultra high accuracy, laser controlled optical delay line, and it was also used to develop an operational procedure for rapid switching between stars. The Mark II was, in turn, followed by the Mark III(2) from 1986 to 1993 which had a maximum baseline of 32 m and operated in two spectral bands from about 400 to 600 nm and from about 600 to 900 nm. Like the Mark II, the Mark III was designed to test out new techniques and technologies. Its aim was to be reliable and easy to operate, whilst being capable of extremely accurate astronomical measurements of stellar positions and diameters. Lessons learnt in operating the Mark III were applied to the design of the US Naval Observatory's NPOI (Navy Prototype Optical Interferometer, see Section 14.6) located in the mountains near Flagstaff, Arizona which became operational in 1996.

Berkeley Infrared Spatial Interferometer (ISI)

Instead of using the Michelson homodyne or direct detection system of the above instruments, Charles Townes, Mike Johnson and Al Betz of the University of

California (UC) at Berkeley(3) used a heterodyne system in 1974 to produce the first mid-infrared interferometer based on techniques used in radio astronomy. In the Michelson system visible starlight received by two apertures is mixed and the interference pattern observed, whereas in the heterodyne system of Townes and colleagues the mid-infrared starlight received by each aperture is first mixed with a carbon dioxide laser local oscillator. The signal is then downconverted to microwave frequencies, followed by path-length matching and fringe detection in a correlator. The prototype system used the two 81 cm McMath auxiliary telescopes on Kitt Peak with a baseline of 5.5 m and operated at 11.1 microns in the infrared.(4)

Townes' team used their experience with this prototype system to design the UC at Berkeley's Infrared Spatial Interferometer (ISI) at their Space Sciences Laboratory. In 1988 they moved the system to Mount Wilson where it was situated about 100 m from the 2.5 m (100 inch) telescope.(5) The moveable part of this ISI system was installed on two trailers, each of which had a 2 m diameter altazimuth-mounted siderostat which fed a fixed 1.65 m parabolic mirror made of Zerodur.(6) In 1988 first fringes were achieved with a baseline of 4 m with the carbon star IRC + 10216.(7) Subsequently the ISI was operated in the atmospheric window from 9 to 12 microns with baselines of up to 32 m. Then in 1999 the baseline was increased to 70 m, improving the resolving power of the system to about 3 mas (milliarcsec) at 11 microns, with a third telescope being added a little later increasing the maximum baseline to 85 m.

Center for High Angular Resolution Astronomy (CHARA)

In the meantime Georgia State University had set up their Center for High Angular Resolution Astronomy (CHARA) in 1984 with the aim of producing a major new optical interferometer capable of achieving unprecedented angular resolution. In the following year the National Science Foundation (NSF) provided an initial grant to enable the project to get started. A number of possible sites were considered for building this CHARA array including Mount Wilson, Mount Palomar and Kitt Peak. In the event Mount Wilson was chosen because of its excellent astronomical seeing, its large number of clear nights, and its good existing infrastructure. Light pollution was expected to be negligible because of the instrument's extremely small field of view.(8)

Over the next ten years the NSF's total contribution to the project exceeded $6 million, with Georgia State University providing matching funds. The early interferometer design had consisted of an array of five 1 m diameter telescopes, but in 1998 the W. M. Keck Foundation awarded the university a grant of $1.5 million to add a sixth telescope and to improve the beam combining instrumentation. In the same year a sum of over $500,000 was donated by the David and Lucile Packard Foundation.

As built the CHARA interferometer consisted of six 1 m telescopes arranged in three pairs on Y-shaped arms allowing telescope separations of from 34 m to 331 m. The primary and secondary optics of each telescope were manufactured as matched sets by LOMO (Leningrad Optical Mechanical Association) in St. Petersburg, Russia using low-CTE (Coefficient of Thermal Expansion) Sitall and Zerodur, respectively. The telescopes' altazimuth mounts were exceptionally stiff and massive for interferometric stability. In addition they incorporated an Invar metering truss to provide a temperature-insensitive focus which enabled the telescopes to stay in focus for up to a year. All enclosure and telescope functions were operable remotely, and most of the telescope and enclosure control electronics were located in adjacent cement block bunkers to avoid heat production within the enclosures. The delay lines, beam management and beam combination subsystems were contained in an L-shaped building with a longest dimension of 94 m. Five different vacuum delay lines were available in discrete lengths up to 143.1 m. These were supplemented by continuous delay lines in air. The array operated over the visible and near-infrared wavebands up to 2.2 microns. Construction was begun in 1996 and first fringes obtained with two telescopes three years later. It became fully operational in 2004.

In addition to operating as a scientific instrument, the CHARA Array was also used as a testbed for new technologies. As a result the Paris Observatory installed their FLUOR[1] fibre optics beam combiner at CHARA. In addition the University of Michigan, the Côte d'Azur Observatory and the University of Sydney also tested beam combiner equipment at CHARA.

The CHARA Array produced the first image of a single main sequence star, ALTAIR, in 2007, showing its distorted shape due to its rapid rotation.(9) It also showed that it was hotter at the poles than at the equator. This image in the near infrared had a resolution of better than 1 mas. At about the same time the CHARA Array was the first to measure the diameter of an exoplanet directly. The planet, which was orbiting its parent star[2] in 2.2 days, was found to have a diameter of 0.38 mas, which equated to about 185,000 km (30% larger than Jupiter) at the star's known distance from earth. The CHARA Array also provided the first detection of gravity darkening in a rapidly rotating star, Regulus, that is not a member of an eclipsing binary system. Then in 2009 the Array helped to unravel the configuration of ε Aurigae when it detected a dark cloud partially occulting the visible source.(10)

14.2 Interféromètre à 2 Télescopes (I2T)

Antoine Labeyrie had been, in 1974, the first to observe stellar sources using a two telescope Michelson interferometer in the optical band. His two 25 cm (10 inch)

[1] FLUOR is an acronym of 'Fiber Linked Unit for Optical Recombination'.
[2] Its parent star was HD 189733.

diameter telescopes with servo-driven altazimuth mounts were separated by 12 m (40 ft). To maintain equal path lengths in the two interfering beams Labeyrie used a movable beam-combiner table and observed the fringes both visually and with a photon-counting television camera.(11) After initial experiments at the Nice observatory in France the interferometer was moved to CERGA[3] on the Plateau de Calern in the Alpes Maritimes where it had a variable baseline of up to 20 m.(12) Over the years the baseline of this I2T (Interféromètre à 2 Télescopes) system was progressively increased to 137 m (450 ft). It was used to measure the angular sizes of a number of stars and to study the orbit of the binary star Capella. The discs of both stars in the binary were resolved with baselines of from 15 to 18 metres to yield diameters of 7 and 4 mas.(13)

The I2T was replaced in 1985 by the GI2T (Grand I2T) which consisted of two 1.5 m (60 inch) Cassegrain-Coudé telescopes operating in the visible and infrared wavebands. The telescopes were each mounted in spherical concrete structures with a baseline that could be extended up to 65 m. Fringe tracking was undertaken visually, but in 1994 an automatic fringe tracker was made available. Subsequently the central beam combiner was redesigned in the REGAIN project to improve the quality and quantity of data. This GI2T/REGAIN interferometer, as it was called, saw first light in 1999. It was closed down seven years later.

14.3 Cambridge Optical Aperture Synthesis Telescope (COAST)

The COAST (Cambridge Optical Aperture Synthesis Telescope) interferometer had been designed in 1986 in Cambridge, UK to provide very high resolution images (not just fringes) of a wide variety of astronomical sources down to a red magnitude of 10 using Michelson interferometry. These sources included the envelopes of pre-main sequence stars, circumstellar shells, compact planetary nebulae and close binary systems. COAST initially consisted of a coherent array of four 40 cm (16 inch) diameter Cassegrain telescopes, each fed by a 50 cm siderostat, operating over the red/near infrared waveband from 600 to 2,300 nm. COAST's beam combining system is shown in Figure 14.1. A fifth telescope was added in 1998 and the maximum baseline of the array was eventually extended to 100 m.

The first stellar fringes were obtained in 1991 and in September 1995 COAST produced the first synthesis image ever produced from separated optical telescopes. This was of the 55 mas spectroscopic binary Capella which was imaged at 830 nm. Just three telescopes were used stopped down to 14 cm (5.5 inches) to match the local seeing conditions. The array was in a close-packed arrangement giving a maximum baseline of 6.1 m (20 ft).(14) The COAST images clearly showed the

[3] CERGA is an acronym of 'Centre d'Etudes et de Recherches Géodynamiques at Astronomiques'.

Figure 14.1. The interior of COAST's beam combining laboratory in which the light paths from each of its telescopes are made to be exactly the same. (Courtesy Cmglee via Wikimedia Commons.)

movement of the two stars in the binary over the 15 days between the two sets of observations. COAST imaged Betelgeuse shortly afterwards, showing strong limb darkening, and in 1998 it detected a periodic modulation in the size of the Mira-type star R Leonis.

The COAST team used their experience to assist the Magdalena Ridge Observatory Consortium to design the Magdalena Ridge Observatory Interferometer (MROI) which is currently (in 2015) being built at an altitude of about 3,200 m (10,500 ft) in the Magdalena mountains of New Mexico. COAST is now mainly used as a testbed for subsystems and components intended for MROI and for the VLTI interferometer of the European Southern Observatory (see Section 7.2).

14.4 Infrared/Optical Telescope Array (IOTA)

The IOTA (Infrared/Optical Telescope Array) Michelson interferometer had been built by a consortium of American academic institutions[4] in the early 1990s at the Smithsonian Astrophysical Observatory's Whipple Observatory on Mount

[4] Harvard University, MIT Lincoln Laboratory, Smithsonian Astrophysical Observatory, University of Massachusetts at Amherst and the University of Wyoming.

Hopkins. Its aim had been similar to that of COAST, namely to produce high-resolution images of astronomical objects using an interferometer array. Scheduled operation started in 1995 with its first two 45 cm diameter telescopes in place on an L-shaped track[5] that allowed spacings of between 5 and 38 m. The interferometer was operational in the visible and near infrared wavebands. It used the FLUOR[6] fibre optics beam combiner, developed by the Paris Observatory, with fluoride glass single-mode waveguides for observations at K band (2.0–2.4 microns).

A pair of discrete InSb photovoltaic detectors was initially used for near infrared observations, but these were not very sensitive having a limiting magnitude of only 2 in K band. As a result a new infrared instrument based on Rockwell International's NICMOS3[7] array was developed by the University of Massachusetts at Amherst and the Harvard-Smithsonian Center for Astrophysics. NICMOS3 arrays were used as they were known to have low noise, high quantum efficiency, low dark current and high uniformity. The resulting focal plane array(15) consisted of 256 × 256 HgCdTe integrating detectors organised in four 128 × 128 quadrants. Its magnitude limits were 8.4 in J band (1.1–1.4 microns), 8.5 in H band (1.5–1.8 microns) and 6.5 in K band. The instrument saw first fringes in 1997.

The visible light detector was a thinned back-illuminated SITe 512 × 512 CCD of Scientific Imaging Technologies with improved short wavelength quantum efficiency. A third telescope was incorporated into the array over the period 1998–2000 enabling aperture synthesis to be performed. But seven years later the interferometer was closed for financial reasons.

14.5 Palomar Testbed Interferometer

The Palomar Testbed Interferometer (PTI)(16) (see Figure 14.2) was a long-baseline interferometer operating in the J, H, and K infrared bands. It had been designed in the early 1990s to test interferometric techniques for the Keck Interferometer (see Section 4.2). These included high-sensitivity direct-detection interferometry with array detectors, and dual-star interferometry with phase referencing for narrow-angle astrometry. The PTI built on experience gained with the Mount Wilson Mark III which had been used with active fringe tracking for, amongst other things, wide-angle astrometry in the visible waveband. But the PTI differed from the Mark III in that it operated in the infrared and provided narrow-angle interferometric astrometry using phase referencing. Major development of the PTI was started by the Jet Propulsion Laboratory in 1992 with funding provided by NASA. It was later installed at the Palomar Observatory where first fringes were detected in 1995.

[5] Thirty five metres oriented north-east and fifteen metres south-east.

[6] It demonstrated for the first time the coherent recombination of stellar signals by guided optics.

[7] NICMOS3 is an acronym of 'Near Infrared Camera and Multi-Object Spectrometer 3'.

Figure 14.2. Aerial view of the Y-shaped Palomar Testbed Interferometer which is to the left of the dome of the 5 m telescope. (Courtesy Gerard T. van Belle via Wikimedia Commons.)

The PTI, which used three 40 cm (16 inch) telescopes each fed by a 50 cm siderostat, was operated as a two-element interferometer, combining light from two of its three siderostats at any one time. The maximum separation of 110 m gave an absolute accuracy of 1 mas. But by tracking two stars simultaneously, even with separations of several hundred milliarcseconds, it could achieve relative accuracies of ~20 microarcsec. This phase-referenced mode, which was used, amongst other things, to search for extrasolar planets orbiting stars in binary systems, was developed over the period 2002–2003. It was the basis for the Palomar High-precision Astrometric Search for Exoplanet Systems (PHASES) programme.(17)

In 2001 the PTI measured the distortion of the 3 mas diameter, A-type star, ALTAIR, due to its rapid rotation. This showed that its equatorial diameter was about 14% greater than its polar diameter.(18) The interferometer was closed down in 2008.

14.6 Navy Prototype Optical Interferometer

The Naval Research Laboratory (NRL) and the US Naval Observatory (USNO) began work on two separate optical interferometers in the late 1980s, namely the

NRL Big Optical Array (BOA) and the USNO Astrometric Interferometer.(19) The NRL imaging array was to be funded by the Office of Naval Research (ONR) whilst the USNO astrometric array was to be funded by the Oceanographer of the Navy.(20) But in 1990 it was decided to combine the two arrays into what was then called the Navy Prototype Optical Interferometer (NPOI). Four possible sites were considered for building the astrometric array, but only one of these was large enough to accommodate the larger imaging interferometer.(21) This was at the Lowell Observatory at a height of 2,200 m (7,200 ft) on Anderson Mesa near Flagstaff, Arizona. Consequently two years later work was started on building the NPOI array there as a joint project of NRL and USNO in cooperation with Lowell Observatory.

The design of this new interferometer took into account experience gained with the Mark III system on Mount Wilson. When completed the NPOI imaging subarray was planned to consist of six movable siderostats feeding 12 cm (5 inch) apertures in a Y-shaped array with three 250 m arms allowing separations of up to 437 m. The astrometric subarray was to consist of four fixed siderostats feeding apertures on the same Y-shaped system, with baselines of from 19 to 38 m. Each siderostat in both subarrays was 50 cm (20 inch) in diameter. It fed a 20 cm tip-tilt mirror which in turn reflected starlight through a window into a 12 cm diameter vacuum pipe that conveyed light to the optics laboratory. This was where pathlength compensation systems and beam combiners were located. Both subarrays were planned to use the same receiving system which was expected to cover 450–850 nm in 32 spectral channels.[8]

The first stellar fringes were observed in 1994. Two years later the first phase of the NPOI interferometer became operational when it produced its first stellar images (of the 6.5 mas Mizar A binary) using three of the astrometric siderostats. This was followed later that year by NPOI's first observations of stellar limb darkening. In the same year concrete piers for the three 250 m long arms of the imaging interferometer were also completed and work started on the long delay lines.

The first station beyond the initial three astronometric stations was commissioned in 2001 to give the NPOI a maximum baseline length of 64 m which produced fringe spacings of 2 mas at 600 nm. Then in the following year routine observations were started with six siderostats operating simultaneously (the first interferometer to do so), four being fixed astrometric siderostats and two movable imaging siderostats to give a maximum baseline of 79 m.(22) Since then all six of the movable imaging siderostats have been built but only a limited number of their originally planned 30 stations have been completed.

NPOI's primary goal was the creation of an astrometric catalogue of stars with a precision of about 2 mas over large angles. This would update the Hipparcos

[8] This requirement was later relaxed to 550–850 nm in 16 channels.

spacecraft catalogue as these stars would have moved in the years since their position had been measured by the spacecraft. At present (2015) the best astrometric accuracy achieved over large angles is 5–10 mas.

The NPOI was renamed the Navy Optical Interferometer in 2011. More recently it was renamed the Navy Precision Optical Interferometer, so keeping its original acronym NPOI. There are currently plans to incorporate the four 1.8 m diameter Keck outrigger telescopes into the array.

Note Other interferometers have been covered elsewhere in this book with their parent observatories. For example the Narrabri Intensity Interferometer and Sydney University Stellar Interferometer (SUSI) have been covered in Section 9.4, the Keck Interferometer in Section 4.2 and the European VLT Interferometer in Section 7.2.

References

1. Fehrenbach, Charles, Twentieth-century instrumentation, in Gingerich, Owen, (ed.), *Astrophysics and Twentieth-century Astronomy to 1950; Part A*, Cambridge University Press, 1984, pp. 182–184.
2. Shao, M., et al., The Mark III Stellar Interferometer, *Astronomy and Astrophysics*, 193, 1988, pp. 357–371.
3. Johnson, M. A., Betz, A. L., and Townes, C. H., Results from the 5.5 Meter Infrared Stellar Interferometer, *Bulletin of the American Astronomical Society*, 6, 1974, pp. 450–451.
4. Shao, M., and Colavita, M. M., Long-Baseline Optical and Infrared Stellar Interferometry, *Annual Review of Astronomy and Astrophysics*, 30, 1992, pp. 457–498.
5. Danchi, W. C., Bester, M., and Townes, C. H., The U. C. Berkeley Infrared Heterodyne Interferometer, *ESO Conference Workshop Proceedings* No. 29, Part 2, pp. 867–877.
6. Hale, D. D. S., et al., The Berkeley Infrared Spatial Interferometer: A Heterodyne Stellar Interferometer for the Mid-Infrared, *Astrophysical Journal*, 537, 2000, pp. 998–1012.
7. McCullough, P. R., et al., *Bulletin of the American Astronomical Society*, 20, 1988, p. 1007.
8. ten Brummelaar, T. A., et al., First Results from the CHARA Array. II. A Description of the Instrument, *Astrophysical Journal*, 628, 2005, pp. 453–465.
9. Monnier, John D., et al., Imaging the Surface of Altair, *Science*, 317, 2007, pp. 342–345.
10. Kloppenborg, Brian, et al., Infrared Images of the Transiting Disk in the ε Aurigae System, *Nature*, 464, 2010, pp. 870–872.
11. Labeyrie, Antoine, Interference Fringes Obtained on Vega with Two Optical Telescopes, *Astrophysical Journal*, 196, 1975, pp. L71–L75.
12. Blazit, A., et al., The Angular Diameters of Capella A and B from Two-Telescope Interferometry, *Astrophysical Journal*, 217, 1977, pp. L55–L57.
13. Labeyrie, Antoine, Stellar Interferometry: A Widening Frontier, *Sky and Telescope*, April 1982, pp. 334–338.
14. Baldwin, J. E., The First Images from the Optical Aperture Synthesis Array: Mapping of Capella with COAST at Two Epochs, *Astronomy and Astrophysics*, 306, 1996, pp. L13–L16.

15. Millan-Gabet, R., et al., A NICMOS3 Camera for Fringe Detection at the IOTA Interferometer, *Publications of the Astronomical Society of the Pacific*, 111, 1999, pp. 238–245.
16. Colavita, M. M., The Palomar Testbed Interferometer, *Astrophysical Journal*, 510, 1999, pp. 505–521.
17. Muterspaugh, Matthew W., et al., Scientific Results from High-precision Astrometry at the Palomar Testbed Interferometer in Monnier, John D., et al., (eds.), *Advances in Stellar Interferometry, Proc. SPIE*, 6268, 2006.
18. van Belle, Gerard T., et al., Altair's Oblateness and Rotation Velocity from Long-Baseline Interferometry, *Astrophysical Journal*, 559, 2001, pp. 1155–1164.
19. Simon, R. S., et al., Imaging Optical Interferometry, in Cornwell, T. J., and Perley, R. A., (eds.), *Radio Interferometry: Theory, Techniques and Applications,* IAU Coll. 131, *ASP Conference Series*, 19, 1991, pp. 358–367.
20. Armstrong, J. T., et al., The Navy Prototype Optical Interferometer, *Astrophysical Journal*, 496, 1998, pp. 550–571.
21. Hutter, D. J., et al., Seeing Tests at Four Sites in Support of the NPOI Project, *Astronomical Journal*, 114, 1997, pp. 2822–2833.
22. Armstrong, J. T., et al., The Navy Precision Optical Interferometer (NPOI): An Update, *Journal of Astronomical Instrumentation*, 2, Issue 2, 2013, pp. 1–8.

15

Solar Observatories

15.1 Climax Observatory and the Sacramento Peak Solar Observatory

Donald Menzel of Harvard College Observatory had been instrumental in establishing the solar observatory at Climax, Colorado, or more properly the Climax observing station of Harvard College Observatory, in 1940. Unfortunately it soon became clear to Menzel and Walter Roberts, a former student of Menzel and the station's superintendent, that this Climax station suffered from long periods of cloudiness, especially during the winter. Consequently they concluded that a second solar observatory should be built as soon as possible after the Second World War had ended.(1)

At about the same time H. H. (Hap) Arnold, the commanding general of the U S Army Air Forces (AAF),[1] had asked his scientific advisor Theodore Von Karman to draw up a long-range research and development plan for the AAF after the war. Arnold had been particularly interested for some time in meteorology, especially as it affected the air force. So it was no great surprise to find that Von Karman included, in his proposed long-range plan, research into the influence of the Sun on the Earth's ionosphere and atmosphere. Arnold and the AAF were not only interested in the effect of the Sun on the ionosphere and radio communications, but also its effect on the upper atmosphere through which guided missiles and supersonic aircraft would travel.

After the war Menzel happened to meet Marcus O'Day of the AAF's Cambridge Field Station who had been given responsibility for upper atmospheric research using captured V2 rockets. During discussions O'Day told Menzel that he also had access to funding to set up a ground-based solar observatory. As a result O'Day mentioned that there was a possibility that the Air Force might be able to support his proposed solar observatory. This case for support would be significantly strengthened if a suitable location could be found on the Sacramento Mountain

[1] After September 18 1947 the US Army Air Forces were known as the US Air Force.

Range close to the White Sands Proving Grounds from which O'Day was planning to launch his V2s.

In the meantime, following a proposal from Menzel and Roberts, in 1946 the Climax observatory became an independent research institution in its own right. (2) Called the High Altitude Observatory (HAO) it was affiliated with Harvard University and the University of Colorado with Roberts as its first director.(3) Its headquarters were at Colorado's Boulder campus.

Roberts found over the years that the association of the HAO with Harvard was no longer such an advantage in raising funds. Consequently in 1954 he persuaded the University of Colorado to take over full responsibility for the observatory. Shortly afterwards he convinced the organising committee for the International Geophysical Year (IGY) to designate the HAO as a World Data Center for Solar Activity.[2]

Returning to the establishment of the Sacramento Peak Observatory, in the summer of 1947 O'Day, Menzel, Roberts and John (Jack) Evans of the HAO, another former student of Menzel, undertook an aerial reconnaissance of the area around White Sands. As a result they picked out the 9,200 ft (2,800 m) high Sacramento Peak adjacent to the Air Force's Holloman Field and overlooking White Sands as the most promising site. Evans then installed two observers at a temporary observing station near the summit of Sacramento Peak to undertake field tests.

In December 1947 the HAO received a formal contract, essentially under Menzel's direction, to examine the possibility of establishing a solar observatory in the Alamogordo region, which included Sacramento Peak. Although other sites were examined in the area and elsewhere, it was concluded by April of the following year that Sacramento Peak was the most suitable. In the following month the Air Force gave a contract to Harvard College Observatory, with Menzel as prime investigator, to design and equip this solar observatory. Unfortunately for Menzel, who hoped that the new observatory would be run directly by Harvard College Observatory, Harvard was not prepared to provide tenured faculty positions for the senior scientists. As a result in 1952 the Air Force took over the observatory and renamed it the Upper Air Research Observatory with an Air Force officer as installation commander and Evans as superintendent. It had a staff of civilian and military scientists with logistical support provided by Holloman Air Force Base.

A regular programme of solar research had been started in 1949 on the site of the new observatory using small outdoor instruments. Then in the following year the first indoor facility, the Grain Bin Dome, was constructed using a small grain silo as the observatory shelter. It housed a 6 inch (15 cm) prominence telescope and

[2] Following a number of name changes, the Climax Observatory was eventually closed in 1974.

a flare patrol camera. The Big Dome facility, now called the Evans Solar Facility, was built shortly afterwards with a rotating turret.

In the meantime, Roberts at Boulder and Evans at Sacramento Peak had received contracts from the Office of Naval Research and the Air Force to develop two 16 inch (40 cm) coronagraphs, one for Climax and one for Sacramento Peak. These would be twice the size of Bernard Lyot's coronagraph at the Pic du Midi Observatory, which was, at the time, still the largest in the world. The Sacramento Peak coronagraph was installed in the Big Dome in 1953 followed two years later by a Littrow spectrograph. In 1956 Evans persuaded the Air Force to change the name of the observatory to the Sacramento Peak Observatory.

Frequent, regular images were produced by the flare patrol camera in the Grain Bin Dome starting in 1950. Then in 1963 these were replaced by more extensive observations from instruments located in the newly built Hilltop Dome. These included whole-sun images taken every minute in both visible light and the monochromatic light of hydrogen alpha at 656.3 nm.

In 1966 the Army Corps of Engineers began building the observatory structure for a radically new design of telescope. This Vacuum Tower Telescope (VTT),(4) as it was called, consisted of a vertical telescope whose light path was evacuated down to a residual pressure of 0.8 millibar. This was to eliminate image distortion caused by air turbulence. Sunlight entered the 41 m (136 ft) high tower at the top through a 76 cm (30 inch) fused quartz entrance window mounted in an altazimuth turret. It was then reflected down the evacuated tube by two 112 cm diameter fused quartz flat mirrors. Fifty seven metres below ground level it was reflected by a 162 cm spherical primary mirror back to ground level where it produced a 50 cm diameter image of the Sun. There the light left the evacuated tube through any one of five quartz windows located around the tube, where it entered one of a number of scientific instruments. The 98 m long central tube and everything connected to it was rotated to counteract image rotation. This was achieved by suspending the complete system from a mercury float bearing at the top of the tower. The VTT cost about $3.3 million (in ~1968 dollars) to build. In 1976 responsibility for the Sacramento Peak Solar Observatory was transferred from the US Air Force to the National Science Foundation.

Robert Smithson and colleagues at the Lockheed Research Laboratories built a Shack-Hartmann adaptive optics system (See Section 5.2) in 1983 using a deformable mirror of nineteen segments arranged in a circular array. They tested this system by imaging sunspots and solar granulation on the VTT, initially achieving an angular resolution of about 2 to 3 arcsec in 6 arcsec seeing conditions. But ten years later they improved this to about 0.33 arcsec in 1 to 3 arcsec seeing.(5)

Meanwhile, in the mid 1980s the Sacramento Peak Solar Observatory and the McMath solar telescope on Kitt Peak had been joined together administratively to form the National Solar Observatory (NSO). The NSO was a division of the National Optical Astronomy Observatory (NOAO) operated by AURA (see Chapter 2). In 1998 the VTT was renamed the Richard B. Dunn Solar Telescope after the name of its original project scientist. Like a number of people in this story of the Sacramento Peak Observatory Dunn had originally been a Menzel student.

15.2 Big Bear Solar Observatory

Robert Leighton supervised an extensive series of site surveys in the mid 1960s to find a suitable location for a Caltech (California Institute of Technology) solar telescope within reasonable travelling distance of their campus at Pasadena, California. Unfortunately, although seeing on the nearby Mount Wilson was good in the morning, it deteriorated towards midday, so a more suitable location was required for daytime observing.(6)

Leighton's site survey showed that all mountains had poor seeing during the day. But almost all sizeable lakes had good seeing, and mountain lakes, in particular, combined good seeing with transparency. In the event Big Bear Lake in California's San Bernardino Mountains was chosen as it had the best combination of seeing and transparency, with 80% of all daylight hours being clear. The lake, which was about 3 miles (5 km) long and one mile (1½ km) wide, was 2,070 m (6,780 ft) above sea level.

Seeing at Big Bear Lake dropped off markedly inland away from the lake's shore. So Caltech decided to build their solar telescope, designed by Harold Zirin, on an island in the lake as the surrounding water would produce much lower thermals than if the observatory had been surrounded by land. Funding for the construction was provided by the National Science Foundation (NSF), the Max Fleischmann Foundation of Nevada, NASA and Caltech.

Construction work on the Big Bear Solar Observatory (BBSO) was started in 1968 by first building an artificial island and an access causeway.[3] The observatory initially had four telescopes all mounted in a 42 inch (1.06 m) tube. These were two 10 inch (25 cm) refractors, a 9 inch reflecting coronagraph, and a 16 inch Cassegrain reflector. Both the Cassegrain reflector and the reflecting coronagraph fed through a coudé system to a vertically mounted spectrograph in a long vertical tube set into the observatory floor. Solar sensors installed on the vertically sliding shutter of the observatory dome kept the shutter at a level just above the Sun line as seen from the telescopes.

[3] When the lease was signed for the land in mid 1968 the causeway was then a few feet (~1 m) above the lake's water level. But when work on the observatory began in September the causeway was under water because the lake was at its highest level for three decades. Consequently, the builders increased the height of the causeway and island. But in February 1969 torrential rain caused the lake to rise 9 ft (2.7 m) and flood the new causeway and island, requiring them both to be increased in height once again.

Figure 15.1. Big Bear Solar Observatory situated on an artificial island in Big Bear Lake. The small dome to the right houses the full disc Hα patrol and Earthshine telescopes. (Courtesy Alla Shumko, Big Bear Solar Observatory.)

An observing station of the Global Oscillation Network Group (GONG) (see Section 15.3 below) was built at the BBSO in 1995 at the shore end of the causeway. Then two years later responsibility for the BBSO was transferred to the New Jersey Institute of Technology (NJIT) as Caltech had decided to concentrate on night-time astronomy.

In the meantime the initial suite of telescopes had been changed to a 65 cm vacuum reflector and a 25 cm vacuum refractor to study sunspots, together with a 20 cm full disc telescope. The BBSO and the NSO then jointly developed a 97 actuator adaptive optics system for the NSO Dunn Solar Telescope at Sacramento Peak and the BBSO 65 cm.(7)

Over the period 2006–2007 the 65 cm and 25 cm BBSO telescopes were removed and the old dome was replaced by a larger, ventilated dome (see Figure 15.1). Then a new 1.6 m clear aperture, open frame, off-axis Gregorian telescope, called the New Solar Telescope (NST), was installed in the new dome. This NST was a joint project of the NJIT, the University of Hawai'i Institute for Astronomy (UH IfA), and the Korea Astronomy and Space Science Institute (KASI).

The f/52 NST telescope was set on an equatorial mount attached to the existing pier of the 65 cm. The new f/2.4 Zerodur primary mirror blank was figured by the

Steward Observatory Mirror Lab using a stressed-lap system and measured using a computer-generated hologram.(8) This mirror acted as a proof-of-concept for the 8.4 m off-axis mirrors of the Giant Magellan Telescope (see Section 12.2).(9) An adaptive optics system, based on the 97 actuator system of the 65 cm telescope, was then added, achieving first light in 2010. Shortly afterwards this was replaced by a 357 actuator system developed, like the 97 actuator system, as a joint project with the NSO to produce nearly diffraction limited observations over the NST's principal operating wavelengths from 0.4 to 1.7 microns. A 10 cm full disc Hα patrol telescope, co-mounted with another small telescope,[4] were also installed nearby in an Ash Dome, replacing the 20 cm full disc telescope.

15.3 The GONG Helioseismology Network

The GONG (Global Oscillation Network Group) network project had been started by the NSO in 1984 to observe oscillations of the surface of the Sun on a continuous basis. They were encouraged in this by the success of the earlier BiSON network organised by the University of Birmingham in the UK (see Section 6.1). The NSO calculated that at least six solar telescopes, approximately equally spaced in longitude, were required for GONG to allow for weather problems and equipment breakdowns.(10) So fifteen different sites were tested,(11) resulting in six sites being chosen in 1991. They were at Mauna Loa, Hawaii; Big Bear, California; Cerro Tololo, Chile; El Teide, Canaries; Udaipur, India; and Learmonth, Australia.

The GONG network of six identical, automatic observing stations was inaugurated when its final station was completed in late 1995. These stations monitored the change in wavelength of a single solar absorption line all over the Sun using a Fourier tachometer that recorded the intensity, radial velocity and spectral-line depth of each pixelated patch of solar surface once per minute. Magnetic activity was also recorded hourly. Data was collected on tape and sent to the project's Tucson, Arizona headquarters for processing.(12) The solar disc was initially imaged on to a 256 × 242 pixel detector but in 2001 this was changed to a 1024 × 1024 pixel device. In the following year GONG was used to image the surface of the far side of the Sun. This was the first time that Earth-based observations had been used to produce such an image.

It had initially been planned to operate the GONG network for three years, but it was so successful that in 2003 the NSO decided to operate it indefinitely. Unfortunately GONG has been under threat of closure for the last few years to save money, but at the time of writing (2015) it is still operational.

[4] To observe earthshine.

Note The McMath-Pierce Solar Telescope and the Kitt Peak Vacuum Telescope are both covered in Section 2.3 and the solar telescopes on the Canary Islands in Section 6.1.

References

1. Liebowitz, Ruth Prelowski, Donald Menzel and the creation of the Sacramento Peak Observatory, *Journal for the History of Astronomy*, 33, Part 2, No. 111, 2002, pp. 193–211.
2. Hufbauer, Karl, *Exploring the Sun; Solar Science since Galileo*, Johns Hopkins University Press, 1991, p. 130.
3. Newkirk, Gordon, and Chronic, Halka, The High Altitude Observatory of the National Center for Atmospheric Research, Boulder, Colo., *Solar Physics*, 7, 1969, pp. 331–338.
4. Dunn, Richard B., Sacramento Peak's New Solar Telescope, *Sky and Telescope*, December 1969, pp. 368–375.
5. Acton, D. S., and Smithson, R. C., Solar Imaging with a Segmented Adaptive Mirror, *Applied Optics*, 31, 1992, pp. 3161–3169.
6. Zirin, Harold, The Big Bear Solar Observatory, *Sky and Telescope*, April 1970, pp. 215–219.
7. Cao, W., et al., Scientific Instrumentation for the 1.6 m New Solar Telescope in Big Bear, *Astronomische Nachrichten*, 331, No. 6, 2010, pp. 636–639.
8. Denker, C., Gary, D. E., and Rimmele, T. R., Ground-based Solar Facilities in the U.S.A., in Kneer, Franz, Puschmann, Klaus G., and Wittmann, Axel D., (eds.), *Modern Solar Facilities – Advanced Solar Science; Proceedings of a Workshop held at Göttingen, September 2006*, Universitätsverlag Göttingen 2007, pp. 34–35.
9. Martin, H. M., et al., Manufacture of a 1.7 m Prototype of the GMT Primary Mirror Segments, in Atad-Ettedgui, E., Antebi, J., and Lemke, D., (eds.), *Optomechanical Technologies for Astronomy, Proc. SPIE*, 6273, 2006.
10. Harvey, J. W., Kennedy, J. R., and Leibacher, J. W., GONG: To See Inside Our Sun, *Sky and Telescope*, November 1987, pp. 470–476.
11. Hill, Frank, et al., The Global Oscillation Network Group Site Survey, II Results, *Solar Physics*, 152, 1994, pp. 351–379.
12. Kennedy, James R., GONG: Probing the Sun's Hidden Heart, *Sky and Telescope*, October 1996, pp. 20–24.

Part 2

Radio Observatories

16
Australian Radio Observatories

16.1 Early Australian Radio Astronomy

Solar Observations

The Australian Radiophysics Laboratory was created at Sydney in 1939 as a secret branch of the Council for Scientific and Industrial Research (CSIR). It was to undertake radar research and development during the Second World War in collaboration with British laboratories and the Australian military.(1) The laboratory was highly successful, but as the war was coming to an end in 1945 the Australian government began to consider how its role should be changed. To assist in the decision Edward G. (Taffy) Bowen,[1] who was soon to take over as the head of the reconstituted Radiophysics Laboratory, assembled a series of papers for a meeting of the CSIR Council in July of that year. These resulted in a decision to concentrate future research in three areas, namely radio, vacuum, and rain and cloud physics. Bowen was to lead the rain and cloud physics group whilst Joe Pawsey, who had studied under John Ratcliffe at Cambridge in the 1930s, led the radio group. In October 1945, the radio group began to study the effect of the Sun on radio communications. This followed receipt of a report that a New Zealand military radar station on Norfolk Island had detected noise at 200 MHz (wavelength λ 1.5 m) that appeared to be connected with the Sun.

To study the Sun, Pawsey, Ruby Payne-Scott and Lindsay McCready initially used a Royal Australian Air Force wartime radar antenna on the coast at Collaroy 400 ft (120 m) above the sea near Sydney. It consisted of 40 half-wave dipoles operating at 200 MHz. After only three weeks of data they concluded that there was a close correlation between the total area of the Sun covered by sunspots and the magnitude of the radio noise which had varied by a factor of 30. Unfortunately the antenna could not be used to track the source in altitude but measurements of

[1] Bowen, a Welshman, had been working in the USA until he moved to Australia in 1944, see Section 16.2.

the noise along the horizon at sunrise and sunset clearly showed that it was associated with the Sun.(2) They also pointed out that for a thermally emitting disc the size of the Sun the detected levels implied equivalent temperatures of ~10^6–10^7K, which was much higher than the Sun's photospheric temperature of ~6,000 K. As a result they concluded that the radio waves were not of thermal origin but were caused by electrical disturbances in the solar atmosphere.

The measurements at Collaroy continued up to March 1946. But in the meantime Pawsey had arranged for parallel observations to be made with an identical radar antenna at Dover Heights which was on a 280 ft (85 m) high cliff overlooking the sea about 10 miles (16 km) south of Collaroy. Unfortunately the resolution of the Collaroy and Dover Heights radar antennae were so poor that it was impossible to determine where on the Sun the radio signals originated. It was thought that they may be coming from disturbances in the solar atmosphere linked with sunspot activity, or from beneath the optical surface of the Sun, or possibly from the solar corona. In these early measurements interference between radio waves coming directly from the Sun with those coming from the Sun via a reflection off the sea had been a nuisance. But Pawsey realised that this interference effect, which was in a way analogous to that of the Lloyd's mirror's optical arrangement of the nineteenth century, could be used to measure the position of the radio source in altitude. So he obtained some chart recorders to provide a permanent record of the effect of the Sun's passage through the interferometer fringes. Then in February 1946, a very large sunspot group started to cross the Sun's disc causing widespread disruption to radio communications. Measurements with both the Dover Heights and Collaroy interferometers clearly indicated that the sunspot group was the cause of the radio interference.[2] McCready, Payne-Scott and Pawsey also reported that the solar radio radiation consisted of two components. One component had a slowly changing intensity and the other consisted of more intense bursts lasting from less than a second up to a minute. But, although the cliff interferometers had been successful, they suffered from the fact it was not easy to change their height above the sea to produce different Fourier components in the interference pattern. As McCready and colleagues pointed out in 1947 maybe a different interference system would be more practicable.

In the meantime, Pawsey had been reanalysing the measurements of the solar radio intensity from October 1945 to May 1946 made using the Dover Heights and Collaroy interferometers. As a result he found that there was a marked lower limit of intensity corresponding to a brightness temperature of about 1×10^6 K. It so happened that a few years earlier, work at optical wavelengths by Walter

[2] They could not actually prove that the sunspot group was the cause of the radio interference as these interferometers had no significant resolution in the azimuth direction.

Grotrian and Bengt Edlén had indicated that the solar corona had a temperature of at least 2×10^6 K. Subsequently David Martyn who was working at Mount Stromlo Observatory began to analyse the possible effect of a solar corona at this sort of temperature on the Sun's radio emission. In 1946 he found that such a corona would be opaque at Pawsey's observational frequency of 200 MHz. So Pawsey's observed radio waves could not be coming from the solar photosphere but must be coming from the outer regions of the corona, although at shorter wavelengths the radio emission could come from lower down in the corona.[3]

A 200 MHz Yagi antenna that could track the Sun throughout the day had been installed at Dover Heights in March 1946. Similar 200 MHz Yagi antennae were also installed at the North Head radar station just outside Sydney and at the Commonwealth Solar Observatory on Mount Stromlo. Shortly afterwards the Dover Heights Yagi was joined by 60 MHz and 75 MHz Yagis which enabled Payne-Scott to undertake multi-wavelength observations of solar bursts. (3) In November 1946, she was joined by John Bolton and Bruce Slee and the 75 MHz antenna was replaced by a 100 MHz Yagi. On March 8 1947 a very large sunspot appeared on the limb of the Sun and an outburst was observed at all three frequencies with the time of onset getting later the lower the radio frequency.

Non-Solar Observations

Stanley Hey, John Parsons and James Phillips of the UK's Army Operational Research Group announced in 1946 that they had discovered a variable radio source in Cygnus when observing at 64 MHz (λ 4.7 m). Hey suspected that this source, now called Cygnus A, was relatively small as it was observed to fluctuate in intensity over the course of a few minutes. In the following year John Bolton and Gordon Stanley used two 100 MHz Yagi antennae for sea interferometry observations at Dover Heights to try to observe this Cygnus source. They detected it almost immediately and continued to observe it over the next few months at various frequencies from 60 to 200 MHz. These observations allowed them to conclude that the source was less than 8 arcmin in size.(4)

By mid 1948 Bolton had announced the detection of six new sources, the three most obvious he named Taurus A, Coma Berenices A and Centaurus A. To improve the measurements of source positions, Bolton and Stanley then built

[3] It has been a subject of some controversy as to whether Martyn analysed the effect of a high temperature corona on the Sun's radio emission after hearing of Pawsey's reanalysis, or whether Pawsey reanalysed his previous observations after hearing of Martyn's conclusion about the high temperature corona. For details see Woodruff T. Sullivan, *Cosmic Noise; A History of Early Radio Astronomy*, Cambridge University Press, 2009, pp. 135–136.

sea interferometers at a height of about 1000 ft (300 m) on the coast of New Zealand, one facing northwards and one facing westwards. This complemented the Dover Heights system, which was operated by Bruce Slee, that pointed eastwards. Using these interferometers over the period from June to August 1948 Bolton, Stanley and Slee found that their earlier position estimates, except for that of Cygnus A, were highly inaccurate being in error by ~10°. In particular it turned out that their previous position estimate of the source that they had called Coma Berenices A was so much in error that they had placed it in the wrong constellation. It was, in fact, in the constellation of Virgo, so its name was changed to Virgo A.

Bolton's group was now able to deduce the position of the four named sources to within ~10 arcmin. As a result they found that the position of Taurus A seemed to correlate with that of the Crab nebula. In addition they were able to put an upper limit of 1.5 arcmin on the size of Cygnus A and of about 6 arcmin on that of Taurus A.

Other Radiophysics Laboratory Observatory Stations of the Late 1940s

In the late 1940s Radiophysics Laboratory staff were carrying out radio astronomy observations at many sites around Sydney, in addition to those at Collaroy and Dover Heights, under Pawsey's general supervision.(5) For example Pawsey assigned two young radio engineers Fred Lehany and Donald Yabsley to carry out solar research at Georges Heights near Sydney from August 1947. They used a 18 × 16 ft (5.5 × 4.9 m) experimental wartime parabolic antenna to make their observations at 200, 600 and 1,200 MHz (λ 150, 50 and 25 cm). This was to extend the observations of solar bursts being made by Payne-Scott and colleagues at Dover Heights to higher frequencies. In 1948 this 18 × 16 ft antenna was refurbished and moved to Potts Hill, about 12 miles (20 km) from Sydney, to observe the partial solar eclipse of November 1 of that year, and the Georges Heights field station was closed down.

In the meantime Jack Piddington and Harry Minnett had undertaken solar and lunar research in 1948 at 24 GHz (λ 1.25 cm) from the top of the main Radiophysics Laboratory building at the University of Sydney. Their antenna was a modified 3.7 ft (110 cm) diameter Second World War searchlight mirror. With this they measured a solar chromospheric temperature of about 10,000 ± 500 K and a lunar surface temperature of 250 K.(6) From their observations of the variation of lunar radio emissions with phase they deduced that the Moon's surface was covered by a thin layer of dust. The 24 GHz antenna was then moved to Potts Hill where Minnett and Norman Labrum used it to observe the partial solar eclipse of November 1948 at 9428 MHz (λ 3.18 cm).

(a) Hornsby Valley

Over the period 1947–1948 Frank Kerr and Alex Shain had used a radar system to undertake ionospheric research and observations of the lunar surface. The signals were transmitted to the Moon by a 50kW Radio Australia transmitter located in Shepparton, Victoria operating at 17.84 and 21.54 MHz. The signals were received back from the Moon by two rhombic antennae set up in the well-shielded Hornsby Valley on the outskirts of Sydney.[4](6,7) Towards the end of 1947 Payne-Scott moved to Hornsby Valley as the Dover Heights Yagis were then being used extensively for non-solar work. At Hornsby she set up Yagi antennae to operate at 60, 65 and 85 MHz and a broadside array to operate at 18.3 MHz. She also used Kerr and Shain's rhombic antennae. Her observations of solar bursts lasted from January to September 1948 before she moved to Potts Hill. Shain stayed on at Hornsby Valley after Kerr and Payne-Scott had left. Then during 1949 and the early 1950s he and Charlie Higgins built 9.15 and 18.3 MHz horizontal arrays which were used to produce the first maps of emission from the Milky Way at low frequencies. The Hornsby Valley field station was closed down in 1955 after Shain had moved his low frequency research to Fleurs.

(b) Penrith and Dapto

The observations of solar radio bursts in 1948 and earlier had been rather confusing as noise storms, enhanced emission, and polarised and unpolarised bursts had been detected, without a clear-cut delineation between each of them. Clearly what was required to sort out these different sorts of solar radio emission was a radiospectrograph which could provide an instant profile of solar emission across a wide range of frequencies. Consequently McCready and Pawsey produced the design of such an instrument and McCready and Paul Wild turned this into a working unit.

Wild now set up a new field station in a relatively radio-quiet location at Penrith about 30 miles (50 km) from Sydney. There he installed a rhombic antenna in late 1948 which was operated with a swept-frequency spectrograph over the range 70 to 130 MHz (λ 4.3 to 2.3 m). Although the equipment did a good job of enabling Wild to separate the solar bursts into three types[5] it was closed down after only six months of observations. This was because, although it had proved the benefits of the radio spectrograph, a more radio-quiet site was required where further antennae could be located.

Following a site search Dapto was chosen, which was shielded by a steep escarpment. It was also twice the distance from Sydney as Penrith which reduced

[4] Reference (1), pp. 274–277, and F. J. Kerr, Early Days in Radio and Radar Astronomy in Australia, in W. T. Sullivan, (ed.), *The Early Years of Radio Astronomy; Reflections Fifty Years after Jansky's Discovery*, Cambridge University Press, 1984, pp. 136–137.

[5] The division into Type I, II and III bursts was based on the change of their radio spectra with time.

Figure 16.1. Rhombic antennae of the Dapto radio spectrometer. (© CSIRO Radio Astronomy Image Archive.)

the level of man-made radio interference. Subsequently in 1951 a completely new system was installed there by Wild, John Murray and William Rowe. It used three separate rhombic antennae (see Figure 16.1) and swept-frequency receivers to cover the frequency range from 40 to 240 MHz. With this equipment Wild found that that two of his three types of solar bursts, namely Types II and III, exhibited bursts in two distinct frequency bands with a frequency ratio of about 2.0.(8)

Potts Hill

As his Radiophysics group expanded Pawsey tried to locate new outstations at sites with minimum radio interference from manmade sources. In April 1948, Payne-Scott and Alec Little reported that a new site had been found at Potts Hill about 12 miles (20 km) from Sydney owned by the Sydney Water Board. Subsequently, as mentioned previously, the 18 × 16 ft antenna from Georges Heights was refurbished and moved to Potts Hill (see Figure 16.2).

Figure 16.2. Potts Hill radio observatory in 1953. The 18 × 16 ft rectangular parabolic antenna that had been moved from Georges Heights is seen top right. Other telescopes shown include a 10 ft dish, top centre, which was used for solar observations, and an early model, 120 × 120 ft in size, of a Mills cross antenna over most of the image. (© CSIRO Radio Astronomy Image Archive.)

At about the same time Chris Christiansen, who had taken over from Fred Lehany, started working at Potts Hill.(9) His first task was to use the 18 × 16 ft antenna to investigate the distribution of 600 MHz (λ 50 cm) radiation across the solar disc during the Sun's November 1948 eclipse.[6] This antenna was then

[6] In parallel eclipse observations were undertaken at the same frequency using portable 10 ft (3 m) diameter parabolic dishes at Rockbank, near Melbourne and Strahan, Tasmania, see W. N. Christiansen, D. E. Yabsley and B. Y. Mills, Measurements of Solar Radiation at a Wavelength of 50 Centimetres during the Eclipse of November 1, 1948, *Australian Journal of Scientific Research A*, 2, 1949, pp. 506–523.

used from 1949 to 1953 at Potts Hill as part of a multifrequency solar observing programme.

Piddington and Minnett also used the 18 × 16 ft antenna and a portable 10 ft antenna to conduct surveys of the Milky Way at 1210 and 3000 MHz (λ 25 to 10 cm) from 1948 to 1950. As a result they detected a source of radio waves in the approximate direction of the galactic centre.(10) Then in 1951 the 18 × 16 ft antenna was used by Christiansen and Jim Hindman to confirm 21 cm (1420 MHz) line emission of neutral hydrogen in the Milky Way.(11) The antenna was subsequently used to undertake a preliminary 21 cm hydrogen line survey of the southern sky which showed structure in the Milky Way. Shortly afterwards Frank Kerr and Hindman supervised the building of a meridian 36 ft (11 m) paraboloid to make a more detailed map of neutral hydrogen in the southern sky to match that produced by Jan Oort and colleagues in the Netherlands for the northern sky. In the process they produced the first map of hydrogen in the Magellanic Clouds.(12)

The other research team operational in the early days of Potts Hill was that of Payne-Scott and Little who built their so-called 'swept lobe' interferometer to observe the Sun from there in early 1949. It operated at 97 MHz and consisted of three Yagi antennae in an east–west line separated by 790 and 130 ft (240 and 40 m), with the two different baselines being used to eliminate ambiguity in the location of the main lobe. The phase of the local oscillator was changed to sweep the interferometer lobes which enabled an interference pattern to be produced from a very short duration source. This swept lobe system produced positions accurate to about 2 arcmin.

In 1950 Christiansen came up with the idea of his radio grating interferometer whilst reading of Bernard Lyot's optical filter in which narrow pass bands can be produced at widely different frequencies. But Bowen and Pawsey only authorised Christiansen to build this solar interferometer provided the cost could be kept to below £500 (about $1,400). In the meantime he was diverted by Pawsey to observe the 21 cm line of hydrogen, as mentioned previously, before he could put his design into effect.(13)

Construction of Christiansen's first solar grating array was begun in 1951 along the east–west wall of a reservoir at Potts Hill. When completed in February of the following year it consisted of thirty two 6 ft (1.8 m) diameter equally spaced dishes, equatorially mounted on a 700 ft (210 m) baseline. The dishes, which were mounted on wooden posts, had to be moved by hand to change right ascension.[7] The system, which operated at a wavelength of 21 cm, showed substantial solar limb brightening.

[7] The changes in right ascension were implemented by having someone run down the line of 32 antennae every 15 minutes, adjusting each one by moving a locking peg in each of the posts.

Unfortunately the Sun, and in particular its corona, is not spherically symmetric so, after undertaking a number of observations with this first grating array Christiansen decided that he needed a second array perpendicular to the first one. So he designed another interferometer to be built along the north–south wall of the reservoir. It consisted of sixteen 10 ft (3 m) wire mesh steerable paraboloids on a 520 ft (160 m) baseline. But instead of the rough and ready posts of the first array the antennae of the second array were mounted on concrete supports. Daily observations were made using both arrays starting in September 1953 to produce maps of the Sun with a resolution of 3 × 4 arcmin. In the following year, Christiansen moved to the Meudon Observatory near Paris on secondment from Radiophysics for a year. In his absence Govind Swarup and R. Parthasarathy, who had been working at the observatory under the Colombo Plan,[8] suggested to Pawsey that the east–west grating array should be converted to operate at 500 MHz (λ 60 cm) to see whether the quiet Sun exhibited limb brightening at that frequency.[9] With Pawsey's and Christiansen's enthusiastic support they made the appropriate modifications and found that it did.

Badgery's Creek

It had been clear to Pawsey that to get a more accurate position estimate of Cygnus A and similar radio sources a better interferometer was required than that at Dover Heights. So in 1949 he suggested to Bernard Mills that he study radio sources using a different design of interferometer at a new location. Initially Mills and Adin Thomas used Payne-Scott's swept lobe interferometer of Yagi antennae at Potts Hill operating at 97 MHz. With it they produced a position estimate of Cygnus A to within about 2 arcmin. Mills later built a three element interferometer at Badgery's Creek about 30 miles (50 km) from Sydney with antenna spacings of 200 ft (60 m) and 900 ft (270 m). These three antennae each consisted of an array of 24 dipoles mounted on a large flat reflector that could be tilted about its horizontal axis to observe the whole sky.(14) Using this interferometer at 101 MHz he was able to published a list of 77 sources in 1952.

Mills also tried to measure the angular sizes of the most intense sources using a mobile 2-Yagi 101 MHz antenna which was connected to one of the large antennae at Badgery's Creek by radio. Interferometer fringes were obtained for the four strongest sources with antenna spacings of up to 6 miles (10 km). These indicated

[8] The Colombo Plan for Cooperative Economic and Social Development in Asia and the Pacific was conceived at the British Commonwealth Conference on Foreign Affairs held in Colombo, Ceylon (now Sri Lanka) in 1950 as a cooperative venture for the development of nations in the area.

[9] At the time, S. F. Smerd and others had predicted theoretically that it should have limb brightening at frequencies of 500 MHz and higher but measurements made H. M. Stanier at Cambridge indicated otherwise.

that Taurus A, Virgo A and Centaurus A all had diameters of the order of 4.0 to 5.0 arcmin, whilst that of Cygnus A was less than 1.0 arcmin.(15) These were the first measurements to clearly show that at least some of the radio sources were extended objects.

Fleurs

In 1953 Bernard Mills, in discussion with Christiansen, came up with the idea of an interferometer with two orthogonal arms with a common centre. To prove that the system would work a prototype was built at Potts Hill by Alec Little and Kevin Sheridan.(16) It had two arms each 120 ft (35 m) long consisting of dipoles operating at 97 MHz (see Figure 16.2 earlier).

Following the success of this prototype, Mills, Little and colleagues built the so-called Mills Cross unfilled aperture system which was completed in 1954 at a cost of A£2,500 ($5,600).[10] It was constructed at Fleurs, about 25 miles (40 km) from Sydney, as Fleurs did not suffer as much radio interference as Potts Hill. The Mills Cross consisted of north–south and east–west arms each 1,500 ft (450 m) in length containing 250 half-wave dipoles backed by a wire mesh reflector. The array operated at 85.5 MHz (λ 3.5 m) producing a 50 arc minute pencil beam.(17)

This Mills Cross design enabled high resolution to be achieved in two directions at once. Although it was a transit instrument, pointing in declination could be achieved by changing the phases of the dipoles in the north–south arm. It was used by Mills, Slee and Eric Hill over the period 1954–1957 to produce a survey of galactic and extragalactic sources. But when they compared the sources listed in their MSH (Mills Slee Hill) catalogue[11] with those of the 2C catalogue of Cambridge, UK there seemed to be no correlation. Eventually it transpired that the Cambridge catalogue had accidentally recorded a large number of spurious sources (see Section 17.1). Then in 1960 Mills joined the University of Sydney after his SuperCross proposal had been rejected (see Section 16.3) and his research project at Fleurs had been closed down.

In the meantime Shain had built a new cross at Fleurs in 1956 to undertake a low frequency survey of the Milky Way at 19.7 MHz (λ 15 m). This Shain Cross, which was built alongside the Mills Cross, had north–south and east–west arms of 3,625 ft (1105 m) and 3,400 ft (1036 m), respectively. The low frequency survey map was found to be completely different from maps at high frequencies. This was because radio emission produced by ionised hydrogen, which was very

[10] This was the last time that the Radiophysics Laboratory funded a major antenna from its own resources.
[11] Published in the Australian Journal of Physics from 1958 to 1961.

evident at high frequencies, was being absorbed by the intervening material at low frequencies.

Bernard Burke and Kenneth Franklin in the USA had discovered decametric radio bursts from Jupiter in 1955. On hearing of this Shain re-examined observations that he had made five years earlier at 18.3 MHz at the Hornsby site and found that what he had thought at the time was terrestrial interference was, in fact, radio emission from Jupiter.(18) By the early 1960s it had become clear that the position of Io relative to Jupiter was important in trying to determine the exact source of these radio emissions. Unfortunately the Shain Cross was not large enough to locate it. So in 1963 the Radiophysics Laboratory decided to undertake observations with a permanent antenna and a mobile one some distance away. As a result Slee and colleagues set up a small field station on the top of a mountain near Jamberoo 50 miles (80 km) south of Fleurs where they installed a large part of Shain's 19.7 MHz equipment. They then modified an old bus to provide a mobile antenna facility which was connected to Jamberoo by radio and which was used at Fleurs, Dapto and Heaton, providing a maximum baseline of 125 miles (200 km). Although this system failed to pinpoint the exact source of Jupiter's emission it did show that the moving solar wind across the Earth was modifying the intensity of the signals.

When Christiansen had returned from France in early 1955 he decided to build a new cross-type interferometer at Fleurs in place of his Potts Hill interferometer.[12] Fleurs was the chosen location as not only did it suffer less from electronic interference than Potts Hill there was simply not enough room at Potts Hill to build Christiansen's proposed new interferometer. Generally called the Chris Cross this interferometer consisted of thirty-two 19 ft (5.8 m) diameter paraboloids on equatorial mounts on each of two 1,250 ft (380 m) arms, oriented north–south and east–west, that could operate up to 1.4 GHz (λ 21 cm).(19) It was used to produce daily two-dimensional high-resolution images of the Sun and to investigate major solar bursts. In 1960 a 60 ft (18 m) prefabricated American antenna, the 'Kennedy Dish', was placed 80 ft (24 m) from the eastern end of the east–west arm at double the spacing of the other antennae. This Fleurs Compound Interferometer, as it was called, was used to measure the right ascension and angular size of eight well-known discrete sources from August 1961 to October of the following year. The Kennedy Dish was moved to Parkes (see Section 16.2) in 1963.

Chris Christiansen, like Mills, had left the Radiophysics Laboratory in 1960 and joined the University of Sydney. At that time Radiophysics had planned

[12] As the thirty-two 6 ft dishes of the Potts Hill interferometer, and their associated electronics, were then no longer required they were donated to India (see Govind Swarup, From Potts Hill (Australia) to Pune (India): The Journey of a Radio Astronomer, *Journal of Astronomical History and Heritage*, 9, No. 1, 2006, pp. 21–33).

to demolish his unused Chris Cross at Fleurs. But they agreed to transfer the ownership of the Fleurs site to the University of Sydney. Subsequently Christiansen added two 45 ft (13.7 m) parabolic dishes to each arm, extending the baseline to 2,600 ft (790 m) and increasing the telescope's resolving power to 20 arcsec at its operational frequency of 1.4 GHz. This so-called Fleurs Synthesis Telescope was used to observe the structure of large radio galaxies, supernova remnants and emission nebulae. In 1984 the east–west arm was extended to 5,200 ft (1,585 m) by the addition of two more 45 ft paraboloids. The Fleurs Synthesis Telescope was closed down in 1988, on the same day as the inauguration of the Australia Telescope (see Section 16.4).

Hole-in-the-ground Antenna, Dover Heights

Bernard Lovell and John Clegg had built a fixed, upward-looking, 218 ft (66 m) diameter dish at Jodrell Bank in the UK in 1946–1947 to try to detect cosmic ray ionisation trails. The wire mesh dish of this very low cost design was formed by suspending it at its periphery from a series of posts driven into the ground (see Section 18.1). Although the cosmic ray search was unsuccessful, Lovell decided to use the dish as a transit telescope operating at 72 MHz (λ 4.16 m) to look for cosmic radio sources instead.

A few years later this Jodrell Bank design inspired John Bolton who wanted to build an inexpensive parabolic antenna at Dover Heights. In Bolton's case his solution was to lay the surface of the dish directly onto the surface of a suitably shaped hole in the ground.(20) Money was very tight at Radiophysics at the time so Bolton and Slee, with the assistance of Gordon Stanley and Kevin Westfold, decided to build the dish in their spare time.

Fortunately for Bolton and his team, the surface at Dover Heights was made of sand, so it was relatively easy to remove it with just shovels and a wheelbarrow. Interestingly, although Taffy Bowen was aware that the work going on, Pawsey, Bolton's immediate superior and Bowen's junior was deliberately kept in the dark in case he intervened and stopped the project. When the excavations had been completed, the surface of the 72 ft (21.9 m) diameter hole was consolidated with ash and partially covered with half inch (1.3 cm) metal strips taken from packing cases to reflect the radio signals to a dipole at the top of a mast. This dipole was then connected to a modified Second World War receiver tuned to 160 MHz (λ 188 cm). As in the case of the Jodrell Bank transit instrument, areas around the zenith were observed by tilting the central mast by adjusting its supporting guy ropes. The first series of observations, which were carried out in late 1951 and early 1952, were successful clearly showing the location of the Sagittarius A source, which seemed to be coincident with the centre of the Milky Way. It was at this stage that

Figure 16.3. The 80 ft diameter hole-in-the-ground antenna at Dover Heights. (© CSIRO Radio Astronomy Image Archive.)

Pawsey was informed of the work which, much to everyone's surprise and relief, he applauded.[13]

Having shown its usefulness, it was then decided to increase the diameter of the hole-in-the-ground dish to 80 ft (24.4 m) (see Figure 16.3) and to increase its operating frequency to 400 MHz (λ 75 cm) to improve resolution. When the further excavations had been completed the new surface was coated in concrete with embedded half-inch wire mesh. The beam width of this new system was only 2°, compared with the 6° beam width of the original system.

A new sky survey was carried out by Richard McGee, Slee and Stanley as soon as the modified dish had been completed in 1953. It clearly showed the location of the Sagittarius A source and the structure of the radio isophotes surrounding it. It was also reported that about 13 other sources had been found, but it turned out that about half of them were not real sources but the result of beam confusion. Radio astronomy ceased at Dover Heights at the end of 1954.

[13] This is, at least, according to the face that he presented to the outside world.

16.2 Parkes Radio Telescope

Taffy Bowen, who was to mastermind the approval of the Parkes Radio Telescope programme,[14] relied in the early approval stages on the contacts that he had made in the USA during the Second World War. During the war he had accompanied Henry Tizard in delivering the top secret magnetron valve[15] from the UK to the USA. Bowen had then remained in the USA working at MIT's Radiation Laboratory as an advisor on the development of microwave radar. But in January 1944 he had moved to Australia at the invitation of Frederick White, the Chief of the Radiophysics Laboratory, to become Assistant Chief in charge of research. Shortly afterwards, as already mentioned in Section 16.1 above, Bowen had become head of the Radiophysics Laboratory.

By the late 1940s it was becoming evident that, if Australia wanted to retain a leading role in radio astronomy, it would have to build a much more capable telescope than any it had built so far. At that stage Bowen had been considering the problem for some time, and over the period 1949–1952 he came up with a number of possible designs for what he called the 'Giant Radio Telescope' to be built in Australia, none of which came to fruition. In the meantime a number of important and influential people in the USA had become increasingly concerned about the lack of a major American radio astronomy observatory to match those in Australia, the UK and the Netherlands. Fortunately in his wartime sojourn in the USA, Bowen had become acquainted with a number of these people. Consequently in early 1952, one of these, Lee DuBridge, president of Caltech, asked him to produce a detailed proposal for American radio telescopes to be built at what he envisaged as the radio equivalent of the optical observatories at Mount Wilson and Palomar Mountain in California. Vannevar Bush, president of the Carnegie Institution in Washington and Alfred Loomis, a trustee of both the Carnegie Corporation and the Rockefeller Foundation agreed to promote the idea of such a radio observatory and seek sources of funding. The plan was for Bowen to be the director of the observatory, with Bush's and Loomis' enthusiastic support, and with Bolton as his deputy.

Although Bowen was interested in taking over as the head of such an American observatory he had some reservations. Frederick White also had reservations as he did not wish to see a 'brain drain', as it was called at the time, of high level Australian scientists going to the USA.[16] So before committing himself to the American venture Bowen looked into the possibility of obtaining funding for a

[14] Various aspects of the origin and development of the Parkes Radio Telescope are covered in D. E. Goddard and D. K. Milne, (eds.), *Parkes: Thirty Years of Radio Astronomy*, CSIRO Publishing, 1994, in Peter Robertson, *Beyond Southern Skies; Radio Astronomy and the Parkes Telescope*, Cambridge University Press, 1992, and in Raymond Haynes, et al., *Explorers of the Southern Sky; A History of Australian Astronomy*, Cambridge University Press, 1996, pp. 238–279.

[15] The magnetron valve was essential to the development of high frequency radar.

[16] There was a similar concern in the UK at the time about losing its best technical talent to the USA.

similar modern radio observatory in Australia. As a result he asked Tizard, who had just retired as chief scientific advisor to the UK's Ministry of Defence, if he thought that the Nuffield Foundation, who had just agreed to pay 50% of the costs of Jodrell Bank's 250 ft (76 m) telescope (see Section 18.2) would be interested in funding a similar sized telescope in Australia. Tizard said that he thought this unlikely and suggested instead that Bowen contact the British Dominion and Colonies Fund, which was managed by the Carnegie Corporation in New York.

In the meantime Bowen had told DuBridge that, whilst he was still interested in the American post that he had been offered, he felt duty bound to also examine the possibility of building a modern radio observatory with a large radio telescope in Australia. Then in August 1952, Bowen wrote to Vannevar Bush to ask whether Carnegie would consider funding both the proposed Caltech radio telescope and a similar telescope in the southern hemisphere as a collaborative development. After a number of twists and turns, in May 1954 the trustees of the Carnegie Corporation approved a sum of $US247,500 as a contribution to a large radio telescope in Australia. In the following year Caltech went their own way and began construction of an interferometer consisting of two 90 ft (27 m) dishes at Owens Valley, California (see Section 21.1) funded by the US Office of Naval Research. At about the same time the United States also decided to build a National Radio Astronomy Observatory at Green Bank, West Virginia (see Chapter 20).

In 1954 Radiophysics thought that a telescope similar in size to that at Jodrell Bank would cost about $US1 million (£400,000 or £A500,000), or four times the sum promised by Carnegie, so a large amount of funding was still required. To assist further fund raising Bowen obtained the scientific endorsement of the International Astronomical Union and the Royal Astronomical Society in London and with these in hand he approached the Australian government for funding via Richard Casey the CSIRO[17] minister. Then in April 1955 Robert Menzies, the Australian prime minister, agreed to government funding for the telescope provided half the cost was guaranteed by private sponsors before any government grant was made available.

As a result Bowen went to the USA in June 1955 to try to raise funding from other large philanthropic organisations there. Unfortunately he drew a blank with all but the Rockefeller Foundation. There he knew Warren Weaver, director of the Natural Science Division, who happened to be extremely interested in radio astronomy. Bowen's approach to the Rockefeller Foundation was followed by a meeting between Casey, in his role as Australian minister for foreign affairs, and Dean Rusk, who was president of the Foundation. This resulted in an announcement

[17] CSIRO (Commonwealth Scientific and Industrial Research Organisation) was Australia's national science agency.

made in December 1955 that the Rockefeller Foundation would provide a grant of $US250,000 towards the telescope. So by the end of the year the project had $US1 million funds approved, 25% from each of the Carnegie Corporation and the Rockefeller Foundation and 50% from the Australian government.

Even though $US1 million had been raised by the end of 1955 the design of the telescope was still unclear. Four years earlier Bolton had suggested building a 'standing paraboloid' 20 ft high in an arc 200 ft (60 m) long with a focus 150 ft in front of the curve. Bowen, on the other hand, had suggested that the new telescope could consist of a cylindrical paraboloid lying on its back, 1,000 ft (300 m) long in an east–west direction, by 200 ft wide.(21) It would be made of five adjoining elements 200 ft square and would be steerable in the north–south direction by a series of cables and winches. There were other telescope designs, including an improved version of the hole-in-the ground-antenna.

Bowen and Pawsey set up a planning committee in 1954 to evaluate the various competing design ideas of the time[18] and to draw up a set of general technical specifications for the telescope. By the following year the preferred configuration was based on a fully steerable 250 ft dish, or larger if possible,[19] with a surface accurate enough to enable it to make observations at the hydrogen line wavelength of 21 cm. There were seven possible designs differing only in the type of mount used to support and steer the dish.

Bowen experienced a great deal of opposition within the Radiophysics team against his proposed Giant Radio Telescope, including that from Pawsey himself, whilst its various configurations were being evaluated. This was because many of the staff anticipated that their own radio telescopes would have to be closed down to pay for this one large instrument, and Pawsey was still very much wedded to the idea of interferometers. Bowen, on the other hand, thought that interferometers had had their day for observing all except the Sun and some of the stronger sources. Pawsey also preferred the 'string and sealing wax' type of approach that he had learnt in his early days at Cambridge University's Cavendish Laboratory,[20] and believed that using one big telescope would stifle originality of thought. In many ways he and Bowen were playing out the same arguments that existed at this time between Bernard Lovell at Jodrell Bank with his 250 ft dish and Martin Ryle at Cambridge with his interferometers.

In June 1954 Lovell suggested to Bowen, in view of the technical problems he had had with building the Jodrell Bank 250 ft, that Bowen should consider

[18] See, for example, Frank Kerr, The Proposal for a Giant Radio Telescope, in D. E. Goddard and D. K. Milne (eds.), *Parkes: Thirty Years of Radio Astronomy*, CSIRO Publishing, 1994, pp. 7–11, and Peter Robertson, *Beyond Southern Skies; Radio Astronomy and the Parkes Telescope*, Cambridge University Press, 1992, p. 135.

[19] To make the Parkes dish larger than the one at Jodrell Bank.

[20] Pawsey obtained his doctorate at the Cavendish in 1934 on the effect of the ionosphere on radio fading.

building a smaller telescope of 150 or 200 ft with very much tighter tolerances. In March 1955 Merle Tuve, who was then involved in establishing the National Radio Astronomy Observatory in the USA, made the same suggestion.[21] By this time Pawsey was prepared to back a dish-type of radio telescope but only if it was smaller than the 250 ft envisaged by Bowen as a smaller telescope would allow much higher surface and pointing accuracies and it could also be brought into operation much sooner. It would also release funds for other antennae that could be linked to the main dish and be used for interferometry.

During a visit to London in 1954 Pawsey had outlined to Tizard the problems then being experienced with the proposed large radio telescope project. Tizard suggested that Pawsey discussed the technical problems with the distinguished British engineer Barnes Wallis who was then head of research in Vickers Armstrong's aircraft division. Wallis, who was rather critical of the design of Jodrell Bank's 250 ft telescope, said that he would be very happy to help. Then in April 1955, he suggested that he may be able to design a larger dish than 250 ft. At about the same time Wallis also came up with the revolutionary idea of positioning the master equatorial control unit, which would control the telescope's position, at the intersection of the telescope's two axes of rotation.

In the meantime at a meeting of the planning committee in March 1955 with Pawsey in the chair and Bowen absent it was agreed that a 150 ft dish was the preferred size as it would be good enough for most observations. At about the same time Bowen himself was beginning to have second thoughts about Radiophysics building a large telescope on their own and he seriously considered the idea of collaborating with the Americans in building a smaller dish which could then be copied in Australia. But in April 1955, Bowen received Barnes Wallis' note in which he signified his strong interest in designing a 250 ft dish, or larger. Then, as mentioned previously, a few days later the Australian government agreed to match pound-for-pound all funds raised privately for the large telescope project. Clearly, the project was back on track.

In June, an advisory committee was formed at Radiophysics which included a number of Australia's best engineers. It met to decide how and where to carry out the preliminary design study. The idea of undertaking this study in Australia was rejected by the committee as it had doubts that a sufficiently experienced team could be assembled there. The possibility of undertaking it in the USA was also rejected as the potential contractors were overloaded with work and it was thought that the subsequent manufacturing costs would be high there. So the committee recommended undertaking the preliminary design study in the UK where three

[21] The advice given by both Lovell and Tuve was technically sound. But if Bowen followed it, it would leave Lovell with the largest fully steerable radio telescope dish in the world. Bowen really wanted to have the largest dish, but he saw the merit of their technical advice.

firms were approached. Of these by far the best offer was made by Freeman Fox and Partners of London who had designed Sydney Harbour Bridge. They quoted a price of £10,000 (£A12,500) which included £1000 for Barnes Wallis who would act as a design consultant.

Radiophysics requested Freeman Fox, in the preliminary design study, to design as large a dish as possible given the total funding constraints. The dish was to be able to operate down to a wavelength of at least 21 cm and preferably lower, and its tracking error, including all error sources, was to be no larger than 1 arcmin. It was also decided to restrict the movement of the dish to a maximum angle of 60° from the vertical to simplify the design and reduce the costs. This would still allow the telescope to reach as far north as +27° declination, which was considered to be good enough bearing in mind the number of telescopes in the northern hemisphere already covering northerly declinations. And even with the 60° limitation, it would be able to view the south celestial pole from the latitude of Sydney.

Radiophysics also accepted that the dish would not be required to operate in wind speeds of more than 30 mph (50 km/hr) and said that they would find a site totally free from ice and snow all year round. The detailed Radiophysics specification for the preliminary design study was completed in November 1955. It was then taken to the UK by Jack Roderick, professor of civil engineering at the University of Sydney, who spent two weeks at Freeman Fox finalising arrangements for the study. Harry Minnett, one of the most experienced antenna designers at Radiophysics, was posted to Freeman Fox for the duration of the study because of their lack of experience in radio engineering and to facilitate the flow of information between Radiophysics and the contractor.

To start with Freeman Fox compared the advantages and disadvantages of either making the dish rigid enough to avoid unacceptable surface deviations or allowing the dish to be distorted but be brought back within specification by some form of automatic compensation mechanism. In the event they found that a rigid structure was preferable up to at least 300 ft. Steel was chosen for the skeleton of the dish because of its superior rigidity compared with aluminium or other light alloys. Freeman Fox also chose an altazimuth mount as it was estimated to be better than an equatorial for dishes larger than about 150 ft. This was a rather controversial decision at the time, particularly with optical astronomers who were used to equatorial mounts. But Freeman Fox thought that the designs of electronic drive systems had matured sufficiently to allow an altazimuth system to be used.

Freeman Fox had initially estimated the preliminary design contract should be complete within about six months, but if that proved to be too optimistic they promised to deliver a detailed interim report by June 1956. In the event the final report, which was very favourably received, was not completed until November of the following year. The main issue still unresolved in that report was the size of the dish as Freeman Fox was confident of meeting all the specifications with a 250 ft diameter structure. Consequently it was simply a matter of what size telescope Radiophysics could afford. In January 1958 they decided that this was one with a 210 ft (64 m) diameter dish, assuming that costs increased as the 5/2 power of dish diameter. Then in the following month they decided that the minimum operating wavelength could be reduced from 21 cm to 10 cm. Subsequently, Freeman Fox was given the full design contract and authorised to engage contractors for the telescope's construction.

Earlier Freeman Fox had experienced difficulty in finding firms in the UK interested in building the telescope, or even parts of it, as the most suitable firms were far more interested at the time in chasing lucrative defence contracts. In addition the idea of building a challenging one-off design on the other side of the world with no expectation of repeat orders was not very attractive. So with such little interest shown by potential contractors Freeman Fox decided there was no scope for asking for competitive tenders. Instead they decided to try to tempt potential contractors by dividing the building contract into three parts based on their knowledge of the competence of UK companies. As a result they chose Metropolitan-Vickers to build the mechanical and electrical drive systems and to act as the main contractor, Grubb Parsons to develop the master control system, and Sir William Arrols and Partners to make the heavy structural components and to assemble the telescope in Australia. The first two of these companies had already been involved in the preliminary design contract working for Freeman Fox so they seemed a logical choice. But unfortunately Metropolitan-Vickers and Freeman Fox were unable to agree on contractual terms and Metropolitan-Vickers were reluctant to provide a price quotation for the main manufacturing contract.

Grubb Parsons, in the meantime, had suggested that the hydraulic error-sensing mechanism of Wallis' master equatorial unit should be replaced by an electro-optical error detector system.[22](22,23) This electro-optical device, like Wallis' proposed system, would be located at the intersection of the telescope's two axes of rotation. It could be accurately pointed in any chosen direction and the dish made to follow it by a master-slave system. An optical error detector within the telescope's hub would detect when the dish was not pointing in the required direction. An error signal would then be produced which was fed to the telescope's altitude and azimuth motors to correct the orientation of the dish. Unfortunately Grubb Parsons quoted an unacceptable manufacturing timescale for this equatorial unit, although its design was retained in the final telescope design.

[22] Harry Minnett, The Construction of the Parkes 210-ft Radio Telescope in D. E. Goddard and D. K. Milne, (eds.), *Parkes: Thirty Years of Radio Astronomy*, CSIRO Publishing, 1994, pp. 12–18, and H. C. Minnett, The Australian 210-foot Radio Telescope, *Sky and Telescope*, October 1962, pp. 184–189.

The problems with Metropolitan-Vickers and Grubb Parsons were not the only problems experienced by Freeman Fox at this time as the change in wavelength from 21 to 10 cm was also causing problems of its own. In addition, Gilbert Roberts, the man responsible at Freeman Fox for this and other engineering contracts, was forced by rising costs to make major simplifications to the mechanical drive system. In the end, after much pressure Metropolitan-Vickers produced a cost estimate for the telescope in March 1959. But this was of £A715,000 (£570,000), which was much higher than the funds available. In addition they also quoted an unacceptable construction period of almost four years.

Bowen then persuaded Freeman Fox that the only sensible solution to the problem experienced with their subcontractors was to open the construction contract to international tenders. As a result the tender documents were issued by Freeman Fox in April 1959 to a shortlist of companies who had expressed a strong interest in the project. This resulted in a winning bid from MAN[23] in West Germany as prime contractor which was for £A611,000 with a construction period of 21 months, ending in April 1961. This was appreciably lower than the Metropolitan-Vickers cost estimate, but once the costs of building an on-site laboratory, residential buildings and roads, and of providing essential services were added there was a still a substantial cash shortfall. So Bowen made another attempt to raise money in the USA.

Carnegie declined to increase their original grant, but the Rockefeller Foundation agreed to provide another $130,000 (£46,000 or £A58,000). At one stage it seemed as if the Sloan Foundation would also support the project but this proved not to be the case. So eventually the Australian government agreed to make up the difference.(24)

Press reports of Carnegie's grant in May 1954 had generated an avalanche of communications with Radiophysics from various organisations including local councils and land owners who wanted the telescope built in their locality. Subsequently, by early 1956 the list of possible sites had been whittled down to a shortlist of ten. The requirements for deciding the best site included low radio interference, no ice or snow, low average wind speed, and good site substrate to take the weight of the telescope. Three general areas seemed to fit the bill both technically and politically, namely somewhere relatively close to Sydney, somewhere relatively close to Australia's capital city of Canberra, or somewhere over the mountains well to the west of Sydney. A location not too far from Sydney would suite Radiophysics, but it would be unlikely to have a low enough level of radio interference. A site in the general area of Canberra could be politically beneficial, especially if it was linked to the Mount Stromlo optical observatory in some way. But Canberra's increasing light pollution was forcing the optical community

[23] Maschinenfabrik Augsburg Nürnberg.

to find a new location some distance away from Mount Stromlo (see Section 9.1). As a result of these problems with potential sites near Sydney and Canberra, a radio site survey was carried out in 1957 of an area appreciably west of Sydney that was protected by the Great Dividing Range. The best site surveyed proved to be the flat floor of the Goobang Valley surrounded by low foothills(25) 15 miles (24 km) north of the small town of Parkes and about 200 miles (320 km) over the mountains from Sydney. The site had an exceptionally low level of radio interference and, consequently, it was selected as the location for the large new radio telescope. With a minimum temperature of 4°C icing of the dish could be ignored, and the site was also found to have very low average wind speeds. The Parkes town council even agreed to restrict industrial development within 6 miles (10 km) of the telescope in order to keep future electromagnetic interference to a minimum.

The telescope contract had been awarded to MAN in July 1959 as prime contractor. MAN had three main subcontractors, these were AEI of Manchester, UK for servo-control, cabling and communication systems, Askania of West Germany for the master equatorial system, and Concrete Constructions of Sydney for the foundations and the reinforced concrete tower which would support the telescope. Construction of the telescope started just two months later and by May 1960 MAN had built the huge turret, hub and counterweight structures. Concrete Constructions had finished building the reinforced concrete tower two months earlier, leaving it to settle for a few months before loading it with the 1000 tons of the telescope's steel superstructure. The radial ribs were then lifted into place and in August 1961 work began on laying the dish reflecting surface, the central 55 ft (17 m) diameter section of which was made of welded steel plates, with the remainder of the 210 ft (64 m) diameter being made of steel mesh. The reflecting surface was completed and the dish moved in elevation for the first time just a few days before the telescope's inauguration on October 31 1961. Although this was six month's late, MAN and everyone connected with building the telescope (see Figure 16.4) could be justifiably proud of their work, having completed it far quicker than any other large radio telescope of that time. The telescope was then subjected to an extensive set of commissioning tests before its main astronomical programme was started in April 1962. By the end of the following year it had achieved a pointing accuracy ranging from about 20 arcsec generally to 6 arcsec for objects in the zenith field, which was much better than that of any other large radio telescope of the period.

One of the main reasons behind building the Parkes telescope had been to produce a radio survey of the southern sky at around 21 cm wavelength. In fact when construction of the telescope had been completed the only large scale catalogue of southern sources was the MSH catalogue (see Section 16.1 above) which had been produced by Mills, Slee and Hill around the end of the 1950s using the Mills Cross at 85.5 MHz (λ 3.5 m).

Figure 16.4. The 210 ft (64 m) diameter Parkes Radio Telescope imaged in 2009. (Courtesy Ian Sutton via Wikimedia Commons.)

The first Parkes survey was started in 1962 by Bolton, Frank Gardner and Brian Mackey at frequencies of 408 MHz (λ 74 cm), 1410 MHz (λ 21 cm) and 2650 MHz (λ 11 cm). The first results, which were published in 1964, listed 297 sources, 51 of which had not been listed in the MSH catalogue for various reasons.(26) The position estimates of those sources that were listed in the two catalogue differed by a median value of 8 arcmin. In the following year Bolton, Margaret Clarke and Jennifer Elkers published 111 possible optical identifications for Parkes sources between declinations of –20° and –33° where plate survey material existed.

In the meantime Mackey, Cyril Hazard and John Shimmins had used the Parkes antenna in 1962 to observe the occultation of the source 3C 273 by the Moon.(27) They found that it was double, with a separation of 20 arcsec. This enabled Maarten Schmidt to identify 3C 273's optical equivalent as what appeared to be a faint blue star and the end of a faint bluish jet coming from the star. The 'star' was found to have an unexpectedly large redshift of 0.158. It was the first quasar to be discovered. Then in 1968 Brian Robinson and colleagues used the Parkes antenna to observe CP 1919, the first pulsar to be discovered, just two weeks after its discovery had been announced. They observed the pulsar at 150 MHz using the first solid state receiver to be constructed on site.

In 1963 the 60 ft (18 m) prefabricated American antenna, the 'Kennedy Dish', (see Section 16.1) was moved from Fleurs to Parkes to be used as an interferometer with the 210 ft Parkes dish.(28) There the smaller antenna was put on rails to provide a variable baseline. Unfortunately phase instability problems with the unburied cable linking the two dishes meant that it was not very useful for its original purpose of accurately measuring the position of radio sources. But it was able to provide very useful data on source sizes and brightness distribution. In particular it was used by Bolton and Ron Ekers to study the fine structure of southern radio sources.

The original Parkes mesh panels between 17 m and 37 m diameter were replaced by perforated aluminium panels over the period 1970–1972.[24] This doubled the aperture efficiency at a wavelength of 3.4 cm and created a 37 m diameter zone that provided a useful performance down to 1.35 cm.(29) The original solid surface of the central 17 m diameter, which had been welded into place, had distorted badly. So three years later a new surface of aluminium panels was provided a few millimetres above the original surface of this central zone which enabled observations at wavelengths as short as 3.5 mm. The screw adjusters of the remainder of the dish's surface were adjusted to remove the step at the edge of this central zone. Some years later the large Parkes antenna became part of a large radio interferometric array called the Australia Telescope (see Section 16.4 below).

In 2003 the 64 m Parkes antenna discovered the first known double pulsar system PSR J0737–3039A/B. The A star, which was a 23 millisecond pulsar, was first detected in April 2003 by Burgay and colleagues using the Parkes antenna. Its pulsar signal was found to have a significant Doppler effect indicating the presence of a nearby binary companion. Follow up observations made using the Parkes antenna and the 76 m Lovell Telescope at Jodrell Bank indicated that the A star's companion, the B star, had a pulsar period of about 2.8 seconds. The orbital period of the system was found to be about 2.4 hours.[25]

16.3 Culgoora and the Molonglo Cross

The arrival of the large Parkes telescope signified a major change in the way radio astronomy was undertaken in Australia. No longer would research be limited to small teams building their own equipment, as Parkes was a major facility requiring

[24] This type of panel was later fitted out to a diameter of 54 m, see Brian Robinson, Spectral Line Astronomy at Parkes in D. E. Goddard and D. K, Milne, (eds.), *Parkes: Thirty Years of Radio Astronomy*, CSIRO Publishing, 1994, p. 109.

[25] See A. Possenti, et al., *Memorie Società Astronomica Italiana*, Supplementi, 5, 2004, pp. 142–147 for an excellent early overview of the discovery.

its own team of technicians to maintain and operate it and with a queue of radio astronomers wanting to use it. Parkes was also an expensive facility to operate, maintain and develop. Consequently the funding available to other radio astronomy facilities in Australia had to be strictly limited.

In preparation for the Parkes era, the Radiophysics Laboratory started to consider in the late 1950s what other new radio telescopes could be afforded in the next decade. Two instruments were considered. One was a Radioheliograph which had been proposed by Paul Wild to study radio emission from the Sun. The other, proposed by Bernard Mills, was the so-called SuperCross to continue the work of the Mills Cross at Fleurs in detecting and cataloguing galactic and extragalactic sources at a wavelength ~1 m. In 1959 the Radiophysics Laboratory chose the Radioheliograph to enable them to continue with their important work on the Sun which was highly regarded internationally. Otherwise the solar group would have had to have been disbanded, leaving the Laboratory to concentrate on galactic and extragalactic radio astronomy.

Wild had initially proposed a Radioheliograph in the form of a cross but he eventually decided on a circular array about two miles (3 km) in diameter. Unfortunately the size of the array proved a particular problem as it was too large to be built at the preferred site of Parkes. Other sites in vicinity of Parkes were considered, but eventually it was decided to build the array at Culgoora, near Narrabri in New South Wales, about 335 miles (540 km) from Sydney. Culgoora was chosen as there was a suitable very large plot of land available there at a reasonable price. In addition the CSIRO Division of Applied Physics decided to locate their solar optical observatory equipment[26] at Culgoora because of its excellent optical seeing conditions. Grants totalling $630,000 (£A1.4 million) were obtained from the Ford Foundation for the array, which was designed and built by Radiophysics staff. It was completed in 1967.

As built the Radioheliograph array consisted of ninety-six identical 45 ft (14 m) diameter dishes positioned at equal intervals around the circle's circumference. The dishes each had a surface of 2 inch (5 cm) galvanised mesh designed to detect radio waves emitted from within the solar corona at a frequency of about 80 MHz. (30) The array produced an image of the Sun every second in two polarisations enabling astronomers to study rapidly varying solar radio bursts. In the event the array, and the associated optical telescopes at Culgoora, enabled the Radiophysics group to remain at the forefront of solar radio research for another decade or so. In 1973 the Radioheliograph array was modified to also work at 160 MHz(31) but the facility was eventually closed down eleven years later when it had come to the end of its useful life.

[26] These included a 12 inch (30 cm) refractor to take time-lapse photographs of selected areas of the solar disc and a 5 inch (13 cm) flare patrol telescope to photograph the Sun at 30 s intervals.

With the rejection of the SuperCross idea, Mills had left the Radiophysics Laboratory in 1960. He joined the University of Sydney where Harry Messel, head of the physics department, had decided to set up a small radio astronomy group. Messel had promised Mills that he would fund the SuperCross as he already had £A80,000 (later increased to £A100,000) from the Science Foundation for Physics. Although more money was required, Messel was aware that the National Science Foundation (NSF) in Washington was, for the first time, willing to consider making grants to scientists outside the United States. Consequently he encouraged Mills to apply to the NSF for funding and, in spite of opposition from Bowen,[27] in 1962 they agreed to provide $746,000 (£A335,000) in return for allowing American astronomers to use the new telescope. In addition Mills also agreed to help train American graduate students in radio astronomy at the University of Sydney. Two years later the NSF agreed to increase this grant by $107,500 (£A241,000).(32)

Subsequently the SuperCross was built in a wide flat valley at Molonglo, near Canberra. This was one of the sites that had been investigated for the large Parkes antenna but had been rejected as the height of the Parkes antenna would have put it in line of sight over the hills to transmitters on Black Mountain. However this was not a problem with the SuperCross, or Molonglo Cross as it was now known, as its antennae were so much lower. Construction of the cross was carried out by staff and students of the university's physics department, supported by many other university and industrial colleagues.

When completed in 1967 this Earth-rotation transit telescope consisted of two 1 mile (1.6 km) long arms, each in the form of a parabolic cylindrical reflector 40 ft (12 m) wide. It was a single frequency telescope with an operational frequency of 408 MHz (λ 73 cm). The east–west arm could be titled mechanically and the other arm could be phased to enable a scan of the southern sky to be produced. The telescope had a collecting area of ten times that of the large Parkes dish and a high angular resolution, for those days, of 2.8 arcmin at 408 MHz. It was used to produce the Molonglo Reference Catalogue of more than 12,000 strong sources, which was published in 1981, as well as a number of other major catalogues of weaker sources.

By July 1968, the Molonglo Cross had discovered the first two pulsars in the southern sky followed in October by the discovery of a pulsar with the shortest pulse rate then known of 89 milliseconds. In fact the Molonglo Cross proved to be an excellent instrument for detecting pulsars, finding 50% of the number of pulsars known by the end of 1971.

[27] This and a number of other events caused a significant rift in the relationship between Radiophysics and the University of Sydney radio astronomy group in the early 1960s. In fact the bad blood between the two groups only stopped with the retirement of the chief antagonists, see Raymond Haynes, et al., *Explorers of the Southern Sky; A History of Australian Astronomy*, Cambridge University Press, 1996, p. 307.

Figure 16.5. Looking along the east–west arm of the Molonglo Observatory Synthesis Telescope (MOST) from its western end. (Courtesy Peter Ellis via Wikimedia Commons.)

The Molonglo Cross had finally broken the monopoly of the Radiophysics Laboratory on radio astronomy in Australia. But it was a one-frequency telescope with a limited steering capability. So in 1978 the north–south arm was closed down, and the east–west arm was converted over the next three years into a new steerable instrument called the Molonglo Observatory Synthesis Telescope (MOST) (see Figure 16.5) operating at the higher frequency of 843 MHz (λ 36 cm) with a resolution of about 40 arcsec.(33) Because of its location in Australia it has been used to study the Magellanic Clouds and the centre of the Milky Way, as well as spiral galaxies, pulsars, and planetary nebulae. In 1987 it, like many other telescopes worldwide, was used to observe the supernova SN1987A which appeared in February of that year. By April, MOST was the only radio telescope with sufficient sensitivity to continue to be able to detect it.

16.4 The Australia Telescope

It had become increasingly clear by the mid 1970s that Australia was gradually slipping further and further behind other countries in the development of modern astronomical facilities. One way of addressing this problem was to build a state-of-the-art aperture synthesis radio telescope, as there was no such instrument available at that time in the southern hemisphere. It would also take advantage of Australia's previous expertise in building and operating radio interferometers. So in the mid 1970s Australian radio astronomers considered a number of possible designs and in 1978 settled on one based on the Parkes 64 m antenna as the prime instrument, together with a number of new, smaller, movable dishes, also located at Parkes, mounted on rails. Unlike the cases of the Parkes 64 m, the Radioheliograph and the Molonglo Cross the Australian government was asked to fund the whole of this new project, then called the Australian Synthesis Telescope (AST). Unfortunately the design chosen, which had been heavily constrained by costs, had too many problems associated with it to produce anything like unanimous backing from the radio astronomy community.(34) And as time went on the backing that there was gradually began to fade.

Robert Frater, in particular, who became Chief of Radiophysics in 1981, considered that the project was severely under-budgeted and offered no scope for industrial development. He also thought that the AST's scope and useful life were too limited and that its design was not innovative enough to promote applied research in Australia. In addition Frater had learned in May 1981 that a company was considering developing a new mining facility very near the Parkes facility which would severely compromise the AST if it was to be located there. Consequently it came as no great surprise when the government rejected the project's funding request.

It was clear at that time that unless an alternative design to the AST was proposed rather quickly radio astronomy in Australia would enter a long period of decline. So in June 1981 a new scheme was outlined in a discussion between Frater and Brian Robinson, a senior radio astronomer at Radiophysics. This was based on a new telescope array to be built, not at Parkes, but at the Culgoora site of the Radioheliograph that was already destined to close. The proposed new array would be able to operate over the very wide wavelength range from about 3 mm to 1 m (100 GHz to 300 MHz), and be linked to a new radio telescope dish near Siding Spring as well as to the Parkes 64 m. This solution had the advantage of largely avoiding problems caused by possible mining operations near Parkes, whilst accommodating the diverse research interests of many different astronomy groups in Australia. These included molecular line work, wide-field mapping and Very Long Baseline Array observing. It would also provide an impressive example of modern Australian design and engineering expertise, as well as involving industrial contractors in a number of different localities. And in a masterpiece of public relations it was called the Australia Telescope and was proposed as a Bicentenary project helping to celebrate, in 1988, the bicentenary of the first British settlement in Australia.[28]

[28] This was a very clever idea as governments tend to have difficulty turning down major visible projects linked to such important periods in their nations' history.

Figure 16.6. Five 22 m diameter dishes of the Australia Telescope's Compact Array. (Courtesy John Masterson, CSIRO via Wikimedia Commons.)

This Australia Telescope (AT) was approved by the Australian Federal Government in 1982 with funding up to $A25 million ($US24 million), indexed for inflation. It was to consist of a Compact Array at Culgoora, now called the Paul Wild Observatory, plus a new radio dish at Mopra, near Siding Spring,[29] about 115 km south of Culgoora, and the 64 m Parkes antenna to provide a total baseline of about 300 km. As built[30] the Compact Array consisted of six 22 m dishes on an east–west baseline of 6 km.(35) Five of the dishes (see Figure 16.6) were mounted on a 3 km railway track, with the sixth being 3 km from the western end of this track on a short length of track of its own at either end of which were two observing stations 75 m apart. The whole 22 m surface of each dish could be used for observations down to 6 mm, whilst the inner 15 m, which was built to a higher standard, could be used for observations down to 3 mm. The 22 m dish at Mopra was identical in design to the Culgoora dishes to which it was linked by a

[29] It had originally been intended to build this dish on Siding Spring Mountain near the UK Schmidt but radio interference, high winds and environmental problems caused it to be built about 4 km away.

[30] Initially the Compact Array was planned to consist of five dishes, four on a 3 km track and one fixed 3 km away, see David H. Smith, Australia's Bicentennial Bonanza, *Sky and Telescope*, February 1983, pp. 120–121.

microwave link. Frater was Project Director of the AT project with John Brooks, who had worked on the upgrade of the 64 m Parkes radio telescope, responsible for day-to-day management.

The surface of the new dishes consisted of aluminium panels forced into shape by attaching them to their supporting ribs with bolts of precisely controlled different lengths. This saved the extra expense of producing the dishes in precisely machined moulds. Usually coaxial cables would have been used to transfer data between the dishes of the Compact Array, but coaxial cables normally exhibit considerable loss of signal with distance and variations of performance with temperature. Consequently the designers of the AT, in a break with tradition, decided to use optical fibres.

As part of the Bicentenary celebrations the AT had to be completed in 1988 and it was, although it took another two years to become fully operational. Ron Ekers, who had been Foundation Director of the Very Large Array in the USA, returned to Australia in 1988 and was appointed Foundation Director of the Australia Telescope National Facility (ATNF). Then in September of that year, when the AT was formally opened by Bob Hawke, the Prime Minister, it was separated from the Division of Radiophysics to become a Division of CSIRO.

Since completion other radio telescopes around Australia have, from time to time, been connected to the ATNF to provide a very long baseline interferometer with baselines of up to 3,000 km. They have included the NASA 70 m Deep Space Network dish at Tidbinbilla, and antennae at Hobart, Ceduna and Perth. Initially this was very much an ad hoc system but in later years the Australian Long Baseline Array (LBA) was formally established which consisted of the ATNF telescopes at Parkes, Culgoora and Mopra and antennae at Hobart and Ceduna operated by the University of Tasmania. Since 2008 the main interferometer correlation for the LBA has been undertaken at the Curtin University of Technology based in Perth.

About ten years after becoming operational, the Compact Array was modified to add a 200 m long north spur perpendicular to the existing 6 km baseline. Initially the surface area of the dishes between 15 m and 22 m diameter had been constructed of perforated panels to reduce wind resistance. But now this outer surface area was changed from perforated to solid panels to improve observations down to 3 mm. In addition the receivers were replaced by those jointly developed by the ATNF and the CSIRO Division of Telecommunications and Industrial Physics (CTIP) to operate at very short wavelengths.[31] This upgrade work was completed in 2004.

Starting in October 2012, funding for the 22 m radio telescope at Mopra was no longer the sole responsibility of CSIRO. From that date additional funding was

[31] These new receivers used indium phosphide chips cooled to 20 K (–253°C).

provided by the National Astronomical Observatory of Japan, the University of New South Wales and the University of Adelaide. In return these organisations were provided with guaranteed observing time. Unfortunately in January 2013 a severe bushfire broke out in the vicinity of the observatory. Although the antenna itself was unharmed, the bushfire caused damage to the local infrastructure causing a four month shutdown of the observing facility.

16.5 The Australian Square Kilometre Array Pathfinder and Murchison Widefield Array

The Australian Square Kilometre Array Pathfinder (ASKAP) was built at the very remote Murchison Radio-astronomy Observatory (MRO) in Western Australia over the period 2009–2012. As its name implies it was built, not just as a telescope in its own right, but as a testbed for the future international Square Kilometre Array (SKA) project. The Murchison site was chosen as it had an extremely radio-quiet environment which the Australian government were prepared to preserve by passing appropriate legislation. The ASKAP consisted of thirty-six 12 m diameter antennae at fixed locations operating at frequencies from 700 MHz to 1.8 GHz (λ 43 to 17 cm) to provide an interferometer with a maximum baseline of 6 km. Although CSIRO led the development of the ASKAP, it was supported by scientists and engineers in the Netherlands, Canada, the USA and China. In 2012 the SKA Organisation announced that Australia and South Africa would co-host the SKA telescope. In the case of Australia this would use the ASKAP as a base, extending its number of antennas from 36 to 96.

In addition to the ASKAP the MRO is also home to the Murchison Widefield Array (MWA) which was an international project supported by institutions in the USA, New Zealand and India as well as Australia. Its main scientific goal was to detect neutral atomic hydrogen from the cosmological Epoch of Reionization (EoR). The MWA was a wide-field, low-frequency dipole array operating between 80 and 300 MHz. Like the ASKAP it was also a precursor for the SKA.

References

1. Sullivan, Woodruff T., *Cosmic Noise; A History of Early Radio Astronomy*, Cambridge University Press, 2009, p. 120.
2. Orchiston, Wayne, Slee, Bruce and Burman, Ron, The Genesis of Solar Radio Astronomy in Australia, *Journal of Astronomical History and Heritage*, 9, No. 1, 2006, pp. 35–56.
3. Stewart, Ronald, et al., The Radiophysics Field Station at Penrith, New South Wales, and the World's First Solar Radiospectrograph, *Journal of Astronomical History and Heritage*, 13, No. 1, 2010, pp. 2–15.

4. Bolton, J. G., and Stanley, G. J., Observations on the Variable Source of Cosmic Radio Frequency Radiation in the Constellation of Cygnus, *Australian Journal of Scientific Research*, 1, 1948, pp. 58–69.
5. Orchiston, Wayne, and Slee, Bruce, The Radiophysics Field Stations and the Early Development of Radio Astronomy, in Orchiston, Wayne, (ed.), *The New Astronomy: Opening the Electromagnetic Window and Expanding our View of Planet Earth*, Springer, 2005, pp. 119–168.
6. Sullivan, Woodruff T., *Cosmic Noise; A History of Early Radio Astronomy*, Cambridge University Press, 2009, pp. 277–280.
7. Kerr, F. J., Early Days in Radio and Radar Astronomy in Australia, in Sullivan, W. T., (ed.), *The Early Years of Radio Astronomy; Reflections Fifty Years after Jansky's Discovery*, Cambridge University Press, 1984, pp. 136–137.
8. Wild, J. P., Murray, J. D., and Rowe, W. C., Harmonics in the Spectra of Solar Radio Disturbances, *Australian Journal of Physics*, 7, 1954, pp. 439–459.
9. Wendt, Harry, Orchiston, Wayne, and Slee, Bruce, W. N. Christiansen and the Development of the Solar Grating Array, *Journal of Astronomical History and Heritage*, 11, No. 3, 2008, pp. 173–184.
10. Piddington, J. H., and Minnett, H. C., Observations of Galactic Radiation at Frequencies of 1210 and 3000 Mc/s, *Australian Journal of Scientific Research*, 4, 1951, pp. 459–475.
11. Wendt, Harry, Orchiston, Wayne, and Slee, Bruce, W. N. Christiansen and the Initial Australian Investigation of the 21 cm Hydrogen Line, *Journal of Astronomical History and Heritage*, 11, No. 3, 2008, pp. 185–193.
12. Kerr, F. J., Early Days in Radio and Radar Astronomy in Australia, in Sullivan, W. T., (ed.), *The Early Years of Radio Astronomy; Reflections Fifty Years after Jansky's Discovery*, Cambridge University Press, 1984, p. 139.
13. Wendt, Harry, Orchiston, Wayne, and Slee, Bruce, W. N. Christiansen and the Development of the Solar Grating Array, *Journal of Astronomical History and Heritage*, 11, No. 3, 2008, p. 175.
14. Sullivan, Woodruff T., *Cosmic Noise; A History of Early Radio Astronomy*, Cambridge University Press, 2009, pp. 330–331.
15. Slee, Bruce, Early Australian Measurements of Angular Structure in Discrete Radio Sources, *Journal of Astronomical History and Heritage*, 8, No. 2, 2005, pp. 97–106.
16. Davies, R. D., A History of the Potts Hill Radio Astronomy Field Station, *Journal of Astronomical History and Heritage*, 8, No. 2, 2005, p. 93.
17. Orchiston, Wayne, From the Solar Corona to Clusters of Galaxies: The Radio Astronomy of Bruce Slee, *Publications of the Astronomical Society of Australia*, 21, 2004, pp. 23–71.
18. Slee, Bruce, Early Australian Measurements of Angular Structure in Discrete Radio Sources, *Journal of Astronomical History and Heritage*, 8, No. 2, 2005, p. 103.
19. Orchiston, Wayne, and Mathewson, Don, Chris Christiansen and the Chris Cross, *Journal of Astronomical History and Heritage*, 12, No. 1, 2009, pp. 11–32.
20. Orchiston, Wayne, and Slee, Bruce, Ingenuity and Initiative in Australian Radio Astronomy: the Dover Heights 'Hole-in-the-Ground' antenna, *Journal of Astronomical History and Heritage*, 5, No. 1, 2002, pp. 21–34.
21. Munns, David P. D., *A Single Sky; How an International Community Forged the Science of Radio Astronomy*, The MIT Press, 2013, p. 124.
22. Minnett, Harry, The Construction of the Parkes 210-ft Radio Telescope in Goddard, D. E., and Milne, D. K., (eds.), *Parkes: Thirty Years of Radio Astronomy*, CSIRO Publishing, 1994, pp. 12–18.

23. Minnett, H. C., The Australian 210-foot Radio Telescope, *Sky and Telescope*, October 1962, pp. 184–189.
24. Kerr, Frank, The Proposal for a Giant Radio Telescope, in Goddard, D. E., and Milne, D. K. (eds.), *Parkes: Thirty Years of Radio Astronomy*, CSIRO Publishing, 1994, pp. 7–11.
25. Kerr, Frank J., Australia's 210-foot Radio Telescope Project, *Sky and Telescope*, October 1959, pp. 666–668.
26. Bolton, J. G., Gardner, F. F., and Mackey, M. B., The Parkes Catalogue of Radio Sources Declination Zone – 20° to – 60°, *Australian Journal of Physics*, 17, 1964, pp. 340–372.
27. Hazard, C., Mackey, M. B., and Shimmins, A. J., Investigation of the Radio Source 3C 273 by the Method of Lunar Occultations, *Nature*, 197, 1963, pp. 1037–1039.
28. Orchiston, Wayne, The Parkes 18-m Antenna: A Brief Historical Evaluation, *Journal of Astronomical History and Heritage*, 15, No. 2, 2012, pp. 96–99.
29. Yabsley, D. E., Millimetre Wavelength Performance of the New 16.7 m Surface of the Parkes Radio Telescope, *Proceedings of the Astronomical Society of Australia*, 3, No. 2, September 1977, pp. 113–115.
30. Higgs, A. J., The Australian Radioheliograph Project, *Sky and Telescope*, July 1968, pp. 11–14.
31. Maxwell, Alan, Radio Emissions from Solar Flares, *Sky and Telescope*, July 1973, pp. 4–9.
32. McAdam, Bruce, Molonglo Observatory: Building the Cross and MOST, *Journal of Astronomical History and Heritage*, 11, No. 1, 2008, pp. 63–70.
33. McAdam, Bruce, Molonglo Observatory: Building the Cross and MOST, *Journal of Astronomical History and Heritage*, 11, No. 1, 2008, pp. 68–69.
34. Haynes, Raymond, et al., *Explorers of the Southern Sky; A History of Australian Astronomy*, Cambridge University Press, 1996, pp. 414–418.
35. Norris, Ray, The Australia Telescope, *Sky and Telescope*, December 1988, pp. 615–620.

17

Cambridge Mullard Radio Observatory

17.1 The Early Years

Martin Ryle, Bernard Lovell and a number of other people who became leaders in British radio astronomy had worked for the British Telecommunications Research Establishment (TRE) during the Second World War. John (Jack) Ratcliffe, who had previously led a research group at Cambridge University's Cavendish Laboratory also worked at TRE as one of its top administrators. But in early 1945 Ratcliffe left TRE to rebuild his radio research group at Cambridge where he was able to attract Ryle, Derek Vonberg and F. Graham Smith amongst others.

When Ryle joined the Cavendish laboratory the sunspot cycle was on its way towards solar maximum. As a result Ratcliffe suggested to Ryle that it may be a good time to investigate the source of radio noise associated with sunspot activity that had been observed by Stanley Hey during the Second World War.(1) But unfortunately at that time a typical radar antenna, operating at a wavelength of 1.5 m, had a beamwidth of the order of 10° making it impossible to find the exact source of radio waves on the Sun. So in the winter of 1945–1946 Ryle decided to build a radio interferometer[1] which should be able to locate the sources much more accurately. It consisted of two antennae working at 175 MHz (λ 1.7 m) at separations of up to 240 m or 140 wavelengths. Each antenna was made of eight half-wave dipoles mounted over a wire mesh reflector. With this system Ryle and Vonberg were able to show that short duration radio bursts from the Sun were often circularly polarised and came from discrete areas on the Sun's disc, and not from the disc as a whole.

Two years later Ryle and Smith used another Michelson interferometer at Cambridge to study the cosmic source Cyg A, which Hey, Parsons and Phillips had found in 1946 fluctuated over a period of a few seconds. Ryle and Smith's

[1] Ryle maintained that his work on aircraft radar during the war gave him this idea, and that it was not a conscious attempt to produce the radio equivalent of a Michelson interferometer, although that is what it turned out to be (see Woodruff T. Sullivan, *Cosmic Noise; A History of Early Radio Astronomy*, Cambridge University Press, 2009, p. 160).

interferometer consisted of two groups of four Yagi antennae spaced 500 m apart, operating at 80 MHz (λ 3.75 m).(2) On their first night of observations they not only observed Cyg A but found another radio source, this time in the constellation of Cassiopeia. This turned out to be Cas A which had been discovered some years earlier by Grote Reber. In addition Ryle and Smith also found evidence of other, weaker sources.

In order to detect these weaker sources Ryle, Smith and Bruce Elsmore constructed the so-called Long Michelson interferometer. It consisted of two broadside arrays aligned approximately east–west, about 400 m apart, of 40 half-wave dipoles operating at 80 MHz. Observations which began in December 1948 produced about 20 sources, many of them faint. But it was difficult to establish their positions with any accuracy as their fringe patterns varied from day to day with some even disappearing from time to time. So Ryle decided to double the east–west length of both these arrays, so narrowing the beam and increasing the intensity of the signals. Ryle also used a phase-switching receiver, which he had invented, that enabled small radio sources to be discriminated from large ones, which were rejected. Subsequently Ryle, Smith and Elsmore used this system to produce the first Cambridge (or 1C) catalogue of 50 sources which was published in 1950. In addition Graham Smith and Antony Hewish also showed, through a long series of observations in collaboration with Gordon Little and Bernard Lovell at Jodrell Bank, that most of the short-term fluctuations that they observed were produced by the Earth's ionosphere.

At about this time Ryle and Smith also built an interferometer consisting of two 7.5 m diameter parabolic German (Würzburg) radar antennae 280 m apart, operating mostly at 214 MHz (λ 1.4 m). This was to try to produce accurate position estimates of the strongest radio sources to enable their optical equivalents to be found. In 1951 Smith, using this system, managed to reduce the positional errors of bright radio sources to 1 arcmin, which enabled Walter Baade and Rudolph Minkowski to use the 200 inch (5 m) Palomar and 100 inch Mount Wilson reflectors to find the optical sources of Cas A and Cyg A. The first was identified as a faint dispersed nebula which had fast-moving gas filaments, whilst the second was an inconspicuous 16th magnitude galaxy with a redshift of about 15,000 km/ s. This and similar discoveries over the next few years finally showed that many radio sources were not radio stars, as originally thought, but were galaxies at some distance from the Milky Way.

The number of radio sources detected at Cambridge, Sydney and elsewhere steadily increased in the early 1950s. As mentioned above, 50 northern sources had been listed in the Cambridge 1C catalogue in 1950. At about the same time 22 southern sources had been listed by Gordon Stanley and Bruce Slee of the Sydney Radiophysics Laboratory in Australia. But the Cambridge team then built a new interferometer over the period 1952–3 of four parabolic cylindrical antennae each 12 × 98 m in size. With this they were able to publish the second Cambridge Catalogue (2C) in 1955 which listed an amazing 1,936 sources.

It is interesting to note the considered views of a number of senior Cambridge staff on the complementary management styles of Ryle and Ratcliffe, his supervisor, as reported by Sullivan.(3) Ryle's personality, not just his undoubted intelligence, had an enormous influence on the development of radio astronomy at Cambridge immediately after the war. Although operating more in the background, Ratcliffe also played an important role. According to their colleagues, Ratcliffe took most of the administrative load off Ryle by raising money, recruiting top-quality personnel, legalising 'looted' equipment, and so on leaving Ryle to concentrate on designing and building the hardware, as 'one of the boys', and supervising the research work. Ratcliffe was apparently a very conventional manager of the period operating largely from his office, whilst Ryle was happy building his interferometers at Cambridge out of all sorts of army surplus equipment. But Ratcliffe, although not so obviously a presence, was also prone to come up with awkward questions. He edited all the published articles of the group and took as much of the lecturing load as he could off Ryle whilst he, himself, produced some of the best lectures of the period at the university.

In his review of radio astronomy in 1950 Ryle had concluded that most radio sources were unusual stars in the Milky Way. But it gradually became clear over the next few years that there were far too many faint sources outside the plane of the Milky Way to sustain this point of view. Then in 1955, following the publication of the 2C catalogue, Ryle suggested that the large number of faint sources in this catalogue indicated that the density of sources in the universe had reduced over time. As a result, the steady state cosmological theory, then favoured by a number of his colleagues in Cambridge, must be incorrect. But the number of sources published in 2C was many more than listed by the Sydney group at that time, and this caused a major disagreement between the two groups. Minkowski, acting as unofficial mediator, eventually concluded that many of the Cambridge sources were spurious. Ryle disagreed,[2] but it eventually became clear that the beamwidth of the Cambridge interferometer was too large to cope with the high density of sources, resulting in source confusion. The Cambridge group then modified their interferometer to operate at a higher frequency. This improved the resolution and reduced the source confusion, so that when the third Cambridge Catalogue was published in 1959 it contained only a quarter of the sources of the second catalogue.

[2] The Cambridge group was very insular at that time. They tended to guard their unpublished material very closely and Ryle tended to act, when dealing with outside groups, in a very arrogant way which, in this case, did not help to sort out which group was right.

17.2 Aperture Synthesis

In the late 1940s/early 1950s Ryle had developed his idea of aperture synthesis to produce radio images, for which he received the Nobel Prize in 1974. In principle, this required just two antennae to be positioned in two locations of a two-dimensional ground-based array to sample the amplitude and phase of the signals received from a source. The two antennae were then placed in various other positions in the two dimensional array, and the image of the radio sky produced by a computer which carried out the Fourier inversion involved in combining all the observations.(4) The first one-dimensional synthesis was undertaken by Harold Stanier at Cambridge in 1949. He did this when observing the Sun as it could be assumed, to a first approximation, to be symmetrical. Stanier used two 3 m diameter Würzburg dishes at 500 MHz (λ 60 cm) with 17 spacings up to a maximum of 220 m. But the first aperture synthesis instrument capable of mapping an arbitrary distribution of sources in two dimensions was built by John Blythe at Cambridge in 1954. This consisted of 48 full-wave dipoles operating at 38 MHz (λ 7.9 m) arranged in a line 365 m (1200 ft) long, plus a single dipole that was moved step by step in a perpendicular direction.(5) By this time Maurice Wilkes and staff at the university's mathematical laboratory had built the EDSAC (Electronic Delay Storage Automatic Calculator) 1 computer which enabled Blythe to produce his map of the sky.

Blythe's antenna was the prototype for the 178 MHz (λ 1.7 m) interferometer that was built in 1957 and used at Cambridge to produce the almost 5,000 source, 4C catalogue in the mid 1960s. This new interferometer consisted of a 440 m long, 20 m wide fixed east–west array of antennae representing slices of a cylindrical paraboloid, and a moveable array which was a 58 m section of a similar cylindrical paraboloid on 305 m long north–south rails 780 m east of the fixed antennae. This interferometer (see Figure 17.1) was built at a new radio observatory site at Lord's Bridge outside Cambridge, which was later called the Mullard Radio Astronomy Observatory.[3] The main constraints on the performance of this, and other early aperture synthesis systems, was the limited capability of the early computers needed to perform the Fourier inversion required.

Antony Hewish and colleagues designed and built a radio telescope at Cambridge in the mid 1960s to detect scintillating radio sources, which were assumed to be quasars. This telescope, which was designed to detect rapidly fluctuating signals at 80 MHz, consisted of an array of permanently fixed wires connected to over 1,000 posts spread out over an area of about 2 hectares (5 acres). Just after this telescope was completed in 1967, Jocelyn Bell discovered a source with it that showed a regular pulse of 1.34 seconds. A little later she discovered three more

[3] The change in name followed a donation of £100,000 ($280,000) by the Mullard electronics company.

Figure 17.1. The 440 m long fixed east–west array of the Cambridge 4C interferometer. (Courtesy Astrophysics Group, Cavendish Laboratory, University of Cambridge.)

sources that also had regular short duration pulses. These were the first pulsars to be discovered, for which Antony Hewish was awarded the Nobel Prize for Physics in 1974.

Three interferometers were built at the Mullard Radio Astronomy Observatory in the 1960s and early 1970s. These were the One Mile telescope built in 1962–1964, followed by the Half-Mile telescope, and finally the Five Kilometre or Ryle telescope, which was completed in 1971. Each of these consisted of one line of dishes, aligned east–west, which used the rotation of the earth to provide a two dimensional scan using earth-rotation aperture synthesis.

Ryle had submitted his request for a grant of £368,000 ($1.0 million) to the Department of Scientific and Industrial Research (DSIR) in April 1960 for the One Mile telescope. Two months later the DSIR awarded Ryle a £3,000 ($8,400) grant for a design study, but budgetary problems at DSIR meant that the construction grant was not approved until December 1961. Ryle and his group at Cambridge had operated on a shoestring for many years, but now he was spending a significant amount of money.

In fact, in his early work at Cambridge Ryle had followed the Cavendish tradition of using simple, inexpensive apparatus of elegant design.(6) But in the 1950s he was gradually forced to accept that this approach would not enable his radio astronomy group to remain one of the subject leaders. So although the 1C interferometer had been built in the late 1940s for a cost of just £350 ($980), the 2C interferometer which had been built a few years later cost about £6,400 ($18,000). Now he had requested over fifty times this amount to build the One Mile telescope.

When completed the One Mile Telescope consisted of three 18 m dishes in line, two of which were fixed about 750 m apart with the third being mounted on an 800 m length of rails, so providing a maximum separation of up to 1550 m.(7) The telescope was designed to detect much fainter and therefore more distant sources than previously observed and to determine the structure of individual sources. By detecting distant sources Ryle was hoping to be able to distinguish between different cosmological models. The first tests with the One Mile telescope were carried out at 408 MHz (λ 74 cm) in the spring of 1964 and the first observations were undertaken at 1407 MHz (λ 21 cm) later that year.

Once the One Mile telescope was in operation, John Shakeshaft and John Baldwin produced plans for a Half-Mile telescope mounted on the same rails as the One Mile. The Half-Mile used four 9 m dishes which were purchased at scrap metal prices from a discontinued radio link service. It was used mainly as a radio spectrometer operating over the 21 cm band of neutral hydrogen to map the distribution and velocity of neutral hydrogen in nearby galaxies.

The Five Kilometre telescope was designed in the mid 1960s by Martin Ryle to map individual sources at wavelengths as short as 2 cm (15 GHz frequency) with a resolution of about 1 arcsec. It consisted of eight 13 m dishes in a line, four fixed at intervals of 1.2 km and four movable on a 1.2 km rail track to provide a maximum east–west baseline of 4.8 km. The telescope was designed as a survey instrument to detect the faintest sources, without ambiguity, and to provided accurate celestial coordinates to facilitate optical identification. It also enabled detailed maps to be produced of the complex structure of radio emission from distant galaxies and quasars. In the late 1980s its sensitivity was increased to enable it to observe the 'shadowing' of the cosmic microwave background (CMB) by foreground clusters of galaxies by the so-called Sunyaev-Zel'dovich (SZ) effect, and its name was changed to the Ryle Telescope.

17.3 Modern Instruments

In the last decade or so the eight 13 m Ryle antennae have been relocated to form a two dimensional array, called the Arcminute Microkelvin Imager (AMI) Large Array (see Figure 17.2), with baselines of up to 100 m. Another array, the so-called

Figure 17.2. The eight 13 m diameter Ryle antennae relocated to form the Arcminute Microkelvin Imager Large Array at the Mullard Radio Astronomy Observatory near Cambridge, UK. (Courtesy Cmglee via Wikimedia Commons.)

AMI Small Array (see Figure 17.3), was also built consisting of ten 3.7 m antennae situated within a small, 4.5 m high metal enclosure. These 3.7 m antennae had rimmed primaries and shaped secondaries to minimise crosstalk and groundspill. The Large Array had a much higher resolution than the Small Array. Consequently the Large Array was used to subtract the effect of foreground sources from the Small Array's observations of galaxy clusters. Essentially the AMI as a whole, which operated over the 13.5–18 GHz range, was designed to observe already-known galaxy clusters and to detect new such clusters.

The Cambridge Low-Frequency Synthesis Telescope (CLFST) had been built in 1980 very much in the Cavendish tradition of the early radio telescopes, that is it was of a relatively simple design with great capability. This design was based on that of the 6C array which consisted of 40 Yagis on a 1.8 km baseline operating at 151 MHz (λ 2.0 m). The CLFST had 60 of the same Yagis on a 4.6 km baseline also operating at 151 MHz. It was used to produce the 7C survey. Then in 1984 an array of 58 eight metre Yagis was built parallel to the 4.6 km array to operate at 38 MHz (λ 7.9 m) during a period of solar minimum when disturbances of the ionosphere should be at a minimum. This was used to produce the 8C survey. The CLFST ceased operating in 2000.

Figure 17.3. The Arcminute Microkelvin Imager Small Array in a shielded enclosure at the Mullard Radio Astronomy Observatory near Cambridge. (Courtesy Matthewldavies via Wikimedia Commons.)

In the meantime a 32 m dish-type radio telescope had been built at Cambridge in 1991 as a node of MERLIN (see Section 18.5), extending the MERLIN baseline to 217 km. This was followed a few years later by the Cosmic Anisotropy Telescope (CAT)(8) which was designed to measure fluctuations in the CMB. It was an interferometer consisting of three 70 cm diameter antennae with a baseline from 1 to 5 m which operated over the frequency range 13–17 GHz. The telescope was surrounded by a 5 m high earth bank lined with aluminium to reduce the effect of spillover and terrestrial radio interference which limited observations to elevations over about 30°. In 1996 CAT was able to provide evidence of the first acoustic peak at about 1° ($l \sim 200$) in the CMB power spectrum after the Ryle telescope had been used to remove the effect of discrete sources.(9) The telescope was turned off in 2000 but experience gained with CAT was used in building the next CMB telescope, called the Very Small Array (VSA), at Cambridge.

The VSA, which was the result of a collaboration between the Astrophysics Group at the Cavendish Laboratories at Cambridge, the Nuffield Radio Astronomy Laboratories at Jodrell Bank, and the Instituto de Astrofisica de Canarias at Tenerife, was built to make high resolution observations of the CMB between 26 and 36 GHz. The telescope was a 14 element interferometer which could be operated in two complementary configurations. It was initially tested at Cambridge before being located at a more suitable site at an altitude of 2400 m in Tenerife. The better site together with the advantage of more antennae and better receivers than the CAT enabled the VSA to collect data over 100 times faster than its predecessor. In 2003 Paul Scott and colleagues reported the clear detection of the first and second acoustic peaks at $l \sim 200$ and 550.(10) Later work showed another peak at $l \sim 850$ in excellent agreement with similar measurements made by the Cosmic Background Imager located at an altitude of 5000 m in Chile's Atacama Desert.(11)

References

1. Hey, J. S., *The Evolution of Radio Astronomy*, Science History Publications, 1973, pp. 14–19 and pp. 36–37.
2. Smith, F. G., Early Work on Radio Stars at Cambridge, in Sullivan, W. T., (ed.), *The Early Years of Radio Astronomy; Reflections Fifty Years after Jansky's Discovery*, Cambridge University Press, 1984, pp. 236–248.
3. Sullivan, Woodruff T., *Cosmic Noise; A History of Early Radio Astronomy*, Cambridge University Press, 2009, pp. 169–172.
4. Ryle, Martin, *Radio Telescopes of Large Resolving Power*, Nobel Lecture 1974, Nobel website.
5. Scheuer, P. A. G., The Development of Aperture Synthesis at Cambridge in Sullivan, W. T., (ed.), *The Early Years of Radio Astronomy; Reflections Fifty Years after Jansky's Discovery*, Cambridge University Press, 1984, p. 251.
6. Munns, David P. D., *A Single Sky: How an International Community Forged the Science of Radio Astronomy*, The MIT Press, 2013, pp. 124–125 and p. 135.
7. Elsmore, B., Kenderdine, S., and Ryle, Sir Martin, The Operation of the Cambridge One-Mile Diameter Radio Telescope, *Monthly Notices of the Royal Astronomical Society*, 134, 1966, pp. 87–95.
8. Robson, M., et al., The Cosmic Anisotropy Telescope, *Astronomy and Astrophysics*, 277, 1993, pp. 314–320.
9. Scott, P. F., et al., Measurements of Structure in the Cosmic Background Radiation with the Cambridge Cosmic Anisotropy Telescope, *Astrophysical Journal*, 461, 1996, pp. L1–L4.
10. Scott, Paul F., et al., First results from the Very Small Array – III. The Cosmic Microwave Background Power Spectrum, *Monthly Notices of the Royal Astronomical Society*, 341, 2003, pp. 1076–1083.
11. Rajguru, Nutan, Cosmic Microwave Background Observations from the Cosmic Background Imager and Very Small Array: a Comparison of Coincident Maps and Parameter Estimation Methods, *Monthly Notices of the Royal Astronomical Society*, 363, 2005, pp. 1125–1135.

18

Jodrell Bank

18.1 From Radar to Radio Astronomy

Bernard Lovell had begun his investigations into cosmic rays in 1937 at the University of Manchester in the UK where P.M.S. (Patrick) Blackett was professor of physics.(1) But two years later, at the outbreak of the Second World War, Lovell was posted to an operational radar station which was assigned to track enemy aircraft. Shortly after arriving he noticed that, in addition to observing reflections from aircraft, there were numerous short-lived echoes which the radar operators attributed to the ionosphere. Lovell wondered what the cause of these signals could be, and suggested to Blackett that they may be being produced by radar reflections from the ionisation produced by highly energetic cosmic ray showers. But the exigencies of war stopped him from investigating them further.

Lovell returned to Manchester immediately the war with the intension of investigating the transient radar echoes. To do this he borrowed an army 4.2 m wavelength mobile radar with a Yagi antenna that had been used to detect V2 rockets.(2) He immediately set up the system outside the university's physics department in the middle of Manchester. There, unfortunately, any radar signals were completely overwhelmed by interference from electric trams that ran nearby. Clearly a quiet location outside the city was required. Lovell considered a number of options before settling on a plot of land at Jodrell Bank about 25 miles (40 km) south of Manchester that belonged to the university's Botanical Gardens. He was given permission to locate his equipment there for a few weeks in December 1945.

Amazingly, Lovell observed several short-lived radar echoes from various distances on his first day of observations.(3) This was repeated on the next two days but, instead of the expected one or two echoes per day, he was detecting several per hour. This led Blackett to wonder whether Lovell was really detecting echoes caused by cosmic ray ionization or whether the echoes were caused by some other effect. So he suggested that Lovell should go to see Stanley Hey, who had experience with using the V2 detection equipment during the war, as he must also have detected these short-lived signals. Hey confirmed that he had indeed detected them and suggested that at least some of the signals were caused by meteors, rather than cosmic rays, as they were detected too high in the atmosphere, were too frequent, and had longer durations than those expected to be produced by cosmic rays.

At the beginning of 1946 Lovell was working on his own but over the next few months he was joined by John Clegg, an antenna expert, and John Banwell, an expert on receiver electronics. He was also put in touch with Manning Prentice, the director of the meteor section of the British Astronomical Association, who suggested to Lovell that he should observe the short-lived radar returns during the next large meteor shower which was that of the Perseids in August. Prentice came up to Jodrell Bank to observe the shower visually and Lovell and Banwell undertook the radar observations. These observations indicated that there was a good correlation between radar echoes lasting more than half a second and meteors, although the correlation with shorter echoes was poor. Clegg then designed a new Yagi antenna and with this Lovell, Banwell and Clegg detected a spectacular peak rate of 168 echoes per minute(4) from the Giacobinid meteor shower of October 1946.

At about this time it became clear that Lovell's calculations of the expected intensity of radar echoes from ionisation trails caused by cosmic rays were far too optimistic. Consequently he required a much larger antenna to detect them. To begin with he and Clegg started to build a large horizontally directed array on a 100 ft (30 m) high tower, but abandoned the construction when the tower was only about 20 ft high. Instead they decided to build a large fixed paraboloid looking vertically upwards, as it would be easier to build. In addition it would be easier to change wavelengths and to incorporate a powerful transmitter. The result was a fixed 218 ft (66 m) diameter dish made of a complex mesh of wire built over the period 1946–1947. The main framework of the dish consisted of 24 heavy duty steel cables each radiating from the centre, which was at ground level, and each supported at the edge by 23 ft (7 m) high scaffolding poles with guy wires attached. These wires were strained to the ground at three intermediate points between the centre and the edge to produce an approximately parabolic profile. A circular array of wire was then wound at 8 inch (20 cm) intervals on top of these heavy duty cables to form the reflecting surface. The feed was placed at the top of a 126 ft (38 m) high mast which was pivoted at the base and secured by guy wires. Again the search for cosmic ray showers was unsuccessful. So Lovell decided to use the dish as a transit telescope, operating at 72 MHz (λ 4.16 m), to search for cosmic radio sources instead, as its beamwidth was much smaller, and its collecting area much larger, than any other comparable antenna of the time.

In 1949 Robert Hanbury Brown and Cyril Hazard joined Lovell's group at Jodrell Bank and began to modify the 218 ft radio telescope to operate at 158.5 MHz (λ 1.89 m). To do this they redesigned the primary feed, designed a state-of-the-art receiver, and replaced the lossy open-wire feeder lines running up the central mast with polyethylene coaxial cable. In addition they decided to make observations over more than just the zenith by tilting the central mast as far from the vertical as possible by adjusting its 18 supporting guy wires. This took two hours of delicate work, to avoid kinking the mast, to move one beam width of 2° from the vertical. (5) Eventually they were able to tilt the mast by up to 15°. Fortunately, the area of sky observable included the Andromeda nebula and in 1950 Hanbury Brown and Hazard were able to prove, after ninety nights of observations, that this was a radio source.[1]

Over the next two years Hanbury Brown and Hazard undertook a survey of the Milky Way visible from Jodrell Bank between 38°N and 68°N declinations, finding a concentration of the brighter radio sources around the plane of our galaxy. This implied that some, at least, of these sources must be members of the Milky Way. Bernard Mills in Australia had already come to the same conclusion, whilst Martin Ryle at Cambridge disagreed. It later became evident that Ryle was mistaken, however, because of unforeseen problems in interpreting the results of the Cambridge interferometer.

In 1953 Hanbury Brown, Henry Palmer and Richard Thompson built an interferometer at Jodrell Bank using a small movable array and the 218 ft paraboloid. Over the next three years they progressively increased the interferometer's baseline up to 12 miles (20 km) using a radio link, but even at this separation three of the sources that they had detected were unresolvable, implying that their angular diameters were less than 12 arcsec. Four years later Rudolph Minkowski of the Mount Wilson and Palomar Observatories identified one of these sources, 3C 295, as a peculiar galaxy with a redshift of 0.46.

18.2 The 250 ft Mark I

The initial success of the 218 ft fixed dish led Bernard Lovell in 1948 to suggest to Patrick Blackett that a fully steerable dish should be built of at least the same diameter.(6) Blackett was highly supportive of the idea and asked Lovell to find someone who could build it and estimate its likely cost. Unfortunately the large UK engineering companies who could probably make such a large telescope had full order books to rebuild Britain after the war. In addition some of them thought that the telescope was not technically feasible, and even if it did turn out to be so they were doubtful that they could make any money from such a challenging project.

[1] An initial paper was published by Hanbury Brown and Hazard in *Nature* in 1950, followed by a more complete paper with corrections in the *Monthly Notices of the Royal Astronomical Society (MNRAS)* in the following year. See Radio-frequency Radiation from the Great Nebula in Andromeda (M.31), *Nature*, 166, 1950, pp. 901–902, and Radio Emission from the Andromeda Nebula, *MNRAS*, 111, 1951, pp. 357–367.

Faced with this lack of interest, in July 1949 Lovell asked George Sisson of Grubb Parsons, who had made a number of large optical telescopes, for his advice.(7) Sisson said that he did not think that his firm would be interested in building the large radio telescope but put Lovell in contact with Head Wrightson and Aldean Ltd (HWA), a firm specialising in structural aluminium work. Their special projects department appeared to be interested in the project and in May the following year HWA came up with a figure of at least half a million pounds ($1.4 million) for a 300 ft (90 m) diameter telescope. This compared with a price range of £60,000 to £70,000 (about $170,000 to $200,000) that the university had been expecting.[2] Blackett made it clear to Lovell that the university could not afford anything like HWA's estimated cost and suggested that it would be better to start with a very much smaller and cheaper telescope. Lovell argued that the telescope diameter should be at least 200 ft (60 m) and asked HWA for a more detailed estimate for a 200 ft telescope. But before such an estimate could be produced HWA decided that they were not interested in the project.

In the meantime Lovell had been put in touch with H. Charles Husband, the head of an engineering consulting firm, by F. D. Roberts of Coubro and Scrutton who had supplied the steel mast for the 218 ft telescope.[3] Husband, who visited Jodrell Bank for the first time in September 1949, was not put off by the proposed size of Lovell's telescope nor the low budget that the university had in mind. In November Roberts suggested that the large telescope could probably be built for around £45,000 ($125,000) using Husband's preliminary design. Husband's structure was made of steel as he had pointed out that steel was heavy and better able to withstand wind forces, whilst that of HWA was made of light alloy.

At this stage Blackett suggested to the council of the Royal Astronomical Society (RAS) that, in view of the rapid progress in radio astronomy in the UK and elsewhere, it might be timely to appoint a committee to consider the future development of radio astronomy in the UK. This it agreed to do. Blackett then suggested that Lovell, together with the leaders of the other two important radio astronomy groups in the UK, namely Martin Ryle of Cambridge and Stanley Hey of the Army Operational Research Group, should prepare a paper for the committee outlining the possible research that the UK could undertake in radio astronomy over the next five years. As a supplement Lovell also produced a paper outlining the design and possible research programme for his proposed large radio telescope. The RAS committee considered Lovell's paper at its second meeting in

[2] These figures from the May meeting with HWA are as given by Lovell on p. 26 of *Voice of the Universe* published in 1987. Figures quoted by Lovell on p. 194 of *Astronomer by Chance* of 1992 are somewhat different when discussing the same meeting, although the gist is the same.

[3] This is according to Lovell in *Astronomer by Chance*, Oxford University Press, 1992. But Blackett says that Husband was recommended by the Ministry of Works, see Jon Agar, *Science and Spectacle; The Work of Jodrell Bank in Post-war British Culture*, Harwood Academic Publishers, 1998, p. 50.

February 1950. Its attendance included William Greaves (the chairman) who was Astronomer Royal for Scotland and past president of the RAS, Harold Spencer Jones who was Astronomer Royal, Edward Appleton, who was principal and vice-chancellor of the University of Edinburgh, Blackett, Hey and Lovell. For some reason Ryle did not attend. Although Blackett had deliberately told Lovell not to assume, in his presentation, that the new telescope would be built at Jodrell Bank, that is what the meeting recommended along with suggesting that Lovell should manage it.

Subsequently Blackett and Lovell had a meeting with Husband on June 14 at which Husband suggested that it should be possible to build the telescope within a ceiling figure of £100,000 ($280,000). But he recommended that a more detailed design should be produced before formal estimates were obtained. Consequently Lovell submitted a request to the Department of Scientific and Industrial Research (DSIR) for £120,000 ($335,000) to cover the design, manufacture and erection of the telescope. Blackett's previous lobbying of senior members of the RAS establishment resulted in the RAS strongly supporting Lovell's request. Consequently on June 22 1950, just one week after receiving the estimate, DSIR agreed to provide £3,300 ($9,250) to enable Husband to produce a detailed design so that tenders could be invited for the manufacture and erection of the telescope. This by now was agreed to be 250 ft (76 m) in diameter with a reflecting surface composed of 4 inch (10 cm) wire mesh to enable operation down to wavelength of 1 m (300 MHz frequency) with a beamwidth of ±½°.

It was decided to have an altazimuth mounting for the telescope. The 250 ft diameter bowl would be suspended from two points 180° apart on the periphery, on towers about 186 ft (57 m) above the ground, to enable it to be moved in altitude (or elevation). These towers would then travel on a circular railway track of 320 ft (98 m) internal diameter to change azimuth. It was also decided to have a deep bowl (see Figure 18.1) with a relatively short focus to shield the focus from local radio interference and to make its access easier. Even so the transmitting and receiving systems still had to be mounted on a tower 62 ft (19 m) above the centre of the bowl.

Husband was concerned about the design of the telescope drive. Blackett observed that the requirements on its design were probably similar to those on the gun aim and control system for a battleship. Consequently he suggested that Husband and Lovell should visit the Admiralty Gunnery Establishment at Teddington where they discussed their problem with J. M. Ford. He advised them to talk to Metropolitan-Vickers about their metadyne drive control system. Husband was also concerned about the size of the gear racks needed to drive the elevation. Here Ford suggested that Husband obtain some of the 25 ft (7.6 m) diameter racks used to drive a battleship's 15 inch (38 cm) gun turrets. It so happened that at that time the battleships *Royal Sovereign*

Figure 18.1. The 250 ft radio telescope dish under construction at Jodrell Bank in the 1950s. (Courtesy Jodrell Bank Centre for Astrophysics, University of Manchester.)

and *Revenge* were in the process of being broken up for scrap and so Husband visited the ship-breakers to discuss the possibility of purchasing some of their gun turret racks. Fortunately, he was able to interrupt the salvage operation sufficiently quickly to enable Jodrell Bank to buy three of these racks for just £250 ($700) each.

At about this time Lovell's team investigated the reflectivity of the dish's planned 4 inch (10 cm) mesh at a wavelength of about 1 m, the lowest wavelength that they anticipated using. To their surprise they found that about 20% of the incoming radiation was lost through this mesh, whereas only 2% was lost through a 2 inch mesh. Consequently it was decided to have the finer mesh.

By early 1951 Husband had sufficient information to provide an updated cost estimate to Jodrell Bank. Based on this Lovell made a formal application to DSIR in March of that year for £259,000 ($725,000) to build the telescope. But this was over double Husband's previous ceiling figure, after accounting for inflation.[4]

[4] £20,000 of the increase was due to an initial underestimate in the cost of the foundations, and £20,000 was due to an underestimate in the cost of the driving system. In addition, £20,000 was required to cover the extra steel needed to strengthen the telescope structure following the decision to reduce the dish's surface mesh size from 4 inches to 2 inches.

Unfortunately there was also a financial crisis in the country at the time and there was no chance that the DSIR would be able to approve an early grant of this magnitude. In addition soon after the request was made the price of steel increased by 20%, making even this new high financial estimate invalid.

In February 1952, Lovell was asked by DSIR to bring the cost estimates up to date assuming a start on April 1 of that year. This sounded hopeful but the new figure turned out to be £335,000 ($940,000) for a construction phase lasting three and a half years, a year longer than originally thought.

In the meantime, however, a number of powerful men had been involved in the background in trying to get the programme approved(8) including:

- Sir Ben Lockspeiser, secretary of DSIR
- Lord Ernest Simon, chairman of the council of the University of Manchester, and chairman of the BBC
- Sir John Stopford, vice-chancellor of the University of Manchester, and Fellow of the Royal Society
- Sir Henry Tizard, Chief Scientific Advisor to the Ministry of Defence, and rector of Imperial College, London
- Lord Frederick Woolton, chancellor of the University of Manchester, and chairman of the Conservative Party

in addition to Patrick Blackett himself, Nobel prize-winner and member of the National Research and Development Corporation.

With these sort of people supporting the telescope case it is hardly surprising that in March 1952 DSIR approved a grant for the construction of the telescope, although they only agreed to provide 50% of the costs. But following the intervention of Tizard the Nuffield Foundation agreed to provide the other 50%.

The first task of construction was to build a circle of reinforced concrete 20 ft (7 m) wide, 320 ft (98 m) internal diameter, and 6 ft (1.8 m) deep supported on 160 reinforced concrete piles to take the railway track. This track was to carry more than a thousand tons of steel and it had to remain level to a high tolerance. The time estimate for sinking the piles was 2.5 months. In reality it took nine months partly due to the fact that a proper site survey had not been carried out on the land used for the telescope.[5]

Meanwhile in 1951 the 21 cm (1.4 GHz) emission line of neutral interstellar hydrogen had been discovered by Harold Ewen and Edward Purcell in the United States. As a result Lovell asked Husband in September of the following year to

[5] Initially, it had been agreed to build the telescope on a field at Jodrell Bank owned by the university and that is where the trial site drillings took place. But under pressure from Lovell a nearby field was subsequently purchased by the university and the telescope built there. Unfortunately, no trial drillings had been undertaken at this new location so the contractor was continuously faced with unpleasant surprises when he came to build the telescope foundations.

look at the effect of increasing the mesh size in the central 100 ft (30 m) diameter of the antenna bowl from 2 × 2 inches to 2 × 1 inch to enable the telescope to undertake observations at 21 cm. When Lovell presented his latest financial estimates to DSIR in April 1953 they included the effect of this change, and of underestimates in the amount of steel required in the 1952 design, underestimates in the costs of the central pivot on which the telescope was to rotate, underestimates in the costs of the railway bogies and driving system, and an increase in the price of steel. The net result was that the cost to completion had increased from £335,000 to £439,000[6] ($ 1.23 million) in a year, of which the Nuffield Foundation would agree to provide only another £32,000 ($90,000). Then in May 1953 DSIR agreed to fund the uncovered extra cost, but warned that if the costs continued to escalate, the University of Manchester may have to find other sources of funding to cover them.

Early in the following year, the Public Accounts Committee (PAC) of the UK parliament expressed concern about the university's lack of financial control. They then used this to raise the more general question as to whether universities were the correct institutions to build and maintain such complex instruments as the 250 ft telescope.(9)

In the meantime things had not remained stationary on the telescope's finances for very long as by December 1953 the cost of the drive systems for both the azimuth and elevation motors had doubled. At about the same time Lovell became a member of the Aeronautical Research Council which was concerned about what effect centimetre wavelength emissions from space may have on air navigation and the defence against guided missiles. Bob Cockburn, scientific advisor at the Air Ministry, asked Lovell for an assessment of the expected performance of the telescope at various wavelengths and the cost of modifications to enable it to operate at high efficiency at a wavelength of 10 cm. As a result Lovell proposed a series of modifications, including a further reduction in mesh size to 0.75 inches (1.9 cm) and a stiffening of the bowl framework at a total additional cost of £46,000 ($129,000). But a little later Cockburn, who had by then moved to be director of guided weapons research in the Ministry of Supply, changed his mind and told Lovell in April 1954 that he was unable to justify the extra costs. Unfortunately Husband had already started making the modifications! Consequently Lovell carried on with them convinced, as he was, of the national importance of the telescope. He argued to himself that their costs could be covered by the 10% contingency for the steelwork.[7]

6 This is according to Lovell on p. 228 of his *Astronomer by Chance*. On p. 51 of his *Voice of the Universe* Lovell quotes £459,896.

7 This looks highly irresponsible as there was hardly a contingency anywhere in the telescope budget which was continually being overspent. A subsequent committee of enquiry into the telescope project (see later) was highly critical of Lovell, but he said that he was unable to explain the reason for the changes because of the defence nature of the additional work.

Shortly afterwards Lovell also became aware, because of his membership of the Air Warfare Committee, that the defence establishment was having trouble tracking high-level supersonic bombers and were anticipating even more severe problems in tracking future ballistic missiles. As a result he suggested in August 1954 that Jodrell Bank and the Telecommunications Research Establishment should undertake a joint investigation of the problem. Lovell argued that the 250 ft telescope needed a more accurate drive system if it was to be used to track high velocity bombers and ballistic missiles. Although this was true, Lovell saw this as a way of getting the extra money he required for the currently designed system. But this way of getting the extra funding eventually fell foul of the Treasury's insistence that project funding could come from only one government department, in this case the DSIR and not the Ministry of Supply, and DSIR said that they had already committed all their money for the next few years.

As time went on it was clear that not only were the Metropolitan-Vickers' drive systems twice their original price estimate, the company was unable to get anywhere near the delivery date required.[8] But Husband, who had visited Siemens in Germany, found that they were able to meet an acceptable delivery date at lower cost. However it then transpired that Siemens' estimate would be subject to a 20% import surcharge wiping out their financial advantage. Finally, just as the contract was ready to be awarded to Siemens in June 1956, Lovell was told that the British firm Brush had quoted an acceptable price and delivery date, which he was forced to accept. Lovell suspected that this was the result of pressure on Brush from someone of influence to avoid having to give the contract to a foreign company.(10)

In the meantime concern had begun to be expressed in early 1954 about the problem of providing good mechanical and electrical connections across the thousands of joints of the wire mesh. Later that year Husband suggested that it may be best to replace the mesh surface of the parabolic dish by metal plates welded together. Not only would this solve the continuity problem but it would also improve the dish stiffness. The wind resistance of such a solid-surfaced bowl would be no more of a problem than that of a bowl with an iced-up mesh. But in January 1955, wind tunnel tests at the National Physical Laboratory showed that a gusty wind could be a real problem for a solid bowl by causing structural resonances. This required Husband to design a damping system which consisted of something that looked like the segment of a huge bicycle wheel attached to the perimeter of the bowl framework and bearing on large tractor wheels. The friction on these wheels damped out any incipient oscillations.

At about this time there was an unsuccessful attempt by DSIR to pass the financial responsibility of the Jodrell Bank telescope to the University Grants Committee

[8] They quoted a delivery time of three years, against a requirement of nine months.

(UGC) and the University of Manchester. In addition DSIR and the UGC discussed the possibility of cancelling the project. But they concluded that too much money had already been spent and the project had had so much good publicity that cancellation was not really an option.(11)

By October 1955, Husband's damping system and other changes had pushed the cost of the telescope to £630,000 ($1.76 million). Whilst DSIR announced that they may be prepared to provide half of the extra funding required, they would only do so if the University of Manchester would find the other half.

It is always difficult to estimate the cost of a civil engineering project which is completely different from that of any other, and the Jodrell Bank dish was no exception. Added to that was the inexperience of anyone at the University of Manchester in managing such a complex engineering project, and the rapid increase in raw material prices particularly that of steel. The net result was a telescope which, in this case, looked as though it would cost about an order of magnitude more than its original estimate. Not surprisingly the DSIR Advisory Council insisted that there should be an inquiry into the management of the project under an independent chairman before they would formally commit to providing half of the extra funding required. The committee of inquiry's first meeting was held in January 1956, by which time the *extra* funding to be found had been increased to £260,000 ($730,000), of which the university had to find half.

The report of the DSIR enquiry, which was published in March 1957 by Sir Frank Tribe, Comptroller and Auditor General at the Treasury, was critical of both Husband and Lovell. It was followed by an investigation by the PAC who published their report on August 13 1957 which was also highly critical of Husband and Lovell. Husband was accused of altering the design without consulting Lovell, and Lovell was accused of irresponsibility in using public funds. Things looked as they were about to turn ugly with Lovell forbidden to publicly criticise either report and Husband threatening to prosecute Lovell unless he put the record straight.

In the meantime on August 2 1957 the first astronomical observations had been made with the new Jodrell Bank telescope in its partially completed condition. Then on October 4 the first artificial earth satellite, Sputnik I, was launched by the Soviet Union. It was eventually to prove the saving of Lovell's, Husband's and Jodrell Bank's reputation.

The United States and the Soviet Union had both offered to launch earth satellites for research purposes during the International Geophysical Year (IGY) that started on July 1 1957. The Americans were hoping to launch their first satellite later that year, and it was expected that the Soviet Union would launch theirs somewhat later. The fact that the Soviets had now beaten the Americans to achieve a successful first launch was astonishing to people in the West and was a humiliation for the Americans who were thought to have been many years in advance of the Soviet

Union technically. The Soviet satellite also had widespread military implications as it had been launched by one of the Soviet Union's modified Intercontinental Ballistic Missiles (ICBMs).

Bernard Lovell had previously made plans to compare radar reflections from the first American satellite with those from the Moon to determine the electron density in the region of space between the Earth and the Moon. Although radio emissions from the cosmos had first been observed by the 250 ft telescope in August, it was not fully operational when Sputnik I was launched in October. For example it could not be driven automatically in elevation and the radar system had not yet been installed. Not that it mattered that this radar system was not available as Lovell considered that the experiment was not that important, and it could be undertaken later in the IGY with one of the other satellites. But the general public in the UK and elsewhere, particularly in the USA, were keen to find out more about Sputnik. How big was it and was it really in orbit around the Earth? Many people were aware that there was an enormous new radio telescope at Jodrell Bank. Surely it could pick up signals from Sputnik. So the press descended on Jodrell Bank to find out. But more seriously it also transpired that neither the USA nor the UK had a powerful enough radar system, other than the 250 ft Jodrell Bank telescope acting in its radar mode, to track Sputnik's spent carrier rocket, which was an ICBM. Nothing much seemed to have changed in the UK since 1954 when Lovell had been involved in an abortive attempt to modify the 250 ft telescope to track ballistic missiles.

By the early evening of October 9, after a hectic three days of work, the telescope was able to be moved automatically under remote control for the first time. Then shortly after midnight it was able to observe the Moon by radar. Unfortunately this radar system developed a fault before it could be used to track Sputnik's rocket. But Jodrell Bank had another, smaller radar transmitter that had a wider transmitted beam, which should make it easier to find the rocket. It did. On October 12, in the full blare of publicity Jodrell Bank was the first facility in the West to track a Soviet carrier rocket. A few weeks later Jodrell Bank's radar also tracked the rocket's final orbits as it burnt up in the Earth's atmosphere on December 1.

But in spite of this success, and of a concerted campaign by the newspapers to shame the government into funding the telescope, the government held out and still required the University of Manchester to find the missing £130,000. Consequently the university launched a public appeal, the Radio Telescope Fund, in November 1958. Within a year this appeal, together with earlier donations from industrial companies, had reached about half-way towards its target.

In the meantime Jodrell Bank had been approached in early 1958 by the US Air Force to ask if they were willing to track their satellites that were due to be launched to the Moon later that year. Lovell, who was in need of good publicity, jumped at the idea. The first of these satellites launched in August exploded just over a minute after lift-off, but the second reached just over 70,000 miles (110,000 km) from Earth before falling back. It was successfully tracked by Jodrell Bank and its scientific data recorded. About a year later the Russians contacted Jodrell Bank an hour after lift-off to ask if they could track the second of their lunar satellites, Lunik (or Luna) 2, that they hoped would reach the Moon on the following day, September 13 1959.[9] The Soviet Union provided the exact position of their satellite as seen from Jodrell Bank and the expected time of impact. In the event Jodrell Bank picked up the satellite telemetry signals exactly as predicted by Moscow. They stopped only a minute after the predicted impact time. Subsequent analysis of their Doppler shift by J. G. Davies at Jodrell Bank clearly proved that the satellite had impacted the Moon, so scotching doubts expressed by various senior Americans including the former president Harry S. Truman and the vice president Richard M. Nixon. Jodrell Bank also provided the Soviet Union with copies of the telemetry tapes from Lunik 3, which in October 1959 was the first satellite to take photographs of the far side of the Moon. Finally in 1960, on the back of this publicity, Jodrell Bank's money problem was solved by a donation of £50,000 from Lord Nuffield, on condition that the Jodrell Bank radio astronomical laboratories were named after him.

At this time the position and sizes of radio sources were only known very approximately. But a number of people suggested that the occultation of these sources by the Moon should enable their positions to be determined much more accurately. Consequently on December 8 1960 Cyril Hazard used the 250 ft radio telescope to observe the occultation of the radio source 3C 212 by the Moon. This resulted in estimates of its position within ±0.3s in right ascension, an improvement by a factor of 20, and within ±2 arcsec in declination, an improvement by a factor of 240.(12)

In 1964 a 50 ft (15 m) antenna was built at Jodrell Bank to track some of the early Soviet and American planetary and lunar probes as well as to undertake astronomical research. It was removed in 1982 and replaced by a 42 ft (13 m) antenna which had been previously located at the Woomera Rocket Testing Range in Australia. The 42 ft was used mainly for observing pulsars.

18.3 Later Parabolic Radio Telescopes

Although Blackett had left Manchester in 1953 to take up a position at Imperial College, London[10] he still followed developments at Jodrell Bank with great interest. He was naturally concerned about the many technical and financial problems that had

[9] The early launch attempts by the Soviet Union were shrouded in secrecy, with the authorities only announcing them after they had been successful. It has since been found that, prior to the launch of Lunik 2, the Russians had attempted to launch no less than five satellites to the Moon, only one of which, Lunik (or Luna) 1 had reached the vicinity of the Moon. It flew past at a distance of 3,700 miles (5955 km) in January 1959.

[10] As head of the Physics Department.

dogged the 250 ft telescope's design and construction phase but was delighted that, in the end, the telescope proved to be a very valuable research tool. Consequently in the autumn of 1958 he asked the DSIR's Sub-committee on Astronomy to consider how radio astronomy in the UK should develop over the next five to ten years to enable the UK to keep its leading position created by the Cambridge and Jodrell Bank observatories.(13) The sub-committee duly asked Lovell and Martin Ryle, the director of the Cambridge radio observatory, to produce a joint paper outlining their ideas taking into account the known plans of other countries. Not surprisingly it took some time for them to produce such a paper as Ryle was interested in developing radio interferometers at Cambridge whilst Lovell wanted to build new large parabolic antennae at Jodrell Bank. But when the paper was finally presented to the sub-committee in March 1960 they agreed to support the development of radio astronomy facilities at both Jodrell Bank and Cambridge.

Whilst working on the joint paper with Ryle, Lovell had started to consider the possibility of building a much larger paraboloid than the 250 ft, or Mark I, and in June 1960 he asked DSIR for $5000 to pay for the study of such a larger telescope. This larger telescope was later called the Mark IV[11] which Lovell envisaged as having a dish extending 1,500 ft (450 m) to 15,000 ft (!) horizontally with a height limitation of 500 ft (150 m). The £5000 study funding was approved in July but in the meantime Husband, the proposed designer, had put in a request to Jodrell Bank for double the amount for the study.

Whilst this study funding was under consideration, discussions between Lovell and Husband resulted in the idea of the Mark IV having an oval rather than a circular dish to simplify its aerodynamics and to reduce constructional problems. Then in September 1960 Husband suggested building a 125 ft × 83 ft (38 m × 25 m) telescope as a prototype, which was later called the Mark II. Lovell was very supportive of the idea of building such a telescope as he could use it to operate at a higher frequency than the 250 ft taking advantage of the recent developments in high frequency amplifiers. He also envisaged a portable version of the Mark II, which was to become the Mark III, that could be located some distance away from the 250 ft to form an interferometer with it. In 1963 the planned Mark IV was abandoned as the UK Treasury had ruled that the DSIR should not finance a design study unless it could foresee a means of finishing the subsequent device and, at this stage, the DSIR could certainly not afford anything like the Mark IV.

In the meantime the approval to build the Mark II had been given in 1962 after Lovell had been forced by DSIR to decide on priorities between the Mark II and Mark III. The Mark II, which was built at Jodrell Bank on the site of the 218 ft

[11] The designation of Mark IV for this large telescope and of Marks II and III for the smaller telescopes discussed later were suggested by Christopher Jolliffe of the DSIR in 1961.

transit telescope, was completed in 1964. It had a solid surface to operate down to 6 cm supported by a steel structure which was carried on a 42 ft (13 m) diameter circular track by a prestressed concrete platform 36 ft high and 85 ft long. In the same year approval was given to build the Mark III at Wardle about 15 miles (24 km) from Jodrell Bank. Its surface, which was the same size and shape as the Mark II, was made of wire mesh that limited its operation down to about 21 cm. When completed two years later it was able to form an interferometer with the Jodrell Bank telescopes. Although it was made to be transportable it was never moved until it was decommissioned in 1996.

Lovell had seen the writing on the wall for the Mark IV in early 1963. So before it was finally killed off by DSIR he was already considering an alternative less expensive design for what came to be called the Mark V. What he had in mind was a 150 to 250 ft telescope at a total cost of about £1–1.5 million ($2.8–4.2 million).(14) By the end of the year Lovell suggested to Husband that something like £1.5–2 million (at 1962 prices) may be available from the DSIR. In addition a financial contribution may be available from the Americans which could increase the Mark V budget to about £4 million ($11.2 million). Consequently Husband was asked by Lovell to carry out a feasibility study that should include an option for an instrument costing £1.5 million at 1962 prices, together with options costing up to about £4 million which could have either a solid or an open mesh surface. The bowl could deviate from the best fit paraboloid by up to ±1 inch (2.5 cm) over the central region out to two-thirds of the radius, and by ±2 inches over the remainder. Husband concluded in March 1965, with these constraints, that it was possible to build a 260 ft (79 m) diameter telescope for £1.76 million (at 1965 prices) or a 400 ft (122 m) telescope for £4.06 million (at 1965 prices).

At this stage the Science Research Council (SRC) had taken over from DSIR the responsibility of funding the physical sciences. So Lovell had to deal with a new organisation when wanting government funding. Not only that, but he was put under pressure by SRC to obtain competitive estimates as Husband's company was no longer the only company with expertise in designing large radio dishes. For example Freeman Fox had built the 210 ft (64 m) Parkes radio telescope in Australia (see Section 16.2) and was undertaking the design and construction of a 150 ft (46 m) radio telescope at Algonquin in Canada (see Section 22.7). Because of this Lovell deferred his request in 1965 for funding from the SRC to build the Mark V. Instead the SRC decided to fund the Anglo-Australian optical telescope at Siding Spring. Then in 1967 he again deferred his funding request to build the Mark V as he needed money to repair and improve the Jodrell Bank 250 ft (see below), and Martin Ryle required funding from the SRC for an aperture synthesis radio telescope at Cambridge.

Meanwhile in January 1967 Husband & Company and Freeman Fox had been invited to undertake independent feasibility studies into the design of the Mark V which was defined as a steerable radio telescope of 400 ft diameter whose price should not exceed £4.25 million (at December 1967 prices). The whole dish was required to operate at full efficiency at a wavelength of 21 cm, and preferably at least 50% of its aperture should be able to operate at high efficiency at shorter wavelengths.[12] The conclusions of the feasibility studies were requested in three months time. Husband almost met this timescale whereas Freeman Fox was four months late, which did not augur well for the future if they were to be the chosen contractor. With their design, which was based on that of their Parkes telescope, Freeman Fox was unable to meet the 21 cm requirement over the whole surface of the dish, whereas Husband could – at least on paper. The construction schedule for Freeman Fox was also twice as long as that of Husband. For these and other reasons(15) it was decided to invite Husband & Company to undertake a detailed design study at an estimated cost of £55,000 ($130,000). This study was begun in February 1968 but rather than taking the expected 9 months it took 15.

Finally Husband revealed his design of 400 ft diameter radio telescope in May 1969. It had the minimum surface deviation from a paraboloid of 2.5 mm at 30° elevation with no wind. But, although his cost estimate was within the required budget, his poor financial estimates for the Mark I project caused SRC to wonder what the final cost would really be. Unfortunately by now the Germans were building a 100 m (330 ft) diameter telescope at Effelsberg (see Section 23.1) which appeared to have a substantially better surface accuracy than Husband's design, which caused additional concern. However shortly afterwards Husband proposed a modification to his design to produced a maximum deviation of 2 mm at all elevations within the specified wind limits for only a relatively small increase in cost. Consequently at the end of 1969 Jodrell Bank submitted a request to SRC for £6.2 million ($14.9 million)[13] at then-current prices to build the Mark V telescope (see Figure 18.2). But at this very moment there was a change in government in the UK which was faced with severe financial problems. The new government's squeeze on the SRC budget and a continued high level of inflation, particularly in the price of steel, forced Lovell to reduce the size of the Mark V telescope to 375 ft (114 m) to become what was called the Mark VA. But by 1974 the estimated cost of this smaller telescope had risen to over £20 million ($47 million) in just four years. Hardly

[12] The efficiency requirements were specified more clearly in an appendix to the bidding document.

[13] Just under half of the increase from £4.25 million to £6.2 million was due to inflation which was running at a very high level at the time. Part of the remainder was due to the modifications required to meet the higher tolerance for the parabolic dish which Husband had estimated would be about £300,000 at end 1969 prices.

Figure 18.2. Model of the 400 ft diameter Mark V telescope proposed by Jodrell Bank but never built. (Courtesy Mike Peel, Jodrell Bank Centre for Astrophysics, University of Manchester via Wikimedia Commons.)

surprisingly this caused it to be abandoned even though £750,000 ($1.8 million) had already been spent on its design.

18.4 Modifications to the Mark I

The 250 ft radio telescope had been renamed the Mark I telescope in 1961. Two years later Husband was contracted to carry out an investigation into the possibility of improving its performance, particularly at higher frequencies. Lovell did nothing with Husband's subsequent report, which he received in April 1964, as he was concerned about getting funding for other telescopes at the time. The Mark II was not then operational and Lovell did not want to have to take the Mark I out action to allow the improvements to be implemented before he had a backup telescope.

Figure 18.3. The 250 ft (76 m) diameter Lovell Telescope at Jodrell Bank as seen in 2007. The vertical double bicycle wheel like structure which improved stability of the dish is clearly seen. (Courtesy Mike Peel, Jodrell Bank Centre for Astrophysics, University of Manchester via Wikimedia Commons.)

But things changed dramatically in September 1967 when cracks were discovered in the Mark I's elevation drive system. As a result Lovell decided to ask the SRC for money to rectify the faults and, at the same time, to improve its design. This resulted in the SRC approving a £400,000 grant in July 1968 for the repair and improvement of the Mark I.

As part of the Mark I telescope's improvement programme a more accurate bowl surface was constructed above the old surface to enable the telescope to be used down to 6 cm (5 GHz). The foundations of the railway track were repaired and the railway track relaid. The single stabilising bicycle wheel was replaced by two load bearing wheels (see Figure 18.3) running on a new inner railway track to reduce the load on the azimuth track and on the trunnion bearings. The repair and improvement work was basically completed by November 1971 to produce what came to be called the Mark IA telescope at a cost of £665,000 ($1.60 million).(16)

In 1980 this Mark IA was used as part of what is now called the MERLIN array (see below), and in 1987 the Mark IA was renamed the Lovell Telescope. The 1971 surface of the dish was replaced by another surface in 2001–2003 to expand the useable frequency range and provided an increase in sensitivity at 6 cm by a factor of five.

18.5 MERLIN

As mentioned previously the 218 ft transit telescope had been used with a small array in the early 1950s to carry out interferometric measurements at baselines of up to 12 miles (20 km). These were followed over the period from 1958 to 1961 by interferometric measurements at a wavelength of 1.89 m (frequency 158 MHz) using the Mark I telescope at Jodrell Bank and a small array at baselines of up to 70 miles (115 km). Five of the 384 sources studied were still unresolved at the longest baseline implying diameters of less than 0.8 arcsec.(17) Shortly afterwards one of these unresolved sources, 3C 48, was found by Jesse Greenstein of the Mount Wilson and Palomar Observatories and Thomas Matthews of Owens Valley Radio Astronomy Observatory to have a redshift of 0.37. It was the second quasar to be discovered.

In 1965, the Mark I and an 82 ft (25 m) radio telescope operated by the Royal Radar Establishment (RRE) 79 miles (127 km) away at Defford were linked by a triple-hop microwave radio link. A number of sources were found to be smaller than 0.1 arcsec at a wavelength of 21 cm (1.4 GHz).(18) In the following year the Mark I was replaced in the interferometer by the Mark II at Jodrell Bank and observations were made at 11 cm (2.7 GHz) and 6 cm (5 GHz), to produce resolutions of 0.05 and 0.025 arcsec, respectively. Again a number of sources were still unresolved.

A little earlier Lovell had asked Henry Palmer and his team to look into the possibility of undertaking interferometric radio observations between Jodrell Bank and the Soviet Union's deep space tracking stations 1,370 miles (2,200 km) away in the Crimea to produce even better resolutions.[14] But something like 50 microwave links would be required to link the two sites, so it was proposed to make tape recordings separately at each site and bring them together using precise timing information to correlate the signals. Clearly it was essential to test out such a system in the UK before making the link to the Crimea. But in May 1967, while planning of this experiment was still underway, it was announced that the Canadians had been the first to successfully carry out long baseline interferometer observations using tape recorders and atomic clocks. Their test was between Algonquin and Penticton (both in Canada) separated by a distance of 1910 miles (3074 km), where they detected fringes for 3C 273 at a wavelength of 67 cm (448 MHz) with a resolution of 0.02 arcsec. But some of the sources were still unresolved and so must have been ≤0.01 arcsec in size. Almost immediately the Americans were also successful with a system of similar resolution operating at 18 m (1.67 GHz). This linked the 140 ft (43 m) antenna at Green Bank, West Virginia to the 85 ft (26 m) antenna at Maryland Point about 525 miles (845 km) away.

[14] This followed a visit by Lovell to the Soviet Union in 1963 as a guest of their Academy of Sciences.

In the meantime the Mark III 38 × 25 m radio telescope had been completed in 1966 at Wardle about 15 miles (24 km) from Jodrell Bank. It was designed to be used, like the Defford 25 m, to form an interferometer with the Jodrell Bank instruments. Seven years later Palmer suggested adding more telescopes to this system to improve the observation of small-scale structure of extended objects when it looked as though the Mark VA telescope project was about to be rejected. He was concerned that the existing telescopes of the Jodrell Bank interferometer were predominantly on a north–south axis which provided good resolution along that axis but poor resolution east–west. So he suggested locating one of the new telescopes to the west of the existing telescopes and one to the east.

After discussing his proposal with Lovell and Graham Smith, professor of radio astronomy at Manchester, it was agreed that three new identical 24 m (80 ft) telescopes should be built. One of these should be located about 60 miles (100 km) south west of Jodrell Bank at Meifod (where the university had already purchased a site for the Mark V telescope), one should be northwest of Manchester on a new site, and a third should be at Jodrell Bank to be used at the shortest wavelengths beyond the capabilities of the Mark IA and Mark II.(19) Lovell asked Husband if he would design such a 24 m telescope that could be used down to a wavelength of 1.3 cm (23 GHz). Husband said that whilst he was happy to do so he suggested that it would be far cheaper to ask E-Systems Inc. if they could supply three 25 m radio telescope dishes of the sort they were in the process of building for the American Very Large Array (VLA, see Section 20.9). In reply to Jodrell Bank's subsequent enquiry E-Systems confirmed that they could accommodate their request without adversely affecting VLA deliveries. So in February 1974, Lovell applied for a grant of £3.65 million ($8.54 million) from the SRC for a Multi-Telescope Radio Linked Interferometer (MTRLI) system, emphasising that this request was to be considered only if the Mark VA were to be cancelled.[15] In the event the Mark VA application was rejected in June 1974. Consequently, the Jodrell Bank proposal for the MTRLI system was considered by the ASR (Astronomy Space and Radio) board of the SRC in September. But doubts were expressed about the potential performance of the proposed system which caused the SRC to approve only the first phase of the project in May 1975 which included only one new 25 m E-Systems dish.

So Phase 1 of this MTRLI system, which was to be controlled from Jodrell Bank, consisted of the Mark IA or Mark II at Jodrell Bank, the 25 m telescope at Defford, the Mark III at Wardle, and the new E-Systems 25 m radio telescope. This new

[15] This approach seems very odd as it clearly gives the impression that Lovell had given up on the Mark VA being approved. This was true (see Bernard Lovell, *Voice of the Universe; Building the Jodrell Bank Telescope*, Praeger, Rev. and Updated Ed., 1987, p. 277.) but surely it would have been better for him to have withdrawn the Mark VA application completely and sent the application for the MTRLI in its place, rather than wait for the SRC to formally reject the Mark VA application some months later. In the event the application for the MTRLI leaked out, as was predictable, causing some embarrassment whilst the Mark VA application was still being considered.

Figure 18.4. The 25 m diameter radio telescope installed at Knockin about 40 miles from Jodrell Bank by E-Systems Inc. in the mid 1970s. (Courtesy Oosoom via Wikimedia Commons.)

telescope was to be installed at Knockin, 42 miles (68 km) south west of Jodrell Bank, instead of at Meifod because of problems with planning permission at the latter site. Project financing was approved by the SRC and UK Treasury in June 1975 followed shortly afterwards by planning permission for the Knockin antenna.

The contract for the supply and construction of the Knockin radio telescope (see Figure 18.4) was signed with E-Systems in August 1975. Amazingly they delivered the hardware three months early and completed the installation five months early in October 1976. Lovell was taken by surprise as, in his experience, contractors were almost always late, sometimes by very large amounts. Local control of the Knockin telescope was achieved in January 1977 followed by tests at 1.2 cm in March which showed that it was within specification. Unfortunately Ferranti was over a year late in installing the microwave link between Knockin and Jodrell Bank and engineers at Jodrell Bank itself were late in providing the L-band local oscillator. But finally in January 1980 the phase 1 system was able to record distant astronomical sources for the first time. The system had cost £2.18 million (about \$4 million) to build which was very close to the original estimate, after taking inflation into account.[16]

In the meantime doubts about the potential performance of the complete MTRLI system had been largely resolved. Consequently the University of Manchester submitted a proposal to the SRC in December 1977 to complete the system by the addition of two more E-Systems antennae and their associated communication links in phase 2 of the project. Phase 2 was approved by the SRC in June 1978 at a cost of £3.08 million.

The two new antennae were installed at Tabley/Pickmere 11 km north west of Jodrell Bank and at Darnhall 18 km south west of Jodrell Bank by the end of October 1979. They were linked into the existing array of radio telescopes by the end of the following year. When the MTRLI became fully operational in 1981 the system, with a maximum separation of 134 km, had a better angular resolution (0.08 arcsec at 6 cm) and dynamic range than the VLA, although the VLA had greater sensitivity. The MTRLI system then became known as MERLIN (Multi-Element Radio Linked Interferometer Network).

The surface of the Jodrell Bank Mark II was upgraded in 1987 to allow it to operate with the three E-Systems telescopes at 22 GHz (λ 1.36 cm). Then in 1991 a 32 m radio telescope was built at Cambridge and added to the system, increasing the maximum baseline to 217 km. The existing MERLIN telescopes were also improved, and in 1997 the first dual frequency observations were made at 5 and 22 GHz. In the meantime the Mark III radio telescope had been decommissioned in 1996.

In 2004 a £8 million (\$12 million) programme of work was begun to transform MERLIN into e-MERLIN.[17] The microwave links were replaced by a fibre optic network to increase the bandwidth of the system from 30 MHz to 4 GHz, the electronics was updated, and a new correlator was installed to handle the large

[16] The \$/£ exchange rate varied widely during the project from a high of 2.21 in 1975 to 1.74 in 1977 to \$2.12 to the £ in 1979 which played havoc with the project budget which was in pounds sterling. This, together with a high inflation rate in the UK, also made it difficult to compare the actual cost of the project with the original budget.

[17] The modifications were paid for by the University of Manchester, the North West Development Agency, the Science and Technology Facilities Council, the University of Cambridge and Liverpool John Moores University.

amount of data of the new system. But in early 2008 it was announced that the UK's Science and Technology Facilities Council (STFC) was considering withdrawing its planned £2.7 million ($4 million) annual funding of e-MERLIN even though the upgrade had almost been completed. However following a media campaign the STFC reversed its position in July 2008 which allowed e-MERLIN to be completed. First fringes were obtained with a prototype correlator in 2009 and first images with the full network were produced in the following year. Commissioning of the system was completed in 2013. It provided L-Band (1.33–1.73 GHz), C-Band (4.5–6.8 GHz) and K-Band (22–24 GHz) observations with a resolution of about 150 mas (milliarcsec), 40 mas and 10 mas, respectively.

References

1. Lovell, Bernard, The Origins and Early History of Jodrell Bank, in Sullivan, W. T., (ed.), *The Early Years of Radio Astronomy; Reflections Fifty Years after Jansky's Discovery*, Cambridge University Press, 1984, pp. 192–211.
2. Sullivan, Woodruff T., *Cosmic Noise; A History of Early Radio Astronomy*, Cambridge University Press, 2009, p. 179.
3. Lovell, Bernard, *Astronomer by Chance*, Oxford University Press, 1992, p. 116.
4. Lovell, A. C. B., Banwell, C. J., and Clegg, J. A., Radio Echo Observations of the Giacobinid Meteors 1946, *Monthly Notices of the Royal Astronomical Society*, 107, 1947, pp. 164–175.
5. Hanbury Brown, R., *Boffin; A Personal Story of the Early Days of Radar, Radio Astronomy and Quantum Optics*, Adam Hilger, 1991, p. 97.
6. Lovell, Bernard, *Voice of the Universe; Building the Jodrell Bank Telescope*, rev. ed., Praeger Publishers, 1987.
7. Lovell, Bernard, *Astronomer by chance*, Oxford University Press, 1992, p. 193.
8. Ibid., pp. 212–219.
9. Agar, Jon, *Science and Spectacle; The Work of Jodrell Bank in Post-war British Culture*, Harwood Academic Publishers, 1998, pp. 63–64.
10. Lovell, Bernard, *Astronomer by Chance*, Oxford University Press, 1992, p. 241.
11. Agar, Jon, *Science and Spectacle; The Work of Jodrell Bank in Post-war British Culture*, Harwood Academic Publishers, 1998, pp. 65–67.
12. Lovell, Bernard, *Out of the Zenith: Jodrell Bank, 1957–1970*, Oxford University Press, 1973, p. 37.
13. Lovell, Bernard, *The Jodrell Bank Telescopes*, Oxford University Press, 1985, p. 3 et seq.
14. Ibid., p. 35.
15. Ibid., pp. 57–58.
16. Lovell, Bernard, *Voice of the Universe; Building the Jodrell Bank Telescope*, Praeger, Revised and Updated Ed., 1987, pp. 266–276.
17. Allen, L. R., et al., Observations of 384 Radio Sources at a Frequency of 158 Mc/s with a Long Baseline Interferometer, *Monthly Notices of the Royal Astronomical Society*, 124, 1962, pp. 477–499.
18. Lovell, Bernard, *Out of the Zenith: Jodrell Bank, 1957–1970*, Oxford University Press, 1973, pp. 50–51.
19. Lovell, Bernard, *The Jodrell Bank Telescopes*, Oxford University Press, 1985, p. 185.

19

Early Radio Observatories Away from the Australian–British Axis

19.1 The Soviet Union

The possibility of using radio to determine the distance of the Moon from Earth had been seriously considered by the Soviet radio physicists Leonid Mandel'shtam and Nikolai Papaleski as long ago as 1925, before accepting that it was not possible with the equipment available at that time.(1) However in 1943 they revisited the situation and concluded that radar observations of the Moon were then feasible. But researchers in the USA and Hungary were the first to succeed in making them three years later.

Papaleski had also considered the possibility of carrying out radar observations of the Sun and around the end of 1945 he asked Vitaly Ginzburg, of the P. N. Lebedev Physical Institute (LPI)[1], to theoretically analyse the reflection of radio waves by the Sun. In the following year Ginzburg concluded from his subsequent analysis that radio waves from Earth would not reach the Sun's photosphere as they would be absorbed by either its chromosphere or corona.(2) Simultaneously and independently Iosif Shklovskii, of the Sternberg Astronomical Institute of Moscow State University, showed that solar thermal radiation in the metre waveband, discovered by the British army during the war, could not be emitted by the solar photosphere or chromosphere but must be emitted by the solar corona.(3) Also independently, in Australia David Martyn concluded in the same year that the solar emission measured by Joe Pawsey in the metre waveband must be coming from high in the solar corona as the corona would be opaque at those wavelengths, so it could not be coming from lower down in the solar atmosphere (see Section 16.1). These theoretical conclusions of Ginzburg, Shklovskii and Martyn were proved to be correct in the following year by a Soviet expedition led by A. A. Mikhailov and Semion Khaikin[2] to observe a total solar eclipse in Brazil.[3] The expedition found

[1] Often called FIAN.

[2] Khaikin had been a student of Mandel'shtam.

[3] Papaleski, who had been due to lead the expedition, died suddenly shortly before it was due to leave the Soviet Union.

that the intensity of radio emission at a wavelength of 1.5 m (frequency 200 MHz) was, at totality, still about 30% of its level out of eclipse.

On his return Khaikin submitted a proposal to study radio wave propagation in the Earth's atmosphere using extraterrestrial sources, such as the Sun, Moon and other discrete radio sources covering the wavelength range from 3 m to 3 cm. This information was required by the military for the radio navigation of rockets,(4) and consequently the proposal was rapidly approved. The observational programme was to be carried out by radio stations in the Crimea and Gorki. That in the Crimea would be undertaken by LPI and led by Khaikin, and that at Gorki by Gorki State University and led by the head of its Statistical Radiophysics Department, Gabriel Gorelik. The military then handed over a number of old radar stations, that needed to be heavily modified, to both LPI and Gorki University.

Lebedev Physical Institute's (LPI's) Radio Observatory in the Crimea

In 1948 the LPI radio astronomy team used two stations in the Crimea overlooking the Black Sea that had been previously set up by Papaleski. It was clear, however, that these two sites at Alupka and Alushta were not sufficient to carry out serious work on a large scale. So with the agreement of Grigorii Shain, the director of the Crimean Astrophysical Observatory, the LPI Crimean station was provided with a site adjacent to the Astrophysical Observatory on Mount Koshka near Simeiz. There over the period 1948–1949 a team installed the phased dipole array that had been used on the Brazil expedition operating at a wavelength of 1.5 m.

The LPI had also been provided with a number of partially destroyed 7.5 m diameter Würzburg German radar antennae which were designed to operate at a wavelength of 50 cm. From these they managed to put together one operational antenna at their Mount Koshka station. Then in the following year a team headed by Pavel Kalachev covered its normal mesh surface with aluminium sheets to enable it to be used at a wavelength of 10 cm (frequency 3 GHz). As a result this so-called RT-7.5 instrument became the first Soviet reflector radio telescope.(5) At about the same time, a 3 m diameter Würzburg radar antenna was also modified and erected at the Alushta site to operate as a sea interferometer, initially at a wavelength of 50 cm. Viktor Vitkevich and his group used this small Würzburg in 1949 to make observations of solar emission. It was later modified so it could be used at wavelengths as low as 3 cm.

Over the period 1951–1953 Vitkevich and his team attempted to observe the structure of the solar corona by observing its occultation of the Crab's radio source, Taurus A. They carried out their first set of observations at a wavelength of 4 m using a sea interferometer with a baseline of 420 m and resolution of 16 arcmin.

This resolution would enable the antenna to distinguish between radio emission from the Sun as a whole and that from Taurus A which had an angular size of 6 arcmin. Unfortunately the Sun was active at the time and the interferometer was unable to distinguish between radio emission from local regions on the Sun and that from Taurus A. Then in the following year Vitkevich and his team used 700 m and 1200 m baseline interferometers at wavelengths of 6 m and 12 m. Unfortunately only the 6 m observations were usable, as the 12 m observations suffered from radio interference. But these 6 m observations appeared to show that the solar corona, or supercorona as Vitkevich called it, extended to at least 4.5 solar radii from the centre of the Sun. However Vitkevich did not publish his results as they were compromised by torrential rain and thunderstorms. Then finally in 1953 his luck changed and observations at 3.5 and 6 m showed that the corona was much larger than previously thought, influencing radio signals as far out as fifteen solar radii from the Sun.

At about this time Khaikin and Vitkevich suggested building 4 m diameter reflectors to observe fainter objects operating down to a wavelength of 3 cm. One was built on Mount Koshka and one near Kaluga, about 100 miles (160 km) southwest of Moscow. In 1953 the Mount Koshka 4 m was moved to Alushta. In the same year a new observing station was built near Katsiveli, also in the Crimea, where it was possible to locate larger interferometers.

Kalachev and colleagues designed a fixed 31 m diameter dish formed of concrete and covered with a zinc film at the Katsiveli station. In 1957 this RT-31 telescope was used to obtain the first two-dimensional radio image of the Sun at 3 cm. In the same year, Arkadii Kuz'min and V. A. Udal'tsov used this antenna to observe Taurus A at a wavelength of 9.6 cm. They found that its radio emission was linearly polarised which confirmed that it was caused by the synchrotron process. In 1959 Udal'tsov also used the RT-31 telescope to make observations of Taurus A at 21, 9.6 and 3.15 cm. He found that the polarisation angle varied as the square of the wavelength.

Gorki State University's Radio Observatory

In the meantime over the period 1947–1948 Vsevolod Troitskii, a young radiophysicist in Gorelik's radiophysics group at Gorki State University, had built a chopping-type radiometer for a wavelength of 4 m, and I. L. Bershtein had developed a Dicke-type radiometer for a wavelength of 10 cm. Then in 1948 Troitskii developed a 1.5 m radio telescope and worked out the theory of chopping radiometers. In the following year this 1.5 m radio telescope was moved to a new field station near Zimenski, 20 miles (30 km) from Gorki on the banks of the Volga. At about the same time S. I. Averkov developed a radiometer to operate at 3 cm which was used with a 60 cm dish antenna in 1950 to measure the radio emission of the Sun and the Earth's atmosphere.

A group of radio astronomers at Gorki's State University, led by Troitskii, undertook radio observations of the Moon in the early 1950s to try to determine the physical properties of its surface. They found that its mean brightness temperature increased with wavelength, which they interpreted as showing that there was a constant flow of heat from the lunar interior possibly caused by radioactive processes. They also concluded that the surface was not covered by the deep layer of dust proposed at the time by some researchers.

LPI's Radio Observatory at Pushchino

In 1951, at Khaikin's suggestion, the USSR Academy of Sciences looked into the possibility of building large reflecting radio telescopes to observe the Sun and other celestial bodies at centimetre and millimetre wavelengths. Consequently, in the following year, Kalachev was asked to develop a radio telescope operating at millimetre wavelengths. This resulted in the construction of RT-2, a 2 m diameter radio telescope operating at 8 mm wavelength (frequency 40 GHz) which used an old searchlight mirror attached to an equatorial mount. Unfortunately in 1954 the surface of the mirror cracked as it was being mounted onto its support structure just before the telescope was due to be shipped to observe a solar eclipse.[4] So a slightly smaller replacement telescope was sent. But in 1955 the RT-2 was installed at the Solar Radio Laboratory in Krasnaya Pakhra, near Moscow. There it was used by Alexander Salomonovich to observe the Sun for the next three years, during one of which he was able to observe a partial solar eclipse.

In the meantime a team led by Salomonovich and Kalachev had produced the draft design of a steerable 16 m radio telescope in 1952 to operate at very short wavelengths. A few years later it was decided to increase the diameter of this proposed telescope to 22 m (72 ft).(6) It was initially proposed to build this so-called RT-22 on a small plot of land near Moscow, but at that time Vitkevich and B. M. Chikhachev were also considering building a large cross-type radio telescope (the DKR-1000, see below) that needed a large amount of land. Consequently a site survey was carried out to try to find a suitable site that could be used for both telescopes. This resulted in the choice of Pushchino on the banks of the Oka River about 100 km south of Moscow (and 60 km from Krasnaya Pakhra). The LPI Radio Astronomy Station was established there in 1956 with the start of construction of the RT-22.

[4] The eclipse was to be observed near Novomoskovsk in the Ukraine.

With remarkable speed the 22 m diameter dish, with an rms surface error of just 0.3 mm, was completed by November 1958 and the telescope became operational in May of the following year. It could be used down to a wavelength of 8 mm where it had an angular resolution of 2 arcmin. In this it outperformed any such instrument in the West at the time.

The RT-22 was used to identify radio hotspots on the Sun and to help determine the physical characteristics of the lunar surface. In early 1960 Kuz'min, Salomonovich and colleagues carried out a survey of discrete radio sources at 9.6 cm, which resulted in the publication of the first radio catalogue in the Soviet Union. It contained 56 radio sources, 34 of which were being observed for the first time at centimetre wavelengths. Then in 1964 Roman Sorochenko and Eduard Borodzich used the RT-22 to detect the 3 cm line resulting from transitions between levels 91 and 90 in the spectrum of the Omega nebula. Four years later Sorochenko and colleagues became the first to observe a radio line in the millimetre waveband. Meanwhile in 1966 an improved RT-22 had become operational at the Crimean astrophysical observatory near Simeiz.[5] Both of the RT-22 telescopes were later used together to form a very long baseline interferometer.

In the meantime construction of another radio telescope, the DKR-1000, had been started at Pushchino. It consisted of two 1000 × 40 m parabolic cylinders operating in the metre waveband, to provide both high sensitivity and high resolution. Its design was a development of that of the Mills Cross in Australia (see Section 16.1) but, unlike the Mills Cross itself, the DKR-1000 allowed simultaneous observations at a relatively wide range of operating frequencies; in this case the wavelengths covered were from 2.5 to 10 m.(7) The telescope's east–west arm became operational in 1964, with the north–south arm a year later. Shortly afterwards a system of 18 amplifiers phased over a wide bandwidth was fitted to the DKR-1000 to increase its sensitivity by about an order of magnitude.

In the late 1960s the east–west arm of the DKR-1000 was used in conjunction with 280 × 20 m cylindrical antennae situated near Pereslavl'-Zalessky and Staritsa to form a 200 mile (320 km) baseline interferometer. This enabled measurements to be made for the first time of solar wind velocities relatively close to the Sun and at high solar latitudes. The characteristic size of the inhomogeneities responsible for the scintillation of radio sources was also determined.

Vitkevich decided in the late 1960s to build what is now known as the Large Phased Array (BSA) at Pushchino consisting of about 16,000 dipoles covering an area of 18 acres (7 hectares). Its very high sensitivity in the metre waveband enabled it to make daily observations of interplanetary scintillations which provided

[5] The improvements included adding a central reference axis to the telescope's rotating platform and installing a radically new automated control system.

valuable information on the variability of the solar wind. BSA has also been used to observe and accurately time the pulse arrival time of pulsars.

In more recent years facilities at the Pushchino Radio Astronomy Observatory, as it is now known, have been used to undertake very long baseline interferometry, including ground-space interferometry.

Pulkovo

From the early days of radio astronomy Khaikin had been convinced that it should be developed alongside optical astronomy. Consequently in 1954 he set up a new department of radio astronomy alongside that of the main optical observatory of the USSR Academy of Sciences at Pulkovo, near Leningrad. The main task of this Pulkovo radio observatory was to develop a new type of large microwave radio telescope, now called a Variable Profile Antenna (VPA), whose design had been proposed by Khaikin and N. L. Kaidanovskii. This so-called Large Pulkovo Radio Telescope (LPRT), which became operational two years later, consisted of 90 tiltable, flat steel plates, each 1.5 m wide × 3 m high, arranged in the arc of a 200 m diameter circle. The beamwidth was just 1 arcmin at 3 cm wavelength in an east-west direction. Later this was reduced to 15 arcsec at 8 mm.(8)

RATAN-600

By the early 1960's the Pulkovo group was convinced that a much larger VPA-type antenna could be built, but their enthusiasm was not echoed throughout the Soviet radio astronomy community. Some people favoured building a large steerable parabolic reflector maybe a little larger than the one at Jodrell Bank, whereas others preferred a much larger, bowl-type antenna built directly onto the ground. Then in 1961 the Radio Astronomy Committee of the Astronomical Council of the Academy of Sciences recommended that the Pulkovo Observatory develop the VPA scheme, whilst other institutions develop competing schemes. Shortly afterwards the newly organised Radio Astronomy Council of the Academy of Sciences endorsed one of the versions of a large VPA. This so-called Academy version consisted of 20 m high reflectors in a 3 km diameter ring. But in 1965 it was decided to build a short-wavelength version of the VPA combined with a Kraus-type system.[6] This became known as RATAN-600.[7]

The next question was where should this general purpose instrument, RATAN-600, be located? At about the same time the Soviet Union were also looking for a

[6] Like that built by John Kraus at Ohio Radio Observatory in the 1950s, see Section 22.5.

[7] RATAN is an acronym of 'Radio Astronomy Telescope of the Academy of sciences (Nauk)'.

site for their proposed large BTA-6 optical telescope (see Section 5.3). Eventually it was decided to build both RATAN-600 and BTA-6 at a new observatory, called the Special Astrophysical Observatory, near Zelenchukskaya in the northern Caucasus.

The main reflector of the RATAN-600 consisted of 895 panels each 1.94 m wide by 7.4 m high arranged in a 576 m diameter ring.(9) Each of the panels, whose orientation could be individually adjusted, was slightly curved, with a depth of about 2 mm, about a vertical axis. Inside the southern part of the ring was a flat 400 m long, 8.3 m high reflector and a number of secondary feeds, enabling the telescope to be operated in a number of different modes. One of these modes used the whole ring, with a very large inverted conical secondary mirror, but others used various parts of the ring.

RATAN-600 operated between 8 mm and 30 cm, with a best angular resolution of 2 arcsec. First light, using just the 225 reflecting panels of the northern sector and one feed, was in 1974, and the whole array was completed in 1977 for manual settings. Since then RATAN-600 has been progressively modified to make its adjustments and its limited tracking automatic.

VLBI

In common with other countries worldwide Russia has built a Very Long Baseline Interferometer (VLBI) network in recent years for astrometry and geodesy research.(10) In Russia the stations of their VLBI Quasar network were located at Svetloe about 100 km from St Petersburgh, which was operational in 2003, at Zelenchukskaya (2005), and at Badary (2006) in East Siberia about 130 km east of Lake Baikal. Amongst other equipment each of these stations had a 32 m radio telescope (RT-32), operational from 1.4 to 22 GHz (λ 21 to 1.36 cm), equipped with special facilities for VLBI observations. The three Quasar stations were recently linked by a fibre optics system, and new satellite laser ranging systems were installed at each station in 2011.

19.2 France

Yves Rocard had worked on radar in the UK during the Second World War following his escape from occupied France in 1943.(11) When he returned to France after the war he was appointed Director of the Physics Laboratory at the École Normale Supérieure (ENS) where, in 1946, he set up a radio astronomy group under the leadership of Jean-François Denisse and Jean-Louis Steinberg.(12) Members of the group had no experience of radio astronomy and most had not even heard of the subject. So to get started they erected a modified equatorially mounted 1.5 m diameter US Air Force searchlight mirror and an array of six Yagi antennae on the roof of the physics laboratory to act as radio telescopes. In addition the ENS group also placed a 3 m Würzburg dish on the roof of the physics laboratory to undertake solar observations at 1.2 GHz (λ 25 cm), including those during the partial solar eclipse of April 1949. The resolution of this dish was not good enough on its own to locate sources of radio emission on the Sun, but the occultation of the sources by the Moon during this eclipse helped to provide approximate locations.

Rocard also suggested that Steinberg arrange for the erection of a 7.5 m Würzburg dish, captured and donated by the British, on the roof of the physics laboratory. But because of its weight (about 25 tons) it was installed, instead, on a wagon on a short length of railway track at the Naval Research Laboratory at Marcoussis, about 20 km south of Paris. It was used there to observe the 1949 partial solar eclipse at 158 MHz (λ 1.9 m). ENS later erected another 7.5 m Würzburg at Marcoussis on a concrete block several hundred metres away from the first dish. In the meantime Marius Laffineur and colleagues of the Institut d'Astrophysique de Paris (IAP) had erected another 7.5 m Würzburg on an altazimuth mount at the Meudon Observatory on the outskirts of Paris which they used to observe the 1949, 1952 and 1954 partial solar eclipses.[8]

The first 7.5 m dish at Marcoussis was used by Steinberg and colleagues between 1954 and 1956 to observe the plane of the Milky Way at 900 MHz. It detected a number of discrete sources which included Sag A, Cyg A, Cyg X, Cas A and Tau A, along with a number of new sources. Denisse, Lequeux and Le Roux also observed the North Polar Spur, and initially concluded that the brightness temperature of the sky was less than 3 K. This is interesting in view of Arno Penzias and Robert Wilson's later 1965 discovery of the microwave background radiation at a temperature of 3.5 K.(13) But in 1957 Delannoy, Denisse, Le Roux and Morlet modified their conclusion to say that the brightness temperature was not greater than 20 K.(14)

Laffineur's IAP group also wanted to up-grade their radio telescope facilities in the mid 1950s following the discovery of the hydrogen 21 cm line by Harold Ewen and Edward Purcell in the United States. In the event Laffineur and his colleagues decided to build a two-element transit interferometer to operate at 300 MHz at the Saint Michel field station of the Haute Provence Observatory. Each element consisted of a large fixed horizontal, cylindrical-parabolic reflector of wire mesh supported by wooden posts. The two elements, which were separated by 1.1 km, were each 60 m in length and 32 m in width. Phase coherence between the signals from the two elements was achieved using a microwave link, the first time

[8] In 1962, this Meudon Würzburg antenna was moved to Bordeaux Observatory where it was provided with an equatorial mounting. It continued to provide daily solar flux measurements up to 1990.

that such a system had been used for a French radio telescope. This Saint Michel Interferometer was used between 1959 and 1967 to survey sources > 1.6 arcmin in diameter that passed through the beam. Subsequently Laffineur and P. Coupiac published their catalogue of sources, but the interferometer was not used again.

Nançay

Steinberg of the ENS group and Marius Laffineur from the IAP had visited various Australian radio observatories in 1952 when they were in the country to attend the International Union of Radio Science (URSI). They were particularly impressed by Christiansen's grating array of 32 small dishes at Potts Hill (see Section 16.1). On his return Steinberg approached Rocard about the possibility of building a new state-of-the-art radio observatory on a large radio-quiet site in France. Not only did Rocard agree with Steinberg's suggestion but in the course of only a few weeks he was able to obtain a commitment from the Ministry of National Education for the 25 million francs (~$US100,000, or ~1.0 million in 2015 US dollars) estimated cost of the facility.[9] Consequently the Physics Laboratory of the ENS were able to begin building a radio observatory in 1953 on a 150 hectare (380 acre) site near the village of Nançay about 190 km (120 miles) south of Paris.(15) It was mostly free from locally generated radio interference as the nearest town of Vierzon was some 25 km (15 miles) away.

At about this time André Danjon, the Director of the Paris Observatory, wanted to set up a radio astronomy group at the Meudon branch of the Observatory whilst Rocard wanted to develop semiconductor research at the ENS. To do this Rocard needed more space in the Physics Laboratory. Rocard then told Steinberg that he wanted to erect military equipment at Nançay to detect atomic bomb tests. Steinberg was strongly opposed to this idea as it would have restricted access to the site for foreign radio astronomers. In the end Rocard decided to transfer his radio astronomy group and the Nançay field station from ENS to the Paris Observatory, allowing him to focus on his other interests.(16)

Initially it had been planned to erect both a cross-type array with 1,500 m long east–west and north–south arms to observe the Sun from the Nançay field station, and a variable-baseline interferometer for non-solar work. But the development of these facilities would take some time. As a result it was decided to make a start with smaller radio telescopes.

One of the simplest of these new radio telescopes at Nançay consisted of two modified 1.5 m diameter searchlight antennae, one of which was fixed and the other moveable on a railway track. Simultaneous observations were carried out in

[9] By 1957, about 120 million francs had been committed on this new facility.

the mid 1950s with both antennae at 9.35 GHz (λ 3.2 cm) to investigate solar scintillations, but these antennae were never used as an interferometer.

At about the same time Jacques Arsac of ENS had designed a completely new type of radio telescope that consisted of four 1.1 m dishes operating at 9.35 GHz. Its antennae were located at positions 0, 1, 4 and 6 (in arbitrary units) to provide six spatial frequencies of equal amplitude without redundancy. This Arsac interferometer was initially installed at Marcoussis in 1955 but shortly afterwards it was moved to Nançay. Although ingenious in design, it was difficult to use in practice and produced little in the way of useful results. A far more useful instrument, that was built at Nançay in 1956, consisted of two equatorially mounted 2 m dishes with a separation of 60 m on an east–west baseline. It was used to undertake Earth rotation synthesis. Mukul Kundu used it to study the size, brightness distribution, polarisation and evolution of localised sources of solar radio emission at 9.35 GHz.

In the mid 1950s the two 7.5 m Würzburg antennae were transferred from Marcoussis to Nançay where they were fitted with equatorial mounts. These were then mounted on two railway tracks, one running 1,480 m east–west and the other 380 m north–south. This variable baseline interferometer, which was designed to measure the fine structure of the most intense radio sources, operated at 1415 MHz (λ 21 cm) and had a maximum resolution of 18 arcsec on the east–west baseline. It became operational in 1959 and not only provided valuable information on the structure of various bright radio sources, including Sagittarius A and supernova remnants, but also found a number of sources, later found to be quasars, that were still unresolved at the highest possible resolutions.

Another large radio telescope was built at Nançay in the mid 1950s. Inspired by Christiansen's Australian grating array at Potts Hill it consisted of thirty-two 5 m diameter equally spaced antennae operating at 169 MHz (λ 1.78 m) on a 1,550 m east–west baseline.(17) It had a resolving power of 3.8 arcmin and was used in 1957 and 1958 to examine the structure of the outer solar corona by observing the occultation of the Crab radio source by the Sun. It was also used to locate the sources of solar noise storms and measure their altitudes and dimensions. Then in 1960 eight 10 m antenna aligned north–south at 110 m spacings were added to the array to produce the so-called Grand Interféromètre. This north–south array was built to measure the declination of sources during their transit in the central beam of the east–west array.

In 1963 it was decided to modify every other of the original 32 east–west antennae to produce an array working at 408 MHz (λ 73.5 cm), with the remaining 16 antennae still operating at 169 MHz. This new array, which was completed in 1965, had a resolving power of 1.7 arcmin, which was more than a factor of two better than that of the original 169 MHz array. Three years later the 16 original antennae of the east–west array were replaced by new 3 m diameter low bandwidth antennae with equatorial mountings operating at 408 MHz. The whole 32 element array became operational at 408 MHz in 1972. In the meantime a fire had destroyed the receiver of the 169 MHz array.

In the late 1950s yet another array had been constructed at Nançay. Designed by Steinberg and Monique Pick it was completed in 1959. The array consisted of sixteen 1.1 m diameter antennae attached to a 23 m long metallic girder with a regular spacing of 1.46 m. This girder, supported by four concrete piles, could be rotated around its east–west horizontal axis to reach different declinations. The array, which observed the Sun at 9.3 GHz, had a resolution of about 5 arcmin. It clearly showed that the width of the Sun at this frequency was virtually the same as its optical width indicating that the radio emission at 9.3 GHz did not come from the corona.

Some members of the ENS group had proposed in 1955 to build an interferometer at Nançay consisting of two 25 m dishes on railway tracks to observe the 21 cm line of interstellar hydrogen.(18) But in the following year Denisse pointed out that the subsoil at the Nançay site was not strong enough to support such large mobile antennae, without building substantial and costly foundations, and he also had reservations about the alternative of building a large parabola like that at Jodrell Bank.[10] So he proposed building a transit instrument based on Kraus' Ohio radio telescope (see Section 22.5). He pointed out that this new telescope would be able to detect the 21 cm line in distant galaxies and so determine their distance from their Doppler shifts.

When built this Grand Radiotélescope at Nançay consisted of a 300 m wide, 35 m high curved reflector (see Figure 19.1) that was part of a 560 m radius sphere illuminated by a tiltable plane reflector 200 m wide and 40 m high facing it from a distance of 460 m. The surfaces of both reflectors were made of 12.5 mm metallic mesh to operate at 21 cm, with a minimum operating wavelength of 9 cm. Receivers were mounted on a curved rail track 280 m from the 300 m wide curved reflector allowing coverage of sources up to 7.5° on either side of the meridian plane.

Only one company, the Compagnie Française d'Enterprises (CFE)[11], replied to the Request for Quotations to build the radio telescope. Accordingly they received a contract in December 1958 for 152 million francs to build the first fifth of the telescope which consisted of a fixed 60 m × 35 m central section of the curved reflector and two tiltable panels 40 m × 40 m in total size. This one-fifth section of the telescope was completed in 1961 and the full telescope was finished four years later.

[10] Lequeux, Steinberg and Orchiston suggest that the real reason that a large fixed parabola was not considered was because French industry was unable or unwilling to build such a large, high-precision dish. See James Lequeux, Jean-Louis Steinberg and Wayne Orchiston, Highlighting the History of French Radio Astronomy. 5: The Nançay Large Radio Telescope, *Journal of Astronomical History and Heritage*, 13, No. 1, 2010, p. 30.

[11] CFE was the result of a merger between Enterprises Métropolitaines et Coloniales, founded by Gustav Eiffel in the nineteenth century, and the Moisant-Laurent-Savey Company. It was purchased by Usinor-Sacilor in 1959.

Figure 19.1. The 300 m wide, 35 m high curved reflector of the Grand Radiotélescope at Nançay, France. (Courtesy Julien Descloux via Wikimedia Commons.)

Its focal plane system was redesigned in the late 1990s to completely cover the frequency range from 1.0 to 3.5 GHz with much reduced system noise temperatures.

The original idea behind this Grand Radiotélescope design had been to produce a radio telescope with a large reflecting area at minimum cost, whilst accepting the limitation that it was a transit instrument that could be used only for about 30 minutes on either side of the meridian. There were a number of other operational limitations which became apparent as time went on, some of which were only rectified in the late 1990s. Although the designers did achieve their goal of producing a large radio telescope at relatively low cost per unit area, the Parkes dish (see Section 16.2) that had been completed in Australia in 1961 was only slightly more expensive per unit area. And the Parkes dish was fully steerable and could be operated at a lower wavelength.

In the early 1970s a new radioheliograph (NRH) was built at Nançay to provide images of the Sun at 169 MHz at the rate of 100 images per second.(19) Initially this NRH, which was restricted to imaging in one dimension with an analogue correlator with 55 channels, was operational in 1976. Later the NRH was modified to provide two-dimensional images in the frequency range from 150 to 450 MHz with a digital correlator of 576 channels. It now has 48 antennae, both simple dipoles and dishes up to 10 m in diameter, in a basically T-shaped array with maximum east–west and north–south baselines of 3.2 km and 2.0 m, respectively.(20) It can produce quasi-simultaneous, multi-frequency observations with a maximum number of 10 frequencies and a limit of 200 images per second.

The Nançay Decameter Array (NDA), operating between 10 and 80 MHz (λ 30 to 3.75 m), was designed and built in the mid 1970s to undertake high resolution spectroscopy of solar and Jovian radio bursts. It consisted of two phased arrays of 72 antennae each operating in opposite senses of circular polarisation, each with an effective aperture of about 3,500 m². The NDA is now also used to study radio emissions from other planets and stars.

To continue its low frequency research at Nançay an E-LOFAR station was installed there in 2010, which is now part of the International LOFAR Telescope (ILT) (see Section 19.3 below).

19.3 The Netherlands

The Netherlands had been occupied by Germany in the Second World War. As a result, the Institute of Technology at Delft, which had been the only source of fully qualified professional engineers, had been closed for four years and Dutch universities had been closed for two years.(21) But some astronomers had been able to continue working, albeit in very difficult circumstances. Two of these, namely Jan Oort, assistant director of the Leiden Observatory, and Hendrik van de Hulst were to prove particularly important for the future of astronomy in the Netherlands and internationally. On the theoretical side van de Hulst calculated in early 1944 that a hyperfine line in the spectrum of the hydrogen atom should eventually be detectable at a frequency of 1.4 GHz (λ 21 cm), although the receivers of the time were not sufficiently sensitive at that frequency.

In early 1944 Oort had been aware of Reber's earlier detection of radio signals that appeared to be coming from the Milky Way. Oort realised that available receivers were not sufficiently sensitive to detect van de Hulst's 21 cm line, but he was still interested in detecting the structure of the Milky Way in the radio continuum. Consequently he asked C. J. (Cornelius) Bakker of the Philips Physical Laboratory if he could supply a receiver operating at a wavelength of about 50 cm (frequency 600 MHz).[12] Oort envisaged using this with a radio telescope dish of about 10 to 20 m diameter which he hoped could be built by the Leiden

[12] As Van Woerden and Strom have pointed out on pp. 16–17 of their *Journal of Astronomical History and Heritage* paper (21), the sequence of events that included van de Hulst's theoretical paper and Oort's request to Bakker is not correct in a number of papers. Van de Hulst presented his original paper at a meeting on April 15 1944, not later as claimed elsewhere, whilst Oort wrote to Bakker on April 19 1944. It seems as though van de Hulst's paper galvanised Oort to start radio observations, even though Oort realised that he couldn't undertake 21 cm observations at that time for the reasons given above.

observatory workshop.(22) Bakker replied that such a receiver may be available after the end of the war, but pointed out that a larger dish would be needed or one operating at shorter wavelengths to meet Oort's required resolution of 0.5°. Alternatively Bakker suggested that the radio equivalent of a Michelson interferometer may be better.

In other countries the motive force behind the development of radio telescopes and radio astronomy after the war was usually supplied by radio and radar engineers and physicists who were looking to use the skills that they had acquired during the war in a post-war environment. But in the Netherlands it was Jan Oort, an astronomer, who was pushing the case for a radio telescope. Unfortunately in his country there was a relative shortage of up-to-date radio/radar engineers and physicists because of the closing down of academia during the war, although there were still some in industrial companies like Philips.

In 1945 Oort looked into the possibility of building a radio telescope of 25 m diameter. As a result he contacted a number of people and organisations for advice and information amongst them Werkspoor, a builder of railway bridges, who quickly produced a tentative design. But, unfortunately, it proved impossible to raise the required funds at that time as they were required to put the country back on its feet after the occupation. However some years later, as a result of an initiative taken by Oort and Marcel Minnaert (the director of the Utrecht observatory), the Netherlands Foundation for Radio Astronomy (NRFA)[13] was created to investigate radio radiation coming from the cosmos. Chaired by Oort, this NRFA was supported by the PTT (Post, Telephone and Telegraph Service), Philips Physical Laboratories and the Royal Netherlands Meteorological Institute, as well as by Leiden and Utrecht observatories. The existence of the NRFA naturally gave a boost to Oort's plans for a large radio telescope.

Meanwhile radio astronomy observations had already been started in the Netherlands, as in France and the Soviet Union, using salvaged Würzburg German radar reflectors. In particular a 7.5 m (25 ft) Würzburg antenna located at the Dutch PTT's telecommunications station at Kootwijk had been made available to NRFA for studies of radio emission from the Milky Way. And in 1948 C. J. Gorter from the Kamerlingh Onnes Laboratory at Leiden University provided a war-surplus radio receiver which could be used at a wavelength of 21 cm. H. Hoo, the NRFA's sole employee at the time, unsuccessfully used it to try to detect the 21 cm emission line of neutral hydrogen before the receiver was accidentally destroyed by fire in 1950.(23)

Shortly afterwards C. Alexander (Lex) Muller, Hoo's successor, started to develop a new receiver for the Würzburg antenna in his search for the 21 cm line.

[13] Stichting Radiostraling van Zon en Melkweg (SRZM) in Dutch.

It so happened that at the same time van de Hulst was spending the 1951 spring semester at Harvard College Observatory as a visiting professor. And whilst he was there he found out that Harold Ewen and Edward Purcell were also trying to detect the 21 cm line. Then on March 25 1951 Ewen and Purcell were successful. Fortunately for the Dutch, Ewen and Purcell were happy to swop information on the design of their equipment with them which resulted in Muller being able to detect the 21 cm line just seven weeks after the Americans. Over the next few years Van de Hulst, Muller and Oort continued to use this Würzburg antenna to observe the distribution of neutral hydrogen in the Milky Way. The result was published in map form in 1954, showing part of our galaxy's spiral structure.(24) In the same year Kwee, Muller and Westerhout were able to determine the rotation curve for the inner part of the Milky Way also using observations of the 21 cm line of neutral hydrogen.(25)

In the meantime the detection of the 21 cm line had given a major boost to Oort's plans for a large radio telescope in the Netherlands and in November 1951 Werkspoor was given the contract to produce the design of a 25 m (82 ft) diameter altazimuth instrument. This design, which had a dish surface of 15 mm × 15 mm mesh, was completed in 1954. Meanwhile a search had been carried out to find a suitable site to locate the telescope. This resulted in Dwingeloo being selected at the edge of the State Forest, overlooking a protected piece of moorland where radio interference could be controlled. When regular observations with this 25 m Dwingeloo telescope (see Figure 19.2) were begun in 1956 it was, for a short time, the largest, fully steerable, radio telescope dish in the world. Gart Westerhout first used it(26) to survey the continuous radiation from the Milky Way at 1.39 GHz (λ 21.5 cm) where it had a beamwidth of 0.57°. This was followed by Oort and G. W. Rougoor who carried out 21 cm line observations of the galactic centre region.[14]

Two Würzburg antennae, which had been purchased from the Norwegian government for the symbolic price of 1 guilder each, were also located at Dwingeloo in 1955 where their altazimuth mountings were replaced by equatorial ones. They were used mainly for solar research as a two-element interferometer.

Oort began to consider the design of the next generation Dutch radio telescope just as the 25 m Dwingeloo dish was becoming operational. As a result he asked Charles Seeger to produce a conceptual design of a large cross-shaped radio telescope. His resulting design was discussed by NRFA in 1958. The arms of the cross were about 3 km long which at a wavelength of 75 cm would produce a resolution

[14] The 25 m telescope was decommissioned in 1998. But seven years later a group of amateur radio enthusiasts founded the C. A. Muller Radio Astronomy Station (CAMRAS) and began to refurbish the 25 m after it had been leased to them. As a result in 2007 the telescope was operational again, this time as an amateur project. See Yvette Cendes, Tales from the Dwingeloo Radio Observatory, *Sky and Telescope*, January 2013, pp. 70–73.

Figure 19.2. The 25 m diameter radio telescope which was built in the early 1950s at Dwingeloo in the Netherlands. (Courtesy ASTRON.)

of ~1 arcmin. Although construction of the relatively simple antennae in the arms would be relatively straightforward, it was clear that the telescope's cost would be higher than the Netherlands could afford on its own. Consequently they approached Belgium to see if a 50/50 joint project was possible. In fact, Belgium were interested but had no money available, so they approached the OEEC (Organisation for European Economic Cooperation)[15] which in 1960 agreed to support the project. By that time Christiansen had joined the project from Australia bringing with him his wealth of experience in designing cross antennae.

The first design of this so-called Benelux Radio Telescope, as it was initially called, was produced by Christiansen, Jan Högbom and Seeger. It was an advanced version of the Mills Cross design (see Section 16.1) with 5 km arms and was to operate at 75 cm (frequency 400 MHz) to produce an instantaneous beam of 1 arcmin or better. The elements on both arms were 30 m cylindrical paraboloids. Unfortunately its estimated cost of about 25 million Dutch guilders ($US 7 million) was thought to be unaffordable so an alternative design was considered in 1961.

[15] From 1961 called the Organisation for Economic Co-operation and Development (OECD).

This new design, which would operate at 21 cm, was to consist of about 100 parabolic dishes with a wire mesh surface in a cross-shaped array. Although most radio sources were considerably fainter at 21 cm than those observed at 75 cm this was compensated, albeit at the expense of telescope time, by the fact that this new design could track the sources for a considerable time. All but one of the dishes were to be of 30 m diameter whilst one was to be of 70 m diameter. In September 1962 the council for what was now called the Benelux Cross Antenna Project (BCAP) chose this 21 cm design as their preferred radio telescope to be completed in 1967, albeit at an estimated cost of 30 million Dutch guilders. The project proposal was submitted to both the Dutch and Belgian governments in February 1963 for approval.

Whilst project approval was being awaited Högbom began to consider building a one-dimensional east–west array, like the Cambridge (UK) One Mile telescope (see Section 17.2) which relied on earth-rotation aperture synthesis. He suggested a 3.1 km long east–west array of 28 parabolic antennae which would produce a resolution of 17 arcsec. It was to be supplemented by another array of 6 antennae 1600 m further east or west to produce a resolution of 10 arcsec. The smaller number of antennae would make it easier to use this array, compared with the previously chosen BCAP configuration. But the individual exposures would take longer, making it more difficult to observe short-term variability of sources. Although each antenna would be more expensive as they had to be steerable over 12 hours of hour-angle, as opposed to 4 hours for the BCAP configuration, the considerable reduction in the number of antennae should make this alternative array somewhat cheaper to build.

In the meantime the Dutch government had given its approval for the BCAP project as presented to it in February 1963. But the Belgian government was, at that time, considering joining the European Southern Observatory (ESO) (see Section 7.1) and they could not afford to join both BCAP and the ESO. Consequently Oort and Adriaan Blaauw, who were both involved in trying to get the ESO approved, suggested that maybe Belgium could contribute to just the preparation cost of BCAP and its operational cost. This compromise was agreed between the two governments in mid 1964, it being understood that the design of the telescope would have to be simplified compare with the BCAP proposal as only the Dutch government was going to pay for its construction.

As long as the project was to be a 50/50 joint project between the Netherlands and Belgium there was a requirement to locate the telescope, if possible, near the border between the two countries. But with the new financial arrangement this was no longer a requirement. At first two possible sites near Dwingeloo were considered but these proved to be unsatisfactory. Subsequently the Dutch Forestry commission suggested a site near the village of Westerbork about 25 km from

Figure 19.3. A number of 25 m diameter equatorially mounted dishes of the Westerbork Synthesis Radio Telescope which became operational in 1970. (Courtesy AdriaanRenting via Wikimedia Commons.)

Dwingeloo on the site of the old Westerbork transit camp that had been used by the Germans to accommodate Jews in the Second World War. This was accepted as a good location and all the necessary planning permission was obtained by late 1964 to build what was to be called the Westerbork Synthesis Radio Telescope.

The design of this Westerbork Synthesis Radio Telescope was based on the east–west linear array proposed by Högbom but simplified as only 50% of the original finance was now available. The new design (see Figure 19.3) which was on a 1.6 km east–west baseline consisted of ten fixed 25 m diameter equatorially mounted dishes and one similar dish movable along a 300 m length of rail. But when the lowest bid came in from the ship-building company Wilton-Feyenoord it was well within the budget, so the NFRA asked for another movable dish to be added to the contract.

In the event the Westerbork radio telescope became operational in 1970. Although it was initially designed to be used down to 21 cm, the dishes were found to be accurate enough down to 6 cm. So a 6 cm receiver was installed on the telescope in 1972 and two additional movable dishes were added later to extend the baseline to 3 km.

It had become more and more difficult in the mid 1960s to retain the binational character of a project with Belgium's limited involvement. Consequently in mid

1967 Belgium decided that it no longer wished to be considered as a partner in the Westerbork Synthesis Radio Telescope.

During its early operation it was found that all the Westerbork observations were subjected to severe interference. This was traced as being due to a NATO transmitter located at a relatively distant military base. But, in spite of the fact that this transmitter was not breaking any frequency protection rules, the military agreed to change its frequency which they did in 1970 to solve the problem.

The European VLBI (Very Long Baseline Interferometry) Network (EVN) was formed in 1980 by a consortium of five major radio astronomy institutes in Europe. But it subsequently became clear that this rather informal arrangement was the cause of many problems. This included the lack of a state-of-the-art correlator to integrate all the observations. So in 1993 the Joint Institute for VLBI in Europe, or JIVE, as it was called, was formed by the European Consortium for VLBI. The main task of JIVE, which was located in Dwingeloo, was to provide, operate and develop the much needed EVN correlator that became operational in 2000.

In the early 1990s ASTRON, as the NRFA was now called, and a number of Dutch universities became interested in building a low frequency radio telescope. Subsequently a feasibility study was undertaken and international partners sought for such a LOFAR (LOw-Frequency ARray) facility. Initially potential sites were considered in Western Australia, the Netherlands and the south-western United States.(27) Although Western Australia was chosen as the best location, the Netherlands government agreed in November 2003 to provide €52 million ($66 million) to fund LOFAR provided it was located in their country. Not surprisingly the project accepted the Dutch offer and, just over a year later, a supercomputer was installed at Groningen in the Netherlands to undertake LOFAR's data processing. LOFAR's main purpose was to observe the very early universe back to the Epoch of Reionization (EoR) of neutral hydrogen.

The first LOFAR station, which consisted of 96 dual-dipole antennae, was installed in 2006 in four clusters with 48 dipole antennae in the central cluster and 16 in each of the other three. These clusters were distributed over an area of about 500 m in diameter near the village of Exloo in the northern Netherlands. Then in the following year the first international LOFAR station was built next to the Effelsberg 100 m radio telescope dish in Germany. Since then 37 more stations have been completed (by 2015) in the Netherlands, 4 more in Germany, and one station each in the UK (Chilbolton), France (Nançay) and Sweden (Onsala). All the stations were linked to the supercomputer at Groningen by a fast fibre-optic network. The antennae were a mixture of high-band antennae operating in the 120 to 240 MHz range and low-band antennae in the 30 to 80 MHz range (see Figure 19.4).(28) By 2015 there were 18 stations in the LOFAR core covering about 400 hectares (1000 acres) centred near Exloo. In addition there were 18 remote stations located within a distance of

Figure 19.4. The central part of the LOFAR station near Exloo in the Netherlands in 2010. The large panels (in groups of 24) enclose high-band antennae whilst the more numerous smaller pale rectangles show the locations of the low-band antennae. (Courtesy LOFAR/ASTRON via Wikimedia Commons.)

100 km of the core, and finally there were 8 international LOFAR stations. LOFAR was not only a telescope system in its own right but it was also considered to be a pathfinder for the Square Kilometre Array (see Section 16.5).

References

1. Salomonovich, A. E., The first steps of Soviet Radio Astronomy, in Sullivan, W. T., (ed.), *The Early Years of Radio Astronomy; Reflections Fifty Years After Jansky's Discovery*, Cambridge University Press, 1984, pp. 268–288.
2. Ginzburg, V. L., Remarks on my work in Radio Astronomy, in Sullivan, W. T., (ed.), *The Early Years of Radio Astronomy; Reflections Fifty Years After Jansky's Discovery*, Cambridge University Press, 1984, pp. 289–302.
3. Strelnitski, Vladimir S., The Early Post-war History of Soviet Radio Astronomy, *Journal for the History of Astronomy*, 26, 1995, pp. 349–362.
4. Ibid., p. 353.
5. Dogel, B. A., et al., Radio Astronomy Studies at the Lebedev Physical Institute, in Braude, S. Y., et al., (eds.), Kellermann, K. I., (trans.), *A Brief History of Radio Astronomy in the USSR*, Springer, 2012, p. 11.

6. Ibid., pp. 28–34.
7. Dagkesamanskii, R. D., The Pushchino Radio Astronomy Observatory of the P N Lebedev Physical Institute Astro Space Center: Yesterday, Today and Tomorrow, *Physics-Uspekhi*, 52, 2009, 1159–1167.
8. Salomonovich, A. E., The first steps of Soviet Radio Astronomy, in Sullivan, W. T., (ed.), *The Early Years of Radio Astronomy; Reflections Fifty Years After Jansky's Discovery*, Cambridge University Press, 1984, p. 281.
9. Parijskij, Yurij N., and Korol'kov, Dmitrij V., Experiment COLD: The First Deep Sky Survey with the RATAN-600 Radio Telescope, *Astrophysics and Space Physics Review*, 5, 1986, pp 39–179.
10. Finkelstein, Andrey, Ipatov, Alexander, and Smolentsev, Sergey, Radio Astronomy Observatories Svetloe, Zelenchukskaya and Badary of VLBI Network QUASAR, *IVS 2004 General Meeting Proceedings*, pp. 161–165.
11. Orchiston, Wayne, et al., Highlighting the History of French Radio Astronomy. 3: The Würzburg Antennas at Marcoussis, Meudon and Nançay, *Journal of Astronomical History and Heritage*, 10, No. 3, 2007, pp. 221–245.
12. Orchiston, Wayne, et al., Highlighting the History of French Radio Astronomy. 4: Early Solar Research at the École Normale Supérieure, Marcoussis and Nançay, *Journal of Astronomical History and Heritage*, 12, No. 3, 2009, pp. 175–188.
13. Denisse, J.-F., Lequeux, J., and Le Roux, E., Nouvelles Observations du Rayonnement du Ciel sur la Longueur d'Onde 33 cm, *Comptes Rendus de l'Académie des Sciences*, 244, 1957, pp. 3030–3033.
14. Delannoy, J., et al., Mesures Absolues de Faibles Densités de Flux de Rayonnement à 900 MHz, *Annales d'Astrophysique*, 20, 1957, pp. 222–237.
15. Steinberg, J.-L., La Station de Recherches de Nançay, Observatoire Radioastronomique Français, *L'Astronomie*, 71, 1957, pp. 89–99.
16. Orchiston, Wayne, et al., Highlighting the History of French Radio Astronomy. 3: The Würzburg Antennas at Marcoussis, Meudon and Nançay, *Journal of Astronomical History and Heritage*, 10, No. 3, 2007, p. 226.
17. Pick, Monique, et al., Highlighting the History of French Radio Astronomy. 6: The Multi-Element Grating Arrays at Nançay, *Journal of Astronomical History and Heritage*, 14, No. 1, 2011, pp. 57–77.
18. Lequeux, James, Steinberg, Jean-Louis, and Orchiston, Wayne, Highlighting the History of French Radio Astronomy. 5: The Nançay Large Radio Telescope, *Journal of Astronomical History and Heritage*, 13, No. 1, 2010, pp. 29–42.
19. Bonmartin, J., et al., The Mark II Nançay Radioheliograph, *Solar Physics*, 55, 1977, pp. 251–261.
20. Kerdraon, Alain, and Delouis, Jean-Marc, The Nançay Radioheliograph, in *Coronal Physics from Radio and Space Observations*, pp. 192–201, in Trottet, Gérard, (ed.), Proceedings of the CESRA Workshop held in Nouan le Fuzelier, France, 3–7 June, 1996, *Lecture Notes in Physics*, 483, 1997, Springer.
21. Van Woerden, Hugo, and Strom, Richard G., The Beginnings of Radio Astronomy in the Netherlands, *Journal of Astronomical History and Heritage*, 9, No. 1, 2006, pp. 3–20.
22. Raimond, Ernst, and Genee, René, *The Westerbork Observatory, Continuing Adventure in Radio Astronomy*, Kluwer Academic Publishers, 1996.
23. Sullivan, Woodruff T., *Cosmic Noise; A History of Early Radio Astronomy*, Cambridge University Press, 2009, p. 406.
24. Van de Hulst, H. C., Muller, C. A., and Oort, J. H., The Spiral Structure of the Outer Part of the Galactic System Derived from the Hydrogen Emission at 21 cm Wavelength, *Bulletin of the Astronomical Institutes of the Netherlands*, 12, 1954, pp. 117–149.

25. Kwee, K. K., Muller, C. A., and Westerhout, G., The Rotation of the Inner Parts of the Galactic System, *Bulletin of the Astronomical Institutes of the Netherlands*, 12, 1954, pp. 211–222.
26. Westerhout, Gart, A Survey of the Continuous Radiation from the Galactic System at a Frequency of 1390 Mc/s, *Bulletin of the Astronomical Institutes of the Netherlands*, 14, 1958, pp. 215–260.
27. Schilling, Govert, LOFAR: Growing a Giant Radio Telescope, *Sky and Telescope*, December 2004, p. 24.
28. Schilling, Govert, End of the Cosmic Dark Ages, *Sky and Telescope*, May 2011, pp. 26–31.

20

The American National Radio Astronomy Observatory

Radio astronomy took some time to take off in America after the pioneering observations of Karl Jansky in 1932 and of Grote Reber in the late 1930s and early 1940s. This was in spite of the tremendous advances in radio technology achieved during the Second World war. Although the US Army Signal Corps had detected radar signals from the Moon in 1946, this was not for astronomical purposes but was a test of their system to detect incoming missiles from the Soviet Union.

At the end of the Second World War American astronomers had access to the best optical telescopes in the world and the Palomar 200 inch was also shortly to become available. In addition the climate in the United States, particularly on the west coast, was generally good for optical observations. So observational astronomers in America saw no need to get involved in the new-fangled field of radio astronomy which required new technical skills that they generally did not have. They were also generally sceptical that radio observations would yield much in the way of useful astronomical results. In Europe, on the other hand, the climate was generally not as conducive for optical observations as in America and large optical telescopes in Europe were not as good. But the climate was no limitation for radio observations. So a new set of observational astronomers, largely ex radio and radar engineers and physicists, grew up in Europe after the Second World War devoted to radio observations. A similar situation prevailed in Australia which also suffered from a lack of top quality optical telescopes, although they did not have the same climate problem as in Europe.

By the early 1950s radio astronomy was beginning to produce interesting results. At that time it became clear in the United States that, unless a bold approach was taken, it would continue to fall further and further behind other countries, particularly the United Kingdom and Australia, in radio astronomy. As a result, a number of American organisations had either started to develop radio telescopes or were considering doing so.(1) Then in January 1954 an interdisciplinary conference took place on radio astronomy, jointly sponsored by the American National

Science Foundation (NSF), the California Institute of Technology (Caltech) and the Department of Terrestrial Magnetism of the Carnegie Institution. This conference, which included radio astronomers from the UK, Australia, the Netherlands and Canada, helped to persuade many of the American participants that a new approach was required to the provision of radio astronomy facilities in the United States. Then on May 3 1954 Alan Waterman, director of the NSF, appointed an Advisory Panel for Radio Astronomy chaired by Merle Tuve.(2)

20.1 AUI Feasibility Study and Early Programme of the 140 ft Telescope

In the meantime Donald Menzel, director of the Harvard College Observatory, had begun discussions in early 1954 on the future of United States involvement in radio astronomy. In the process Lloyd Berkner, president of AUI (Associated Universities Incorporated),[1] told Menzel that AUI may be interested in managing a national radio astronomy observatory. Encouraged by this response Menzel in collaboration with Julius Stratton, the Provost of MIT, Jerome Weisner, also of MIT, and Bart Bok and colleagues of Harvard produced a report outlining possible future cooperation in radio astronomy. The report said that, unless a *major* facility was built, the United States would fall further and further behind in radio astronomy research. But the high cost of such a major facility meant that it would be beyond the means of any one institution.

The Harvard/MIT report also emphasised the interdisciplinary nature of radio astronomy in which astronomy, physics, and electrical and mechanical engineering all had a role to play. Consequently it suggested that radio astronomy should be treated in a similar way to nuclear physics research after the war. In that case AUI had established and operated the Brookhaven National Laboratory as a national user institute. As a result, the Harvard/MIT report specifically suggested that AUI would be the preferred operating institution for the large radio astronomy facility under consideration.

In April 1954, Berkner received authorisation from the AUI board of trustees to investigate the possibility of AUI operating a major radio astronomy facility for the benefit of the entire American research community. Consequently, he invited thirty-seven scientists with an interest in radio astronomy from twenty-eight institutions to a meeting on May 20 to discuss cooperation in radio astronomy.[2] In the

[1] AUI was formally incorporated in 1946 and began by developing the Brookhaven Graphite Research Reactor. It was the first time a consortium of universities had been formally created in the United States to support a large-scale scientific research facility. In the early 1950s, AUI were looking to expand into other areas of basic scientific research, one of which included radio astronomy.

[2] *Planning Document for the Establishment and Operation of a Radio Astronomy Observatory*, prepared for the National Science Foundation by Associated Universities, Inc., August 1956, p. 7.

invitation, he mentioned that the meeting was in response to the suggestions made during the January NSF/Caltech/Carnegie conference and indicated that AUI's involvement was in response to the request of the Harvard/MIT group.

Berkner's and Menzel's background work resulted in the May 20 meeting recommending that AUI should request funds from the NSF to conduct a feasibility study considering the planning, design and construction of a national radio astronomy facility. AUI had a draft proposal ready by the end of June. An ad-hoc committee, which included Tuve and other members of the NSF Advisory Panel, was then set up to guide the study.[3] At its first meeting on July 26, a slightly revised proposal for a grant of $105,000 was approved by the committee and submitted directly to the NSF.(3) On receipt Waterman sent the proposal to Tuve who was chairman of the NSF subpanel that advised the agency on what proposals deserved funding. Tuve sent out copies to members of the subpanel and referees for review.

Tuve was very much against the post-war trend in science to ever larger research facilities, often supported by government money, whilst Berkner was completely the opposite being more of a 'big science' man. So when a number of referees and others voiced concerns about the management of the proposed radio astronomy facility by AUI, Tuve used them to support his personal views. He emphasised that there were no radio astronomers on the AUI board of trustees and no-one at AUI was active in the field of radio astronomy. So the planned facility was likely to be institution-led rather than research-led. In particular he suspected, with some justification, that AUI were determined to build a larger dish than the 250 ft (76 m) at Jodrell Bank to be the largest dish in the world, even if it was not the most appropriate facility for American radio astronomers at that time. He was also concerned that the amount of money required to build this large national facility would inevitably drain money away from smaller, traditional astronomy research groups. In addition three southern members of the NSF board were concerned that AUI was too north-east oriented as there were no astronomers from universities outside the north-east on the AUI board. There was also a suggestion that there should be just one national astronomy facility covering both radio and optical research. But all that needed to be decided in late 1954 was whether to approve the proposed AUI feasibility study, not decide on which organisation should build and operate any subsequent facility.

The NSF Radio Astronomy Panel discussed the AUI proposal at its first meeting in November 1954 with Tuve in the chair and recommended a grant of $85,000 but with conditions. The feasibility study was to be limited to considering instruments of modest size and the search for a site for the proposed observatory was to

[3] The presence of Tuve and other members of the NSF Advisory Panel on this ad hoc committee was only allowed after consultation with NSF lawyers. They accepted such overlapping responsibilities in this case as there were so few people with the necessary expertise in America at that time.

be limited to a region within a 300 mile (480 km) radius of Washington, DC. The decision to restrict the site location to the Washington area in the east of the United States was to achieve a balance with the best optical observatories which were in the west.

On March 25 1955 AUI held a meeting with people familiar with building large radio telescopes in the USA and abroad.[4] The result of this meeting was considered on the following day by the AUI ad-hoc committee, which had now been reconstituted as a steering committee for the project. The steering committee decided at that meeting that the first large telescope should be a reflector with a diameter of about 150 ft (46 m). It was suggested that existing designs in the 50 to 80 foot range could be easily extrapolated up to 150 ft. They also decided that the largest practicable diameter of parabolic dish with then-current technology would be about 600 ft (180 m).[5] Finally, they called for site selection and acquisition before the end of 1955 with the beginning of operations with the 150 ft reflector in July of the following year. After the meeting AUI contracted Jacob Feld, a consulting engineer from New York, to undertake a design study of a 600 ft. This study, which he completed in July 1955, showed that, although it was technically feasible to build a 600 ft, such an undertaking would be exceptionally challenging with the consequent risk of serious timescale delays.

Following the March 26 meeting Richard Emberson, Berkner's assistant and acting director of the project, contracted Eggers and Higgins of New York to prepare a hypothetical site development plan to cover all foreseen telescopes and facilities. This was to be used to determine the likely site development costs, excluding those for the telescopes themselves.

The first AUI report, which was submitted to NSF on May 6 1955,[6] outlined a preliminary view on the mode of operating the facility, its organisational plans and approximate estimates of the funds required for construction and operation of the facility over the next five years. In response the NSF requested that AUI should now concentrate on a more modest plan, with particular emphasis on the design of a 40 m (131 ft) radio telescope. The NSF's Advisory Panel for Radio Astronomy also proposed a five year budget for such a facility.

The AUI's steering committee met on May 28 to consider the comments made by the NSF and the NSF's Advisory Panel on the AUI report. The committee discussed the five year budget proposed by the Advisory Panel and decided that it was too small. Instead they agreed to develop a minimum operating budget for

[4] *Planning Document for the Establishment and Operation of a Radio Astronomy Observatory*, prepared for the National Science Foundation by Associated Universities, Inc., August 1956, p. 9.

[5] This may seem to be an enormous antenna to be considering at that time. But much larger dishes were under investigation as it was not realised then how difficult they would be to build and operate.

[6] A draft had been delivered on April 22.

an observatory with just a 40 m telescope. They also reviewed the specifications which were to be sent out to interested industrial companies as part of AUI's request for proposals covering both a 40 m telescope and a larger telescope up to about 600 ft (180 m) diameter. The requests for proposals, which asked for indicative costs, were sent out on the following day with a request that replies be submitted by July 1. Quotations were required for both altazimuth and equatorial mounts for the 40 m which was to operate down to a wavelength of about 10 cm.

Twenty companies submitted proposals or expressions of interest for at least some of the task, with five of these companies offering to do the whole job. The steering committee considered these five proposals for the design and construction of the 40 m telescope at their meeting of July 11–12. Initially AUI had hoped to let a fixed price contract to design, make and erect the telescope. But following the advice of Charles Husband, who was working on the Jodrell Bank 250 ft dish at the time (see Section 18.2), AUI decided to limit the first contract to producing a detailed design for a 140 ft.[7] The surface was now required to be accurate enough to allow observations at wavelengths down to 5 or 6 cm with some capability at 3 cm. The steering committee also suggested that the decision on the choice of site, which was still to be made, should allow the accommodation at some future time of more radio telescopes including a much larger dish of around 600 ft in diameter. Subsequently AUI requested NSF for an additional $234,500 to support their studies on the 140 ft up to June 30 1956.

The NSF cut this new requested grant to $140,500 before approving it in October 1955. Now that these extra funds had been agreed, AUI awarded contracts to Jacob Feld, the D. S. Kennedy Company of Massachusetts (builders of various radio antennae, both civil and military, up to 84 ft), and Husband and Company of Sheffield, England to undertake preliminary designs of the 140 ft. The three companies were asked to concentrate on an altazimuth design as engineers had advised AUI that an equatorial design would be more complex and more costly. In addition AUI wanted the 140 ft telescope to act as a prototype for a much larger telescope which would have to have an altazimuth mount.

The question of the site for the proposed national radio observatory within 300 miles of Washington, DC had been considered early in the AUI feasibility study by a team led by Emberson. A number of criteria had been proposed to assist in making a choice, the most restrictive being the requirement to avoid areas subject to significant radio interference and subject to hurricanes and tornadoes.

[7] At about this time, according to David Heeschen, an 'unidentified government official' decided that the size of an American telescope should be expressed in feet, so the 40 m was rounded upwards to 140 ft., see David S. Heeschen, The First Ten Years, 1955–1965, in F. J. Lockman, F. D. Ghigo and D. S. Balser, (eds.), *But It Was Fun: The First Forty Years of Radio Astronomy at Green Bank*, National Radio Astronomy Observatory, 2007, p. 268.

Eliminating these parts of the country produced an oval area about 300 miles long by about 100 to 150 miles wide. Engineers from Jansky and Bailey, Inc. were contracted to make radio interference measurements at selected potential sites in this area with equipment supplied by the Naval Research Laboratory. These measurements, which did not include those of white noise, showed that one site was clearly better than all the rest. This was at Green Bank, West Virginia which was also best on the basis of population studies (an indicator of the likely amount of white noise), limited nearby urban and industrial areas and infrequent aviation activity. In addition Green Bank was virtually surrounded by mountains which would help to protect the site from future radio interference. Consequently Green Bank was recommended as the location of the proposed radio observatory at a meeting of the AUI steering committee and others on December 13 1955. The steering committee also recommended that all of the land in the Green Bank valley be purchased or, failing that, restrictions be put in place to ensure that the local radio environment around the radio astronomy site was controlled. Shortly afterwards NSF accepted these recommendations.

The steering committee also discussed the altazimuth-mounted 140 ft telescope at their meeting of December 1955. The committee suggested that, initially, two smaller telescopes of about 25 and 60 ft diameter should be provided at the observatory. In addition this time some members of the committee were not happy with the decision to limit the design of the 140 ft to an altazimuth-mounted telescope. Consequently in March 1956 R. J. Seeger, assistant director of the NSF, asked AUI to get at least one of the three contractors to produce a design and cost estimates for an equatorially mounted 140 ft. In the event all three contractors agreed to do this.

AUI issued a 'Planning Document for the Establishment and Operation of a Radio Astronomy Observatory'[8] in August 1956 in which they described the provisional results of the three design contracts. This was to assist in the writing of a contract (see Section 20.3 below) between the NSF and AUI for the establishment and operation of the National Radio Astronomy Observatory (NRAO). In it AUI outlined an initial programme that would include a 28 ft telescope, two 60 ft telescopes and the 140 ft.

The steering committee reviewed all the designs of the 140 ft in October 1956 and found that none of them was satisfactory. So N. L. (Ned) Ashton from the University of Iowa, the designer of the Naval Research Laboratory's (NRL's) 50 ft high precision telescope (see Section 22.1), was asked to develop an equatorial design under the supervision of an ad hoc panel chaired by T. C. Kavanagh, a consulting engineer. Subsequently in January 1957 the question of whether to use an

[8] *Planning Document for the Establishment and Operation of a Radio Astronomy Observatory*, prepared for the National Science Foundation by Associated Universities, Inc., August 1956.

equatorial or altazimuth mount was finally settled in favour of the equatorial design after reviewing once more the relative advantages and disadvantages of both types of mount. Ashton's final design was approved by the committee later that year and a request for fixed price proposals issued in August (see Section 20.7 below).

20.2 Role of the NSF in Funding Large Facilities

In parallel with the above Waterman had begun to consider the fundamental question as to whether the NSF should be in the business of supporting the creation and operation of large facilities like a national radio astronomy observatory. As a result he asked AUI in April 1955 to provide a general planning document for an NSF-sponsored radio astronomy facility to the National Science Board Committee on the Mathematical, Physical and Engineering (MPE) Sciences before the May meeting of the full National Science Board (NSB).[9] In response AUI provided the requested planning document together with design proposals for a 140 ft instrument. In addition the document, partly drafted by Berkner, drew attention to the advantages that the technology behind designing and building radio telescopes, which had to detect very weak signals, would have in military applications. As mentioned above, Tuve was already sceptical about building a large national radio observatory facility. But mention of the military caused even more problems not only with Tuve but with many other radio astronomers who were very much against linking peaceful scientific research with military applications.

At their May meeting, the NSB formally discussed the NSF role in support of large-scale scientific research. As a result it passed a resolution committing the NSF to supporting large-scale basic scientific facilities such as 'a national [optical] astronomical observatory, a major radio astronomy facility, and university installations of computers, accelerators and reactors' when the funds were not readily available from other sources.(4) The NSB also decided that financial support for these facilities should be provided by special budgets to avoid money being diverted from 'the regular established programs of the Foundation'.

20.3 Choice of AUI to Manage the National Radio Astronomy Observatory

Tuve was particularly concerned about Berkner's comments about the possible military applications of technology developed for a radio astronomy facility. For these and other reasons Tuve looked into possible alternatives to AUI managing the proposed radio astronomy observatory. At that time a number of universities were

[9] The National Science Board was the governing body of the NSF.

examining the possibility of forming a consortium to manage the planned national optical astronomy observatory, then called the National Astronomical Observatory (see Section 2.2). So Tuve suggested that this consortium could manage both the radio and optical observatories. Various alternative management arrangements were discussed at a meeting between AUI and NSF in April 1956 without coming to a conclusion.

Subsequently in June 1956 Paul Gross, dean at Duke University, was contacted by the Mathematical, Physical and Engineering division of the NSF to ask whether he could suggest a possible alternative to AUI managing the facility. Gross then asked William Pollard, executive director of the Oak Ridge Institute of Nuclear Studies (ORINS) to talk to Raymond Seeger, assistant director of the NSF, to discuss the situation and see if ORINS would be prepared to facilitate a suitable management consortium. Pollard said that ORINS was not willing to enter into competition with AUI but he was willing to see if he could organise an alternative consortium under a temporary ORINS umbrella. Seeger suggested that Pollard discussed the matter with Tuve, which he did. After a number of discussions with potential interested parties a possible university consortium called AURA[10], for Association of Universities for Radio Astronomy, was then assembled by Pollard and Tuve.

By mid July, Waterman had reduced the management choices for the radio observatory to three: management by a single university, management by AUI, or management by a new organisation (the Pollard-Tuve university consortium). But over the next few weeks Waterman tried, unsuccessfully, to interest West Virginia University (as the local university to Green Bank) in managing the facility, and the idea of a new multi-university organisation fell by the wayside, leaving AUI as the preferred option. In response to the criticism that AUI was managed by nuclear physicists, Berkner agreed to have astronomers appointed to its board of trustees as vacancies became available. And in response to concerns that the universities represented by the AUI were based in the northeast,[11] AUI indicated that they would consider adding up to three 'trustees at large' to the AUI board who would come from outside the area. With these assurances the NSB decided on August 24 1956 to award a contract to AUI to manage the national facility for research in radio astronomy, called the National Radio Astronomy Observatory (NRAO), for a period of five years. This was 'subject to the clear understanding with the AUI that the [National Science] Foundation will give serious consideration to the possibility of establishing at the end of that time a common management for the

[10] Not to be confused with AURA (Association of Universities for Research in Astronomy) which was later set up to manage the American National Optical Observatory (see Section 2.2).

[11] These universities were Columbia, Cornell, Harvard, Johns Hopkins, MIT, Pennsylvania, Princeton, Rochester and Yale.

Radio Astronomy Observatory and for the Optical Astronomy Observatory.'(5) The formal contract was signed by Waterman and Berkner on November 17 to cover construction and operation of the 140 ft telescope.(6) Berkner was nominated as interim director.

With the signature of the main contract for the observatory in place, site procurement was the next priority. The NSF asked the US Army Corps of Engineers, who were highly experienced in this sort of work, to undertake the task. Subsequently they acquired a 2,700 acre site for $550,000 at Green Bank. Initial site development plans were prepared by Eggers and Higgins whilst Irving Bowman and Associates of Charleston, West Virginia were contracted to provide the architectural and engineering work for the local infrastructure. Construction of the site infrastructure was begun in May 1957.

20.4 The First Radio Telescopes at Green Bank

In the meantime the first telescope, a simple radio interferometer, had been installed at Green Bank in late 1956 to observe bright radio sources. A temporary instrument, it consisted of two 38 × 38 × 50 ft (12 × 12 × 15 m) corner reflectors set up with the assistance of the radio astronomy branch of the NRL on a 2000 ft (600 m) baseline. It was initially used to measure the local noise environment, but in the following year its operational frequency was increased to 40 MHz to track the first earth-orbiting satellite, Sputnik I, that had been launched on October 4 1957. Ground breaking ceremonies were held at Green Bank on October 17. In the meantime Emberson had been made project manager for the development of the site and David Heeschen had been appointed as head of the astronomy department.(7)

Earlier Heeschen and John Findlay had expressed concern in 1956 at the difficulties in producing an acceptable design for the 140 ft and suggested, as a result, that a smaller and simpler telescope should be constructed as a matter of urgency to enable research to start as soon as possible. The NSF agreed and a 85 ft (26 m) diameter telescope was purchased (see Figure 20.1) with a structural design that was basically identical to that of the University of Michigan's 85 ft telescope.

The basic design concept behind these NSF/Michigan 85 ft telescopes was produced by the Carnegie Institution's Department of Terrestrial Magnetism headed by Merle Tuve. One of the members of this department, Howard Tatel, came up with the idea of using large diameter gears for the two axes of rotation to provide high precision at relatively low cost. Unfortunately Tatel was to die suddenly in 1957. As a result in the following year the Green Bank 85 ft telescope was officially named the Howard Tatel telescope in his memory.

In the meantime the Blaw-Knox Company had been selected to build both the Green Bank and Michigan 85 ft telescopes. Their contract was signed for the Green

Figure 20.1. The first 85 ft (26 m) diameter radio telescope built at Green Bank, West Virginia in the late 1950s. (Courtesy NRAO/AUI/NSF.)

Bank telescope in October 1957, erection of the telescope was started in the summer of the following year, and the first observations were carried out in February 1959.

This Howard Tatel 85 ft was an equatorially mounted telescope with a solid surface made of aluminium sheets. Its focal length/diameter (*f*/*d*) ratio of 0.43 was chosen to be the same as that of the planned 140 ft to enable the feeds to be interchangeable. A dual feed operating at 21 and 3.75 cm wavelengths (1.4 and 8 GHz) was supplied virtually off-the-shelf by Jasik Laboratories. The receivers for both wavelengths showed little cross interference which allowed simultaneous observations. Shortly afterwards a 75 cm (400 MHz) feed and receiver were designed and built by NRAO staff to allow simultaneous observations at three wavelengths.

One of the earliest investigations carried out with the Howard Tatel telescope was undertaken by Frank Drake, a senior astronomer at the NRAO, and colleagues to determine the positions of 30 sources to an accuracy of ~10 arcsec. At about this time Drake and colleagues found that they could resolve the nucleus of the Milky Way into at least four components, two of which were not located at the galaxy's centre of rotation.(8) Several sky surveys were then begun, including one by Heeschen which took advantage of the telescope's ability to perform simultaneous observations at frequencies up to 8 GHz.(9) Planets were also observed and the telescope was used from 1960, somewhat controversially, in Project OZMA[12] to search for signals emitted by extraterrestrial civilisations.

Various other smaller telescopes were located at Green Bank in the late 1950s/early 1960s including a 12 ft (3.6 m) reflector that was obtained in 1959 with the same *f/d* ratio as the 85 ft. It had an off-the-shelf dish made of spun aluminium mounted on a surplus radar turret.(10) Three years later a 20 ft wire mesh paraboloid was purchased. Two other systems were installed in this period, the so-called Little Big Horn and a 40 ft (12 m) dish.

The Little Big Horn had a 120 ft (35 m) long horn with an aperture of 13 × 17.5 ft (4 × 5.3 m) that reduced to 3 × 6 inches (7.5 × 15 cm) at the detector end. It was used to regularly observe the radio source Cas A starting in April 1960 to determine its flux density very accurately for standardisation purposes. This was observed to reduce at the rate of 1.75 ± 0.52% per year over the period from 1960 to 1964.(11)

The 40 ft radio telescope was specifically designed to automatically monitor a small number of sources every day for a period of years to determine if there was any long-term variation in their intensities. It was not practical to tie up one of the larger telescopes for this period of time, so the 40 ft, with a wire mesh surface, was built for the purpose. As such it was one of the first fully automated radio telescopes in the world. Unfortunately the monitoring programme, which lasted from 1962 to 1968, found no long-term variability in the eight sources measured. The 40 ft telescope was refurbished in 1968 and was completely modernised twenty years later with state-of-the-art front-end equipment. It is now used as an educational tool.

20.5 Green Bank Interferometer

A second 85 ft telescope, essentially identical to the Howard Tatel telescope, was supplied by Blaw-Knox to Green Bank in 1964 to operate as part of an interferometer with the original 85 ft. The new dish, designated 85–2, was designed to

[12] Project OZMA was named after queen Ozma the ruler of the land of Oz, a place 'very far away, difficult to reach, and populated by strange and exotic beings'.

Figure 20.2. Three 85 ft dishes of the Green Bank Interferometer. (Courtesy NRAO/AUI/NSF.)

be moved by an 80-wheel dolly along a roadway and to be placed on one of six concrete pads at distances ranging from 0.75 to 1.7 miles (1.2 to 2.7 km) from the Howard Tatel telescope, which was designated 85–1. Operations of this two-element NRAO Green Bank Interferometer (GBI) began in October 1964. Three years later a third 85 ft telescope was installed and extra concrete pads added whilst the maximum baseline remained at 1.7 miles.(12)

In addition to operating as a research facility in its own right, this GBI (see Figure 20.2) was also used as a prototype instrument for the Very Large Array (VLA, see Section 20.9), whose design studies had been started in 1961. As a result, a portable 42 ft (13 m) telescope, located 22 miles (35 km) away from Green Bank, was incorporated into the Green Bank interferometer system in 1969. It operated with one of the 85 ft telescopes using a microwave link to provide a two-element interferometer. A few years later the 42 ft was replaced by a portable 45 ft (14 m) telescope which was linked to all three 85 ft telescopes to form a four element interferometer. This had a resolution of 0.2 arcsec at 3.75 cm wavelength which allowed observation of the fine structure of radio sources. Then in 1982 a portable 45 m telescope was added to the interferometer some 20 miles (32 km) from Green Bank to provide a Y-shaped arrangement with 85–1 and 85–2, mimicking the Y-shaped arrangement planed for the VLA. This enabled the image synthesis processes to be tried out in advance of the VLA's operation.

In the meantime new developments in receiver technology had been incorporated into the GBI to allow simultaneous observations at 3.75 cm, 6 cm and 11.1 cm (8, 5, and 2.7 GHz), the same wavelengths as the VLA. In 1970 Hjellming and Wade used this newly improved interferometer to see if they could detect radio emission from stars, particularly red supergiants which they thought were the most likely to be detectable. Early on in their search they detected radio emission from two novae, Nova Delphini 1967 and Nova Serpentis 1970, followed shortly afterwards by that from the red supergiant Antares.(13) Subsequently the GBI was also used to observe pulsars and quasars as well as other radio sources.

From 1978 to 1996 the GBI was operated in support of the United States Naval Observatory's (USNO's) studies of Earth rotation and monitoring of variable radio sources. Then from 1996 to 2000 it was operated by NRAO as a two telescope interferometer working simultaneously at two frequencies. Partly funded by NASA it was used to monitor the radio emission from transient X-ray and gamma-ray sources. In 2000 the GBI was closed down for financial reasons.

20.6 The 300 ft and its Replacement

In June 1958 the contract had been signed to build the 140 ft, but shortly afterwards it became clear that its completion would be severely delayed (see Section 20.7 below). Although the 85 ft would allow radio observations to begin relatively soon at the NRAO with a reasonably sized telescope, this size was no longer at the cutting edge in world terms. So with a badly delayed 140 ft and with the 600 ft still in the distant future senior members of the NRAO, including John Findlay, Dave Heeschen and Frank Drake, saw a requirement for a new large telescope that could be built relatively quickly. Findlay and his colleagues also felt that they had had little involvement in the design and construction of the 140 ft which was being handled by AUI headquarters in New York. In their view the key to their new large telescope, which they suggested should be about 300 ft (90 m) in diameter, was to build something as quickly and inexpensively as possible and not let it turn into an all-singing/all-dancing machine, with the inevitable delays and cost increases that such an approach would entail.

An initial request was made in 1958 to the NSF for $300,000 for FY1960 to build a 300 ft fixed zenith antenna, but the request was rejected.(14) The request was resubmitted in May 1959 for FY1961. Then, two months later, Otto Struve became the first director of NRAO and now, with his enthusiastic support, the NSF decided to reverse their earlier decision and approve the grant. Finally in January 1960, the 300 ft was included in the President's FY1961 budget.

In the meantime in 1959 the NRAO staff, under the supervision of Findlay who was to become the project manager, had begun a preliminary design of the

300 ft radio telescope. By the following year a set of target specifications had been defined. These included the requirement that this inexpensive telescope should have a target lifetime of five years whilst being able to operate at a wavelength as low as 21 cm. In October 1960, it was decided that the 300 ft should be a transit telescope.

Late in 1960 Bob Hall and colleagues, who had constituted the antenna division at Blaw-Knox which had built the 85 ft, were in the process of transferring to Rohr Aircraft. Whilst they were in this process, consulting contracts were signed with them for the detailed design of the 300 ft. The manufacturing drawings were to be produced by Ed Faelten, a consulting engineer in Buffalo, New York. Subsequently a request for proposals was sent out in February 1961 to a number of companies for building and erecting the telescope, twelve of which submitted bids. In the event the prime contract was awarded to Bristol Steel and Iron Works of Bristol, Virginia in April 1961. In the same month the telescope's groundbreaking ceremony was carried out by Otto Struve.

As built the 300 ft diameter parabolic dish (see Figure 20.3) was mounted on two 87 ft (27 m) towers which carried the elevation bearings. These were attached to the dish superstructure about half way from the centre of the dish to its circumference. The dish surface consisted of ⅝ inch (16 mm) Squarex aluminium mesh, which was relatively inexpensive and lightweight, supported by a steel superstructure which was bolted together for speed and simplicity. The telescope was designed to withstand wind speeds of up to 65 mph (105 km/h) stowed. But in high precision operation it was limited to wind speeds of up to 15 mph (24 km/h), although it could be used at lower precision in winds of between 15 and 25 mph (40 km/h). Declination movement was achieved to within 30° of the horizontal north and south using a semicircular bicycle wheel type structure similar to that used for the 250 ft Jodrell Bank dish. The feed, at the focus of the 300 ft dish, was supported some 128 ft (39 m) above the dish surface by two legs attached to the steel superstructure. The legs were braced by guy wires attached to the dish surface.

Amazingly the design and construction of this 300 ft telescope took just 23 months, being completed in September 1962, and cost only $850,000.(15) In 1966 an upgrade programme was undertaken to improve the strength and integrity of the superstructure and new, stronger feed support legs were fitted. Attempts to repair the aluminium surface panels of the dish, which had become distorted over time, were only partially successful, however.[13] So in 1970 they were replaced by aluminium panels made of ⅜ inch mesh, at a cost of about $650,000,[14]

[13] They were removed from the dish and reflattened with a road roller! Unfortunately, they did not stay flattened for very long.

[14] *Sky and Telescope*, November 1971, p. 272.

Figure 20.3. The 300 ft (90 m) diameter radio telescope at Green Bank which was completed in 1962. (Courtesy NRAO/AUI/NSF.)

which allowed the telescope to be operated down to 6 cm (5 GHz). In the meantime a travelling feed had been installed in 1969 on a 50 ft (15 m) track to provide tracking observations for several minutes at frequencies of up to 500 MHz. Two years later this system was improved to allow tracking of sources at frequencies of up to 5 GHz for up to 4 minutes. Various other improvements were installed over the years adding cryogenic systems and improving the feeds, receivers, computers and so on to make a dramatic improvement in the telescope's performance.

The 300 ft radio telescope was a very important survey instrument observing hundreds of sources, primarily extragalactic objects, for the first time. It discovered numerous pulsars, and was used extensively for observations of neutral hydrogen in the Milky Way and beyond. It was also used for planetary observations.[15]

[15] For an excellent summary of the research carried out with the 300 ft see Benjamin Malphrus, *The History of Radio Astronomy and the National Radio Astronomy Observatory; Evolution Toward Big Science*, Krieger Publishing, 1996, pp. 130–140.

Figure 20.4. The 300 ft following its collapse in November 1988. (Courtesy NRAO/AUI/NSF.)

The problem with buildings and large structures of all types is that people get used to them and forget that they do not have an infinite lifetime. So it was something of a shock when on November 15 1988 on a cool windless night the 300 ft radio telescope collapsed under its own weight (see Figure 20.4). Fortunately no-one was injured. The 300 ft had originally been built as an inexpensive temporary facility with an expected lifetime of about 5 years to fill the gap left with the delayed 140 ft. So it should really not have been a surprise that it collapsed after 26 years of use, even though it had been structurally strengthened in the meantime and had been regularly inspected.

Three days after the collapse, the NSF and AUI established a Technical Assessment Panel to investigate the cause of the failure.[16] This panel found, using a computer-based finite element stress analysis, that many structural members had been overstressed by as much as 100%, with some being overstressed by over 200%. Many of the gusset plates connecting structural members were found to be overstressed by secondary forces not considered in the original design. It was

[16] Report of the Technical Assessment Panel for the 300 Foot Radio Telescope at Green Bank, WV, March 1989, in F. J. Lockman, F. D. Ghigo and D. S. Balser, (eds.), *But It Was Fun: The First Forty Years of Radio Astronomy at Green Bank*, National Radio Astronomy Observatory, 2007, pp. 252–261.

clear that many overstressed structural members would eventually fail due to metal fatigue, and that is what had happened. A gusset plate, which could not have been seen during routine inspections, had failed due to overstress and consequent metal fatigue. Unfortunately when the telescope was designed some 26 years or so before, it had been impossible to undertake the complex computer stress analysis that had been carried out as part of the failure review.[17] So, according to the Technical Assessment Panel, the failure could not have been predicted using the design methods available when it was built initially. In addition, they concluded that its failure was not the subsequent result of inadequate maintenance or inappropriate operation of the telescope.[18]

The 300 ft was replaced by a 100 × 110 m altazimuth dish (see Figure 20.5), the Robert C. Byrd Green Bank Telescope (GBT), that was named after the West Virginia senator who pushed its funding of $75 million through Congress. Amazingly this funding had been approved less than a year after the collapse of the 300 ft. Unlike the case of similar radio telescopes, the feed of the GBT was not supported by a symmetrical array of legs over the centre of the dish. Instead it was carried by a structure that projected upwards from one edge of the dish. This produced an unblocked aperture that gave the dish a higher efficiency, lower sidelobes, and reduced pick-up of ground radiation. The offset feed support arm had a retractable prime focus feed horn in front of an 8 m subreflector that could focus the beam onto an array of eight feed horns on a revolving turret at the Gregorian focus. This enabled a rapid change of observing wavelengths. The GBT had a computer-controlled active surface of over 2000 panels that allowed operation down to about 3 mm (100 GHz). It was formally dedicated in 2000 and is currently the largest fully steerable radio telescope dish in the world.

In 2012 the NSF's Division of Astronomical Sciences Portfolio Review Committee recommended that the GBT should lose its funding over a five year period. As a result the GBT's future became unclear and the NRAO made a major effort to find partners to help fund its $10 million annual operating costs.[19] So it was with some relief that in 2015 it was agreed that about 20% of these would be

[17] Interestingly, a year before the 300 ft's collapse Bernie Burke had written a letter to NRAO suggesting that a new large steerable radio telescope should be built to replace both the aging 140 ft and 300 ft Green Bank telescopes, see James J. Condon, Zapped! … by Hostile Space Aliens, in Alan H. Bridle, James J. Condon and Gareth C. Hunt, (eds.), *Frontiers of Astrophysics; A Celebration of the NRAO's 50th Anniversary*, Astronomical Society of the Pacific Conference Series Vol. 395, 2008, pp. 323–332.

[18] A later analysis of radio observations taken with the telescope over the few years before the 300 ft had collapsed had shown that pointing hysteresis had increased from about 10 arcsec in 1986 and 1987 to 30 to 50 arcsec in 1988, indicating a failure of the telescope structure, see James J. Condon, Zapped! … by Hostile Space Aliens, in Alan H. Bridle, James J. Condon and Gareth C. Hunt, *Frontiers of Astrophysics; A Celebration of the NRAO's 50th Anniversary*, Astronomical Society of the Pacific Conference Series Vol. 395, 2008, pp. 323–332.

[19] Bryan Bumgardner, Too Big to Fail? The Green Bank Telescope's Uncertain Future, Scientific American website (www.scientificamerican.com/article/too-big-to-fail-the-green-bank-telescopes-uncertain-future) September 4 2013.

Figure 20.5. The 100 × 110 m Robert C. Byrd Green Bank Telescope that was built to replace the 300 ft. (Courtesy NRAO/AUI/NSF.)

covered for a five year period by the Breakthrough Prize Foundation. The foundation wanted to use it to detect radio signals transmitted by intelligent life elsewhere in the universe.

20.7 The 140 ft Telescope

We return now to the troubled 140 ft that had been left in August 1957 when AUI had issued a request for fixed price proposals (see Section 20.1). This was for construction of the telescope to Ashton's design. About 40 proposals were received, eight of which were for a fixed price contract for the whole task with prices ranging from \$3.96 million to \$12.02 million. This wide range of estimates should have raised serious concerns, but it didn't. Even the lowest bid was about twice the budget and twice the latest estimates provided by AUI's consultants. Then after negotiating with four of the bidders AUI awarded a contract in June 1958 to the E. W. Bliss company of Canton, Ohio for \$4.75 million, down from their initial estimate of \$5.75 million, for completion of the 140 ft within 24 months. Problems soon arose, however, and it became obvious that the telescope would be severely delayed.

The telescope was designed to have a high precision surface to enable operations down its design wavelength of 3 cm (frequency 10 GHz) with some capability down to 1 cm. The combination of large surface area for high sensitivity and accurate surface for operations at a short wavelength caused considerable difficulties. The decision to have an equatorial mount also caused problems. So, although construction of the foundations had started in 1958, completion of the telescope was considerably delayed beyond its target completion date of 1960.

The first serious problem appeared in the spring of 1960 with the Ashton-designed construction of the 22 ft (7 m) diameter welded spherical bearing at E. W. Bliss.(16) Many small cracks had appeared in the steel whilst the sphere was being welded. On investigation it transpired that these cracks could propagate rapidly under low temperature and certain stress conditions causing a catastrophic failure of the bearing. Metallurgists advised that there was a small but finite chance that the spherical bearing could collapse but were unable to quantify the risk. This so-called 'brittle fracture' effect had first been observed over a decade earlier but neither Ashton nor the Bliss company seemed to be aware of it. Apparently it was due to the type of steel that Bliss had used. Bliss had chosen this type of steel as they had used it before and had a supply available. Unfortunately they had also used the same type for the polar shaft and yoke which were also both welded structures and which they were in the process of making when the problem was found with the spherical bearing. Three alternative solutions were considered: either (i) accept the build of the sphere, shaft and yoke as is and accept that the risk of a catastrophic failure was small, or (ii) try to repair the damage in some way, or (iii) throw the existing metalwork away and start again using a better steel. The last option was clearly the safest, but it was originally not considered as the Bliss contract was already very late and considerably overspent.

Before a decision could be made on what to do, the responsibilities of the various parties had to be sorted out. Unfortunately these were not clear. Initially Bliss was to have developed shop drawings together with fabrication procedures based on the Ashton design. When these had been accepted by AUI, Bliss would then have been responsible only for materials and workmanship. The problem was that Ashton's design had not been finished when Bliss had been awarded their contract and had started work, although both AUI and Bliss seemed to have assumed that Ashton's design was sufficiently complete so that all that would be required was detailing.

A great deal of hot air was generated over the next few months in trying to resolve both the technical and contractual problems. Eventually in August 1960 AUI, NRAO and NSF agreed to continue with the existing hardware with no modifications. But Emberson, the 140 ft project manager, was not happy with this decision. Consequently Waterman appointed a committee, chaired by the distinguished metallurgist Auguste Kinzel,[20] to review the problem and make recommendations.

In January 1961, the Kinzel committee issued its report. It recommended that the existing sphere and shaft should be heated to a high temperature to reduce the stresses and then bolted together, instead of being welded together as in the original design. Emberson and Ashton were not happy with this solution and suggested that a new sphere and shaft should be produced using better steel and be designed to be bolted together. In the end it was agreed that both solutions should be developed in parallel and in April 1961 AUI contracted Stone and Webster Engineering Co to act as their agents with Bliss.

Stone and Webster decided, first of all, to review the design and build status of all the work in progress. Subsequently in October 1961, they rejected the Kinzel committee recommendations and produced a radical rework of the Ashton design so that it was now virtually unrecognisable from the design that they had inherited. AUI and NSF accepted all of the Stone and Webster changes, so at last the 140 ft was back on track, although it was very late and costing far more than originally foreseen.

When Struve had been made the director of NRAO in 1958 he and Berkner had agreed that AUI would continue with the management of the 140 ft and would hand over the telescope to Struve and the NRAO as a finished product. That was fine as long as things were going well, but as soon as things took a turn for the worse Struve and the NRAO staff at Green Bank said that they should have been much more involved in the project in the first place. In fact the NRAO staff thought that the project should be run by Struve and his staff.

In 1961 Emberson left AUI shortly followed in November of that year by Struve who resigned as director of NRAO. Struve had resigned as he felt that he should have been running the 140 ft project and because of his deteriorating health. Shortly afterwards the Bliss contract was cancelled and AUI appointed Stone and Webster to undertake the telescope's design and supervise its construction. NRAO took over supervision of the 140 ft in 1963 with Max Small as project manager. Finally the telescope was completed in mid 1965 (see Figure 20.6) and formally dedicated in October of that year. It had cost about $14 million, which was over double the original contract value.

By 1967 receivers, which had been developed by the NRAO to operate down to 0.9 cm, were operating on the telescope. Seven years later the 140 ft was converted from a prime focus to a Cassegrain configuration in which the feed horns pointed upwards towards a subreflector, rather than downwards towards the main dish but

[20] Kinzel held a number of high-level positions at the time including that of chairman of the Division of Engineering and Industrial Research of the National Research Council.

Figure 20.6. The 140 ft (43 m) telescope at Green Bank which was completed in 1965. (Courtesy NRAO/AUI/NSF.)

particularly towards the ground which radiated noise.(17) The subreflector was of a nutating design which pointed the beam alternately to the object being observed and to the sky beside it. This significantly increased the effective sensitivity whilst minimising atmospheric effects. Then in 1979 a deformable subreflector was installed to compensate for dish distortions that were produced as the angle of the dish to the vertical was changed. Eight years later a holographic survey was carried out that allowed the antenna's surface panels to be adjusted more accurately than before. This resulted in the dish performance at 35 GHz (λ 8.6 mm) being as accurate as its previous performance at 24 GHz (λ 12.5 mm). Shortly afterwards the first observations were made with a 35 GHz receiver which was the highest frequency used successfully on this telescope.

The 140 ft has been particularly useful in galactic astronomy and in molecular spectroscopy. For example, it was used in 1965 by Bertil Höglund and Peter Mezger to make the first unambiguous detection of a recombination line of ionised hydrogen.(18) This was the N110 to N109 line at 5,009 MHz which they observed in M17, the Orion Nebula and nine other galactic HII regions. Then in 1968 Gerrit Verschuur used the 140 ft to make the first unambiguous detection of the Zeeman effect in neutral hydrogen regions, which clearly proved the existence of interstellar magnetic fields.(19) In the following year the 140 ft made the first detection of the first organic polyatomic molecule, formaldehyde, in the interstellar medium.(20) This was followed by the detection of formic acid and the first of the long-chain carbon compounds, cyanoacetylene, in 1970. Since then the 140 ft has made the first detection of a number of other molecules in interstellar space.

In 2001 the 140 ft was retired from routine astronomical observations. Then in 2013 it got a new lease of life as one of two ground stations for the Russian-made RadioAstron satellite.

20.8 Millimeter-Wave Telescope

At the end of the 1950s Frank Drake and others at the NRAO had thought that telescopes at the observatory should cover the whole radio spectrum down to the microwave region. As a result, Drake visited Texas Instruments in February 1962 to look into the possibility of purchasing one of their new low temperature gallium doped germanium bolometers that had been developed there by Frank Low. Drake wanted to use one of these bolometers to observe the planets and other astronomical objects in the millimetre waveband.

Shortly afterwards an early model of Low's bolometer was used by Drake with a precision 5 ft (1.5 m) diameter Kennedy spun-cast gold-plated dish at Green Bank fixed to an old radar altazimuth mount. First observations were made of the Sun and Moon at a wavelength of 1.2 mm but, unfortunately, the altazimuth mount was not good enough to carry out accurate astronomical observations. In addition the background noise was too high so preventing the telescope from detecting fainter astronomical sources.(21)

Low then joined the NRAO staff in October 1962 and made observations with his bolometer at a wavelength of 11 microns in the infrared and at 1.2 mm. The detector was mounted in a special chamber where it was cooled to its operating temperature of below 2 K by liquid helium. Subsequently it was used with a 3 ft (90 cm) diameter Kennedy spun-cast gold-plated dish that was fixed to a modified equatorial optical telescope mount. In addition it was used with the 5 ft dish that had been remounted onto a similar optical telescope mount, and a new 12 ft (3.6 m) Kennedy dish fixed on a modified surplus radar mount.

A dual-beam radiometer was then tested at 1.2 mm using the 12 ft dish. It showed an order of magnitude reduction in sky brightness fluctuations. Subsequently a dual-beam radiometer was used by Low and Harold Johnson[21] to make stellar

[21] Of the University of Arizona's Lunar and Planetary Laboratory.

observations at the Cassegrain focus of the 82 inch McDonald Observatory telescope at a wavelength of about 10 microns.(22)

It rapidly became clear during the early part of this observing programme that there was too much water vapour in the atmosphere at Green Bank, even on the best observing nights, to make top-quality infrared and millimetre-wave observations. In addition suitable dishes available at Green Bank were too small. So Drake, Low and Heeschen, by then the NRAO director, decided to push for a better site, probably in the drier southwest of the United States, at which to build a new large millimetre-wave radio telescope. Drake thought that it should be possible to build a 36 ft (11 m) diameter telescope there at a reasonable cost.

With the blessing of the NSF, the NRAO decided to examine the possibility of building such a 36 ft radio telescope in more detail. As a result they awarded a feasibility study contract to the Rohr Corporation who had already built a smaller millimetre-wave telescope for the Aerospace Corporation. On completion Rohr concluded that it should be possible to build a 36 ft with a surface accuracy of 0.002 inch (50 microns) rms, which would be satisfactory for observations down to a wavelength of 1 mm (frequency 300 GHz).

NRAO then sent out a request for proposals for the construction of the telescope to a number of companies including Rohr and Newport News Shipbuilding of Norfolk, Virginia who were familiar with producing large structures to such high tolerances. Low had previously visited this shipbuilding company, which built the pressure hulls of nuclear submarines, and had been very impressed by their capabilities. But Findlay, the deputy director of NRAO, thought that the construction contract should go to a high-tech company as they would be more capable of building a high accuracy paraboloid. Consequently in 1964 Rohr were awarded a $600,000 contract to build the telescope.

As far as location was concerned two potential sites were considered, both near Tucson, Arizona in the dry southwest. One of these was the 9,200 ft (2,800 m) Mount Lemmon and the other was the lower altitude Kitt Peak where the Kitt Peak National Observatory was being built (see Section 2.2). Unfortunately although Mount Lemmon with its extra height would have a more transparent atmosphere, there were some commercial radio transmitters nearby. In addition at that time Mount Lemmon had no support facilities available like those on Kitt Peak. So a site at an altitude of 6,300 ft (1,920 m) on the southwest ridge of Kitt Peak was chosen as the location for the 36 ft millimetre-wave radio telescope.

Rohr's dish design consisted of aluminium plates welded together and machined as one piece to produce the accurate surface required. This dish was supported by a steel backup structure which was carried on a computer-driven altazimuth mount. When completed the telescope was enclosed in a 95 ft (29 m) diameter dome with a 41 ft (12.5 m) wide entrance slit which rotated with the telescope (see Figure 20.7).

Figure 20.7. NRAO's 36 ft (11 m) diameter millimetre-wave radio telescope in its dome on Kitt Peak. (Courtesy NRAO/AUI/NSF.)

This dome was to shield the dish from the wind to minimise pointing problems and from direct sunlight to minimise thermal distortions of the dish surface.

During manufacture of the dish the cutting tool accidentally cut a hole through the aluminium surface. Unfortunately although the manufacturers welded the hole shut, the new weld was harder than the original surface causing problems in achieving the high surface tolerance required. Nevertheless, although this repair work caused the dish to be late, when it arrived on Kitt Peak in February 1966 the dome and telescope mount were not ready. In the event the telescope was completed and the first astronomical observations were undertaken just over eighteen months later. In the meantime Findlay, who was then project manager, had left the project to become site director of the 1,000 ft (300 m) Arecibo radio telescope in Puerto Rico (see Section 22.6) and Low, who was frustrated with delays in completing the 36 ft telescope, had joined the University of Arizona at Tucson where he undertook observations in the infrared.

Unfortunately the accuracy of the dish surface was found to be poor and, because of its method of construction, it could not be adjusted. Its aperture efficiency was only about 12% at its design wavelength of 1.2 mm. In addition its focal length changed with temperature because of the difference in materials between the surface of the dish and that of its backup structure requiring the prime focus receiver to be moveable. It turned out that the azimuth and elevations axes were not quite perpendicular and there were also problems with the computer programme used to point the telescope.(23)

At first there was a problem in finding suitably sensitive receivers to operate in the millimetre waveband. But by 1969 receivers had improved sufficiently to observe the stronger continuum distant radio sources. At around this time ammonia and water were discovered in the interstellar medium at a wavelength of about 1.3 cm by A. C. Cheung and colleagues of the University of California at Berkeley using the 20 ft (6 m) Hat Creek radio telescope (see Section 22.9). Lewis Snyder and colleagues discovered interstellar formaldehyde at 6 cm with the NRAO 140 ft, then in 1970 Robert Wilson, Keith Jefferts and Arno Penzias detected carbon monoxide at 2.6 mm using this 36 ft NRAO dish. These discoveries produced a major increase in interest in the 36 ft dish which resulted in the discovery of many more molecules in the interstellar medium including hydrogen cyanide and the formyl ion in 1970, seven other molecules in 1971,[22] and so on. In fact in the 1970s the 36 ft was the instrument of choice for discovering interstellar molecules.[23]

Unfortunately if sunlight struck the telescope dish surface during the day it could take up to four hours to cool down. So with the large increase in observational demand in the 1970s the NRAO team looked into the possibility of protecting the surface from the Sun whilst still allowing it to be used for astronomical observations. The first idea was to use a fabric stretched across the dish opening but this was easily torn by the wind. Instead a conical-shaped 'teepee' of radio-transmitting fabric was built over the dish with the apex just below the height of the dome. Although this could also be damaged by the wind it was much more robust than the previous solution. The main disadvantage was that it took some time to put the teepee up and take it down. Consequently it was often left up for days or weeks at a time, with a consequent impact on the sensitivity of the telescope at night.

By 1980 the 36 ft was showing its age, despite attempts to improve it over the previous ten years or so. It was now time to think of building a better millimetre-wave telescope. Some years earlier NRAO had looked into the possibility of building a copy of Bob Leighton's inexpensive but highly accurate 34 ft (10.4 m) millimetre-wave telescope that he was designing for Caltech (see Section 21.1). But this idea had fallen through. Instead in 1980 NRAO considered replacing the 36 ft dish in its current enclosure. The problem was that NRAO was in the middle of trying to get funding for a new 25 m (82 ft) millimetre-wave radio telescope to

[22] Lines of carbon monosulphide, silicon monoxide, hydrogen isocyanide, carbonyl sulphide, isocyanic acid, methyl cyanide and methyl acetylene were detected at wavelengths between 2.0 and 3.5 mm.

[23] The NRAO 36 ft discovered twenty-nine different molecules in the 1970s, compared with the next best result of five for the 140 ft NRAO telescope and four for the 210 ft Parkes antenna.

be built on Mauna Kea and it was unlikely that they could have both telescopes. Nevertheless, NRAO decided to replace the 36 ft dish in its existing dome whilst still pushing for the 25 m.[24]

The new NRAO dish was 12 m (39 ft) in diameter, the largest that could see through the existing slit. It consisted of individually adjustable pressed aluminium surface panels supported by a rigid space-frame structure. A jig was used to produce the initial setting of the panels, whilst holography was used for the fine adjustment. First observations took place in 1983. A few years later a subreflector was made with a negative image of the dish surface's errors.[25] This finally allowed the telescope to meet the design goal of the original telescope but, unfortunately, a structural fault was discovered in 1989. Mending this caused the dish to warp, requiring a new subreflector which was not able to correct the telescope as well as the previous one. Nevertheless, the telescope's performance was still far better than that of the previous 36 ft.

In 2000 budgetary pressures at NRAO caused primarily by its financial commitment to ALMA (Atacama Large Millimeter Array, see Section 24.1) forced NRAO to close down the 12 m telescope. However shortly afterwards they transferred control of the telescope and its auxiliary buildings on Kitt Peak to the University of Arizona.

20.9 Very Large Array

Studies were begun in 1961 at the NRAO into the possible design of a flexible, general-purpose interferometer array that was significantly better than any currently existing or being built. At that time the angular resolution of radio telescopes was poor, with the result that imaging, where it was possible, was very poor and very slow. In addition interferometers, which generally had better resolution than large dish antennae, suffered from their own problems of complex side lobes and source confusion. In the following year David Heeschen, the director of NRAO, who coordinated the studies, summarised the desirable characteristics of the versatile radio telescope that the NRAO had in mind. It would have a resolution of about 10 arcsec in the centimetre waveband with good declination coverage and sufficient sensitivity to observe several hundred sources. The preferred design was an aperture synthesis array consisting of 80 to 150 ft (24 to 46 m) diameter antennae which may be steerable in hour angle and/or movable on tracks.(24)

[24] The 25 m project was killed off in 1982 by the Astronomy Advisory Committee, see M. A. Gordon, *Recollections of 'Tucson Operations'; The Millimeter-Wave Observatory of the National Radio Astronomy Observatory*, Astrophysics and Space Science Library, Vol. 323, Springer, 2005, pp. 145–146.

[25] The 12 m telescope could also be used as a prime focus as well as a Cassegrain instrument. But the Cassegrain focus had the advantage that it could carry large and heavy instruments, many of which needed to be cooled.

Shortly afterwards the National Academy of Sciences' (NAS') first decennial review entitled 'Ground-Based Astronomy; A Ten-Year Program' was produced by the Whitford committee.(25) It recommended that a radio telescope array of about one hundred 85 ft diameter parabolic antennae should be built to study bright extragalactic sources, the structure of the Milky Way, and so on. The main requirement of the array was a resolution of better than 10 arcsec at centimetre wavelengths with acceptable sidelobes, with a secondary requirement of a resolution of about 1 arcmin at 21 cm. The report further suggested that the array, which was to become the Very Large Array (VLA), should be built and operated by the NRAO as a national facility, but it left open what type of array should be built. The costs were estimated at about $40 m with a construction timescale of about 10 years. The Whitford committee recommended that work should start as soon as possible.

The Green Bank Interferometer (GBI, see Section 20.5 above) began operations at about the same time as the Whitford committee issued their report. This GBI had two antennae at first, then three years later a third antenna was added. In addition to being a useful interferometer in its own right, the GBI provided the VLA designers with experience of operating an interferometer. In addition they could use the GBI to try out a number of their ideas for the design and operation of the VLA.

In 1964 the NRAO organised a VLA design group with George Swenson, who had temporarily joined NRAO from the University of Illinois, in the chair. After about eighteen months of work it became clear that as far as the type of array was concerned, an earth rotation synthesis array was the only realistic choice.(26) Then by 1967 the VLA design group had fleshed out the original design goals to produce more specific requirements. These now included a resolution of 1 arcsec at 10 cm wavelength,[26] a field of view of 1–10 arcmin, declinations down to – 20°, minimum observational wavelength of 3 cm, with spectroscopy and polarisation capabilities. The additional requirement that the first sidelobe should be at – 20 dB implied an array of 36 antennae. The VLA design to meet these requirements was contained in Volumes 1 and 2 of the NRAO proposal sent to the NSF in January 1967. Its thirty-six 82 ft (25 m) diameter, rail-mounted antennae[27] were arranged in a Y-shaped array with arms 13 miles (21 km) long. They were movable along these rail tracks by special transporters. The cost of the VLA was estimated to be about $36 million.[28]

[26] The initial resolution requirement of 10 arcsec had been found to be too conservative and astronomers wanted to have a radio telescope with a resolution similar to that of ground-based optical telescopes, see D. S. Heeschen, The Very Large Array, in G. Burbidge and A. Hewitt, (eds.), *Telescopes for the 1980s*, Annual Reviews Monograph, 1981, p. 6.

[27] The movable large antennae of the Green Bank Interferometer had been moved on wheels with rubber tyres. But the road bed had proved expensive to maintain, the tyres were expensive, and the high antennae were difficult to pull safely. So the VLA antennae were designed to be moved on rails instead.

[28] A Proposed Very Large Radio Telescope Array, *Sky and Telescope*, April 1967, p. 213.

At this time the proposed VLA was regarded by many radio astronomers both in the USA and abroad as being too large and too expensive. In the USA, at least, these views were predicated on the idea that spending a large amount of money on the VLA would divert money and resources from other radio telescopes, so threatening the continuation of many research groups. In addition, a number of people wondered where the radio astronomers would come from to use the tremendous amount of data that the VLA would produce.

After receiving the first two volumes of the NRAO proposal in 1967, the NSF called for a review of the various radio astronomy proposals around at the time by a committee chaired by Robert Dicke of Princeton University. This committee responded favourably to the VLA proposal, but raised a number of queries and recommended additional work. Then in January 1969, Volume 3 of the VLA proposal was sent to the NSF by the NRAO. In this they described technical advances that had occurred in the last two years, the most important of which resulted in the reduction of the number of antennae from 36 to 27.(27) The Dicke committee reviewed this modified proposal later that year.

At the same time, Caltech was planning to build an array of eight 130 ft (40 m) antennae in Owens Valley (see Section 21.1) and a group of universities was planning a 440 ft (135 m) steerable dish inside a radome (see Section 22.2). Pressurised to make a recommendation on which radio telescope project to support, the Dicke committee recommended in late 1969 that the Caltech and radome projects should go ahead immediately, and that the VLA should be built in stages.[29] But it was impracticable to start all these projects at the same time. As a result the NRAO interpreted this Dicke committee recommendation as signalling the end of the VLA project and consequently they stopped any further substantive design work, at least for the time being, and disbanded their hardware design team. As Heeschen pointed out there was no point in starting the detailed design of something that would not be built for a number of years, if at all, as the technology was developing so fast that by the time the array was completed it would be out of date.

The VLA project had been considering where to build the array for some time. Its site requirements were quite constraining as, for example, the array needed to be at high altitude and be in a dry climate to minimise atmospheric effects. The site also had to be at least 22 miles (35 km) in diameter to take all the dishes, be relatively flat and be situated at low latitude to maximise sky coverage.(28) Furthermore it had to be remote from large population centres to minimise man-made interference whilst being reasonably accessible for visiting observers. Seven

[29] Heeschen was unclear as to what this 1969 Dick committee report really meant, see David S. Heeschen, Reminiscences of Early Days of the VLA, in T. J. Cornwell, and R. A. Perley, (eds.), *Radio Interferometry: Theory, Techniques and Applications*, Proceedings of the 131st IAU Colloquium, ASP Conference Series, Vol. 19, 1991, p. 152.

sites were shortlisted but by far the best, meeting all these criteria, was at an altitude of 2,120 m (6,960 ft) on the Plains of San Augustin near Socorro, New Mexico. Site selection was covered in Volume 4 of the NRAO proposal which was published in January 1971.

Things changed for the better for the VLA project when the second NAS decennial report was produced by the Greenstein committee in 1971 (published 1972). At first the Greenstein committee's radio astronomy subpanel had endorsed the 1969 recommendations of the Dicke committee. But William Wright of the National Science Foundation let it be known that they favoured the VLA and Harvey Brooks[30], speaking on behalf of the National Academy of Sciences, pressurised the subpanel to prioritise the various radio astronomy projects.(29) As a result Greenstein's radio astronomy subpanel recommended the VLA as the top priority radio astronomy project and the full Greenstein committee subsequently recommended the VLA as its top priority project for the whole of astronomy.(30) Greenstein quoted an estimated capital cost of $62 million in 1970 dollars, with an estimated construction period of five years.[31]

In 1972 the NSF and the US Congress approved construction of the VLA, whilst the NSF and the Office of Management and Budget agreed that the project should have drip-feed funding of about $10 million per year. This decision not only lengthened the construction period but also increased its cost.(31) On the other hand by 1972 there had been great advances in the design of electronics since the VLA had been designed some years earlier. So the extended timescale was used to undertake detailed redesign and testing.

Unfortunately 1973 saw an oil embargo from the Middle East, a devaluation of the US dollar and the start of a period of rampant inflation. This all put pressure on the VLA team to cut costs even before the project had really been started. As a result they scrapped plans for a local airstrip, reduced the number of antenna transporters from three to two, and dropped the requirement of operating the antennae simultaneously at two frequencies.

The VLA needed 14,000 tons of used rails for the track at an estimated cost of about $90 per ton. But by 1973 the price of used rails had increased to over $250 per ton. Consequently the project team, with the active support of the NSF, looked to see whether it was possible to take spare rails from military bases and from other government facilities. It was eventually agreed that this could be done for the cost of transporting them to the VLA site, which turned out to be just $53 per ton.

NRAO decided that they would be prime contractor for the VLA project themselves so they could keep a tight control on its performance, timescale and costs.

[30] Chair of the Committee of Science and Public Policy of the National Academy of Sciences.

[31] In addition the first five years of operation were estimated to cost about $30 million, worked out as 10% of the capital cost per annum.

Construction using 25 m diameter antennae supplied by E-Systems, Inc., of Dallas, Texas was started on the Plains of San Augustin in 1974. These antennae were altazimuth-mounted, Cassegrain reflectors, with a solid surface. Their subreflector was asymmetric so the Cassegrain focus was laterally displaced from the geometric axis of the 25 m dish. Consequently by rotating this subreflector the focal point could be made to coincide with the position of one of four feed horns which were arranged in a circle around the axis of the main dish. Each of the four feed horns operated at a different wavelength, so changing from one band to another could be achieved in less than a minute. In addition the feeds each had two outputs to provide either opposite circular or orthogonal linear polarisations.(32)

The receiver front ends for each antenna were placed in a Dewar to maintain them at a temperature of 18 K. The signals were then heterodyned down for transmission to the array's central control room along a 60 mm diameter circular waveguide which was buried in the ground alongside each arm of the array's Y-shaped configuration.

The first 25 m antenna was completed on site in 1975. Then in February of the following year interference fringes were produced with the first two antennae operating over a baseline of 1.2 km at a wavelength of 6 cm.[32] The first six dishes were operational in 1977 and the complete Y-shaped array, with arms between about 19 km (11.8 miles) and 21 km (13 miles) in length,[33] was completed in early 1981 (see Figure 20.8). There were 72 antenna stations in the array at each of which there were three concrete piers for mounting an antenna. Each of these stations was at the end of a short spur of rails at right angles to the local arm of the array. The VLA had cost $78 million which, allowing for inflation, was only 3.4% over budget.(33)

The VLA was, literally, a massive enterprise with each of it 27 antennae weighing about 193,000 kg (213 tons). These were moved by a 73,000 kg (80 ton) transporter on pairs of standard gauge rail track 18 ft (5.5 m) apart so that the antenna configuration could be changed. This configuration varied from the tightest, the D configuration, in which all antennae were within about 600 m of the centre, to the widest, the A configuration, using the full length of rails. The smallest array gave the highest sensitivity and the largest array the highest resolution. The VLA initially operated at wavelengths of 21, 6, 2 and 1.3 cm (frequencies 1.43, 5.0, 15 and 23 GHz) with an angular resolution of between 2.1 and 0.13 arcsec.(34) Readings were obtained from one pair of dishes at a time, which resulted in 351 pairs of measurements. Well-known sources were observed typically every 30 minutes for calibration purposes.

[32] Initially, the primary operating wavelength of the VLA was expected to be 11 cm. But this was changed to 6 cm because it was expected that powerful transmitters on the ATS spacecraft would interfere with observations at 11 cm.

[33] Two of the arms were 21 km in length but the north arm, at 19 km, was shorter than the other two as the owner of the land on which the final 2 km would have been built had asked an exorbitant price for it.

Figure 20.8. The VLA on the Plains of San Augustin, New Mexico. The twenty-seven 25 m dishes were movable on rails up to 21 km (13 miles) in length. (Courtesy NRAO/AUI/NSF.)

Twenty-eight antennae had been purchased so that every seven weeks one of the 27 antennae could be taken out of the array for servicing and replaced by the one that had been most-recently refurbished. The data processing system proved to be a real headache as it was found to be incapable of undertaking all the tasks required. Consequently it was progressively improved over time, but even then it struggled to catch up with astronomers' requirements which were also constantly expanding. In a sense the VLA proved to be a victim of its own success.

The VLA was used to produce spectacular radio images of extended objects including supernova remnants, planetary nebulae, emission nebulae, radio galaxies, gravitational lenses, and the central region of the Milky Way. Other observations have included those of the Sun and planets, protoplanetary disks, interstellar hydrogen, astrophysical masers, pulsars, quasars and gamma-ray bursts.

In 1991 Robert Becker, Richard White and David Helfand(35) suggested to the NRAO that the VLA should be used in its C configuration (maximum distance of furthest antenna from the centre of the array of 1.95 km) to undertake a sky survey at a wavelength of 20 cm. In parallel Jim Condon made a proposal to the NRAO to use the VLA in its most compact (D) configuration for a sky survey. As mentioned above the larger configuration gave the better resolution, but at the expense of sensitivity, whereas the most compact configuration, whilst being the most sensitive provided the poorest resolution.

Subsequently NRAO sent out both proposals for independent review and in parallel undertook trial observations in the C and D configurations. This resulted in a decision made in 1993 to carry out two surveys, one in the B configuration (maximum distance of antennae from the centre of the array of 6.4 km) and one in the D configuration. The B configuration was to cover 25% of the sky centred on the North Galactic Pole whereas the D configuration was to cover 85% of the sky. The B survey, which was designed to be the radio equivalent of the Palomar Observatory Sky Survey, was called FIRST (Faint Images of the Radio Sky at Twenty centimeters), whilst the D configuration survey was called NVSS (NRAO VLA Sky Survey).

The FIRST survey, which was completed in 2004, produced images with 5 arcsec resolution and 0.5 arcsec position accuracy. Since then a number of smaller targeted surveys have been carried out in the FIRST project. Up to 2011 FIRST observations at 1.435 GHz (λ 20.8 cm) took 3 minutes each, but since then the improved VLA (the EVLA, see below) has been used at 1.730 GHz (λ 17.3 cm) with exposures of just 1 minute. The NVSS survey detected over 1.8 million sources with position uncertainties of 1 arcsec or better for relatively strong sources (intensity >15 mJy) to 7 arcsec for the faintest detectable sources (intensity 2.3 mJy).(36)

A major upgrading of the VLA, to produce the so-called Expanded VLA (EVLA), was begun in 2001. This was to provide major improvements in the VLA's sensitivity, frequency coverage, bandwidth and spectroscopic capability. This EVLA was completed in 2012 when it was renamed the Karl G. Jansky Very Large Array. It was over ten times more sensitive to faint radio emissions than the VLA was before the latest modifications. The EVLA also covered appreciably more radio frequencies in a range from 75 MHz to 43 GHz (λ 400–0.7 cm) to produce maximum resolutions of from 24 to 0.04 arcsec, respectively. The NSF spent about $59 million on the EVLA project, whilst Canada supplied the $17 million correlator and Mexico contributed $1.8 million.

A proposal had also been made to the NSF in 2004 to increase the resolution of the VLA by adding more antennae around New Mexico in Phase II of the EVLA project. It was also proposed to add a condensed array configuration smaller than the existing D configuration and add low frequency observing bands to the existing antennae. But in the following year the NSF notified the NRAO that they were unable to finance this Phase II project.

20.10 Very Long Baseline Array

Canadian astronomers had shown in 1967 that interferometer observations could be made using tape recorders and atomic clocks to transfer data between telescopes rather than having to rely on microwave links (see Section 18.5). Their very long baseline interferometer, which operated at a wavelength of 67 cm (448 MHz), consisted of two radio telescopes separated by a distance of about 1900 miles (3000 km) or about 5 million wavelengths. Shortly afterwards American astronomers were also successful with a similar system using dishes at Hat Creek, California and Green Bank, West Virginia. Although the distance between these two American antennae was not much more than between the Canadian antennae, the Americans had used a wavelength of 18 cm (1.67 GHZ) to produce an antenna separation of about 20 million wavelengths, with consequently better resolution. These very long baseline interferometry (VLBI) experiments were followed by those using pairs of radio telescopes separated by even larger distances. But, unfortunately, all these radio telescopes already had a full observing schedule in the late 1960s so they could not be dedicated to VLBI work.

The obvious solution was to build an array of two or more radio telescopes thousands of kilometres apart specifically dedicated to VLBI work. But the NRAO was in the process of planning the VLA at the time that these VLBI experiments were being carried out and it was clearly out of the question for them to build a permanent VLBI array in parallel with the VLA. Not only would it be impossible to raise the extra finance, but the NRAO did not have the staff to undertake the task.

However by 1975, as the first VLA antenna was being completed, the NRAO could afford to start planning a permanent array of radio telescopes distributed across the country that would be exclusively devoted to VLBI. This so-called Very Long Baseline Array (VLBA), which was planned to be a multi-purpose instrument,[34] was to consist of ten identical 25 m diameter antennae located at various places around the USA. This would give an effective diameter of several thousand kilometres and produce an angular resolution of the order of a few milliarcsecond (mas). The plan was to control the whole array with a computer located in an operations centre near the VLA, to which all the data tapes would be shipped for correlation and transformation into radio images. The correlator was planned as a 20-station system to support processing of all observations from the ten antenna VLBA array together with up to ten other antennae worldwide.

This VLBA was a key project recommended in 1982 by the influential Field Report of the National Academy of Sciences.[35] It was approved by Congress in 1984 although a reduction in the 1985 appropriation from $15 million to $9 million and a delay in first funding extended the period of construction by one year. The

[34] As a multi-purpose instrument, it was designed to support continuum, spectroscopic, astrometric, multi-frequency, bandwidth synthesis, polarisation and pulsar measurements, see Jonathan D. Romney, The Very Long Baseline Array, in Mark Jonathan Reid and James M. Moran, (eds.), *The Impact of VLBI on Astrophysics and Geophysics*; Proceedings of the 129th IAU Symposium 1987, Kluwer Academic Publishers, 1988, pp. 461–468.

[35] At this time, the VLBA was estimated to cost about $35 million plus about $15 million for operations, see George B. Field, et al., *Astronomy and Astrophysics for the 1980s, Vol. 1*, 1982, National Research Council, National Academy Press, Washington, DC, p. 19.

Figure 20.9. One of the ten identical 25 m diameter antennae of the Very Long Baseline Array. This one was located in Owens Valley. (Courtesy NRAO/AUI/NSF.)

total cost was estimated to be about $70 million with operating costs of about $5 million per year.(37)

Construction of the 25 m diameter antennae (see Figure 20.9, for example) began in 1985 and three years later the first of these was commissioned at Pie Town, New Mexico.[36] In the following year this dish was linked to the VLBA dish at Kitt Peak, Arizona to make the VLBA's first coordinated observation of an astronomical source. This was the radio galaxy 3C 84 which was observed at a wavelength of 6 cm. It had initially been expected that construction of the VLBA would take four years, but financial constraints meant that it was not completed until 1993. However, its final cost of $84 million(38) in 1989 dollars was about the same as approved by Congress in 1984 after allowing for inflation.

When completed, the VLBA extended 8,600 km (5,300 miles) east–west from the Virgin Islands to Hawaii, and 2,900 km (1,800 miles) north–south. It operated at ten frequencies from 330 MHz (λ 90 cm) to 43 GHz (λ 7 mm), and achieved resolutions of from 22 to 0.17 mas. The system was controlled from a joint VLA-VLBA operations centre opened in 1988 on the campus of the New Mexico Institute of Mining and Technology in Socorro. Here broadband digital tapes were

[36] *Sky and Telescope*, September 1988, p. 231.

received from each of the antennae with the recorded data and associated time signals provided by atomic clocks. Non-VLBA radio telescopes including the VLA and telescopes at Arecibo, Green Bank, and Effelsberg (in Germany) were later brought online from time to time to work with the VLBA to complete what was called the High Sensitivity Array.

The VLBA has observed the motion of masers near the centre of the Milky Way and of other galaxies, enabling the masses of their possible black holes to be determined. It has measured the motion of gas in the circumstellar envelope of evolved stars, and has shown the evolution of magnetic field structures in active galaxy jets using high resolution polarimetry.

Since the VLBA's completion the NRAO, in partnership with other radio astronomy institutions both in the USA and elsewhere, has increased the array's available bandwidth by more than an order of magnitude and has improved its sensitivity as well. In addition the tape recorder system at each VLBA site has been replaced by modern one-terabyte hard disc drives.

References

1. Malphrus, Benjamin, *The History of Radio Astronomy and the National Radio Astronomy Observatory; Evolution Toward Big Science*, Krieger Publishing, 1996, p. 32.
2. Edmondson, Frank K., *AURA and its US National Observatories*, Cambridge University Press, 1997, p. 9.
3. Needell, Allan A., Science, *Cold War and the American State; Lloyd V. Berkner and the Balance of Professional Ideals*, Harwood Academic Publishers, 2000, p. 273.
4. Edmondson, Frank K., *AURA and its US National Observatories*, Cambridge University Press, 1997, p. 10.
5. Needell, Allan A., *Science, Cold War and the American State; Lloyd V. Berkner and the Balance of Professional Ideals*, Harwood Academic Publishers, 2000, p. 290.
6. Heeschen, David S., The Origins and Early Years of NRAO, in Bridle, Alan H., Condon, James J., and Hunt, Gareth C., (eds.), *Frontiers of Astrophysics; A Celebration of the NRAO's 50th Anniversary*, Astronomical Society of the Pacific Conference Series, Vol. 395, 2008, pp. 311–321.
7. Lockman, F. J., Ghigo, F. D., and Balser, D. S., (eds.), *But It Was Fun: The First Forty Years of Radio Astronomy at Green Bank*, National Radio Astronomy Observatory, 2007, p. xi.
8. Drake, F. D., A High-resolution Radio Study of the Galactic Center, *Astronomical Journal*, 64, 1959, p. 329.
9. Heeschen, D. S., Observations of Radio Sources at Four Frequencies, *Astrophysical Journal*, 133, 1961, pp. 322–334.
10. Malphrus, Benjamin, *The History of Radio Astronomy and the National Radio Astronomy Observatory; Evolution Toward Big Science*, Krieger Publishing, 1996, p. 69.
11. Findlay, J. W., Hvatum, H., and Waltman, W. B., An Absolute Flux-density Measurement of Cassiopeia A at 1440 MHz, *Astrophysical Journal*, 141, 1965, pp. 873–884.

12. Findlay, John W., The National Radio Astronomy Observatory, *Sky and Telescope*, December 1974, pp. 352–360.
13. Hjellming, R. M. and Wade, C. M., Radio Novae, *Astrophysical Journal*, 162, 1970, pp. L1–L4, and Wade, C. M., and Hjellming, R. M., Detection of Radio Emission from Antares, *Astrophysical Journal*, 163, 1971, pp. L105–L106.
14. Heeschen, D. S., The 300 Foot Telescope and the National Center Concept, in Lockman, F. J., Ghigo, F. D., and Balser, D. S., (eds.), *But It Was Fun: The First Forty Years of Radio Astronomy at Green Bank*, National Radio Astronomy Observatory, 2007, pp. 136–137.
15. Findlay, John W., The 300-foot Radio Telescope at Green Bank, *Sky and Telescope*, February 1963, pp. 68–75.
16. Heeschen, David S., The First Ten Years, 1955–1965, in Lockman, F. J., Ghigo, F. D., and Balser, D. S., (eds.), *But It Was Fun: The First Forty Years of Radio Astronomy at Green Bank*, National Radio Astronomy Observatory, 2007, pp. 265–283.
17. Findlay, John W., The National Radio Astronomy Observatory, *Sky and Telescope*, December 1974, pp. 352–360.
18. Hoglund, B., and Mezger, P. G., Hydrogen Emission Line N110→N109: Detection at 5009 Megahertz in Galactic HII Regions, *Science*, 150, Issue 3694, pp. 339–348.
19. Malphrus, Benjamin, *The History of Radio Astronomy and the National Radio Astronomy Observatory; Evolution Toward Big Science*, Krieger Publishing, 1996, pp. 95–96.
20. Snyder, Lewis E., The NRAO 140 Foot Telescope and the Beginning of Astronomical Spectroscopy of Interstellar Polyatomic Organic Molecules, in Lockman, F. J., Ghigo, F. D., and Balser, D. S., (eds.), *But It Was Fun: The First Forty Years of Radio Astronomy at Green Bank*, National Radio Astronomy Observatory, 2007, pp. 361–373.
21. Drake, Frank D., NRAO's First Millimeter Wave Telescopes, in Lockman, F. J., Ghigo, F. D., and Balser, D. S., (eds.), *But It Was Fun: The First Forty Years of Radio Astronomy at Green Bank*, National Radio Astronomy Observatory, 2007, pp. 506–512.
22. Low, Frank J., and Johnson, Harold L., Stellar Photometry at 10 μ, *Astrophysical Journal*, 139, pp. 1130–1134, 1964.
23. Gordon, M. A., *Recollections of 'Tucson Operations'; The Millimeter-Wave Observatory of the National Radio Astronomy Observatory*, Astrophysics and Space Science Library, Vol. 323, Springer, 2005, pp. 19–27.
24. Heeschen, David S., Reminiscences of Early Days of the VLA, in Cornwell, T. J., and Perley, R. A., (eds.), *Radio Interferometry: Theory, Techniques and Applications*, Proceedings of the 131st IAU Colloquium, ASP Conference Series, Vol. 19, 1991, pp. 150–159.
25. Whitford, A. E., et al., *Ground-Based Astronomy; A Ten-Year Program*, National Academy of Sciences – National Research Council, Washington, DC., 1964, pp. 50–52.
26. Swenson, George W., Jr., The Early Technical Development of the VLA, in Cornwell, T. J., and Perley, R. A., (eds.), *Radio Interferometry: Theory, Techniques and Applications*, Proceedings of the 131st IAU Colloquium, ASP Conference Series, Vol. 19, 1991, pp. 160–164.
27. Thompson, A. R., et al., The Very Large Array, *Astrophysical Journal Supplement Series*, 44, 1980, pp. 151–167.
28. Tucker, Wallace, and Tucker, Karen, *The Cosmic Inquirers; Modern Telescopes and Their Makers*, Harvard University Press, 1986, pp. 20–21.
29. Ibid., pp. 23–24.
30. Greenstein, J., et al., *Astronomy and Astrophysics for the 1970's, National Academy of Sciences*, 1, 1972, pp. 76–80.

31. Lancaster, John H., The Design and Construction of the VLA: The Project Manager's View, in Cornwell, T. J., and Perley, R. A., (eds.), *Radio Interferometry: Theory, Techniques and Applications*, Proceedings of the 131st IAU Colloquium, ASP Conference Series, Vol. 19, 1991, pp. 165–170.
32. Heeschen, D. S., The Very Large Array, in Burbidge, G. and Hewitt, A., (eds.), *Telescopes for the 1980s*, Annual Review Monograph, 1981, pp. 18–20.
33. Federer, C. A., Jr., The VLA: Ears on the Universe, *Sky and Telescope*, December 1980, pp. 472–473.
34. Ibid.
35. Becker, Robert H., White, Richard L., and Helfand, David J., The VLA's FIRST Survey, in Crabtree, D. R., Hanisch, R. J., and Barnes, J., (eds.), *Astronomical Data Analysis Software and Systems III*, ASP Conference Series, Vol. 61, 1994, pp. 165–174.
36. Condon, J. J., et al., The NRAO VLA Sky Survey, *Astronomical Journal*, 115, 1998, pp. 1693–1716.
37. Gordon, Mark A., VLBA – A Continent-size Radio Telescope, *Sky and Telescope*, June 1985, pp. 487–490.
38. Ulvestad, James, and Goss, Miller, Radio Astronomy's Resolution Machine: The Very Long Baseline Array, *Sky and Telescope*, December 1999, pp. 36–46.

21

Owens Valley and Mauna Kea

21.1 Owens Valley Radio Observatory

Edward G. (Taffy) Bowen, head of the Radiophysics Laboratory in Australia, paid a visit to some old wartime friends during a visit to the United States in 1951. These included Lee DuBridge, the president of Caltech, Robert Bacher, head of the physics department at Caltech, Vannevar Bush, president of the Carnegie Institution in Washington, and Alfred Loomis, a trustee of both the Carnegie Corporation and the Rockefeller Foundation.(1) His discussions with these luminaries of America's scientific establishment tended to focus on the tremendous advances in radio astronomy in Australia and the relatively poor situation of radio astronomy in the United States. So in December 1951 Bacher asked Bowen if he would produce a draft specification for a suitable telescope to get the USA back in the game, whilst shortly afterwards DuBridge asked Bowen to outline the sort of instruments that would be required to build the radio equivalent of the superb optical observatories on Mount Wilson and Palomar Mountain. DuBridge wanted Bowen to be the director of the radio observatory with John Bolton as his deputy. But at this stage Bowen was non-committal about his future as he was also interested in building a large radio telescope in Australia.

In August 1952 Bowen wrote to Vannevar Bush to ask whether Carnegie would consider funding both the proposed Caltech radio telescope and a similar instrument in the southern hemisphere as a collaborative development. This eventually resulted in the trustees of the Carnegie Corporation approving funding in May 1954 for what was to become the Parkes Radio Telescope in Australia (see Section 16.2). In the meantime, as mentioned in the previous chapter, an interdisciplinary conference had been held in Washington, DC in January 1954, jointly sponsored by the NSF, Caltech and the Carnegie Institution, to discuss the state of radio astronomy in the USA and what to do about it. This eventually resulted in the NSF funding the National Radio Astronomy Observatory, whilst Caltech decided to go it alone and build their own observatory which was largely funded in the early years by the US Office of Naval Research.[1]

In the meantime Bowen had decided to stay in Australia as he had obtained significant funding for the Parkes radio telescope. Bolton, on the other hand, decided in 1954 to take a two-year leave of absence from Australia's CSIRO and accept a two-year appointment at Caltech as a senior research fellow, with the promise for a longer-term role at the end of that period. Bolton then arrived in the USA in January 1955 to take up his new appointment, followed shortly afterwards by Gordon Stanley who was on secondment to Caltech to help Bolton with the design of the new Caltech radio telescope.

By the time Stanley arrived, Bolton had already decided to build a large, two-element, variable-spacing interferometer.(2) But this would take some time so, in the interim, they decided to build a 32 ft (10 m) diameter prototype antenna on Palomar Mountain to try out a number of ideas that they had for the main telescope. This prototype dish, which was completed in 1956 (see Figure 21.1), was Caltech's first large radio telescope.

Meanwhile Bolton and Stanley had begun to search for a suitable site for the large Caltech radio interferometer. The main requirements were that the site should be flat and have a low level of radio interference. Bolton's initial choice of location near Ojai, California proved to be unsuitable, and whilst he was on mission in Europe, Stanley and Temple Larrabee, who had helped with the construction of the 32 ft dish, began investigating Owens Valley, California and other desert areas. Some areas were too remote and some too near military bases. In the end the best compromise between low radio interference and reasonable accessibility proved to be in the northern part of Owens Valley, five miles (eight kilometres) north of Big Pine. Not only was this area shielded from man-made interference by the nearby Sierra Nevada and Inyo Mountains, but it was owned by the Los Angeles Department of Water and Power (DWP) which could control any unsuitable further development. Caltech then arranged a 300 acre (120 hectare) lease in 1956 with the DWP. In the same year Bolton resigned his position at CSIRO and was appointed director of the Owens Valley Radio Observatory (OVRO) assisted by Stanley.

In the following year a low-frequency interferometer, consisting of two large dipole arrays, was installed at Owens Valley to monitor the Crab nebula's radio emission at a frequency of 25 MHz (λ 12 m) as it passed near the Sun. Then in 1958 the 32 ft dish on Palomar Mountain was dismantled and installed at Owens Valley with a better surface.

[1] The Office of Naval Research (ONR) provided all but \$200,000 of the total cost of \$1.5 million for the initial construction, see Some Radio Telescopes – I, *Sky and Telescope*, April 1959, pp. 302–303. In 1970, however, Congress passed the Mansfield Amendment which restricted the ONR to financing projects with military applications. Subsequently, the NSF took over as the major funding source for the Owens Valley Radio Observatory.

Figure 21.1. John Bolton on the 32 ft (10 m) diameter radio telescope on Palomar Mountain that was completed in 1956. The dome in the background is that of the 200 inch Hale reflector. (Courtesy of the Archives, California Institute of Technology.)

When completed in 1959 the new, large Owens Valley interferometer consisted of two 90 ft (27 m) equatorial dishes (see Figure 21.2), with surfaces of ⅜ inch (9.5 mm) steel mesh.[2] They had a restricted polar motion of ±4 hours, which simplified the design. The dishes were movable on rail tracks, oriented east–west, to give a maximum separation of 1600 ft (500m) with intermediate stations at 200, 400 and 800 ft.[3] It was initially planned to operate the interferometer at a frequency of 400 MHz (λ 75 cm) but, because of technological breakthroughs, by the time it was built it was able to operate at 750 MHz (λ 40 cm). When completed this Owens Valley interferometer was the first operational large interferometer in the United States.[4]

[2] Some Radio Telescopes – I, *Sky and Telescope*, April 1959, pp. 302–303.

[3] The rails, which came from a US Navy depot at Port Hueneme, California, were really old, having been made in Germany in 1880!

[4] Not only that but the first 90 ft dish, when it had been completed in early 1959, was the second largest fully directable radio telescope dish in the world, next only to Jodrell Bank's 250 ft.

Figure 21.2. The two 90 ft (27 m) dishes of Caltech's Owens Valley interferometer built in the late 1950s. (Courtesy of the Archives, California Institute of Technology.)

Almost immediately after completion the 750 MHz system was replaced by a 960 MHz (λ 31 cm) system. This new frequency was in a protected spacecraft communications band. Consequently the receiver's new front-end design, produced by Stanley, could be and was used on the Jet Propulsion Laboratory's (JPL's) Explorer spacecraft. The telescope system was then progressively improved so that by 1963 observations could be made at five frequencies from 480 MHz (λ 63 cm) to 2.8 GHz (λ 10.6 cm). The dishes' supporting structure was subsequently strengthened and the surface of the central part of the dishes was changed to solid aluminium to enable them to operate at an even higher frequency. So that in 1967 observations could be made at a frequency of 10 GHz (λ 3 cm) with an antenna efficiency of about 30%.

In the meantime a 1,600 ft north–south track had been added to the western end of the east–west track in 1960 mostly by the observatory staff undertaking the manual work themselves rather than using contractors. This north–south track, which had the same intermediate station distances as the east–west track, extended north of the original track to form an 'L' shape with it. More stations were later added and the east–west track was extended 600 ft west of the intersection of the two arms.

Whilst all this work had been going on, Caltech had submitted a proposal to the NSF in 1962 to extend the above two-element Owens Valley interferometer by adding four new dishes of about 125 ft (38 m) in diameter and extending the existing track 1,200 ft (390 m) to the south. In response the NSF paid for engineering studies and in 1964 they funded the construction of a prototype 130 ft (40 m) dish (see below). In the same year the National Academy of Sciences' Whitford committee endorsed the proposed extension to the Owens Valley interferometer. They did so as they thought that it could be built relatively quickly and that it would produce high-resolution measurements, albeit on a relatively limited number of radio sources. In fact the committee went further and suggested that eight new 130 ft antennae should be built instead of four.(3)

Subsequently Caltech proposed in 1966 that the interferometer extension of eight new 130 ft telescopes should be built on tracks 9,000 ft (3 km) long east–west and 16,000 ft (5 km) long north–south.(4) Unfortunately at this time the NSF were being asked to fund a number of other large radio astronomy projects. So they appointed a committee, under the chairmanship of Robert Dicke (see Section 20.9), to review them all and to make recommendations as to which should be funded. However, although the Dicke committee endorsed the Caltech array in 1969 as an immediate start, it also endorsed two other major projects, including what was to become the VLA, and the NSF could not afford all three. Consequently the National Academy of Sciences' Greenstein committee was asked to judge between the three potential projects, and in 1971 they recommended the VLA as its top priority project for the whole of astronomy. As a result the proposed Caltech array of eight new 130 ft telescopes was abandoned, although one dish had already been built by Westinghouse Electric Corp. This was on a short length of track 3,500 ft (1 km) east of the centre of the existing interferometer. This single 130 ft telescope was dedicated in 1968.

The inner part of this new 130 ft dish was made of solid aluminium panels whilst the aluminium panels in the outer 25 ft were perforated to reduce wind resistance. The dish, which could also be operated independently from the existing two-element array, was used at wavelengths from about 1 m (300 MHz) to 1 cm (30 GHz).[5] In April 1969, it took part in a very long baseline interferometer experiment

[5] The dish's efficiency of 40% at 10 GHz (λ 3 cm) reduced to 20% at 24 GHz (λ 1.25 cm).

with the NRAO 140 ft (43 m) telescope at Green Bank and the 210 ft (64 m) Parkes antenna in Australia. Unfortunately there was found to be too much phase noise between the Green Bank and Australian dishes, but the OVRO – Australia interferometer worked well, showing that a number of quasars were smaller than the interferometer's 0.4 milliarcsec angular resolution.

Initially the surface panels of the Owens Valley's 130 ft had been adjusted to produce the best paraboloid when pointing at the zenith. But subsequent measurements showed that the antenna efficiency reduced rapidly with angular distance from the zenith. So in 1970 the surface panels were adjusted to produce the best paraboloid with the dish axis inclined at 40° to the zenith. The result was a 24 GHz aperture efficiency of 23% at 40° which fell to about 13% at both 0° and 90°.

By 1979 the two 27 m dish interferometer had been overtaken by larger and better instruments elsewhere, so the decision was taken to devote that interferometer to solar observations. Three 1.8 m antennae were added to the two dishes by 1997 when management of this 27 m/1.8 m interferometer, then called the Owens Valley Solar Array, was transferred to the New Jersey Institute of Technology. Two more 1.8 m antennae were added by 2004. One of the main tasks of this Solar Array has been to observe the development of magnetic fields on the Sun that were associated with solar flares.

Millimetre Arrays

In 1971 Robert Leighton of Caltech had begun to design a 10.4 m dish-type radio telescope that could be used at millimetre wavelengths. Some years later Caltech installed the dish at Owens Valley and in 1984 two more 10.4 m dishes were added to form an interferometer. The antennae, each of which was altazimuth mounted, had a reflecting surface of aluminium honeycomb panels supported by a steel space-frame backstructure. They were movable along a T-shaped railway track 100 m north–south and 200 m east–west. In 1989 the receivers were changed to reduce the operating wavelength to 1.3 mm (230 GHz) where the aperture efficiency was still a highly respectable 45%. Three more 10.4 m dishes were later added to form a six dish millimetre array with funding provided by Caltech, the NSF and the K.T. and E.L. Norris Foundation.(5)

In 2004 it was decided to merge this Owens Valley millimetre array with that of the Berkeley-Illinois-Maryland Association (BIMA, see Section 22.9), which consisted of nine 6.1 m telescopes located at Hat Creek, California. This new array, called CARMA (Combined Array for Research in Millimeter-wave Astronomy), began routine operations at the nearby Cedar Flat location, near Bishop, California in 2007. At an altitude of about 2,200 m (7,200 ft) this was about 1,000 m higher than Owens Valley, and so had typically about 50% less atmospheric water vapour.

Then in 2008 the University of Chicago's Sunyaev-Zel'dovich Array (SZA) consisting of eight 3.5 m antennae was added to CARMA.

CARMA had a baseline up to about a mile long (1.6 km). It operated at wavelengths of 7, 3 and 1.3 mm and was used to detect and image radio emission from molecules and dust. This has enabled research to be undertaken into the formation, evolution and dynamics of galaxies, the formation of stars and any associated planetary systems, and the composition of planetary atmospheres.

21.2 Submillimetre Radio Telescopes on Mauna Kea

James Clerk Maxwell Telescope

Astronomers in the United Kingdom were interested in the late 1960s in building a telescope to study the formation of stars by measuring the variations in density, composition, temperature and gas flow in star forming regions. These regions were opaque to optical radiation but not the submillimetre. So in 1975 the UK's Science Research Council decided to build a 15 m diameter telescope to observe down to a wavelength of 0.75 mm. It was initially intended to build this, the largest submillimetre telescope in the world, on La Palma. Then the Dutch joined the project in 1981 contributing 20% of the costs.

In the following year a decision was made to build the telescope at an altitude of 13,425 ft (4092 m) on Mauna Kea because of the very low water vapour content of its atmosphere. The site was leased from the University of Hawai'i who provided the local infrastructure and observer accommodation; in return they were allowed 10% of the telescope's observer time. Construction started in 1983, and this Cassegrain instrument, subsequently called the James Clerk Maxwell Telescope (JCMT), saw first light just over three years later. The facility was funded by the United Kingdom (55%), the Netherlands (20%) and Canada (25% from 1987) and was operated by the Joint Astronomy Centre (JAC),[6] from Hilo, Hawaii.

The main JCMT dish, which was on an altazimuth mount, consisted of over 270 individually adjustable aluminium panels mounted on a steel frame.[7] It was protected from the Sun, wind and other elements by a transparent PTFE (polytetrafluoroethylene) membrane attached to a co-rotating enclosure which enabled the telescope to be operational in winds of up to 45 miles per hour (70 km/h). The membrane was 90% transparent in the telescope's operational waveband whilst transmitting no more than 20% of the Sun's heating radiation. The original detector, a continuum single pixel bolometer, was replaced by the Submillimetre

[6] The JAC also operated the UK Infrared Telescope (UKIRT, see Section 8.4).

[7] *Astronomy Now*, April–June 1987, pp. 6–7.

Common-User Bolometer Array (SCUBA) in 1996 to detect interstellar dust grains at temperatures ranging from 3 K to 30 K.(6)

SCUBA, which was built by the Royal Observatory Edinburgh, was a dual camera system. It had a 91 pixel short wavelength detector array optimised to work at 450 microns and a 37 pixel long wavelength array designed to work at 850 microns. The detectors were cooled to well below 1 K. Additional pixels were provided alongside the long wavelength array for photometric work at 1.1, 1.35 and 2.0 mm. Both the short and long wavelength arrays had approximately the same field of view of about 2.3 arcmin in diameter. SCUBA, which had been in development for almost ten years, found voids in the distribution of dust grains around nearby stars that may have been caused by newly formed planets. Observations with SCUBA also showed bright knots in molecular clouds where stars were being born. In addition submillimetre spectral lines were observed using heterodyne receivers which provided information on the temperature, density and velocity of interstellar gas. An improved version of SCUBA, called SCUBA-2, was installed on the telescope in 2011 with detectors cooled to 0.1 K.[8] It was two orders of magnitude faster at mapping large areas of the sky compared with its predecessor, and provided wide-field imaging at 450 and 850 microns with arrays of 1280 pixels.

In 2011 the Netherlands announced that they would withdraw from the JCMT project in March 2013. This announcement was followed by Canada and the UK that said shortly afterwards that they would withdraw from the JCMT project in September 2014. In the event the Netherlands and Canada withdrew, as expected, whilst an extension of operations from October 2014 to February 2015 was funded by the UK and some Canadian university funds. Finally on March 1 2015 ownership of the JCMT was transferred from the UK to the University of Hawai'i, whilst the telescope was operated by a consortium headed by the East Asian Observatory (EAO)[9] in partnership with the UK and Canadian research communities.[10]

Caltech Submillimeter Observatory

As mentioned previously[11] Robert Leighton had started work in 1971 on designing a large (~10 m diameter) millimetre-wave radio telescope dish. It basically consisted of a number of hexagonal, lightweight aluminium honeycomb panels

[8] The SCUBA-2 project was led by the UK Astronomy Technology Centre in Edinburgh in collaboration with the universities of British Columbia (Canada), Cardiff (Wales), Edinburgh (Scotland), and Waterloo (Canada), the US National Institute of Standards and Technology, and the Joint Astronomy Centre.

[9] EAO was created by the East Asian Core Observatories Association (EACOA) which was an affiliation of the research institutes of ASIAA (Taiwan), NAOC (China), KASI (South Korea) and NAOJ (Japan).

[10] UK Science and Technology Facilities Council, Press Release, March 1 2015, 'JCMT now under new ownership'.

[11] See Section 21.1.

supported by a steel space-frame backup structure. This supporting structure consisted of a strut and post array based on a lattice of equilateral triangles. The posts were all parallel to each other and to the optic axis of the dish, and all joints were pinned, rather than welded, to avoid the introduction of unknown shear forces.[12]

Al Moffet, a Caltech professor of radio astronomy, was so impressed with Leighton's design that he joined with Gerry Neugebauer[13] and Leighton in submitting a proposal to the NSF to build three of these millimetre-wave radio telescopes together with a fourth which would have a more accurate antenna to operate in the submillimetre waveband. The three millimetre-wave telescopes would form an interferometer and be built at the Owens Valley Radio Observatory (OVRO) whereas the submillimetre telescope would be built at a higher, dryer site.

In response the NSF decided to provide a grant for the OVRO interferometer but deferred making a grant for the submillimetre telescope until the interferometer had been completed. Unfortunately the development of this interferometer was rather slow, consequently holding up the NSF funding for the submillimetre telescope. But in the meantime Leighton had managed to squeeze out some funding for the submillimetre telescope from that already supplied by the NSF for the interferometer. Caltech also received additional financial help from NASA and the Kresge Foundation of Troy, Michigan for the submillimetre telescope.[14]

Even so it was still important to get funding from the NSF for the submillimetre telescope and that would not be forthcoming until the build of the OVRO interferometer had been completed. As a result Thomas Phillips,[15] the director designate for the submillimetre telescope, was put in charge of the interferometer project in 1980 to get that finished as soon as possible. Unfortunately this took almost another four years. So it was not until November 1983 that the NSF finally agreed to provide $3.9 million for the Caltech submillimetre telescope. The money was primarily used for the construction of the 56 ft (17 m) diameter dome which, like that of the JCMT, was designed to rotate with the altazimuth-mounted telescope.

It had been recognised early on that Owens Valley was not high enough to support submillimetre observations but that Mauna Kea probably was. Consequently in 1980 Phillips, who had been taken on by Leighton to work with him on the design of the submillimetre telescope, began an Environmental Impact Study for Mauna Kea which he completed three years later in parallel with his job on the

[12] Walter R. Steiger, *A Brief History of the Caltech Submillimeter Observatory*, www.cso.caltech.edu/cso_history/CSO_History/index.html.

[13] Gerry Neugebauer had previously led Caltech work on infrared and worked with Leighton on the two micron survey.

[14] Additional funds were later provided by a gift from Francis L. Moseley of Flintridge, California and the Prince Charitable Trusts of Chicago whilst later funding was also provided by the NSF. The environmental impact studies and the dome design were initiated with Caltech funding.

[15] As a matter of interest, Phillips had written the first proposal for the JCMT when he spent a year at London University as reader in physics in 1975.

OVRO project. On Mauna Kea he undertook measurements of total column water vapour over various parts of the mountain, which was a crucial factor in deciding where to locate the proposed telescope. In the meantime in 1982 the very influential Field Report of the American National Academy of Sciences had endorsed the project.

Phillips, with a background in superconductivity and low temperature physics, was responsible for the development of the telescope's receiver system, whilst Leighton designed the dish and the dome of what was called the Caltech Submillimeter Observatory (CSO). The dish was constructed so that its surface panels could be easily dismounted, whilst its backup structure could be disassembled into easily transportable pieces. As a result the whole of the dish structure could be (and was) assembled in a controlled environment[16] in Pasadena where it could be tested, adjusted and modified, if necessary, close to various technical services. The dome was similarly erected in Pasadena before being disassembled for the trip to Mauna Kea.

When parts of the telescope finally arrived on the mountain in the mid 1980s they were assembled into a telescope only about 160 m away from the JCMT (see Figure 21.3). This was to facilitate a possible linkage between the two to produce a submillimetre interferometer. Caltech's aim had been to build the most accurate dish possible at about 10 m aperture, whereas the JCMT designers had wanted to build a larger dish, whilst accepting a compromise in dish accuracy. As a result, although the Caltech dish had only half the area of the JCMT, it was able to operate down to shorter wavelengths of about 0.35 mm. In the event the CSO was dedicated in late 1986, two weeks before the JCMT saw first light. First light of the CSO dish then took place in March of the following year with an image of M82.

The initial CSO detectors were heterodyne receivers and a single channel bolometer. The latter was later replaced by SHARC (Submillimeter High Angular Resolution Camera), a 24 element imaging bolometer array optimised for 350 and 450 micron continuum mapping. The CSO has, amongst other things, improved our understanding of the chemistry of the interstellar medium, improved our knowledge of the development of red giant stars, mapped molecular gas and dust in galaxies, and helped to determine the volatile constituents of comets.

The CSO undertook short baseline interferometry with the JCMT to produce a best angular resolution of about 0.5 arcseconds at 345 GHz (0.9 mm wavelength). (7) Experience gained with these observations assisted the design of the subsequent Submillimeter Array (see below).

NSF funding of the CSO ended in March 2013 as it was anticipated that Caltech would begin decommissioning of the observatory starting in 2016, with the site

[16] The building used had been previously occupied by the 200 inch Palomar mirror.

Figure 21.3. This image of Millimeter Valley on Mauna Kea shows the dome of the 10 m diameter Caltech submillimeter telescope in the left foreground and the enclosure for the 15 m diameter JCMT in the centre. (Courtesy A. Woodcraft via Wikimedia Commons.)

being returned to its natural state by 2018. Then in May 2015 Caltech announced that it would end operations ahead of schedule in September of that year to enable decommissioning to begin.

Smithsonian Submillimeter Array

The Smithsonian Submillimeter Array (SMA) was the result of an initiative in 1983 by Irwin Shapiro, the new director of the Harvard-Smithsonian Center for Astrophysics (CfA), who wanted to extend the development of high angular resolution instruments across as wide a range of frequencies as possible.(8) At that time the submillimetre waveband, which was important for the observation of very cool gas clouds in the Milky Way, had been virtually unexplored astronomically, and those telescopes in the process of being built were only single aperture instruments and so would not be able to produce high resolution images. As a result, an array of submillimetre telescopes, producing high resolution images, would be unique.

Shapiro set up a Submillimeter Telescope Committee, with James Moran of the CfA in the chair, to outline the configuration of such a submillimetre array. The committee first examined the idea of building a single radio telescope to work with a planned 10 m telescope of the University of Arizona and the Max Planck Institute for Radio Astronomy. But such a two-element interferometer would have little imaging capability. So the committee decided that the best solution was to build a new submillimetre array of six dishes, as that was the number of dishes required to provide reasonable image quality without multiple reconfigurations.

The requirement placed on the array was that it should be at least as sensitive as that of any of the planned submillimetre telescope of which the JCMT was the largest.(9) Consequently the committee settled on dishes of 6 m diameter to provide a total collecting area virtually the same as that of the JCMT. As far as location was concerned these dishes would have to be situated at a dry high altitude site to produce a sub arcsecond resolution in the required wavelength range from 0.3 to 1.3 mm. A formal proposal was then made to the Smithsonian Institution in 1984 based on this configuration of array. The astronomical community, who were also consulted, reacted favourably to the project.

The Moran committee recognised that the development of appropriate receivers was crucial to the success of the SMA project and recommended that a receiver technology laboratory should be established as a matter of some urgency. Subsequently the CfA set up such a laboratory at the Smithsonian Astrophysical Observatory (SAO). The design work on the array started in 1988 followed by construction four years later.

As far as location of the array was concerned a number of possible sites were considered. These included Mount Graham in Arizona, the Atacama Desert in Chile, the South Pole, and Mauna Kea. Eventually in 1992 Mauna Kea was selected as the preferred site for the SMA because of on-site test results, its excellent infrastructure and because the CSO and the JCMT were already on Mauna Kea operating at similar frequencies. In fact it was hoped to install the six 6 m telescopes of the SMA near to the CSO and JCMT so they could all be linked together to form an extended interferometer. Consequently two years later an agreement was reached with the University of Hawai'i to build the SMA in 'Millimeter Valley', a little below the summit of Mauna Kea where these other two telescopes were located.(10)

The main problem with the site provided for the SMA was that it was rather uneven. Consequently the telescopes could not be moved around on a rail track and a special transporter had to be built to pick up and move them over slopes of up to 15%. In addition, because of the requirement to move the telescopes it became impracticable to cover each of them with radomes similar to those of the CSO and JCMT. So the reflecting surfaces of the antennae (see Figure 21.4) were unprotected from the local environment, resulting in them being subjected to large temperature

Figure 21.4. A 6 m diameter antenna of the Submillimeter Array on Mauna Kea. (Courtesy Harvard-Smithsonian Center for Astrophysics.)

swings. If these surfaces were continuous, this would have resulted in them exhibiting substantial warping, so the surface panels of the dishes were arranged in four rings; 12 panels in each of the inner two rings and 24 panels in each of the outer two.

The reflective surface of the dishes was made of machined cast aluminium panels supported by carbon fibre tubes and steel nodes. Each of the panels was attached to this backup structure by three or four mechanical adjusters per panel.[17]

[17] There were three adjusters per panel for panels in the inner ring and four per panel for the outer three rings.

Figure 21.5. The Submillimeter Array on Mauna Kea. As mentioned in the text a special transporter had to be built to move the antennae over the uneven ground. (Courtesy Harvard-Smithsonian Center for Astrophysics.)

These adjusters were accessed from the front, as the backup structures were covered by an aluminium skin to protect them from the weather.

In 1996 the Academia Sinica Institute of Astronomy and Astrophysics of Taiwan (ASIAA) joined the SMA project, and agreed to add two more telescopes, increasing the number of possible baselines from 15 to 28.

The first prototype telescope was built at the MIT Haystack Observatory in 1996, and two copies of a second prototype operated there two years later as an interferometer. These two telescopes were then installed on Mauna Kea where they produced first fringes in 1999. The production American telescopes were also assembled at Haystack for testing before being disassembled and shipped out to Mauna Kea. In the event the SMA was completed on Mauna Kea with the six American and two ASIAA telescopes by the end of 2003. There were 24 pads for locating the eight individual dishes to provide baselines varying from 9 to 508 m. The array (see Figure 21.5) operated in the atmospheric window between 180 and 900 GHz (λ 1.7 to 0.3 mm), producing maximum resolutions of from 0.5 to 0.1 arcsec, respectively.

In 2008 the SMA was connected to the JCMT and CSO using fibre optic cables to produce a ten element interferometer operating at 230 and 345 GHz (λ 1.3 and

0.9 mm). This connection to the two larger telescopes increased the maximum baseline to 782 m in what was called the eSMA (extended SMA) configuration.

The SMA and eSMA have been used to observe star-forming molecular clouds, highly redshifted galaxies, evolved stars and nearby supermassive black holes, including the source Sgr A* at the centre of the Milky Way. Although Sgr A* was unresolved, as expected, it was found to be linearly polarised and highly time variable. A detailed analysis of the results indicated that the black hole at the centre of the Milky Way was in a quiescent state.

References

1. Robertson, Peter, *Beyond Southern Skies; Radio Astronomy and the Parkes Telescope*, Cambridge University Press, 1992, pp. 115–116.
2. Cohen, Marshall H., The Owens Valley Radio Observatory: Early Years, *Engineering and Science*, Spring 1994, pp. 8–24.
3. Whitford, A. E., et al., *Ground-Based Astronomy; A Ten-Year Program*, National Academy of Sciences – National Research Council, Washington, DC, 1964, pp. 52–53 and p. 75.
4. Cohen, Marshall H., A History of OVRO: Part II, *Engineering and Science*, No. 3, 2007, pp. 33–43, and Giant Radio Telescopes, *Sky and Telescope*, October 1967, p. 203.
5. Scoville, N., et al., The Owens Valley Millimeter Array, in Ishiguro, M. and Welch, Wm. J., (eds.), *Astronomy with Millimeter and Submillimeter Wave Interferometry*, ASP Conference Series, Vol. 59, 1994, pp. 10–17.
6. Holland, W. S., et al., SCUBA: a Common-user Submillimetre Camera Operating on the James Clerk Maxwell Telescope, *Monthly Notices of the Royal Astronomical Society*, 303, 1999, 659–672.
7. Wiedner, M. C., et al., Introduction to the JCMT-CSO Interferometer, in Bender, Ralf, and Davies, Roger, (eds.), *New Light on Galaxy Evolution*, IAU symposium 171, June 1995, Kluwer Academic Publishers, 1996, p. 467.
8. Moran, James M., The Submillimeter Array, *ASP Conference Series*, 356, (2006), 45–57.
9. Blundell, Raymond, The Submillimeter Array, in *Proceedings of the IEEE/MTT-S International Microwave Symposium 2007*, pp. 1857–1860.
10. Ho, Paul T. P., Moran, James M., and Lo, Kwok Yung, The Submillimeter Array, *Astrophysical Journal*, 616, 2004, L1–L6.

22

Further North and Central American Observatories

22.1 US Naval Research Laboratory

Just after the end of the Second World War John Hagan, the head of the Centimeter-Wave Research Branch[1] of the US Naval Research Laboratory (NRL), was looking for a new field of research in which to use his branch's experience. As a result he hit on the idea of observing astronomical sources of radio emission. Consequently Hagen and his deputy, Fred Haddock, decided to observe the Sun, not only because of its effect on the Earth's atmosphere, but because it was probably the only astronomical source observable with their type of equipment.

For their first solar observations Hagen and Haddock used parabolic antennae up to 10 ft (3 m) in diameter with receivers operating at wavelengths of 8.5 mm, 3.2 cm and 9.4 cm (frequencies of 35, 9.4 and 3.2 GHz respectively). Then in 1947 they went further and used an 8 ft diameter antenna to observe a total solar eclipse at a wavelength of 3.2 cm from on board ship in the South Atlantic. As a result they were able to conclude that the Sun, at this wavelength, was only slightly larger than the optical Sun.(1)

Unfortunately Hagen and Haddock's equipment only enabled them to observe the emission from the Sun as a whole as it did not have enough resolution to locate the exact sources of radio emission. It was clear that to locate these sources they would need a much larger dish. And if they managed to procure one they may also be able to observe other cosmic sources, as well as detecting thermal radio emission from the planets. So in the late 1940s Hagen and Haddock managed to persuade the US Navy to provide $100,000 for its purchase and, as a result, they were able to acquire a 50 ft (15 m) parabolic reflector from the Collins Radio Company designed by Ned Ashton of the University of Iowa.

[1] Later called the Radio-Frequency Research Branch.

Figure 22.1. The 50 ft diameter radio telescope dish built on the top of the Naval Research Laboratory's main administration building in 1950. (Courtesy Tim1965 via Wikimedia Commons.)

This 50 ft NRL dish was made of 30 aluminium sector castings which had been bolted together. The surface was then machined to its parabolic shape to enable it to be used at wavelengths as low as 1 cm. The segments were subsequently dismantled and reassembled on an altazimuth mount which used a modified anti-aircraft naval gun mounting installed on the roof of NRL's main administration building in Washington (see Figure 22.1).(2) The dish's beamwidth was only 3 arcminutes, which was very good for the time.

This radio telescope, which was the first large reflector to operate down to centimetre wavelengths, was operational in 1951. Unfortunately there were problems with the mount's pointing and tracking, so in 1953 it was decided to use the dish at the longer wavelength of 9 cm where such problems would not be as important. As a result Haddock, Cornell Mayer and Russell Sloanaker made the first detection and measurement of interstellar ionised hydrogen clouds as discrete radio sources. Subsequently Mayer, Timothy McCullough and Sloanaker detected 3.15 cm radio emissions from Venus and concluded, in 1956, that its surface temperature was very high at about 600 K.

NRL was the leading radio astronomy group in the United States in the mid 1950s. In 1955 they ordered an 84 ft (26 m) equatorially mounted radio telescope from the D. S. Kennedy Company which, when it was erected at Maryland Point in 1957, was the largest equatorial radio telescope in the world.[2] The surface of the dish consisted of ⅜ inch (9.5 mm) square aluminium mesh to enable operations down to a wavelength of about 10 cm (frequency 3 GHz). Maryland Point, on the north shore of the Potomac River, had been chosen as the site for the new observatory as it had an unrestricted southern horizon and had by far the lowest level of radio interference measured within 50 miles (80 km) of Washington D. C. A few years later NRL ordered an 85 ft equatorially mounted parabolic reflector from the Rohr Corporation of San Diego which became available at Maryland Point in 1965. The surface of this dish was made of stretch-formed aluminium to allow operations down to a wavelength of about 1 cm.(3)

In the mid 1950s both the NRAO (see Section 20.1), on the one hand, and the NRL and Office of Naval Research (ONR), on the other, became interested in building a 600 ft (180 m) parabolic radio telescope. But subsequently NRAO decided to concentrate on building smaller more practical and more affordable telescopes. On the other hand in 1956 NRL and the ONR decided to build a fully steerable 600 ft paraboloid to operate down to a wavelength of 21 cm. The site they chose was at Sugar Grove, West Virginia in the remote Allegheny mountains.

This 600 ft Sugar Grove antenna was a highly classified project supported by the National Security Agency (NSA) to detect Soviet radio signals reflected from the surface of the Moon.(4) When the existence of this American project was announced in June 1958 it was suggested that it was primarily to be used for astronomical research. But by the following month it had been admitted that it was being built for defence research, with any astronomical work being undertaken as a by-product.[3] A Navy spokesman giving evidence to a House of Representatives subcommittee further suggested that this 600 ft antenna might be used to study the reflection of radio signals from the Moon for communication purposes, without making it clear that the signals of interest were those emitted by the Soviet Union for various purposes rather than by the United States. In 1959 the NSA and ONR were awarded \$60 million for the project. But just one year later, after construction had begun, they requested authorisation for an additional \$18 million.

In 1962 it was finally decided to abandon the 600 ft project when it became clear how difficult it would be to complete it and how marginal was its military

[2] Edward F. McClain, The Naval Research Laboratory's 84-foot Radio Telescope, *Sky and Telescope*, October 1958, pp. 608–610. The previous largest equatorially-mounted radio telescope was the 60 ft built by the D. S. Kennedy Company for Harvard Observatory's Agassiz Station about 30 miles (50 km) from Boston which had been dedicated in 1956.

[3] *Sky and Telescope*, October 1958, p. 626.

surveillance value.[4] At this stage about $96 million had already been spent but only two 420 ft (130 m) high erection towers and the circular concrete base had been built. Its estimated cost had increased to more than $200 million.

NRL had also built a 60 ft (18 m) paraboloid at Sugar Grove in 1957 as a prototype for the 600 ft. They later had a 150 ft (46 m) built there made partly from materials left over from the 600 ft project. Sugar Grove was a military installation and these, as well as other antennae located there, have been used for military purposes.

As time went on, NRL stopped building its own state-of-the-art radio telescopes and began using US national facilities, particularly those at Green Bank. The NRL 85 ft radio telescope was later used with other radio telescopes worldwide as part of a Very Large Baseline Interferometry (VLBI) network. However in the mid 1990s the Maryland Point observatory was closed down.

22.2 MIT Lincoln Laboratory, Millstone Hill and Haystack

In 1954 MIT's Lincoln Laboratory, who specialised in defence-related contracts, moved out of its buildings on the MIT campus into a newly constructed facility next to the Air Force Cambridge Research Center in Lexington, Massachusetts. Two years later Lincoln Laboratory began construction of an experimental prototype Ballistic Missile Early Warning System (BMEWS) radar at Millstone Hill near Westford, Massachusetts under contract to the US Air Force.(5) This location was chosen as it was not troubled by major air movements or television transmitters whilst it was relatively close to MIT and the Lincoln Laboratory.

This BMEWS radar, which used a 26 m (84 ft) diameter steerable parabolic dish, operated at 440 MHz (λ 68 cm) with signals powered by two specially designed klystrons each having a peak power of 1.25 megawatts. By coincidence the system, which began operation in October 1957, was just in time to track the Soviet Union's first Sputnik spacecraft. It was later used for both military and civilian purposes, including ionospheric research, and radar studies of the aurora, meteors and the Moon. In 1958 Bob Price and colleagues used the Millstone Hill antenna equipped with a new highly sensitive maser to try to detect radar signals reflected by Venus. Even though a klystron failure significantly reduced transmitted power they were convinced of their success(6) until their attempt to repeat the detection in the following year failed.[5] In the event, JPL was the first to *unambiguously* detect Venus

[4] In 1962, when explaining the reason for the cancellation, the Pentagon pointed out 'the need originally established in 1954 for classified research in ionospheric physics, space communications, navigation and radio astronomy has been substantially reduced by major advances in science and technology not foreseen at that time', see *Sky and Telescope*, September 1962, p. 132.

[5] Lincoln Laboratory later withdrew their claim to have detected radar signals from Venus in 1958. But later analysis of their 1959 observations showed that they had actually been the first to detect radar signals from Venus, although their signals had been initially lost in the noise, see Andrew J. Butrica, *To See the Unseen; A History of Planetary Radar Astronomy*, NASA SP – 4218, 1996, p. 42.

Figure 22.2. The Radome of the MIT Lincoln Laboratory's 37 m (120 ft) diameter radar antenna at Haystack Observatory as seen in 2012. (Courtesy Daderot via Wikimedia Commons.)

by radar in 1961. Then in the following year the Millstone antenna was upgraded to operate at 1.32 GHz (23 cm).

With the launch of Sputnik in 1957 it was clear that the United States needed a more powerful radar to track and observe future Soviet spacecraft. At the same time the US also realised that it needed to improve long-range military communications. Consequently in 1958 the Lincoln Laboratory started the design of a 37 m (120 ft) diameter X-band (7.65 GHz; λ 4 cm) radar dish that could also be used for communications research. Unfortunately the Air Force found the initial price estimate of $5 million too high. Consequently Lincoln Laboratory decided to place the dish inside a fibreglass radome (see Figure 22.2) to eliminate the effects of wind and snow loading and solar heating. This enabled the dish and its supporting structure to be made much lighter with a consequent reduction in the power needed to drive it. As a result the estimated costs were reduced to around $1.5 to $2 million.

This 37 m diameter X-band radar antenna was built on Haystack Hill about half a mile (800 m) from Millstone Hill. It had a parabolic main dish of solid aluminium panels in a Cassegrain configuration with a 9.5 ft (2.9 m) diameter

Figure 22.3. The 46 m (150 ft) diameter fully steerable radar antenna of MIT's Haystack Observatory in the distance with, in the foreground, the observatory's 67 m (220 ft) zenith antenna. (Courtesy Daderot via Wikimedia Commons.)

secondary. This secondary focused the received energy to the centre of the main dish where instrument boxes were located.[6] The initial wide-band receivers could operate up to 16 GHz (λ 1.9 cm). On completion in 1964 this altazimuth mounted Haystack antenna was used not only for spacecraft surveillance and for military communications research but it also took over from the nearby Millstone antenna as the Lincoln Laboratory's planetary radar facility. In addition the Haystack facility was used as a ground terminal for space communications and as a radio telescope.

In parallel with the construction of this 37 m Haystack radar antenna, Lincoln Laboratory also built an 18 m (60 ft) radome-protected X-band radar dish for use with the controversial Project West Ford. In this project about 500 million thin copper wires were launched into orbit in 1963 to enable messages to be transmitted from coast to coast across the United States. At about the same time the Laboratory also built a fixed, zenith pointing, 67 m (220 ft) antenna (see Figure 22.3) which was used for ionospheric research using incoherent scatter techniques.

[6] Haystack Radio-radar Telescope, *Sky and Telescope*, December 1964, pp. 349–351.

Harvard College Observatory suggested in 1963 that MIT (including the Lincoln Laboratory), the Smithsonian Astrophysical Observatory (SAO) and themselves join together to undertake a regional effort to build a large antenna free of military limitations. The project would build on Lincoln's strength in radar astronomy and Harvard's work on radio astronomy. In the following year Harvard approached the Lincoln Laboratory to try to obtain access to the Haystack facility, which they were granted by the Air Force in 1965.

In the meantime the Lincoln Laboratory had become increasingly concerned that their 37 m Haystack antenna would be outperformed as a radar facility by the 1,000 ft (305 m) Arecibo antenna and JPL's 210 ft (64 m) Mars station, which were under construction at the time, so they welcomed the idea of a cooperative project. It so happened that in 1963 the National Academy of Sciences (NAS) Whitford Report recommended building two 300 ft (90 m) fully steerable dishes, possibly as regional facilities, which matched the Harvard idea of a joint approach at Haystack. Subsequently Harvard, MIT and the SAO agreed, under the name of CAMROC (Cambridge Radio Observatory Committee), to undertake a joint study of a large radio and radar telescope. Unfortunately although a joint study was carried out the full project funding was not forthcoming. Consequently CAMROC looked into the possibility of changing itself into a regional organisation, called NEROC (Northeast Radio Observatory Corporation), with additional member institutions providing extra technical and financial support. In this they were partly successful as they managed to recruit ten new member institutions, although their financial problem was still not resolved. By now NEROC was seeking $28 million over five years to cover the design and construction of a fully steerable 440 ft (135 m) parabolic dish operating at 6 GHz (λ 5 cm). It would be enclosed in a radome and be used both as a radio telescope and radar facility. Unfortunately at this time the Dicke committee, which had been asked by the NSF to judge between the various claims on its funds for radio telescopes, deferred a decision in August 1967 on the NEROC project.

Two months later the Lincoln Laboratory asked NEROC if they were interested in assuming responsibility for Haystack. NEROC were definitely interested as support for their proposed large new telescope had almost dried up following the Dicke report. So whilst the Smithsonian Institution tried to obtain Congressional funding for the large new NEROC telescope, NEROC itself sought funding from the NSF and other organisations to enable them to take over responsibility for the existing 37 m Haystack antenna as a regional facility. Eventually after much discussion the NSF agreed to fund the 37 m. Then in 1970 the Air Force transferred the Haystack dish to MIT, which already owned the land, whilst NEROC was made responsible for its operations.

In the meantime the Dicke committee, which had been reconvened in 1969, had finally endorsed the large new NEROC antenna project. But subsequent attempts to get its funding through Congress met with a number of difficulties. Then in 1971 the NAS Greenstein committee made the VLA its top priority project (see Section 20.9). At about the same time it became clear to NEROC that the next new frontier of radio research would be in the millimetre waveband. Consequently NEROC decided to cancel their plans for the large new dish and concentrate instead on an NSF-proposed upgrade to the Haystack antenna to enable it to operate at wavelengths of about 3 mm (100 GHz frequency).(7)

Over the years the available observing time devoted to astronomical radar research using the Haystack dish gradually reduced. Then in March 1974 it ceased completely because of diminishing Air Force and NASA interest. There were also problems with internal arcing in the klystron tubes, and there was increasing pressure from radio astronomers for observing time with the dish particularly for molecular line research.

In the event the modifications to enable the Haystack antenna to operate down to 3 mm were undertaken in two phases, the first from 1986 to 1989 to allow it to operate down to 7 mm and the second from 1991 to 1993 to achieve the 3 mm figure.[7] In the first phase the radome membrane was replaced, active thermal control of a critical structural element was introduced and the surface panels of the main dish were adjusted to produce a more accurate surface with a tolerance of about 500 microns rms.(8) In the second phase a deformable subreflector with 19 actuators was installed to help compensate for gravitational distortions of the main dish. Thermal distortions were also minimised by using a modern computerised environmental control system for the radome heater-blowers. The intake air was also pre-heated during winter.(9) Further adjustment of the surface panels of the main dish reduced its surface deviations to about 250 microns rms to produce an aperture efficiency of about 15% at 3 mm wavelength. A 3 mm radiometer was also developed to cover the 84–115 GHz (W-band) frequency range.

By this stage the Haystack dish had essentially reached its maximum possible performance. Then because of financial pressures elsewhere NSF funding for the system as a user facility was withdrawn in 1998. Seven years later control of the telescope was passed to the Air Force and the radiometers were removed in 2007 whilst the Air Force continued to use the antenna's radar capability.

What the Air Force really wanted at this time was a wideband radar system operating in the 92 to 100 GHz frequency band to produce high resolution images of microsatellites in low Earth orbit. This was beyond the capability of the existing Haystack system. As a result the Air Force and the Department of Defense's Advanced Research Projects Agency paid for the existing 37 m dish to be removed

[7] Lynn D. Matthews, of MIT Haystack Observatory, *The Upgrade of the Haystack 37-m Antenna*, www.cfa.harvard.edu/DCDCLXV/presentations/Dense_Cores/L_Matthews.pdf.

in 2010 and be replaced by a brand new, more accurate dish with improved thermal properties to enable it to work more efficiently in the millimetre waveband. The radome skin was also replaced. When completed the 75 micron rms surface accuracy of the main dish was, at the time, the best in the world for an antenna of its diameter. This renamed Haystack Ultrawideband Satellite Imaging Radar (HUSIR) became part of the US Space Surveillance Network, but it was also available part time for astronomical purposes.

In addition to the 37 m dish the MIT Haystack Observatory today (in 2015) operates the 18 m Project West Ford antenna which is mainly used for geodetic research, the 67 m zenith antenna, and a 46 m (150 ft) fully steerable radar antenna that had been built at the Sagamore Hill Air Force facility at Hamilton, Massachusetts, in 1963. It was moved to Millstone Hill in 1978 (see Figure 22.3) where, along with the 67 m, it has been used for ionospheric research.

22.3 Harvard Radio Astronomy Station, Fort Davis, Texas

A key part of the planned International Geophysical Year (IGY) which started in July 1957 was continuous monitoring of the Sun in all possible wavebands including radio. Consequently because of the geographical position of the United States it was clearly important that the US extend their solar observations into the radio frequency range to increase timescale coverage of such observations worldwide. At that time the US Air Force already had an optical solar observatory on Sacramento Peak, New Mexico (see Section 15.1) and they were already interested in extending their solar coverage into radio frequencies. Consequently it was proposed that the Air Force should be responsible for the US radio observations of the Sun during the IGY and longer term. As a result they ordered a 28 ft (8.4 m) diameter parabolic equatorially mounted antenna from the D. S. Kennedy Company to undertake the observations.(10)

Some years before Donald Menzel of the Harvard College Observatory (HCO) had been involved with the Air Force in setting up the Sacramento Peak solar observatory. So it was natural that he should also be involved in setting up its radio equivalent. As a result in 1955 HCO recruited Alan Maxwell, an experienced solar astronomer who had already made radio observations in New Zealand and at Jodrell Bank in the UK, to manage the task. His first job was to find a suitable site for the 28 ft antenna. It was especially important to find a location free from radio interference as it was planned to measure the spectra of solar radio bursts over a wide frequency range. The first idea was to build the radio telescope on Sacramento Peak near the optical solar observatory but this location turned out to have far too much radio interference because of its proximity to the White Sands Proving Grounds and Holloman Air Force Base.(11) The optical solar observatory on the top of a mountain was also completely unshielded from television and radio signals from distant towns. Consequently a number of other possible sites were considered in Texas, New Mexico, Arizona and Colorado.

Henry Smith, a solar observer at Sacramento Peak, suggested locating the radio telescope somewhere near Fort Davis in Texas, whilst Marlyn Krebs of the McDonald Observatory on Mt Locke, which was near Fort Davis, suggested a grassy valley called Cook Flat near the base of Mt Locke. This site, which was at an altitude of 5,200 ft (1,600 m), was surrounded on nearly all sides by mountains about 1,500 ft (450 m) higher. It proved to be an excellent location so Harvard decided to lease a one acre plot there on which to build their 28 ft antenna. They also decided to have an office and laboratories located at Fort Davis which was only about 8 miles (12 km) away.

The 28 ft radio telescope, which was built in 1956, was designed to continuously monitor solar activity over the frequency range from 100 to 580 MHz (λ 3 m to 52 cm). To do this it was provided with receivers which swept through this frequency band about three times per second to produce a radio spectrum. It soon became clear that interesting spectral information was being missed by the low frequency cut-off at 100 MHz, so additional receivers were obtained for the observatory in 1958 to take the cut-off down to 25 MHz using fixed dipoles in front of a plane reflecting screen. Eight years later this cut-off was reduced further to 10 MHz (λ 30 m) using a log-periodic antenna. The observatory's solar monitoring programme was supported financially by the US Air Force as a radio extension of their optical Sacramento Peak Observatory until 1973. Sponsorship was then taken over by NASA, the NSF and the NOAA World Data Center until 1982 when solar observations ceased.

It was clear from the early solar observations at the Fort Davis site that radio interference there was very low. As a result Maxwell suggested that the location would be ideal to carry out non-solar radio observations of sources both in the Milky Way and in the wider universe. This resulted in Harvard ordering an 85 ft (26 m) diameter equatorially mounted antenna from the Blaw-Knox Company of Pittsburgh, Pennsylvania which was delivered in 1962.[8] The Harvard observatory site at Cook Flat was then extended to 4 acres to accommodate both this new antenna and a new laboratory building.

Although it was not intended primarily as a solar telescope the 85 ft was used from 1970 to 1974 to make solar observations from 580 MHz to 4 GHz (λ 7.5 cm). But the main use of the 85 ft was to observe the wider universe. For example it was used to undertake occultation observations of radio sources by the Moon and of the Crab nebula by the Sun. The Galactic Centre and Galactic Plane were also mapped and flux densities measured at 5 GHz of Galactic and extragalactic

[8] The design of the Fort Davis 85 ft was based on that of the Green Bank 85 ft which Blaw-Knox had delivered to the NRAO a few years earlier (see Section 20.4).

sources. Then starting in 1972 the 85 ft was used for VLBI observations with antennae at the Owens Valley Radio Observatory, the National Radio Astronomical Observatory (NRAO) at Green Bank and other radio observatories with the support of the NSF and NASA. In 1982 the Harvard Radio Astronomy Station was renamed the George R. Agassiz Station of the Harvard College Observatory.

The 85 ft was used in the 1980s for geophysical research, sponsored by the National Geodetic Survey, studying the rotation of the Earth and the movement of the Earth's poles. Then in 1987 the 28 ft antenna was moved to Penticton, British Columbia.

The VLBI programme continued at the Agassiz observatory until 1991 when the 85 ft antenna was demolished. By this time the NRAO had completed the construction of a 25 m (82 ft) antenna only a few hundred metres away from the 85 ft. This new NRAO antenna was one of ten identical dishes which made up its Very Long Baseline Array (VLBA, see Section 20.10).

22.4 Vermilion River Observatory

George McVittie, head of the University of Illinois' Astronomy department, was interested in the mid 1950s in building a large radio telescope to help to resolve the disagreements between the Cambridge and Sydney radio source catalogues (see Section 17.1). At that time these two major catalogues disagreed where their celestial coverage overlapped and McVittie thought that the best way of resolving the problem would be to produce a new catalogue using an instrument of completely different design. He discussed the situation with Edward Jordan, head of the university's Electrical Engineering department, who agreed to help in 1954 in designing and building a suitable telescope.

McVittie and Jordan also agreed that this new large telescope should be built within a reasonable distance of the university's Urbana campus to enable it to be used for both research and training purposes.(12) The Office of Naval Research (ONR) indicated that it was willing to fund the project and two years later the university recruited George Swenson a professor of electrical engineering at Michigan State University as project engineer.

Illinois university first sent Swenson on a tour of the world's best radio astronomical observatories in 1957 to see their facilities and to discuss the advantages and disadvantages of various possible radio telescope designs.(13) Subsequently he defined the key characteristics required of the new Illinois instrument. It should have low sidelobes, to reduce source confusion, and operate in a different frequency range from existing instruments. It should also be large enough and have sensitive-enough receivers to detect weaker sources than in the Cambridge and Sydney catalogues.

It was also decided, to keep costs to a minimum, to build a telescope that was steerable in declination only, using the rotation of the Earth to scan in right ascension. One possible solution was to build a fixed parabolic dish, but this would have a strictly limited capability in scanning off-axis. However this would not be the case for a parabolic-cylindrical reflector with its focal line in the meridian. In this case steering in declination could be achieved by phase-adjusting the receiving elements. As a result this was the chosen configuration.

The next question was how to build a very large antenna of this design. One possibility was to build an antenna surface of wire mesh suspended on wooden posts above a roughly shaped piece of ground. But the alternative of laying the mesh directly on to a more accurately shaped ground surface was chosen as it was thought to be a more economical solution. An aerial survey subsequently found a suitable piece of ground only about 45 miles (70 km) from the university campus. This was in the form of a depression, oriented approximately north–south, that had been formed by a tributary of the Vermilion River. Compared with a flat site this natural depression reduced the amount of earth to be removed by about 65%.

Construction of this University of Illinois transit telescope was begun in 1959. It consisted of a parabolic cylindrical reflector 600 ft (180 m) long by 400 ft wide, aligned north–south, and made of galvanised wire mesh. This was placed on sheets of asphalt that had been laid directly on the ground. The telescope's 276 downward-looking conical log-spiral antennae, designed by John Dyson of the university's antenna laboratory, were mounted on a 425 ft (130 m) long horizontal beam carried on 165 ft (50 m) high towers at the linear focus of the reflector. The telescope (see Figure 22.4), which operated at about 610 MHz (λ 49 cm),[9] could detect sources at declinations between +10° and +70° by adjusting the phase of its feed elements.

In modern times what had once been a reasonably sized tributary of the Vermilion River was now no larger than a small stream. So during construction of the telescope this stream was contained in a concrete channel along the centre line of the telescope which had a shallow gradient to facilitate drainage. This was judged to be quite sufficient for normal operation of the telescope as a dam upstream was expected to provide sufficient storage during flash floods. Unfortunately flash floods in 1961 and 1962 caused extensive damage to the reflector's surface for 20 ft (6 m) on each side of the central channel. As a result, a second dam some 37 ft (11 m) high and 100 ft (30 m) long was constructed further upstream.

Over time, however, there was a progressive migration of earth from the sides of the parabolic cylinder to its centre gradually increasing its focal length. As a result by 1969 the focal length could no longer be accommodated within the

[9] Swenson pointed out in a 1986 IEEE article (reference 13) that the frequency band 608–614 MHz was, at that time, unused so it was available for use for radio astronomy. But it took years to get the band formally assigned to radio astronomy and the university could not wait that long. So they took the risk and used that frequency band anyhow, assuming the regulatory authorities would assign the band to radio astronomy in due course, which fortunately they did.

Figure 22.4. The 600 ft (180 m) long University of Illinois' transit telescope at their Vermilion River Observatory. (Courtesy ECE Illinois, University of Illinois at Urbana-Champaign.)

available range of adjustments in the height of the focal line array. In the meantime in November 1967 the ONR had announced that it would not renew the operating budget for the Vermilion River Observatory (VRO).(14) Although the National Science Foundation (NSF) agreed to take over funding responsibility this was at only 74% of the 1967–68 funding level. Consequently for both technical and financial reasons the telescope was closed down in 1970.

By the time construction of the VRO transit telescope had been completed in 1962 the differences between the Cambridge and Sydney catalogues had been largely resolved. But by the time that the VRO telescope was closed down eight years later it had completed its original survey programme. In the process it had detected over 1,000 discrete radio sources beyond the Milky Way. It had also mapped many ionised hydrogen regions of the Milky Way and detected two previously unknown supernova remnants.

Swenson had also planned in the late 1960s to build an interferometer consisting of three 120 ft (37 m) equatorially mounted parabolic antennae that could be moved along two lengths of rails intersecting at right angles. The east–west track was to be one mile (1.6 km) long and the north–south one 0.5 mile long. The first antenna, operating at 18 cm (1.67 GHz) and 49 cm, was built in-house using staff and students and government surplus equipment to keep costs to a minimum.(15) The surface of the dish consisted of a ¼ inch (6 mm) mesh of lightweight aluminium, and to minimise cost it was decided to restrict antenna tracking from 2½ hours before meridian transit to 2½ hours after. In addition only northern declinations were covered. This approach kept the height of the mounting system within reasonable limits and eliminated the need for counterweights to balance the antenna when it was tilted. The telescope was equipped with a 50 channel, frequency-filter spectrometer designed in-house by Kwang-Shi Yang and John Webber.

During construction of the 120 ft, which began in 1967, it became clear that funding that had been provided for this first antenna by the NSF would not be extended to cover the other two antennae. So instead of a three antenna interferometer, astronomers used the one antenna that was completed in 1971 as a radio spectrometer. That was until 1974 when a terminal was built allowing the antenna to also be used as one element of a very large baseline interferometer (VLBI) with other radio telescopes worldwide.

Lightning hit the VRO engineering shop one day destroying all the equipment. (16) Although the university paid for some replacement equipment, on another occasion vandals broke into the observatory at night and seriously damaged the surface of the 120 ft antenna. Although this surface was repaired it was never as accurate as before. In the meantime other facilities had become available for Illinois radio astronomers like the VLA (see Section 20.9). So in 1981 the VRO was closed down.

22.5 Ohio Transit Radio Telescope

The first large radio telescope at Ohio State University was an array of 98 helical antennae designed to be used at 250 MHz (λ 1.2 m). They were mounted on a tiltable ground plane 160 ft (50 m) long by 22 ft (6.7 m) wide.[10] But helical antennae were restricted in bandwidth so when John Kraus looked to build a larger telescope he decided to make one with a large continuous antenna surface. To keep the cost within reasonable bounds his antenna was to be close to the ground and be limited to transit observations only. Kraus designed this radio telescope and in 1954 Robert Nash, a graduate student of Kraus, built a very successful working scale model. Two years later Nash supervised the building of the complete instrument.(17)

This transit radio telescope consisted of a tiltable flat reflector of wires stretched on a 260 ft (80 m) long by 100 ft (30 m) high metal frame. This structure reflected radio waves from the source on to a section of a fixed parabolic

[10] See photo on the cover of *Sky and Telescope*, February 1957.

Figure 22.5. John Kraus' transit telescope of Ohio State University with the tiltable flat reflector on the right and the fixed parabolic reflector on the left. (Courtesy NRAO/AUI/NSF.)

reflector 360 ft (110 m) long by 70 ft (20 m) high[11] with a reflecting surface made of copper-clad steel wires. From there they were reflected to feeds at the focus which was on the ground just in front of the flat reflector. The ground between the two reflectors was covered with aluminium sheet cemented to a concrete floor to reduce thermal radiation from the ground and simplify the design of the feeds at frequencies below 1 GHz. The telescope was operational between 20 MHz and 2 GHz (λ 15 m and 15 cm), having a fan-shaped beam of about 11 by 40 arcmin at 1.4 GHz (λ 21 cm). It could track sources for about 30 minutes near the meridian by using movable feeds fixed to a motor-driven rail car.

This Ohio reflector took almost seven years to build as it relied on manual work by university technicians and students working part time. It was completed in 1962 at a cost of about \$500,000 at the Ohio State – Ohio Wesleyan Radio Observatory in the grounds of Ohio Wesleyan University's Perkins Observatory. The relative simplicity of its design encouraged French radio astronomers to produce a larger, improved version at Nançay (see Section 19.2), which became fully operational a few years later. The Ohio radio telescope, otherwise called 'Big Ear', did not initially use the whole width of the parabolic section as funds had run out before the flat section could be completed. But in 1970 this flat section was extended from 80 m to its full width of 100 m (see Figure 22.5).

[11] Initially Kraus had planned for both the flat and curved reflectors to be 720 ft long, but he was constrained by funding limitations.

One of the first tasks of the Ohio reflector in 1963 had been to produce the most detailed radio map of the Andromeda galaxy available at that time. This was followed by a thorough survey of the sky from declinations 63° N to 36° S, mainly at 1.4 GHz. Sixty percent of the approximately 20,000 sources detected were new. (18) Then in 1973 the telescope was used by the SETI (Search for ExtraTerrestrial Intelligence) project, which continued to use it for a number of years until the telescope was demolished in 1997. This was to make way for residential development and an extension to a golf course.[12]

22.6 Arecibo Radio Telescope

Ionospheric research received a much-needed boost in the second half of the 1950s by the approach of the International Geophysical Year (IGY), which lasted from July 1957 to the end of the following year. The IGY was deliberately planned to include the period of solar maximum, and studies of the ionosphere were included in the planning which began in 1953. Balloons, sounding rockets and satellites were envisaged to provide observations. But in early 1958 William Gordon of Cornell University suggested a new way of studying electrons in the ionosphere by bombarding them with high energy radar signals and observing the incoherent scatter produced.(19) He concluded in May 1958, based on his theoretical calculations, that a dish of about 1,000 ft (305 m) diameter would be required for observations above about 1,000 km in altitude. In addition he pointed out that such a dish could be used to undertake radar observations of the Sun, Moon, Venus and Mars, and possibly Jupiter and Mercury. His Cornell electrical engineering colleague Henry Booker also suggested that it could be used for radio astronomy.(20)

Academics like Gordon were not the only people interested at the time in the ionosphere as intercontinental ballistic missiles (ICBMs) and satellite launchers travel through this region. As a result, the US military was very interested in ionospheric research as they could see the possibilities of tracking ion trails produced by the Soviet Union's ICBMs and satellite launchers. In addition they wanted to understand the effect that high-altitude nuclear explosions would have on the ionosphere and consequently on radio communications.

Following his theoretical analysis Gordon led a study group, which included Booker and Ben Nichols, also of Cornell, to undertake an initial radar design study. This study was funded by the Office of Naval Research, the Rome[13] Air Development Center, and the Aerial Reconnaissance Laboratory at Wright-Patterson Air Force Base, Dayton, Ohio. The study report, which was issued at the

[12] The Big Ear Falls Silent, *Sky and Telescope*, July 1998, p. 22.

[13] Rome is a city in New York State.

end of 1958, assumed the use of existing, state-of-the-art components. It proposed a radar with a peak output of 2.5 megawatts, or a continuous power of 150 kw, that would operate at about 400 MHz (λ 75 cm).[14] The antenna would be a fixed parabolic dish 1,000 ft in diameter with a horn feed at the top of a 500 ft high vertical tower. However, by this time experimental work by Kenneth Bowles of the National Bureau of Standards, had shown that such a large dish was not required for ionospheric research using incoherent scatter. But by then the other possible uses of such a large diameter antenna had been identified, so the diameter remained at 1,000 ft.

In the meantime Gordon had tried to find funds to develop the design of this antenna and to build it. Although the Office of Naval Research (ONR) had helped to provide study funds neither they nor the National Science Foundation (NSF) had the sort of money available to build the large antenna.[15] Eventually he persuaded the newly established Advanced Research Projects Agency (ARPA), an arm of the Department of Defense (DoD), to help. Although Gordon didn't realise it at the time, ARPA was then involved in a programme to research, develop and build a state-of-the-art defence against Soviet missiles. They saw Gordon's proposed radar as one possible element of this.

As far as location for the proposed dish was concerned, it was clear from early on that it should be located in the tropics where the Sun, Moon and planets would cross the sky virtually overhead. Donald Belcher of Cornell University, and an expert in aerial photographic interpretation, led the site search which included locations in Hawaii, Mexico, Cuba, Puerto Rico and some small Caribbean islands. It would clearly simplify the support system for this large antenna if it could be built in a natural bowl on the earth's surface and Puerto Rico had a number of sinkholes that could be suitable (see Figure 22.6). This country was also relatively close to the USA and was stable politically. As a result it was concluded that a site in Puerto Rico was the most suitable, based on geographical, political and technical considerations. At one stage it was suggested that, although a 1,000 ft diameter antenna should be built initially, a site should be found that could eventually take a 1,500 ft diameter dish. The best site to allow for this was located about 18° north of the equator some 12 miles (20 km) south of the city of Arecibo. Consequently that is where the 1,000 ft dish was eventually built.

As mentioned above it had initially been assumed that the 1,000 ft diameter dish would be a fixed paraboloid with a vertical axis. But if the Sun was to be observed

[14] This was later changed to 430 MHz (λ 70 cm) in the final design.

[15] The US Navy were in the process of trying to build the massive 600 ft diameter Sugar Grove antenna at the time which also required substantial funding. In fact a number of people questioned why there was a need for the Cornell dish at all when this large Navy antenna, which was fully steerable, was in the process of being built. But the Sugar Grove antenna was cancelled in 1962, well before its construction was finished, as it had run into too many problems and was too expensive (see Section 22.1).

Figure 22.6. The sinkhole used to build the Arecibo radio telescope in the early 1960s. (Courtesy of the NAIC-Arecibo Observatory, a facility of the NSF.)

by radar, the feed at the focal point would have to move by ±2° off axis to cover the movement of the Sun in the ~16 minute round trip that the radar waves would have to and from the Sun. To observe Jupiter by radar the feed would have to move much more. But even if the feed could be moved laterally, the signal from a paraboloid would be degraded. On the other hand a spherical reflector would not have this problem, although it would suffer from spherical aberration. At this stage Ward Low of the Institute for Defense Analysis, and an ARPA advisor, introduced Gordon to the Air Force Cambridge Research Laboratories (AFCRL) who were to monitor the contract for the ARPA. Their antenna group had been studying spherical reflectors for over a decade. Consequently AFCRL changed the design of the Cornell antenna from a paraboloid with a fixed feed at the top of a high tower to a spherical antenna with a moveable feed mounted on a suspended platform.

In November 1959, Cornell University were contracted by the DoD to further study the radar system and to prioritise the early experiments that the system could be used to undertake. Eight months later, Cornell was awarded the contract to build this Arecibo radio telescope, or DoD's Ionospheric Research Facility as it was called at the time. In turn Cornell awarded separate subcontracts for the antenna, electronics, support towers and so on, whilst the Army Corps of Engineers supervised the telescope's construction. On completion William Gordon was appointed director of the Arecibo observatory with Gordon Pettengill, who had carried out radar astronomy experiments at the Lincoln Laboratory, as associate director. It had cost $8.3 million to build.[16]

The Arecibo telescope, which became operational in 1963, initially had a dish made of half-inch (12 mm) square wire mesh held in place by cables suspended from the rim of the dish.[17] Observations could be made up to 20° from the zenith using a feed system carried by a platform which was suspended above the dish by cables attached to three towers each ~300 ft (90 m) high.[18]

A few years after the telescope had been completed it was subjected to high winds from the edge of a nearby hurricane.(21) Much to everyone's surprise the structure moved very much less than anticipated. As a result it was realised in late 1966 that the telescope's structure was sufficiently stable to enable it to operate up to about 3 GHz (λ 10 cm) provided the dish was resurfaced to make it more accurate. Thus was born the idea of converting the Arecibo radar system to an S band (2 to 4 GHz) radar facility. But Arecibo had been suffering since its completion with an inefficient antenna feed, and, unfortunately, ARPA funding was now in short supply because of other calls on its budget. In addition, ARPA was really only interested in Arecibo's use for ionospheric research. But since the facility had been completed it had been used mainly for astronomical research. Consequently the idea of ARPA providing the funding to improve Arecibo's radar capability fell on stony ground. Unfortunately at that time the ARPA/Cornell contract did not allow Cornell to look for finance elsewhere, although ARPA did indicate that they were prepared to share financing the facility with the NSF.

At this time Thomas Gold was the joint director of the Center for Radiophysics and Space Research (CRSR) who were responsible for the administration of the Arecibo facility. So with ARPA's blessing Gold with Frank Drake, who was then the Arecibo director, approached the NSF to see if they could be persuaded to help finance the Arecibo facility. To help their case Gold, Drake and Rolf Dyce[19] also presented their proposal for Arecibo modifications to the Dicke panel which had been set up by the NSF to advise on which new radio astronomy facilities to finance. In August 1967 the Dicke panel gave the Arecibo modifications high priority for implementation.

As a result the NSF were prepared to consider helping to fund the Arecibo modifications, but there was no point in proceeding with these modifications until the feed problem had been resolved. Consequently Cornell undertook, using in-house money, to see how a suitable feed could be produced. In the meantime the major

[16] Arecibo Observatory in Puerto Rico, *Sky and Telescope*, December 1963, pp. 315 and 323.
[17] Some Current Programs at Arecibo – I, *Sky and Telescope*, July 1964, pp. 4–8.
[18] The actual heights were 365, 265 and 265 ft. Two of the towers were built on hills, hence their shorter height.
[19] Rolf Dyce was an expert Arecibo radar investigator.

Arecibo modification project was put on hold. Then in 1969 a retractable 35 ft (10.7 m) boom with attached feeds was added to the existing feed platform to allow observations up to ± 24° from the zenith. This enabled Arecibo to observe Cygnus A, Cygnus X and the Andromeda galaxy.[20]

Meanwhile in a series of meetings starting in 1967 it had been agreed that the financing and supervision of the Arecibo operational contract would gradually be transferred from ARPA to the NSF with the NSF taking over full control on October 1 1969. As an interim measure, ARPA was prepared to pay one third of Arecibo's operational budget for two years starting from that date, which represented the proportion of the observatory time expected to be devoted to ionospheric research.

There was considerable discussion between the NSF and Cornell about the management structure of the Arecibo facility both before and after responsibility for the facility passed to the NSF in 1969. As a result Tor Hagfors, who had an excellent background in ionospheric and radar research in both Norway and the United States, was appointed director. In addition in 1971 the name of the facility was changed to the National Astronomy and Ionosphere Center (NAIC).

The money to upgrade the Arecibo radar system had still not been found when the NSF took over as management agency for Arecibo in October 1969. But the start of a solution had been found in January of that year when Harry Hess, who was chairman of the Space Science Board of the National Academy of Sciences, had written to John Naugle of NASA urging NASA, who were planning to launch a radar spacecraft to Venus, to fund the upgrade to the Arecibo radar system. This was estimated to cost about $5 million for the dish resurfacing plus about $2 to $3 million for the radar equipment. At that time Arecibo had already been used to produce radar images of Venus,[21] although they were rather crude, and NASA's plans for a Venus radar spacecraft were still very provisional.[22] And, as Hess pointed out, the costs of modifying the Arecibo facility, which would be able to produce images with a resolution of a few kilometres, would be much less than those of building and launching a spacecraft to do the same thing. NASA was also interested in the S-band radar for their own purposes. So after the usual discussions of who should pay for what, and getting the necessary funding through Congress for both agencies, an agreement was signed in June 1971 between the NSF and NASA. Under this agreement NSF agreed to pay for the antenna resurfacing and NASA agreed to pay for the S-band radar system, whilst the NSF would monitor the contract with Cornell for all the work.

[20] Arecibo's Sky Enlarged, *Sky and Telescope*, June 1970, p. 358.

[21] Venus is covered in thick cloud and so optical observation of its surface from the Earth or Venus orbit is impossible.

[22] NASA's first Venus orbiter, called the Pioneer Venus Orbiter (PVO), with a planetary radar on board did not enter orbit around Venus until December 1978, see Andrew J. Butrica, *To See the Unseen; A History of Planetary Radar Astronomy*, NASA SP – 4218, 1996, pp. 162–176.

Cornell appointed LTV Electrosystems of Dallas in November 1971, later called E-Systems, to resurface the reflector with almost 40,000 individually adjustable perforated aluminium panels to enable the telescope to operate down to a wavelength of about 6 cm (5 GHz frequency). This was an order of magnitude better than with the original surface. At about the same time a new power supply was installed together with an S-band radar transmitter operating at 2,380 MHz (λ 12.6 cm) and two refrigerated maser receivers provided by JPL.[23] At the time Varian was developing a 1 megawatt continuous-wave klystron transmitter, but NAIC decided against using it as existing lower power klystrons had a proven reliability and spares were easy to obtain. So it was initially planned to use a 800 kw transmitter using two klystrons, but after further thought NAIC decided that a 450 kw transmitter was sufficient for scientific purposes. The net result of all these modifications was to increase the system's total sensitivity by three orders of magnitude. When they had been completed by 1975 they had cost about $9 million.[24]

In the meantime Tor Hagfors had left Arecibo to return to Scandinavia before returning in 1982 as director of the NAIC. Shortly after his return he commissioned a number of investigations into how to improve the Arecibo facility. One of the first targets was to eliminate the spherical aberration inherent in the initial choice of a spherical main dish. To carry out the design work Hagfors recruited Sebastian von Hoerner from NRAO as a consultant.

The idea of having a Gregorian subreflector to correct for the spherical aberration had been around for some time. But when von Hoerner looked into a possible design he soon realised that it would be beneficial to also have a tertiary reflector. In addition he decided to build a geodesic dome around this secondary/tertiary system to protect the antennae and block any external radio interference. The installation of this subreflector system would also have the advantage of being able to replace the line feeds on the suspended platform by normal horn feeds mounted on a rotating platform at the focus of the tertiary reflector. To try out the proposed Gregorian configuration the NAIC built, at their own expense, a smaller version which illuminated a 350 ft (107 m) diameter area of the main dish.

When completed the full Gregorian system for the 1,000 ft dish had a 72 ft (22 m) diameter secondary and a 30 ft diameter tertiary reflector enclosed in a 100 ft diameter geodesic dome (see Figure 22.7). A complete new system of receivers was also installed in the dome to cover the frequency range from 300 MHz up to 10 GHz (λ 1 m to 3 cm).(22)

Other elements of the Arecibo upgrade included a 16 m high ground screen which was installed around the dish perimeter to reduce noise, and a new 1 megawatt

[23] These maser receivers were spares from JPL's Deep Space Network programme.

[24] Arecibo's Giant Radio Telescope Upgraded, *Sky and Telescope*, March 1975, pp. 140–142 and p. 146.

Figure 22.7. Receivers for the Arecibo radio telescope are in this movable geodesic dome suspended above the telescope's 1,000 ft (305 m) dish. (Courtesy of the NAIC-Arecibo Observatory, a facility of the NSF.)

S-band radar transmitter. New computers and twenty-six new electric motors were installed to control a number of new drive systems.[25] In addition the surface panels of the primary antenna were readjusted to have an RMS accuracy of better than 2 millimetres to improve its efficiency at the highest frequencies. But another idea of increasing the diameter of the main dish[26] from 1,000 to 1,200 ft was scrapped. The total cost of the upgrades, which were completed by 1997, was about $25 million, with the costs split $14 million/$11 million between the NSF and NASA.

The Arecibo radio telescope (see Figure 22.8) has been used for a wide number of research projects, including imaging the Moon, planets and asteroids using radar, tracking near-earth asteroids, detecting and observing pulsars, and the Search for ExtraTerrestrial Intelligence (SETI). In 1965 Gordon Pettengill and Rolf Dyce used the telescope to determine the rotation period of Mercury for the first time, and three years later Richard Lovelace and colleagues used the facility to discover the periodicity of the Crab pulsar. In 1975 Joe Taylor and Russell Hulse discovered

[25] These motors were to drive the Gregorian dome, the tertiary reflector and the rotating floor that carried the feed horns, amongst other things.

[26] *Sky and Telescope*, August 1985, p. 106.

Figure 22.8. The Arecibo radio telescope following modifications in the late 1990s. The new 16 m high ground screen can be seen around the edge of the dish. (Courtesy of the NAIC-Arecibo Observatory, a facility of the NSF.)

the first binary pulsar, PSR 1913+16, and eight years later Don Backer and colleagues discovered the first millisecond pulsar, PSR 1937+214, using Arecibo. Then in 1990/91 Alex Wolszczan and Dale Frail found evidence of two planets orbiting the millisecond pulsar PSR B1257 + 12 by measuring timing variations. More recent observations of Mars with the Arecibo telescope showed lava flows and near-surface geological features not seen in visible images.

Funding of the Arecibo Observatory was gradually reduced in the first decade of the twenty-first century. For example, NASA reduced its support of planetary radar between 2001 and 2006, although it restored its funding in 2010. In the meantime the NSF had reduced its funding starting in 2007. Since then the government of Puerto Rico provided some funding and the NAIC received extra finance subsequent to the American Recovery and Reinvestment Act of 2009.

In 2010 the NSF issued a call for proposals for the management of the NAIC and in May 2011 they advised Cornell University that as from October 1 2011 it would no longer be the operator of the Arecibo facility. Instead the NSF signed a five year cooperative agreement, valued at about $42 million, with a consortium led by SRI International which included Universities Space Research Association (USRA) and Universidad Metropolitana (UMET) de Puerto Rico to manage, operate and maintain the facility.[27]

One of the Arecibo suspension cables was damaged by an earthquake in January 2014 at a splice made during its original construction in 1962. Fortunately, the splice did not part completely and the damage was repaired in just two months.

22.7 Algonquin Radio Observatory

Arthur Covington had started radio frequency observations of the Sun from the Ottawa Radio Field Station of the Canadian National Research Council (NRC) in mid 1946 using ex-radar equipment.(23) He used a 4 ft (1.2 m) parabolic antenna operating at a frequency of 2.8 GHz (λ 10.7 cm) to try to detect Reber's galactic radiation which Reber had detected at lower frequencies. But Covington was unable to detect any radiation from the centre of the Milky Way at 2.8 GHz, neither could he detect any from the Moon, Mars or Jupiter.(24) The Sun, on the other hand, gave a much higher signal than expected. Later that year he used the occultation of the Sun by the Moon during a partial solar eclipse to show that radio emission from the Sun could be divided into two components, one from the Sun as a whole and one from sunspots. Routine daily observations of the Sun were begun the following year, but interference from other equipment at the field station caused Covington to move his equipment in 1948 about 5 miles (8 km) to what became known as the Goth Hill radio observatory. There he had a 150 ft (46 m) long slotted waveguide collector placed in a metal trough to produce an 8 arcmin fan beam which he used to observe the Sun starting in 1951. Shortly afterwards this antenna was extended along its axis by a waveguide collector of the same design which became available in January 1956.(25) Two years later this two-element simple interferometer was replaced by a four element interferometer that had a beamwidth of just 1.2 arcmin.

Although the move to Goth Hill had solved a number of issues, increasing problems with radar and radio interference in the Ottawa area in the 1950s resulted in another new site having to be found for radio astronomy. This was at Lake Traverse in Algonquin Park, about 100 miles (160 km) from Ottawa. In 1959 the Algonquin Radio Observatory was inaugurated there and three years later it became Canada's national radio observatory.

Covington, the first director of the Algonquin Radio Observatory, had installed a 6 ft (1.8 m) parabolic dish at Algonquin in 1960. Six years later a new radio telescope was completed there which consisted of an array of thirty-two 10 ft (3 m) parabolic dishes arranged over a common 700 ft (215 m) long waveguide.

[27] SRI International to Manage Arecibo Observatory, *SRI press release*, June 2 2011.

Figure 22.9. The 150 ft (46 m) diameter dish of the Algonquin Radio Observatory. (Courtesy Bquine via Wikimedia Commons.)

Meanwhile in 1961 Algonquin had been selected as the site for the construction of a 120 ft (37 m) diameter fully steerable radio telescope dish. By the following year the required dish size had been increased to 150 ft (46 m). Construction of the telescope, which was designed to study galactic and extragalactic objects down to a wavelength of 2 m (15 GHz frequency), began in 1964. It was designed for the NRC by Freeman Fox and Partners of London, who had designed the very similar Parkes antenna in Australia (see Section 16.2), and was constructed by Dominion Bridge Co. of Montreal. The central 37 m diameter of the 46 m dish was made of solid metal plates whilst the remainder of the dish consisted of wire mesh (see Figure 22.9). The 46 m dish was later joined by an 18 m telescope which had originally been located at the David Dunlap Observatory outside of Toronto.

Shortly after its completion, the 46 m was used in conjunction with the 26 m dish at the Dominion Radio Astrophysical Observatory (DRAO), see below, about 1900 miles (3100 km) away at Penticton to create the world's first successful very long baseline interferometer (VLBI). The observations from both radio telescopes were recorded on magnetic tapes, and the time signals recorded using atomic clocks.

The NRC looked into the possibility of modifying the 46 m in the 1980s to enable it to operate down to a wavelength of about 3 mm. But in 1987 they decided, instead, to participate in the James Clerk Maxwell Telescope (JCMT) on Mauna Kea (see Section 21.2) which had been designed to operate below 1 mm. At about the same time they offered the IEI use of the 18 m dish and the 32 dish solar observatory for their SETI project called project TARGET.[28] But in 1991 further budget cuts forced the NRC to close down the Algonquin Radio Observatory. The 1.8 m dish and the 32 dish solar observatory were later moved to a new site near Shelburne, Ontario.

Since 2008 the Algonquin observatory has been operated by Thoth Technology to provide telemetry, tracking and control services for spacecraft and geodetic and radio astronomy services using the 46 m antenna. In addition the observatory's hydrogen maser maintains a highly accurate time standard to facilitate VLBI experiments.

22.8 Dominion Radio Astrophysical Observatory

In 1956 Carlyle Beals, the Dominion Astronomer in Canada, suggested that the Observatories Branch of the Department of Mines and Technical Surveys be allowed to enter the new field of radio astronomy. Amazingly, just seven months later he had the approval of the Treasury Board.(26) A site survey carried out by Jack Locke and Ed Argyle soon found an ideal location with very low levels of radio interference at White Lake, near Penticton, British Columbia, just north of the US boarder. This was accepted as the observatory site by Beals in 1957, and two years later the first telescope was delivered, an 84 ft (26 m) diameter fully steerable parabolic dish (see Figure 22.10) designed to operate at the 21 cm (1.4 GHz) line of neutral hydrogen. The telescope was then erected and commissioned and the Dominion Radio Astrophysical Observatory (DRAO) was inaugurated in 1960.

As mentioned above (in Section 22.7) this telescope was used in conjunction with the 150 ft radio telescope of the Algonquin Radio Observatory in 1967 to produce the world's first successful very long baseline interferometer (VLBI). Following later modifications, the Penticton 26 m, since called the John A. Galt Telescope,[29] can now be operated at various frequencies from 408 MHz to 8.4 GHz, including the hydroxyl lines around 1.6 GHz and the methanol line at 6.6 GHz.

A 6 ft (1.8 m) dish, like that at the Algonquin Radio Observatory, was installed at Penticton in 1965 to complement the solar observations obtained at Algonquin.

[28] TARGET is an acronym of 'Telescope Antenna Researching Galactic Extraterrestrial Transmissions'.

[29] Named after the DRAO's first employee who was responsible for commissioning the 26 m and who was the director of the observatory from 1963 to 1980.

Figure 22.10. The 84 ft (26 m) diameter dish of Canada's Dominion Radio Astrophysical Observatory. (Courtesy Jason Nishiyama via Wikimedia Commons.)

Penticton later took over Algonquin's responsibility for solar monitoring in 1991 when the Algonquin observatory was closed down.

Immediately after the inauguration of the Penticton observatory in 1960 Carman Costain, who had recently arrived from Cambridge, suggested that the observatory should have a low frequency telescope. This was to carry out observations during the expected solar minimum of the mid 1960s which would be an optimum period for observing at low frequencies. As a result a T-shaped array was built over the period from 1962 to 1964 covering an area of 65,000 m² to operate at 22 MHz (λ 14 m).(27) Its main task was to produce a map of radio sources in the Milky Way, as well as allowing observations of a number of specific sources, particularly quasars, in more detail.

In the late 1960s an interferometer was built at Penticton consisting of two 8.5 m diameter dishes operating at 1.4 GHz over a baseline of 300 m. This interferometer was later extended in stages to finally have seven similar sized dishes operating at 1.4 GHz and 408 MHz over a baseline of 600 m.(28) Called the Synthesis Telescope this interferometer produced high-resolution imaging of discrete sources and mapping of the galactic plane in the Canadian Galactic Plane Survey from 1995 to 2005.

Since then Canadian engineers began construction in 2013 of a completely new type of telescope at the DRAO which will consist of five side-by-side cylindrical reflectors each 20 m wide by 100 m long. This $11 million CHIME (Canadian Hydrogen Intensity Mapping Experiment) telescope, which will operate over the frequency range from 400 to 800 MHz, has been designed to measure baryon acoustic oscillations at redshifts of between 0.8 and 2.5 to probe dark energy.(29) It is a joint project of the University of British Columbia, McGill University (of Montreal), the University of Toronto and the DRAO.

22.9 Hat Creek

The University of California's Hat Creek radio observatory was founded in 1959 in the Lassen National Forest in northern California. This location was chosen after a search led by Harold Weaver, professor of astronomy at the university's Berkeley campus, who became the observatory's first director. About thirty possible locations were investigated in the northern part of the state before selecting Hat Creek at an altitude of about 990 m (3,200 ft). It was virtually surrounded by mountains which shielded it from radio interference.[30] The Office of Naval Research agreed to provide 70% of the observatory's initial capital cost of $500,000 and also agreed to share its operating costs with the University of California.[31] The first two telescopes, both of which had equatorial mounts and dishes made of sheet aluminium, were a 33 ft (10 m) diameter parabolic antenna, which was completed in 1959 and an 85 ft (26 m) dish that was completed in 1962.[32] The 85 ft telescope, which was

[30] Radio Observatory at Hat Creek, California, *Sky and Telescope*, August 1962, pp. 64–66.
[31] *Sky and Telescope*, November 1959, p. 3.
[32] By then the capital cost of the observatory had increased to $1.2 million.

designed to operate in winds of up to 50 miles/hour and withstand winds of up to 100 miles/hour, could operate down to a wavelength of 1 cm.

Initially both telescopes used the observatory's three receivers, two of which were used to observe frequencies in the region of the 1.4 GHz (21 cm) hydrogen line. The third receiver operated in the 7.5 to 8.5 GHz band but, because of the high surface accuracy required of the dish surface at these frequencies, it could not be operated in windy weather. In 1965 Weaver and colleagues discovered emission at 1665 MHz with the 85 ft telescope(30) which was later shown to be caused by maser radiation emitted by hydroxyl molecules. This was the first astrophysical maser to be discovered. In January 1993, the 85 ft dish collapsed in a wind storm shortly after it had been refurbished.[33] It was not rebuilt.

Meanwhile a 20 ft (6.1 m) diameter antenna had been built at Hat Creek in the 1960s with a highly accurate surface to enable it to operate in the wavelength range from 2 cm to 8 mm. In 1968 it was the first radio telescope to detect interstellar ammonia and water vapour (at a wavelength of 1.3 cm).(31) Then in the early 1980s a three element millimetre-wave interferometer was built at Hat Creek which was progressively modified and extended over the next two decades to produce the Berkeley-Illinois-Maryland Association (BIMA) interferometric array(32) (see Figure 22.11) working down to a wavelength of about 1 mm. BIMA was formally dedicated in 1996 but eight years later it was decided to move it to Cedar Flat, near Bishop, California where it formed part of the CARMA (Combined Array for Research in Millimeter-wave Astronomy) millimetre array (see Section 21.1).

Allen Telescope Array

In 1999 the SETI Institute and the University of California at Berkeley announced plans to build an array of 500 to 1,000 three to five metre diameter radio telescope dishes having a total collecting area of about one hectare (2.5 acres).[34] It was expected that this $25 million One Hectare Telescope (1HT), as it was called at the time, would be used for both SETI and for conventional radio astronomy. The receivers were to be designed to observe at any frequency between 0.3 and 12 GHz (λ 1 m to 2.5 cm) and the array was to be able to observe up to 100 different objects simultaneously. Clearly the challenge here was not the design of the dishes but the computer system required to analyse the observational data. It was expected that the array would be built at Hat Creek.

[33] *Sky and Telescope*, April 1993, p. 8.
[34] Giant SETI Radio Telescope Planned, *Sky and Telescope*, July 1999, p. 22.

Figure 22.11. The BIMA array at Hat Creek before its move to Cedar Flat. (Courtesy Marc Pound, University of Maryland.)

The first significant funding of \$11.5 million was provided in 2001 for this new array by the Paul G. Allen Family Foundation (named after the co-founder of Microsoft). Consequently the 1HT was renamed the Allen Telescope Array (ATA). Over the next few years the design of the system to be built at Hat Creek was gradually modified, although it still remained both a SETI array and one to carry out normal astronomical observations. For example the design of antenna was changed from that of a commercial TV-type dish about 3 to 5 m in diameter to a 6.1 × 7.0 m hydroformed offset Gregorian with a 2.4 m subreflector. The observational waveband was changed to 0.5 to 10 GHz (60 to 3 cm) and the number of dishes expected to form the array was reduced to 350, with an initial complement of 42 antennae being built once the BIMA array had been removed from the site. Unfortunately although this initial 42 dish array was completed at Hat Creek by 2007, only two sources could be monitored simultaneously.(33) Routine observing started two years later although further construction had to be put on hold pending the raising of additional funds. At that stage it had cost about \$50 million to build the 42 antenna system, about 50% of which had come from the Allen Foundation. Another $50 million was required to enable the full 350 antenna system to be completed.

Unfortunately not only was there not enough money to expand the existing ATA, but shortly afterwards it became clear that there was not enough money to operate what was already there. So in April 2011 the ATA had to be shut down. However, following an appeal for private donations, operation was able to be restarted in December 2011. Then in 2012 the University of California at Berkeley withdrew from the project and handed over management and operation of the ATA to SRI International who already managed and operated the Arecibo facility. In the same year Franklin Antonio, co-founder of QUALCOMM, provided $3.6 million to allow the receivers to be updated to provide greater sensitivity over a wider frequency range.

22.10 Five College Radio Astronomy Observatory and the Mexican–American Large Millimeter Telescope

The Five College Radio Astronomy Observatory (FCRAO), located near Amherst, Massachusetts, was founded in 1969 by the state-supported University of Massachusetts and four small private colleges (Amherst, Hampshire, Mount Holyoke and Smith). As its first significant radio telescope Richard Huguenin, who had just joined the University from Harvard, suggested building a telescope 'in house' to observe pulsars at metre wavelengths. It would consist of thirty-two 120 ft (37 m) diameter, fixed, spherical dishes of wire mesh with movable feeds, acting together. This assembly would have about half the sensitivity of the 1,000 ft (305 m) Arecibo radio telescope in Puerto Rico.[35] As a start he proposed building an array of four such 120 ft dishes to provide a sensitivity comparable to that of the 250 ft (76 m) Jodrell Bank radio telescope in the UK. The site chosen was on state-owned land in Massachusetts' Quabbin Reservation and initial funding was provided by the NSF, the Five Colleges and private benefactors.[36] In addition Huguenin, who was appointed director of the FCRAO in 1970, obtained large amounts of government surplus equipment and persuaded a number of companies to provide equipment free of charge. The first antenna, built mainly by students and observatory staff, was completed in November 1970, and the first observations of pulsars were made in the following month. The four antenna pulsar array was

[35] William M. Irvine, *Reflections on the Growth of Astronomy at the University of Massachusetts and the Five College Astronomy Department*, www.astro.umass.edu/media/dept_file/history.pdf.

[36] The NSF provided a grant of $75,200, the Fleishman Foundation donated $53,000, the Sloan Foundation $20,000, and the Land Foundation $1,500. The University of Massachusetts provided more than $30,000, Amherst College about $20,000, Hampshire College $27,900, and Mount Holyoke and Smith Colleges $2,000 each. In addition, the private Research Corporation provided $73,000 in 1970 to Huguenin and Joseph Taylor for their research on pulsars.

completed shortly afterwards. Unfortunately it was never extended beyond four dishes(34) and this pulsar array was decommissioned in 1981.

In the meantime there had been an increasing interest worldwide on astronomical observations in the centimetre and millimetre wavebands to detect complex interstellar molecules. In particular Olof Rydbeck, Director of Chalmers University of Technology's Onsala Space Observatory in Sweden (see Section 23.2) had begun negotiations with the Electronic Space Systems Corporation (ESSCO) of West Concord, Massachusetts to procure a large millimetre-wave radio telescope. At about the same time one of Rydbeck's former PhD students, Sigfrid Yngvesson, an expert in maser receivers operating at these wavelengths, had joined the University of Massachusetts' Electrical and Computer Engineering Department. So the telescope manufacturer (ESSCO) and Rydbeck's former student and receiver expert were both in Massachusetts. Consequently Rydbeck decided to spend a sabbatical year in Amherst during 1972–1973.

By this stage ESSCO had already built a 14 m diameter millimetre wave dish which they offered to Rydbeck's Onsala observatory pending the delivery of the 20 m antenna that Onsala required. This and other options for using the 14 m were considered at a meeting in December 1972 between Rydbeck, Huguenin and Albert Cohen, the president of ESSCO. At this meeting Cohen suggested that ESSCO provide Huguenin with a complete telescope system based on the 14 m dish within a protective radio-transparent radome for observations down to 1 mm (300 GHz frequency). The cost would be about $400,000 of which ESSCO might donate $150,000.

Cohen's offer stopped Huguenin and William Irvine, of the University of Massachusetts' physics and astronomy department, from investigating the building of more fixed 120 m dishes. It persuaded them to look, instead, at purchasing the proposed 14 m millimetre-wave telescope that would be the largest in the United States operating at such short wavelengths. But Huguenin estimated that, in addition to the purchase cost of the ESSCO system, another $400,000 would be required for instrumentation and a two year commissioning phase. As a result a request was submitted to the NSF in April 1973 for $750,000 to include the cost of the telescope, instrumentation and commissioning of the new facility. The request was accepted allowing the FCRAO to purchase the 14 m telescope from ESSCO. Construction was completed at the FCRAO observatory in 1976.

This FCRAO observatory was located at a low altitude, which was normally a problem for millimetre-wave observations because of atmospheric absorption. But it was found that the 14 m could operate at these wavelengths on cold winter days when the water vapour was largely frozen out of the atmosphere. In the event the 14 m radio telescope was used to study the physics and chemistry of interstellar clouds, circumstellar envelopes, planetary atmospheres and comets. It was, in 1987, the first telescope to detect phosphorus nitride in the interstellar medium. In the same year its original radome, which had a transparency of 60% at the telescope's shortest operating wavelength, was replaced by a Gore-Tex radome which had a transparency of 95%.

By this time in the late 1980s the 14 m telescope was becoming non-competitive internationally compared with other, more modern millimetre-wave instruments. As a result, the FCRAO began to consider ways of replacing it with a 50 m diameter, state-of-the-art, radome-protected, millimetre-wave dish, called the Large Millimeter Telescope (LMT), that would be at least as good as the 45 m Nobeyama Radio Observatory instrument in Japan. ESSCO estimated in 1989 that it would cost about $28 million.

In a parallel development the Mexican government had successfully applied for a large loan from the World Bank to significantly strengthen its scientific and technical infrastructure. Stephen Strom, the Chair of the Five College Astronomy Department, subsequently contacted Alfonso Serrano, Director of the Institute of Astronomy at the Universidad Nacional Autónoma de México (UNAM) and Luis Carrasco, Director of the Observatory at San Pedro Martir, and suggested that Mexico might like to collaborate with the University of Massachusetts on either the LMT or on a new optical-infrared telescope. The Mexicans chose the LMT. As a result the FCRAO built a radiometer to test the suitability of the San Pedro site for observations at millimetre wavelengths.

Meanwhile Serrano had been made head of a new division of UNAM called PUIDE[37] concerned with space science which became the lead Mexican institution for the LMT. But then in 1992 he was appointed General Director of the Instituto Nacional de Astrofísica, Óptica y Electrónica (INAOE) in the province of Puebla. Unfortunately INAOE and the Institute of Astronomy at UNAM, who had decided to push for a new Mexican astronomical facility based on a large optical telescope, were rival institutions. As a result UNAM, which had the largest group of astronomers in Mexico, pulled out of the LMT project, leaving INAOE to support it alone. INAOE submitted a proposal to CONACyT,[38] the Mexican equivalent of the NSF, in 1992 for their share of the proposed funding for the joint Mexican–American LMT and were awarded a grant equivalent to $14,680,000 early the following year.

In the meantime the University of Massachusetts had received funding from their state legislature to develop technology relevant to the LMT. Unfortunately, this LMT project had not been supported by the NAS decennial review of 1991 chaired by John Bahcall, so trying to get financial support from the US Congress was an uphill struggle. However Michael Hooker, who had been appointed

[37] PUIDE is an acronym of Programa Universitario de Investigacion y Desarrollo Espacial.

[38] CONACyT is an acronym of COnsejo NAcional de Ciencia y Technologia.

President of the University of Massachusetts in 1992, was fully supportive of the LMT project, promoting it in the higher political levels of the US Congress and the State of Massachusetts. This included meeting with Senator Edward Kennedy in 1993 to discuss the project which was, unfortunately, unsuccessful in receiving his support. Nevertheless, by September 1994 it appeared as though a breakthrough was on the horizon for the project in the form of a contract between the University of Massachusetts and the Department of Defense's Advanced Research Projects Agency (ARPA) who were interested in the possibility of using the LMT as a millimetre wavelength radar.(35)

Unfortunately at this time relations between the University of Massachusetts and ESSCO had broken down not aided by a substantial increase in the ESSCO financial estimate. But, following a meeting of all interested parties, Cohen concluded that ESSCO and the University would be able to complete the project within the ARPA ceiling of $42.8 million (which included the money to be provided by the state, University and Mexico). The project was still not out of the wood, however, as major disagreements existed between the University and ESSCO on the terms of the ESSCO contract which included a definition of the working interface between ESSCO and INAOE.

At this stage Hooker gave responsibility for his role in the LMT project to Allen Sessoms, university Vice-President, with instructions to find an accommodation with ESSCO. Subsequently the University caved in and Sessoms even signed the contract without the agreement of the University's Mexican partners.[39] This naturally upset the Mexicans as it required them to provide detailed technical information to ESSCO which they had not been consulted about. Irvine also resigned as Project Scientist in protest at the signature of the ESSCO contract to be replaced by Steve Strom. In addition Hooker resigned as President of the University followed shortly afterwards by Allen Sessoms.

The relationship between the University and ESSCO had not been helped by the patchy performance of ESSCO in undertaking their previous contract nor by the suggestion that came from some experts that the ESSCO radome be scrapped and the telescope be operated in the open air. Then in the summer of 1995 contracts were let by the University to ESSCO for the preliminary design of the radome and reflector panels and by INAOE to TIW/SGH[40] for a conceptual design of the LMT in a radome. In addition Vertex, who (as Krupp) had with MAN built the 100 m Effelsberg telescope in Germany (see Section 23.1), was given a feasibility study

[39] William M. Irvine, *Reflections on the Growth of Astronomy at the University of Massachusetts and the Five College Astronomy Department*, p. 49, www.astro.umass.edu/history.pdf.

[40] TIW/SGH was an acronym of Toronto Iron Works of Santa Clara, California and Simpson Gumpertz and Heger of Waltham, Massachusetts.

contract for a telescope in the open air. Then in early 1997 MAN submitted an unsolicited proposal for an open air telescope.

Eventually in mid 1997 Requests for Quotations were sent out to ESSCO, TIW, MAN, Vertex and RSI to bid for the critical LMT design contract. They were given the option to propose either an open air telescope or one in a radome. Responses were received from TIW (who had just been taken over by the Vertex Corporation) for an enclosed dish, MAN for an exposed telescope, and Vertex Antennentechnik (the German subsidiary of the Vertex Corporation) also for an exposed telescope. Eventually it was decided that the contract should be awarded to MAN, resulting in contract signature with them in December 1997. The contractual problems with ESSCO were subsequently resolved with ESSCO early the following year.

In the meantime a site had been chosen for the LMT on the 4,600 m (15,000 ft) summit of Sierra Negra, an extinct volcano in Mexico, at a latitude of just 19°. One reason for choosing such a low northern latitude was that it would allow an excellent view of the southern sky including the centre of the Milky Way.(36)

Strom guided the project for a number of difficult years in the 1990s until he also resigned from the University in 1999 following a breakdown in communications between him and the University's Chancellor, David Scott. Unfortunately inter-personnel problems continued to dog the project until the telescope achieved first light at centimetre wavelengths in 2006. Subsequently precision surface segments were installed over the inner 32 m diameter to allow millimetre-wave observations in June 2011.[41] Then Alfonso Serrano died in the following month and shortly afterwards the telescope was named the 'Large Millimeter Telescope Alfonso Serrano'.

Over time the LMT (see Figure 22.12), which was initially sold as a joint Mexican–American project with the costs split equally, had gradually become a largely Mexican telescope with a minority University of Massachusetts financial interest. In addition a telescope which was initially planned to operate over its full 50 m aperture down to 0.8 mm by the year 2000 had by 2015 still not reached that capability, reaching, instead only a wavelength of 2.5 cm over full aperture in 2006, and 3 mm over an aperture of just 32 m in 2011. By 2012 its costs of about $130 million were also much higher than originally estimated, even allowing for inflation. On the other hand, when the LMT is finished, it will be a remarkable telescope for a country which, when the project was envisaged some decades earlier, had no experience in millimetre-wavelength astronomy.[42]

[41] The 14 m FCRAO telescope was also decommissioned in 2011.

[42] Mexico's Large Millimeter Telescope Opens for Business, Science Insider May 8 2013, *American Association for the Advancement of Science*, www.aaas.org/.

Figure 22.12. The Large Millimeter Telescope or Gran Telescopio Milimétrico built on Sierra Negra in Mexico. (Courtesy Luyten via Wikimedia Commons.)

References

1. Sullivan, Woodruff T., *Cosmic Noise; A History of Early Radio Astronomy*, Cambridge University Press, 2009, p. 207.
2. Hey, J. S., *The Evolution of Radio Astronomy*, Science History Publications, 1973, pp. 84–85.
3. McClain, Edward F., A High-Precision 85-foot Radio Telescope, *Sky and Telescope*, July 1966, pp. 4–6.
4. Malphrus, Benjamin, *The History of Radio Astronomy and the National Radio Astronomy Observatory; Evolution Toward Big Science*, Krieger Publishing, 1996, p. 39.

5. Butrica, Andrew J., *To See the Unseen; A History of Planetary Radar Astronomy*, NASA SP – 4218, 1996, p. 29.
6. Price, R., et al., Radar Echoes from Venus, *Science*, 129, No. 3351, pp. 751–753.
7. Butrica, Andrew J., *To See the Unseen; A History of Planetary Radar Astronomy*, NASA SP – 4218, 1996, p. 83.
8. Barvainis, Richard, et al., The Haystack Observatory λ3-mm Upgrade, *Publications of the Astronomical Society of the Pacific*, 105, 1993, pp. 1334–1341.
9. Barvainis, Richard, and Salah, Joseph E., Upgrade Capabilities of the Haystack Radio Telescope, in Clemens, D. P, and Barvainis, R., (eds.), *Clouds, Cores and Low Mass Stars*, Astronomical Society of the Pacific Conference Series, 65, 1994, pp. 411–418.
10. Thompson, A. Richard, The Harvard Radio Astronomy Station at Fort Davis, Texas, *Journal of Astronomical History and Heritage*, 13, No. 1, 2010, pp. 17–27.
11. Maxwell, Alan, Solar Radio Astronomy in Texas, *Sky and Telescope*, June 1958, pp. 388–393.
12. McVittie, G. C., Illinois' Vermilion River Observatory, *Sky and Telescope*, December 1962, pp. 322–325.
13. Swenson, George W., The Illinois 400-foot Radio Telescope, *IEEE Antennas and Propagation Society Newsletter*, 28, 1986, pp. 12–16.
14. McVittie, G. C., University of Illinois Observatory, Urbana, Illinois, Report 1967–1968, *Bulletin of the American Astronomical Society*, 1, 1969, pp. 25–27.
15. Schriefer, Arno H., Yang, Kwang-Shi and Swenson, George W., The Illinois 120-foot Radio Telescope, *Sky and Telescope*, March 1971, pp. 132–138.
16. Hutchinson, Jamie, The Vermilion River Observatory; Two Decades of Radio Astronomy in ECE, *ECE Alumni News*, 39, Spring 2005, pp. 6–11.
17. Kraus, John D., The Large Radio Telescope of Ohio State University, *Sky and Telescope*, July 1963, pp. 12–16.
18. Rinsland, C. P., Dixon, R. S., and Kraus, J. D., Ohio Survey Supplement 2, *Astronomical Journal*, 80, 1975, pp. 759–770.
19. Cohen, Marshall H., Genesis of the 1000-foot Arecibo Dish, *Journal of Astronomical History and Heritage*, 12, No. 2, 2009, pp. 141–152.
20. Butrica, Andrew J., *To See the Unseen; A History of Planetary Radar Astronomy*, NASA SP – 4218, 1996, p. 88.
21. Ibid., p. 94.
22. Altschuler, Daniel R., The National Astronomy and Ionosphere Center's (NAIC) Arecibo Observatory in Puerto Rico, in Stanimirović, Snezana, et al., (eds.), Single-Dish Radio Astronomy: Techniques and Applications, *ASP Conference Series*, 278, 2002, pp. 1–24.
23. Covington, Arthur E., Algonquin Radio Observatory, Lake Traverse, Ont., Canada, *Solar Physics*, 9, 1969, pp. 241–245.
24. Sullivan, Woodruff T., *Cosmic Noise; A History of Early Radio Astronomy*, Cambridge University Press, 2009, pp. 211–213.
25. Covington, Arthur E., Beginnings of Solar Radio Astronomy in Canada, in Sullivan, W. T., (ed.), *The Early Years of Radio Astronomy; Reflections Fifty Years after Jansky's Discovery*, Cambridge University Press, 1984, pp. 317–334.
26. Locke, Jack, The Beginning of the Dominion Radio Astrophysical Observatory, *Journal of the Royal Astronomical Society of Canada*, 92, 1998, pp. 112–114.
27. Costain, C. H., The New 22 Mc/s Radio Telescope of the Dominion Radio Astrophysical Observatory, *Publications of the Astronomical Society of the Pacific*, 74, 1962, pp. 400–401.

28. Landecker, T. L., et al., The Synthesis Telescope at the Dominion Radio Astrophysical Observatory, *Astronomy and Astrophysics Supplement Series*, 145, 2000, pp. 509–524.
29. Newburgh, Laura B., et al., Calibrating CHIME, A New Radio Interferometer to Probe Dark Energy, in Stepp, Larry M., Gilmozzi, Roberto, and Hall, Helen J., (eds.), *Ground-based and Airborne Telescopes V, SPIE Conference Volume 9145*, 2014.
30. Dieter, Nannielou H., Weaver, Harold, and Williams, David R. W., The Interstellar Hydroxyl Radio Emission, *Sky and Telescope*, March 1966, pp. 132–136.
31. Malphrus, Benjamin, *The History of Radio Astronomy and the National Radio Astronomy Observatory; Evolution Toward Big Science*, Krieger Publishing, 1996, pp. 98–99.
32. Welch, W. J., et al., The Berkeley-Illinois-Maryland-Association Millimeter Array, *Publications of the Astronomical Society of the Pacific*, 108, 1996, pp. 93–103.
33. Shostak, Seth, Closing in on E.T., *Sky and Telescope*, November 2010, pp. 22–28.
34. Arny, T., and Valeriani, G., The Five College Radio Astronomy Observatory, *Sky and Telescope*, June 1977, pp. 431–435.
35. Baars, Jacob W. M., *International Radio Telescope Projects; A life among their designers, builders and users*, Jacob W. M. Baars, 2013, p. 112.
36. Irvine, William M., Carrasco, Esperanza and Aretxaga, Itziar, (eds.), *The Large Millimeter Telescope; Neighbors Explore the Cosmos*, University of Massachusetts Press, 2005.

23

Further European and Asian Radio Observatories

23.1 Stockert Observatory and the Effelsberg Radio Telescope

Germany was excluded from the development of radio astronomy in the crucial years immediately after the end of the Second World War as restrictions on radio research were not lifted until 1950. But even then there were still many more important calls on state funding to rebuild the country after the war. So there was no sudden explosion of radio astronomy research in Germany in the early 1950s to match that already underway in Australia, the UK and elsewhere.

But starting in 1952 Friedrich Becker and H. Strahl gave a number of lectures about radio astronomy in local government ministries in Düsseldorf, the state capital of Nordrhein-Westfalen.(1) This resulted in the idea of building a 25 m diameter fully steerable radio telescope which could also be used for radar research. Design studies were carried out by Metallwerk Friedrichshafen and Telefunken who had built the German Würzburg radar antennae during the war. Enthusiastic backing for the proposed project came from Telefunken's former chief engineer Leo Brandt who was then Secretary of State in the Ministry of Economics and Traffic in Nordrhein-Westfalen. With his support 1.2 million DM (about $300,000) was raised to pay for the project which was undertaken by a consortium of companies headed by Telefunken.

The basic requirement of this 25 m radio telescope was that it should be able to detect radio emissions at the 21 cm wavelength (1.4 GHz) of neutral hydrogen and shorter if possible. This meant that its dish should have a surface accurate to about ±5 mm. It was also expected that the dish, which would be supported by an altazimuth mount, would be pointed to an accuracy of 1 arcminute. Naturally, considering the pedigree of the main contractors involved, the actual design was based on that of the Second World War 7.5 m diameter Würzburg radar antennae. The surface panels were of 2 mm thick sheet aluminium with 10 mm square perforations to reduce wind resistence. As far as location of the telescope was concerned,

Figure 23.1. The 25 m diameter telescope (in the background) and the 10 m diameter solar telescope at West Germany's Stockert Observatory. They were built there in the late 1950s and early 1960s, respectively. (Courtesy Stephan Wershoven via Wikimedia Commons.)

its use for radar research required it to be built with a good view of the horizon. Consequently Stockert hill in Nordrhein-Westfalen, about 40 km (25 miles) from Bonn, was chosen for locating the instrument which was to be operated as a radio telescope by the Astronomy Institute of the University of Bonn.[1] It would also be used to carry out radar research by the Institute of High Energy Physics. Construction was started in 1955 and the telescope (see Figure 23.1) became operational two years later.

Initially the conversion from equatorial coordinates to the telescope's altazimuth coordinates was carried out by a mechanical system produced by the firm of Rademacher in the Netherlands. This was based on an invention made at the Collins Radio Company in the United States. But in 1967 the purchase of a Ferranti 400 computer enabled the conversions to be carried out electronically.

In the late 1950s the 25 m telescope was used to produce a survey of the Galactic Plane at a wavelength of 11 cm. Then in 1966 a cooled parametric amplifier, which

[1] The Astronomy Institute of the University of Bonn continued to operate the Stockert site until 1966 when the Max Planck Institute for Radio Astronomy took over the task. Then in 1978 the operation of the site reverted to the University of Bonn.

was purchased from the USA, showed a dramatic reduction in system noise temperature. Initially the dish focused the received radio energy on to a dipole feed but in the mid 1960s this feed was replaced by a four leg support system that allowed either prime focus or secondary focus operations.

Returning to the 1950s, whilst funding was being raised in West Germany for the 25 m, funds had also been found by the Academy in East Berlin, later called the Academy of the German Democratic Republic (GDR), for a 32 m diameter transit radio telescope.(2) This was for the Heinrich Hertz Institute (HHI) at Berlin-Adlershof in East Germany. The telescope, which was the responsibility of Otto Hachenberg, the director of the HHI, could not be designed to sufficient accuracy analytically. So a test rig was built to measure the deformation of the 32 m space frame structure when it was loaded with concrete blocks. It showed that once the deformation of one section was analysed the deflection of the whole symmetrical structure could be predicted.

Then in 1961 the building of the Berlin Wall separated Hachenberg, who lived in West Berlin, from his 32 m telescope whose construction had just been completed in East Germany. Unfortunately Hachenberg was denied access to the 32 m by the East German authorities in spite of representations by the Academy of the GDR. As a result he stayed in the west, and in the following year he was appointed professor of radio astronomy at the University of Bonn and director of the Stockert Observatory with its 25 m radio telescope.

Shortly after his arrival Hachenberg applied for funds to build a 10 m telescope on the Stockert site for millimetre-wave solar research. Construction of this telescope (see Figure 23.1) was completed in 1965. But what Hachenberg really wanted was a very large parabolic radio telescope of high surface accuracy with a design based on his work with the 32 m dish. So, following his move to Bonn, he began to study the design of a fully steerable 80 m to 90 m diameter radio telescope. Initial funding came from the State of Nordrhein-Westfalen.(3) But it was obvious that substantial funds would be required to build such a large telescope which were beyond the resources of the university and the local State. Consequently in 1964 the University of Bonn applied for funding to the Volkswagen Foundation. The foundation's reaction was positive and they granted the university study funding. As a result Hachenberg was able to award study contracts to MAN (the prime contractor for the Parkes Radio Telescope, see Section 16.2) and Krupp for an 80 m to 90 m diameter radio telescope to operate down to a wavelength of 1.3 cm (frequency 23 GHz). Hachenberg was very keen to explain to the contractors the results of his experimental deformation work that had been carried out during the design of the 32 m dish. Since then IBM's FRAN computer programme had become available which indicated that a self-compensating elastic structure could work. Normally telescope dishes of the period were made sufficiently rigid to avoid them deforming as their inclination to the vertical varied. But this 100 m dish would be allowed to deform as, in this case, it would retain its paraboloidal shape with just the focus position changing.

In the meantime Germany's leading research funding organisation, the Deutsche Forschungsgemeinschaft, had issued a report in 1962 on the state of astronomy in West Germany. One of its recommendations was that the country should build a major radio telescope. As a result the Max-Planck-Gesellschaft (MPG) decided to become active in the field of radio astronomy and they set up the Max-Planck-Institut für Radioastronomie (MPIfR) to carry out their programme. The MPG also decided to invite Sebastian von Hoerner, who was then at the NRAO at Green Bank, to become a founding director of this new institute whilst the University of Tübingen decided to offer him a professorship. The MPG also invited Hachenberg to become a founding director of the MPIfR and the University of Tübingen also offered him a professorship if he became a second MPIfR director.

In parallel with the University of Bonn's request of 1964 to the Volkswagen Foundation to fund Hachenberg's proposed large telescope, von Hoerner had also asked the Foundation to fund a 160 m lower frequency telescope.[2] Both Hachenberg and von Hoerner were interested in MPG's offer of a directorship of the MPIfR but MPG and von Hoerner wanted the MPIfR to be in Tübingen, whilst Leo Brandt and Hachenberg wanted it to be in Nordrhein-Westfalen where the University of Bonn's existing Radio Astronomy Observatory was located. Eventually it was decided in 1965 to locate the MPIfR in the Bonn area and although the Volkswagen Foundation were prepared to fund the construction of Hachenberg's proposed 80 to 90 m telescope, they felt that more design work was required before they could commit to funding the construction of von Hoerner's proposed 160 m instrument. This led von Hoerner to withdraw his request for funding. Consequently the Volkswagen Foundation agreed to provide the money set aside for both projects, that is 27 million DM ($6.8 million), to Hachenberg's Bonn project which enabled its diameter to be increased to 100 m. Hachenberg also became the MPIfR's founding director.

The two companies who had carried out the study contracts, namely Krupp and MAN, were then asked in 1966 to combine forces to build the 100 m diameter telescope which had to be capable of making observations at frequencies up to at least 25 GHz (λ 1.2 cm). Assuming an efficiency of 67% this implied an average deviation of the dish surface from the ideal paraboloid of no more than 0.65 mm, rms.(4) In the event Krupp and MAN formed a consortium, called ARGE STAR,

[2] Hachenberg had used a mixture of analysis and experiment to design his 32 m dish. But in 1964 von Hoerner was in the process of designing large dishes using an analysis concept, later called 'homology', which he wanted to use in designing his proposed 160 m diameter dish.

with other companies to build the telescope with Krupp mainly responsible for the steelwork and MAN for the on-site assembly.

The search for a suitable site for this 100 m telescope focused on valleys that would help protect it from man-made radio interference. A number of possible locations were investigated before a north–south valley was chosen in the Eifel mountains near to the village of Bad-Münstereifel-Effelsberg about 40 km (25 miles) from Bonn.

Telescope construction started in 1968 with pouring the concrete ring to carry the 64 m diameter azimuth rail track. This ring was supported by 140 reinforced concrete piles each 1.2 m in diameter and seven to ten metres long resting on the bedrock. The rails were then installed and levelled. A large kingpin was built in the centre of the ring to centre the dish and provide a conduit for electrical cables from the ground to the telescope. The supporting structure for the dish and the equatorial bearings were then constructed and finally the dish was built with 17 rings of panels. The inner rings (numbers 1 to 9) out to 60 m diameter were made of high accuracy aluminium glued honeycomb panels, the next section (rings 10 to 14) from 60 to 80 m was made of aluminium frame panels, whilst the final ring of panels out to 100 m were made of wire mesh on steel frames. Finally a 1.8 m high collar was built at the periphery of the dish to reduced man-made interference. The telescope could be operated with a receiver at the dish's prime focus just 30 m from the dish surface or in a Gregorian configuration with the receiver in the centre of the dish. The Gregorian subreflector was 6.5 m in diameter.

As mentioned above the design of the Effelsberg dish was based on a self-compensating elastic structure. So although its surface could be distorted by up to 10 cm due to gravity, its shape still matched a true paraboloid within about 0.4 mm. As the angle of the dish to the vertical was changed, the position of the receiver was simply moved to match the change in focus position. On completion the dish of the Effelsberg 100 m telescope (see Figure 23.2) was the largest fully steerable radio telescope dish in the world.

First light observations took place in April 1971 at a frequency of 2.7 GHz (λ 11 cm) at the prime focus, whilst shortly afterwards the first astronomical observations were undertaken at 408 MHz. Then in October 1972 operations were undertaken at 10.6 GHz (λ 2.8 cm). In the following year cooled receiver systems were installed to operate at 1.4 GHz (21 cm) and 5 GHz (6 cm), whilst in 1974 a tuneable maser receiver was installed operating in the 23 GHz (1.3 cm) band. The first VLBI experiments involving the Effelsberg antenna took place in 1973.

The glued honeycomb panels in rings 1 to 9 proved problematic over time with failure of the glue seam allowing water ingress which caused damage during frosty weather. So over the period 1980–1982 all of these panels were replaced by panels of a different design. When the new panels had been fitted, measurements of the

Figure 23.2. The 100 m diameter Effelsberg radio telescope which, when it was completed in 1971, was the largest fully steerable radio telescope dish in the world. (Courtesy Wolkenkratzer via Wikimedia Commons.)

dish's surface accuracy by holography and subsequent adjustment resulted in a much more accurate surface. A few years later a more accurate Gregorian subreflector was fitted to improve matters even more. Since then the wire mesh outer panels have been replaced by perforated aluminium surface panels, allowing the telescope to operate at frequencies up to 86 GHz (3.5 mm). A new 6.5 m subreflector was installed in 2006 with active optics to improve system performance even more. In the following year the first international LOFAR (LOw-Frequency ARray) station was built next to the Effelsberg 100 m dish for observations at long wavelengths (see Section 19.3).

As far as the Stockert site is concerned, the 25 m telescope was closed down in 1995 and two years later the University of Bonn sold the site to a private company. In the meantime a group of enthusiasts had set up a foundation to restore the 25 m telescope and in 1999 they managed to get it accepted as a 'historic monument'. The site owner became insolvent in 2004 and in the following year the Nordrhein-Westfalen-Stiftung (NRW Foundation) purchased the site and allocated €300,000 (about $380,000) to enable the telescope to be restored. This restoration was begun in 2007. Three years later the facility was opened to the general public and in the following year astronomical observations were restarted mainly for the education and training of students at schools and universities.

23.2 Chalmers Onsala Space Observatory and the Swedish-ESO Submillimetre Telescope (SEST)

Olof Rydbeck of the Chalmers University of Technology in Gothenburg, Sweden set up a coastal field station in 1949 at Råö, near Onsala, at what later became known as the Onsala Space Observatory. There he undertook radio observations of the Sun at 150 MHz (λ 2 m).(5) In the following year he also undertook radar observations of meteors at the suggestion of Bertil-Anders Lindblad of Lund University also in Sweden. Shortly afterwards Rydbeck managed to get hold of five ex-German Würzburg radar antennae that had been located at remote mountain sites in Norway. In 1954 he used the first of these antennae to observe a solar eclipse at the Onsala observatory where it was 99% total.(6) Subsequently, these Würzburg antennae were used to observe neutral hydrogen in the Milky Way.

Encouraged by this first foray into radio astronomy, Rydbeck examined the possibility of building a 25 m diameter radio telescope. His search for financial support resulted in Skandinaviska telesatellit kommitten and Malmfonden helping to finance the telescope which was built at Onsala in 1963. Four years later a maser receiver, which had been developed in-house, was fitted to the 25 m significantly improving its sensitivity. Then in 1968 this radio telescope was linked to the 43 m (140 ft) radio telescope at the NRAO at Green Bank some 6319 km (3926 miles) away in a VLBI experiment. Observations were made at frequencies of 1.7 and 5 GHz (λ 18 and 6 cm) to give a resolution of 1 milliarcsecond.(7)

After the 25 m became operational Rydbeck hoped to build a replica of the 100 m German Effelsberg dish whose construction had been approved in 1966. But this proved too expensive. So he decided instead to enter the new field of millimetre-wave astronomy using a new, state-of-the-art telescope. Funding, which was provided by the Swedish Natural Science Research Council (NFR) and the Wallenberg Foundation, enabled him to purchase a 20 m diameter radio telescope from ESSCO (see Section 22.10). When it was inaugurated by King Carl Gustaf in 1976, it was the largest telescope in the world operating at millimetre wavelengths. (8) The dish had a radome to protect it from the weather and the heat of the Sun. This radome was optimised for frequencies of about 100 GHz (λ 3 mm) where it had a transparency of about 95%. The telescope was upgraded in 1992, and in 2014 a new radome was fitted (see Figure 23.3) together with a new receiver operating over the 85 to 116 GHz frequency range. The 20 m was used to detect millimetre-wave emission from comets, circumstellar discs, the interstellar medium and extragalactic objects. It was also used as one element of the European and worldwide VLBI networks.

Figure 23.3. Sweden's 20 m diameter millimetre-wave radio telescope inside its new dome at the Onsala Space Observatory in 2015. (Courtesy Peterkz via Wikimedia Commons.)

A few years after the 20 m had become operational, Swedish astronomers began to investigate the possibility of building a telescope operating at even shorter wavelengths. At that time IRAM had already started work on designing an array of 15 m diameter submillimetre/millimetre telescopes which were to be built on the 2,550 m (8,400 ft) altitude Plateau de Bure in France (see Section 23.3 below). So a cost-effective solution to Onsala's desire to have a similar telescope would be to add one telescope to the IRAM production run. Peter de Jonge, the director of IRAM and Roy Booth, the new director of the Onsala Space Observatory, discussed this idea and the possibility of further collaboration between their two organisations at the IAU General Assembly in 1982.(9)

Peter de Jonge, who had previously worked at the European Southern Observatory (ESO), was also keen to establish a close relationship between IRAM and ESO. Consequently, Booth and de Jonge discussed the possibility with ESO's Director General Lodewijk Woltjer of locating the new 15 m submillimetre/millimetre dish at the ESO observatory at an altitude of 2,375 m (7,800 ft) on La Silla (see Section 7.1). Woltjer backed the idea of such a three-way partnership which was ratified by the ESO Council in June 1984. As a result, it was agreed that Onsala and ESO would install and operate this telescope on La Silla and share the expense and observing time over a 15-year period, whilst IRAM would build the telescope under contract to ESO. Onsala also had the technical responsibility for the first receivers and the overall project.(10)

This new 15 m diameter telescope, called the Swedish-ESO Submillimetre Telescope (SEST), which operated down to 0.8 mm, became the first large telescope operating at such a short wavelength in the southern hemisphere. It was also the first ESO-backed facility not dedicated to optical/infrared astronomy.

Construction of the 15 m SEST (see Figure 23.4), which was started in 1986, was marred by an accident that resulted in the Sun damaging part of the subreflector and several panels of the main reflector. This happened when the handset used to rotate the telescope, to avoid it pointing straight at the Sun, failed during daytime operations undertaken to install the subreflector. Smoke was seen coming from the edge of the subreflector but the fire brigade, who arrived quickly, could not reach the source of the fire with their water hoses due to strong winds and low water pressure. And an attempt to rapidly move the telescope manually resulted in the dish hitting some scaffolding. In spite of all this, the repaired telescope was successfully handed over to ESO in March 1987.

SEST was mainly used to carry out a spectral survey of the central region of the Milky Way and the Magellanic Clouds. It was also used for VLBI observations at millimetre wavelengths. Then in 2003 it was decommissioned to be superseded by APEX and ALMA (see Section 24.1).

Figure 23.4. The Swedish-ESO Submillimetre Telescope (SEST) on La Silla. (Courtesy Iztok Boncina/ESO.)

23.3 IRAM

Émile-Jacques Blum and Jean Delannoy were working on the Grand Radiotélescope at Nançay in France when in 1964 they became aware that Bordeaux Observatory and the University of Bordeaux wanted to enter the field of radio astronomy.(11) As a result, they contacted Pierre Sémirot, the director of the observatory, and Roger Servant and André Charru, professors at the university and members of its 'ultra-hertzian' optics laboratory, suggesting that they build a millimetre-wave interferometer at the observatory. This suggestion was accepted and Delannoy moved to the observatory to build such an interferometer whose construction was begun in 1967. When completed six years later the interferometer operated at a wavelength of 8 mm (frequency 38 GHz) and consisted of two fixed 2.5 m diameter antennae on altazimuth mounts. Subsequently it was used to observe the Sun for a number of years. One of the two dishes was then given to the new observatory of Grenoble and installed on the Plateau de Bure about 100 km (60 miles) south of the city in 1986.

In the meantime, Blum, James Lequeux and others had tried in the late 1960s to produce a longer term plan for French radioastronomical research. Unfortunately this ran into insurmountable funding constraints. But in 1973 Jean Brossel took over as Director of the Physics Laboratory at the École Normale Supérieure (ENS), and two years later he set up a new Laboratory for Millimetre Radioastronomy (Institut de Radioastronomie Millimétrique). In the absence of a suitable radio telescope of their own, laboratory staff used American millimetre telescopes for their observations. Meanwhile a small sum of money had been allocated in France for technical developments and site studies for a possible millimetre-wave radio telescope.

At about the same time there had also been a push in parts of Europe led by Jan Oort and others to develop a joint, multi-national approach on millimetre-wave radio astronomy. For example, in February 1973, the Organisation de Coopération et de Développement Économiques (OCDE) suggested that France, Germany and the UK should cooperate in millimetre radio astronomy. As a first step, this resulted in a multi-national European laboratory being set up in Ireland to build diodes for millimetre mixers. The first successful junctions were produced in 1977.

Meanwhile in the autumn of 1973, the OCDE representatives had also decided to set up a Scientific Advisory Group for Millimetre Astronomy (SAGMA). This group defined a European millimetre project that they estimate would cost about 37 million Deutschmarks (about $12 million). It would consist of a 30 m radio telescope and an interferometer consisting of four 10 m antennae, both radio telescopes being located at the same site. Site surveys were due to start in 1974 with both telescopes becoming operational by the end of the decade.

The choice of site was not determined solely by its astronomical usefulness. For example, Germany did not want the telescopes to be sited in France as they were apparently afraid of political instability and strikes, and they thought that France already had more than its fair share of international organisations located there. As a compromise, the site survey committee considered sites in the Atlas Mountains of Morocco, the Sierra Nevada in Spain and the Canary Islands, as well as sites in France. But they concluded that the best site was the Plateau de Bure at an altitude of 2,550 m (8,400 ft) in France where the atmosphere had a precipitable water vapour content of only about 2 mm in winter.(12) The UK withdrew from the project in 1975 for financial reasons and decided, instead, to build what became the 15 m JCMT on Mauna Kea (see Section 21.2 above).

In spite of the recommendation of the site survey team for the Plateau de Bure, Germany continued to oppose a site in France. Instead they pushed for one in Spain, preferably at Calar Alto where there was already a joint German–Spanish optical observatory. But, unfortunately, Calar Alto turned out to be unsuitable for a millimetre-wave telescope. On the other hand, the much higher site of Pico Veleta, at an altitude of 2,870 m (9,400 ft) in the Sierra Nevada, about 45 km (28 miles) from Grenada, appeared to be suitable. So at this stage there seemed to be two possible sites for the observatory, the Plateau de Bure, favoured by France, and

Pico Veleta favoured by Germany. A three-man neutral international committee was then set up to judge between these two sites. They recommended that the interferometer, which needed a relatively large flat area, should be set up on the Plateau de Bure and that the 30 m telescope should be built at Pico Veleta, with the two-site observatory being administered from a headquarters in Grenoble. CNRS (Centre National de la Recherche Scientifique) for France and MPG (Max-Planck-Gesellschaft) for Germany accepted these recommendations. Subsequently they established the provisionally named Joint Institute for Millimetre Radioastronomy (JIMA) in Grenoble to run the observatory.

Unfortunately at this time in 1976–1977 both France and Germany had encountered financial problems. MPG for their part had to delay any expenditure until 1978, although they did give a firm commitment to continuing with the project. France, on the other hand, allocated just 2.5 million French Francs ($500,000) to cover both 1977 and 1978. To make matters worse they also temporarily blocked the actual expenditure of any of this money. Faced with this apparent lack of interest or commitment, particularly in France, Blum and Lequeux, the two main French supporters of the project, resigned bringing matters to a head. MPG was appalled at these developments and put pressure on CNRS to commit proper financing to the project.

The Blum and Lequeux resignations and pressure from MPG had the required effect, with France committing, in November 1977, to a funding of 83 million French Francs (about $16 million), albeit spread over eight years (from 1978 to 1985). Subsequently Blum and Lequeux returned to the project in February 1978. Then in the following year, IRAM (Institut de Radioastronomie Millimétrique) was formally established by CNRS and MPG, acting on behalf of their respective countries, with the requirement that both the 30 m telescope and the interferometer should be able to operate down to a wavelength of 1.8 mm (frequency 170 GHz).

Meanwhile interest in millimetre-wave astronomy in Spain had led the Royal Observatory in Madrid, which was a department of the Instituto Geográfico Nacional (IGN), to order a 14 m millimetre radio telescope in 1974 from a company in the USA. Jesús Gómez-González was made responsible for this telescope which was subsequently installed in 1976–1977 at the Yebes Observatory about 50 miles (80 km) from Madrid. Gómez-González had previously spent some time at the Paris Observatory, Meudon and he had been involved, both before and after he left France, in the IRAM search for a suitable site for their radio telescopes in Spain. So it was no great surprise when the IGN showed an interest in being associated with the IRAM project. This led to an agreement in 1980 that the IGN would provide IRAM with land on Loma de Dilar, near Pico Veleta, where the 30 m telescope would be located and also provide a building in Granada for offices and

Figure 23.5. IRAM's 30 m Millimeter Radio Telescope near Pico Veleta in Spain. (Courtesy IRAM/Nicolas Billot.)

laboratories. In return, IGN would be allowed up to 10% observing time on both the 30 m and interferometer telescopes. Eventually in 1990 Spain became a full member of IRAM contributing 6% of the budget.

Initial design work for a large millimetre-wave radio telescope had been started at the Max-Planck-Institut für Radioastronomie (MPIfR) in 1975. At that time the main requirement had been to design as large a dish as finances would allow that could be used down to about 1 mm (300 GHz) wavelength. After the formation of IRAM, it was formally agreed that the telescope should be of 30 m diameter and the MPIfR was contracted to be responsible for its construction. Subsequently the detailed design and construction of what was to be called the Millimeter Radio Telescope (MRT) was undertaken by the same industrial consortium that had designed the Effelsberg 100 m radio telescope.(13)

The MRT's 30 m dish (see Figure 23.5) consisted of 420 aluminium panels arranged in seven concentric rings. The designers used a self-compensating structure for the dish, similar that used for the Effelsberg telescope, to minimise the effect of gravitational and wind loads. They also minimised thermal effects on the MRT by designing a cavity underneath the dish that was thermally controlled by fans. In addition, the dish's aluminium panels were coated with a heat reflecting paint, and thermal insulation sheets were installed between these panels and the steel support structure. Active thermal control was also used elsewhere on the telescope structure.

Distortion of the antenna surface was measured by two different methods, one using a theodolite and the other using a holographic system. In the latter case, the strong water vapour maser source in Orion was used as the test transmitter at a wavelength of 13.3 mm. An alternative holographic system was also employed using a 3.5 mm transmitter located on the top of Pico Veleta. The final adjustment of the surface panels was then undertaken using a number of iterative steps with intermediate measurements.

Tests of the MRT were begun in 1984 at a wavelength of 3.5 mm (80 GHz), and the telescope became operational two years later, operating down to 1.3 mm. Later improvements enabled it to be used down to the atmospheric window at 0.87 mm (350 GHz). The telescope's excellent thermal design also enabled it to be used in daylight hours.

Meanwhile construction work had started in 1983 on the IRAM interferometer at the Plateau de Bure(14) that was designed to operate down to 0.8 mm. It was built with three 15 m diameter dishes[3] each mounted on fully steerable altazimuth fork mounts that incorporated self-propelled transporters. These allowed the dishes to be placed at any of 26 stations on a T-shaped railway track stretching 160 m approximately north–south, and 288 m approximately east–west.[4]

Each dish was composed of 176 panels, 60 mm thick arranged in six concentric rings. They were made of aluminium honeycomb cores sandwiched between carbon fibre skins, with the reflecting surface on top of the carbon fibre consisting of 40 nanometres of aluminium protected by a 30 micron film of Hostaflon PTFE. The reflecting surface could be heated electrically during icy weather. The antennae could not be pointed within about 35° of the Sun because of the smoothness of this reflecting surface. Two digital correlators were available for the interferometer, one for continuum and one for spectroscopic observations.

First fringes were obtained on 3C 273 and the Orion SiO maser in December 1988 with two dishes operating at 86 GHz (λ 3.4 mm) on a 48 m baseline.(15) The first operation with three dishes was in June 1989. The number of dishes was then increased to six, and both arms of the 'T' were progressively extended to produce a maximum baseline of 760 m.

This interferometer has, amongst other things, been used to examine the dust and gas in the coma of comets, and to measure the wind and thermal profile of Mars' atmosphere. It has also been used to undertake a systematic study of proplyds (protoplanetary discs), and to detect the distribution of dust and gas in distant galaxies.

[3] The increase from the originally foreseen diameter of 10 m to 15 m was because the Owens Valley millimetre array had since been designed with 10 m diameter dishes (see Section 21.1 above).

[4] The two tracks crossed at an angle of 75° not 90°.

Figure 23.6. Seven 15 m diameter dishes of IRAM's NOEMA array at the Plateau de Bure in France. (Courtesy IRAM/André Rambaud.)

In 2014 the first of six new 15 m dishes was added to the IRAM array to form NOEMA (NOrthern Extended Millimeter Array) (see Figure 23.6). It was then planned to extend the track system to increase the maximum antenna separation to 1.6 km and to provide the antennae with completely new instruments.

23.4 Indian Radio Telescopes

India was anxious to develop its scientific base after independence in 1947. But at that time radio astronomy was still a very new science, and so it did not figure in India's plans. But in 1955, following a suggestion by Govind Swarup, the CSIRO (Commonwealth Scientific and Industrial Research Organisation) in Australia, where he was working, agreed to donate thirty-two 6 ft (1.8 m) redundant antennae from Potts Hill (see Section 16.1) to India under the Colombo Plan provided India paid for their transportation.(16) Unfortunately, the Indian Council of Scientific and Industrial Research (CSIR) refused to pay for this and suggested that the Australians should also pay for their transportation because of the shortage of foreign exchange in India. Initially, the Australian CSIRO rejected this suggestion but eventually they relented and paid for the transport of the 32 dishes to New Delhi.

For a time these dishes remained unused. But in 1961 a group of Indian radio astronomers, including Swarup, who were working at the time in the United States wrote to five major scientific organisations and agencies in India. These expatriates said that they were willing to return India to start a solar radio observatory using the 32 dishes if some organisation was willing to finance the project. The most encouraging response came from Homi Bhabha, director of the Tata Institute of Fundamental Research (TIFR) in Bombay (now called Mumbai) who said that they had decided to set up a radio astronomy group. Subsequently Bhabha managed to get the 32 antennae transferred from CSIR to TIFR in mid 1963.

In the meantime Swarup, who had already returned to India, produced a design for a new interferometer using these dishes. He also proposed building a large cylindrical radio telescope to try to measure the distribution of objects like the newly discovered quasars which were a considerable distance from the Milky Way. Swarup was keen to measure this distribution as it should enable astronomers to resolve the controversy between the Big Bang and Steady State cosmologies. He proposed building this cylindrical radio telescope on a suitably inclined hill in southern India near the equator to enable the telescope's axis to be made parallel to the Earth's axis.

It was important at this time for the small group of radio astronomers who were gradually being recruited by TIFR to get some practical experience in radio astronomy. Consequently they set up a grating type interferometer at Kalyan, near Bombay using the 32 dishes from Australia. This interferometric array, which was completed in 1965, consisted of 24 dishes on a 630 m east–west baseline and the remaining 8 on a 256 m north–south baseline. It was used to measure the two-dimensional distribution of radio emission from the quiet Sun at a wavelength of 610 MHz (λ 49 cm).(17)

Over the years since the Second World War radio astronomy in the developed world had gradually progressed from observations at these relatively long wavelengths to observations at ever shorter wavelengths in order to obtain higher resolution. The technology of these developed countries had allowed this transition to take place. But in India labour costs were low and government budgets very low. So it made more sense to build large antennae operating at long wavelengths where the mechanical tolerances were lower. Swarup's proposed large cylindrical radio telescope fitted the bill.

The first task for this project was to find a suitable hill in southern India on which to build the telescope. This was found at Ootacamund, or Ooty, at an altitude of about 6,900 ft (2,100 m) in the Nilgiri Hills. The telescope, which was completed

Figure 23.7. India's Ooty radio telescope that was built on a north–south slope with the same angular gradient as the station's latitude so providing a natural equatorial mount. (Courtesy Ooty Radio Telescope/Razimantv via Wikimedia Commons.)

in December 1969, consisted of a 530 m long, 30 m wide parabolic cylinder, the 'surface' of which consisted of one thousand one hundred 530 m long, thin, stainless steel wires running parallel to each other. They were supported on 24 steerable parabolic frames. There were over a thousand half-wave dipoles located along the telescope's linear focus operating as a phased array.

This Ooty Radio Telescope (ORT) (see Figure 23.7) was built with its long axis along a north–south hill with a slope at the same angle, 11.35°, as the latitude of the station. This meant that the telescope's long axis was parallel to the earth's rotational axis, giving it an equatorial mount. As a result, the ORT was able to track radio sources within 35° of the celestial equator for up to 9.5 hours per day by a simple mechanical rotation of the array. The telescope, which operated at 327 MHz (λ 91.7 cm), could also be steered in the north–south direction by introducing a phase delay along the dipole array.

The ORT was used to observe the lunar occultation of a large number of relatively faint radio sources like distant radio galaxies and quasars. These occultations provided accurate position estimates of the sources and showed their structure to arcsecond resolution. The results supported the Big Bang cosmology.[5] In addition, observations of interplanetary scintillation provided information on the compact structure of some radio sources at subarcsecond resolution. Then in the mid 1980s, seven small parabolic cylindrical antennae[6] were combined with the ORT to form the Ooty Synthesis Radio Telescope that covered an area of about 4 km (east–west) by 2 km (north–south).(18)

In the mid 1970s, India built one of the few large low frequency radio telescopes of the world at Gauribidanur about 80 km north of Bangalore. When it was completed in the early 1980s it consisted of a T-shaped array 1.38 km east–west, 445 m north–south, of 1,000 broad band dipole antennae. These were situated about 1.5 m above a 60,000 m² reflecting screen.[7] The array, which operated at 34.5 MHz (λ 9 m), could be pointed to any direction along the meridian within a declination range of – 45° to +75° in steps of 12 arcminutes.[8] It was used to observe the Sun, to produce a sky survey of radio sources at 34.5 MHz, and to examine large areas of diffuse radio emission like those associated with clusters of galaxies.

The decommissioning of the Culgoora radio heliograph in 1984 followed by the closure of the Clark Lake radio observatory meant the there were no telescopes in the early 1990s to observe radio emission from the Sun's outer corona. As a result, the Indian Institute of Astrophysics decided to build a radioheliograph at Gauribidanur to continue this work. Their new instrument was designed to produce two dimensional images of the Sun's outer corona simultaneously at different frequencies in the range from 40 to 150 MHz (λ 7.5 to 2 m). As with the other low frequency telescope at Gauribidanur the radioheliograph's dipoles were also arranged in a T-shaped configuration, in this case 1.28 km east–west and 441 m north–south. Following completion in 1998, it was used to study coronal mass ejections, coronal holes and active regions, amongst other things. Currently (2015) work is under way to extend the length of both the east–west and north–south arms.

In the meantime, the success of the ORT had encouraged Swarup and his colleagues in 1984 to make a proposal to build what became known as the Giant Metrewave Radio Telescope (GMRT). Three years later the Government of India approved the project and in 1990 its detailed design was finalised. Subsequently, the GMRT was built at Khodad about 80 km (50 miles) from Pune in west central India. It consisted of thirty 45 m fully steerable, altazimuth-mounted, parabolic dishes with reflecting surfaces of wire mesh. About half of these were located in

[5] New Evidence Favors the Big Bang, *Sky and Telescope*, September 1976, pp. 162–163.

[6] There were six 23 m × 9 m parabolic cylinders and one 92 m × 9 m.

[7] India's Low-Frequency Radio Telescopes, *Sky and Telescope*, June 1984, p. 520.

[8] K. R. Subramanian, *Low Frequency Solar Radio Astronomy Research at Indian Institute of Astrophysics*, ursi.org/Proceedings/ProcGA05/pdf-P.19(0764).pdf.

the central 1 km², with the remainder being on three 14 km long arms arranged in an approximate 'Y' formation. The GMRT, which became fully operational in 2000, could be operated between about 40 and 1,450 MHz (7.5 m to 21 cm).(19) The telescope, which cost about $25 million, has been used to observe an enormous variety of objects from the Sun and planets to the centre of the Milky Way, pulsars, supernova remnants, interstellar hydrogen clouds, cosmic masers and quasars. It has even observed the afterglows of gamma-ray bursts.

23.5 Nobeyama Observatory

Solar Telescopes

Early radio astronomy in Japan had been undertaken mainly at Tokyo Astronomical Observatory's (TAO's) observatory at Mitaka. But in the 1960s, TAO undertook a survey to try to find a more suitable site which resulted in Nobeyama being selected. It was on a 1,350 m (4,400 ft) high plateau at the foot of the Yatsugatake Mountains about 150 km (90 miles) north-west of Tokyo.

The building of the Nobeyama Observatory was started in 1967 with the construction of a 160 MHz (λ 1.9 m) compound solar interferometer. It consisted of an east–west array of nine 6 m and two 8 m parabolic antennae with a baseline of 2.3 km, and a north–south array of six 6 m dishes with a baseline of 1.2 km.(20) This was followed by the addition of a 70–220 MHz multi-channel spectrograph using an 8 m diameter parabolic antenna and a 200–600 MHz acousto-optical spectrograph using a 6 m diameter antenna. These spectrographs, which were also built to observe the Sun, could be used with both the 160 MHz and 17 GHz interferometers (see next paragraph). They became operational in 1977.

In the meantime, a 12 element interferometer had been built in 1971 on an east–west baseline using 1.2 m diameter dishes operating at 17 GHz (λ 1.8 cm). Unfortunately, its minimum time resolution of 30 s was too long to observe rapidly changing solar phenomena. Consequently in 1978, this system was modified to produce a time resolution down to 0.8 s and spatial resolution of ~50 arcsec.(21) In the same year, the Nobeyama Radio Observatory was designated the Japanese national radio observatory specialising in microwave observations.

The 160 MHz interferometer was closed down in 1989 and the 17 GHz interferometer was replaced shortly afterwards by a radioheliograph that consisted of eighty-four 80 cm diameter parabolic dishes operating at 17 GHz. They were arranged in a T-shaped configuration with east–west and north–south baselines of 490 m and 220 m, respectively. In 1995 this radioheliograph was modified to enable it to operate at both 17 and 34 MHz by adding a frequency-sensitive subreflector to each of the 84 dishes. This subreflector reflected 17 GHz radio waves into

Figure 23.8. Japan's 45 m diameter millimetre-wave radio telescope at Nobeyama Observatory. (Courtesy of the National Astronomical Observatory of Japan.)

the Cassegrain focus whilst transmitting 34 GHz waves into the primary focus.(22) The system had a temporal resolution of 0.1 s, and a spatial resolution of 5 arcsec at 34 GHz. Its main purpose was to observe solar flares.

Millimetre-Wave Telescopes

The first radio telescope in Japan to operate in the millimetre waveband was built in 1969 at the TAO's observatory at Mitaka to enable Japan to enter the relatively new field of interstellar molecular research. It was a 6 m paraboloid that operated at frequencies from 70 to 100 GHz (λ 4.3 to 3 mm).

Once this radio telescope was operational, the TAO began to look into the possibility of building a large millimetre-wave telescope at their new Nobeyama observatory. This seemed to be almost the ideal location after taking into account the low levels of atmospheric water vapour there, particularly in winter, and the low levels of radio interference assisted by the surrounding mountains.

Consequently, a fully steerable 45 m parabolic telescope was built (see Figure 23.8) at the Nobeyama observatory around the end of the 1970s to operate down to a wavelength of about 2 mm (frequency 150 GHz).(23) Its main dish had a similar self-compensating system to that of the Effelsberg antenna in West Germany in which distortions due to gravity, as the dish was moved from the zenith, were compensated by moving its secondary reflector.

The 600 reflective panels of this 45 m Japanese dish were made of 10 cm thick aluminium honeycomb sandwiched between 1 mm thick sheets of carbon fibre reinforced plastic. This produced a very lightweight and extremely rigid structure. In addition, the plastic had a very low thermal expansion coefficient and the aluminium had a good thermal conductivity. These panels were individually attached to the dish's main substructure by computer-controlled motorised supports to enable the shape of the dish to be adjusted. A closed cavity underneath the dish was thermally controlled by fans to try to maintain the temperature of the supporting system constant, and so reduce thermal distortion.

When it became operational in 1982 this 45 m telescope was the largest radio telescope in the world operating at such short wavelengths. In addition, it had an acousto-optical spectrometer that provided a bandwidth of 2 GHz with 16,000 channels. The telescope has since been used, amongst other things, to study molecular clouds to aid understanding of interstellar chemistry, the formation of stars, and the structure of galaxies.

The Nobeyama Millimeter Array, which was built at Nobeyama in the early 1980s, initially consisted of five 10 m dishes operating down to a wavelength of 2.6 mm. The dishes were arranged on two baselines of just over 500 m each, one east–west and the other intersecting the east–west baseline at an angle of about 60°. The first interference fringes were produced at the end of 1982. Later a sixth 10 m dish was added and the wavelength coverage extended down to 0.8 mm (frequency 365 GHz). This Millimeter Array was used, amongst other things, to study the fine structure of interstellar clouds.

Around the end of the millennium a consortium of Japanese institutions, led by the National Astronomical Observatory of Japan (NAOJ), and supported by the University of Chile, produced a 10 m diameter submillimetre telescope to operate down to a wavelength of about 0.3 mm (frequency 1 THz). It was an early prototype for the ALMA (Atacama Large Millimeter Array, see Section 24.1) antennae. Made by Mitsubishi Electric in Japan, this 10 m dish called ASTE (Atacama Submillimeter Telescope Experiment) was located at an altitude of 4,800 m (15,700 ft) on Pampa la Bola in the Chile's Atacama Desert in 2002. It became operational two years later. Since then a number of different sensor systems have been used with it to support both spectroscopy and continuum observations. For example, the AzTEC camera, with a 144 element bolometer array, was installed on ASTE in 2007 to provide continuum observations at wavelengths of 2.1 and 1.1 mm.[9]

23.6 Heinrich Hertz Submillimeter Telescope

In the mid 1970s, the Max-Planck-Institut für Radioastronomie (MPIfR) were interested in building a 30 m diameter millimetre-wave radio telescope and a 10 m diameter telescope working in the submillimetre band.(24) But they eventually decided, instead, to work with France in building the IRAM (Institut de Radioastronomie Millimétrique) millimetre-wave observatory on the Plateau de Bure in the Alps. Unfortunately, this location was not thought to be suitable for submillimetre observations.

A few years later MPIfR started to look once more at providing a submillimetre telescope, this time in cooperation with the University of Arizona. The initial plan was to build a 10 m diameter submillimeter telescope (SMT) on the summit of Mount Lemmon in Arizona. But it was later decided to locate the SMT on Mount Graham as it was, at 3250 m altitude, about 450 m higher than Mount Lemmon and was more suitable for submillimetre observations. In May 1985, MPIfR gave a contract to a consortium of Krupp Industrietechnik and MAN Technologie to undertake the detailed design and manufacture of the telescope which would operate down to the atmospheric cut-off at about 0.3 mm (frequency 1 THz). The telescope enclosure was to be built by the University of Arizona, whilst the telescope's instruments were to be developed jointly.

Work was due to start on Mount Graham in 1985 but environmental concerns over its protected red squirrel habitat led to a frustrating delay (see Section 13.1). As a result, the site work was not started until 1990. Nevertheless, the telescope was dedicated as the Heinrich Hertz Telescope (HHT) (see Figure 23.9) just three years later, with first light taking place in the following year. The surface of the dish was made of sheets of 1.5 mm thick Carbon Fibre Reinforced Plastic (CFRP) bonded to a 90 mm thick aluminium honeycomb core. The support structure was made of CFRP tubes connected to joints of invar steel to produce a low thermal expansion system of low weight and very high stiffness. The HHT or SMT (it is called both) was operated and maintained by the Arizona Radio Observatory to work over the frequency range from 130 GHz to 1 THz.

The main areas of research undertaken with the HHT included the structure and dynamics of late-type stars, planetary nebulae, molecular clouds, and the early stages of star formation.

[9] The AzTEC camera had been developed by a team led by the University of Massachusetts. It was commissioned on the JCMT in 2005 and, as mentioned above, was installed on the ASTE two years later. In 2011, it saw first light on the Mexican–American Large Millimeter Telescope.

Figure 23.9. The 10 m diameter submillimetre Heinrich Hertz Telescope on Mount Graham. (Courtesy Geremia via Wikimedia Commons.)

References

1. Wielebinski, R., Fifty Years of the Stockert Radio Telescope and What Came Afterwards, *Astronomische Nachrichten*, 328, No. 5, 2007, pp. 388–394.
2. Wielebinski, R., The New Era of Large Paraboloid Antennas: the Life of Prof. Dr. Otto Hachenberg, *Advances in Radio Science*, 1, 2003, pp. 321–324.
3. Wielebinski, Richard, Junkes, Norbert, and Grahl, Berndt H., The Effelsberg 100-m Radio Telescope: Construction and Forty Years of Radio Astronomy, *Journal of Astronomical History and Heritage*, 14, No. 1, March 2011, pp. 3–21.
4. Hachenberg, O., The New Bonn 100-meter Radio Telescope, *Sky and Telescope*, December 1970, pp. 338–343.
5. Sullivan, Woodruff T., *Cosmic Noise; A History of Early Radio Astronomy*, Cambridge University Press, 2009, p. 228.

6. Booth, Roy, Obituary of Olof E. H. Rydbeck 1911–1999, *Bulletin of the American Astronomical Society*, 33, No. 4, 2001, pp. 1580–1581.
7. Kellermann, K. I., et al., High-Resolution Interferometry of Small Radio Sources using Intercontinental Baselines, *Astrophysical Journal*, 153, 1968, pp. L209–L214.
8. Menzel, Donald H., A New Radio Telescope for Sweden, *Sky and Telescope*, October 1976, pp. 240–242.
9. Madsen, Claus, *The Jewel on the Mountaintop; The European Southern Observatory through Fifty Years*, ESO and Wiley-VCH, 2012, pp. 164–166.
10. Booth, R. S., De Jonge, M. J., and Shaver, P. A., The Swedish-ESO Submillimetre Telescope, *ESO Messenger*, No. 48, June 1987, pp. 2–5.
11. Encrenaz, Pierre, et al., Highlighting the History of French Radio Astronomy 7: The Genesis of the Institute of Radioastronomy at Millimeter Wavelengths (IRAM), *Journal of Astronomical History and Heritage*, 14, No. 2, 2011, pp. 83–92.
12. Baars, J. W. M., et al., The IRAM 30-m millimeter radio telescope on Pico Veleta, Spain, *Astronomy and Astrophysics*, 175, 1987, 319–326.
13. Baars, J. W. M., The 30 meter Millimeterwavelength Radiotelescope, *Mitteilungen der Astronomischen Gesellschaft*, 54, 1981, 61–69.
14. Guilloteau, S., et al., The IRAM Interferometer on Plateau de Bure, *Astronomy and Astrophysics*, 262, 1992, pp. 624–633.
15. Lucas, Robert, The IRAM Plateau De Bure Interferometer, in Cornwell, T. J., and Perley, R. A., (eds.), Radio Interferometry: Theory, Techniques and Applications, IAU Colloquium 131, *ASP Conference Series*, 19, 1991, pp. 449–452.
16. Swarup, Govind, From Potts Hill (Australia) to Pune (India): The Journey of a Radio Astronomer, *Journal of Astronomical History and Heritage*, 9, No. 1, June 2006, pp 21–33.
17. Swarup, Govind, Reminiscences Regarding Professor W. N. Christiansen, *Journal of Astronomical History and Heritage*, 11, No. 3, November 2008, pp. 194–202.
18. Swarup, Govind, The Ooty Synthesis Radio Telescope: First Results, *Journal of Astrophysics and Astronomy*, 5, 1984, pp. 139–148.
19. Ananthakrishnan, S., The Giant Metrewave Radio Telescope (GMRT): Salient Features and Recent Results, *29th International Cosmic Ray Conference, Pune*, 10, 2005, pp. 125–136.
20. Takakura, T., et al., A New Compound Interferometer Operating at 160 MHz at Nobeyama, *Proceedings of the Astronomical Society of Australia*, 1, 1967, pp. 56–58.
21. Nakajima, Hiroshi, et al., A New 17-GHz Solar Radio Interferometer at Nobeyama, *Publications of the Astronomical Society of Japan*, 32, 1980, pp. 639–650.
22. Takano, Toshiaki, et al., An Upgrade of Nobeyama Radioheliograph to a Dual-Frequency (17 and 34 GHz) System, in Trottet, Gérard, (ed.) Coronal Physics from Radio and Space Observatories, *Lecture Notes in Physics*, 483, 1997, pp. 183–191.
23. Akabane, Kenji, et al., The Nobeyama Radio Observatory, *Sky and Telescope*, December 1983, pp 495–499.
24. Baars, J. W. M., and Martin, R. N., The Heinrich Hertz Telescope; A New Instrument for Submillimeter-Wavelength Astronomy, *Reviews in Modern Astronomy*, 9, 1996, pp. 111–126.

24

ALMA and the South Pole

24.1 ALMA

As mentioned previously (see Section 20.8) the NRAO had been trying in the late 1970s to get funding for a new 25 m (82 ft) millimeter-wave radio telescope to be built on Mauna Kea, but this had been killed off in 1982 by the Astronomy Advisory Committee. However, in the same year, an internal NRAO proposal was made to build a VLA-type millimetre array of fifteen 10 m diameter dishes on 1 km long arms. The plan was to build it near the VLA(1) at an altitude of about 2,100 m (6,900 ft) on the Plains of San Augustin, New Mexico. A little later, higher sites were considered in Arizona and New Mexico to enable the telescope to operate down to a wavelength of at least 0.85 mm (frequency 350 GHz). Then in 1990 the NRAO proposed what was called the Millimeter Array (MMA) which was to consist of forty 8 m diameter dishes, with a total area of 2,000 m^2, at a cost of at least \$120 million.(2) It would operate in the wavelength range from 10 mm to 0.35 mm (frequencies 30 GHz to 850 GHz)(3) and be built on a site about 3 km in diameter at an altitude of about 2,500 m (8,200 ft) in Arizona. At this stage the NSF, the potential funding source, let it be known that they expected that any large project of this nature should involve a number of international partners. By the mid 1990s, the NRAO were considering much higher sites on Mauna Kea or on Llano de Chajnantor at an altitude of about 5,000 m (16,400 ft) in the Atacama Desert to enable the array to operate well into the submillimetre range.

In Japan, meanwhile, the Tokyo Astronomical Observatory had built the Nobeyama Millimeter Array in the 1980s (see Section 23.5) which consisted of a number of 10 m dishes operating down to 0.8 mm. They followed this with plans to build an array in northern Chile consisting of fifty 8 m or 10 m diameter dishes called the Large Millimeter and Submillimeter Array (LMSA) to operate down to 0.35 mm.

In Europe, Onsala, ESO and IRAM had already cooperated in the 1980s on building the 15 m submillimetre/millimetre SEST radio telescope (see Section 23.2) that operated down to a wavelength of 0.8 mm. In 1991 the SEST group then began to consider building an interferometer. Their first idea was to build an additional telescope near to the SEST site at La Silla.(4) Then at the suggestion of Roy Booth, the director of the Onsala Space Observatory, they began to consider building an interferometer consisting of ten 8 m antennae near Paranal at a cost of 50 million Deutsche Mark (about $30 million). But just as the Onsala, ESO and IRAM working group was about to study this option, news broke that American astronomers had discovered carbon monoxide with a redshift of 2.3. This showed that it should be possible to study the distribution of not only carbon monoxide but also other atoms and molecules in the distant universe, provided the observations were carried out with a telescope of high sensitivity and good resolution in the millimetre and hopefully the submillimetre wavebands. This required more antennae with significantly more total collecting area than the proposed Booth interferometer.

The SEST group consisting of ESO, Onsala and IRAM, together with the Dutch Netherlands Foundation for Radio Astronomy (NFRA, later called ASTRON), now agreed to pool their resources and carry out a two year study into such an array starting in 1995. This Large Southern Array (LSA), with a collecting area of about 10,000 m^2, would consist of fifty 16 m or sixty 15 m diameter antennae with baselines of up to 10 km. It was planned to locate it at an altitude above 3,000 m in Northern Chile, possibly at Pampa San Eulogio, to allow observations down to 0.8 mm. This European LSA would produce high resolution, and whilst it would operate mostly in the millimetre waveband its large collecting area would give it high sensitivity. The American MMA, on the other hand, would operate in both the millimetre and submillimetre wavebands, albeit with lower sensitivity. But to do this the American array would need to be at a higher altitude than the European one.

As mentioned above the NSF were pushing the NRAO to find international partners to develop their MMA array and the Europeans also saw the benefit of a joint project. Consequently, on completion of the two-year European study in 1997, they and the NRAO began a joint study to see if an array could be designed to operate at submillimetre wavelengths, like that of the Americans, whilst having the sensitivity and resolution of the European array. The study evaluated the relative merits of two possible options for the design of a joint LSA/MMA system. Option 1 was a hybrid array of forty 8 m antennae and twenty-five to thirty-five 15 m antennae. The 8 m antennae would be used down to 0.35 mm, whilst the larger antenna would probably be limited to about 0.8 mm. Option 2 was an homogenous array of fifty to sixty antennae of about 12 m diameter that would operate down to 0.45 mm.(5) By the following year, a version of option 2 had been selected for the LSA/MMA array or Large Millimeter Array (LMA), as it was then called, with sixty-four 12 m diameter antennae (total area 7,200 m^2).(6) This LMA would be built with baselines of up to 10 km on the 5,000 m altitude Llano de Chajnantor and would operate in the wavelength range from 3.5 to 0.35 mm. Because of this extreme altitude, it would be operated remotely from San Pedro de Atacama. The LMA's total development and construction cost was estimated to be about $400 million.

The completion of this two-year joint European/American study resulted in a formal agreement in 1999 between the NSF and a number of European institutions, including ESO,[1] to collaborate on the design and development of what was to be called ALMA (Atacama Large Millimeter Array). Clearly, it would be an advantage if Japan were to cancel its development of a similar but smaller array and be involved in ALMA instead. Unfortunately, Japan was unable to commit to joining the ALMA project at that time, but it was agreed that they would interact informally with the project on a temporary basis with a view to hopefully joining at a later date.[2]

In the meantime, the Europeans and Americans had in February 2000 each ordered a prototype 12 m antenna to be tested side by side. The Europeans ordered their antenna from European Industrial Engineering (EIE) of Italy in partnership with Costamasnaga. Unfortunately, the work quickly ran into problems, which resulted in ESO negotiating a new contract with a modified consortium called AEC (Alcatel-EIE-Consortium) led by Alcatel of France.

A key requirement for the 12 m antenna was that it should be able to make long-duration observations of the Sun.(7) Consequently, it not only had to have very accurate surface panels, but these panels had to scatter, rather than reflect infrared light from the Sun to avoid overheating the subreflector and other parts of the antenna. The Italian company Media Lario, who had manufactured a set of mirrors for the European Space Agency's XMM X-ray space telescope, were chosen to supply the European panels. These ALMA antenna panels consisted of an aluminium honeycomb core sandwiched between nickel-deposited front and back plates. The nickel surface was covered with a thin film of rhodium to improve the thermal behaviour when illuminated by the Sun.[3]

The Americans in the meantime had placed their order for a prototype antenna with VertexRSI of Texas. They used their German subsidiary Vertex Antennentechnik, who had worked on both IRAM and on the HHT on Mount Graham, to do most of the work. The Vertex antenna panels were made of machined aluminium which were then chemically etched to scatter the solar heat.

[1] The European institutions who signed the agreement with the NSF in June 1999 were CNRS (France), MPG (Germany), NFRA (the Netherlands), and PPARC (UK), in addition to ESO. Subsequently Sweden agreed to fund more than their contribution through ESO, and Spain, who were not a member of ESO at the time, agreed to join the project in the following year. It was later agreed that ESO would act on behalf of all these organisations when required.

[2] *ESO Annual Report,* 1999, p. 81.

[3] The AEC antenna as a whole turned out to be significantly lighter than the American one because of the AEC's more extensive use of carbon fibre reinforced plastic (CFRP).

In February 2003 ESO and Spain,[4] on the one hand, and the NSF and the National Research Council of Canada, on the other signed an agreement to proceed with the 64 antenna ALMA project. The costs, estimated at about $650 million, would be shared equally between Europe and North America. ESO agreed to provide 50% of the antennae, the cryostats for all the front ends, the receivers for bands 7 and 9 (275–370 GHz and 602–720 GHz, respectively) and a substantial amount of software and other items.[5] The NSF agreed to provide the other 50% of the antennae in addition to receivers, back end oscillators, the system correlator, fibre optics networks and other items.

It was agreed that the 64 ALMA antennae would not be fixed permanently on their 5,000 m altitude observing site on the Chajnantor Plateau, but could be moved between over two hundred fixed stations arranged over a distance of up to 16 km. The signals from the antennae would then be transmitted by fibre optics to a central correlator. Unfortunately, the Chajnantor Plateau is rather too high for human beings to work normally, particularly if physical effort is required, so the observatory was to be operated from and supported by the ALMA Operations Support Facility (OSF) at a much more human-friendly 2,900 m altitude. The antennae would also be assembled and maintained there. Two massive, specially made transporters built by the German company Scheuerle Fahzeugfabrik, and provided by ESO, would then be used to transport the antennae from the OSF to the observatory site and move them between observing stations at that site. Groundbreaking took place at both the OSF and Chajnantor sites in November 2003.

At about this time the prototype European and American antennae were tested side-by-side at the VLA site in New Mexico. Although these tests were completed by the end of May 2004, the final report was not published until December.[6] In the meantime, ESO and NRAO had both issued their request for quotations for their 50% share of the total production antennae. Unfortunately, the antenna costs quoted in reply were much higher than expected.

Two alternatives were then considered to significantly reduce the programme cost; either to reduce the number of antennae or to reduce the number of receiver bands covered. The reduction in the number of antennae could, to some extent, be compensated by extending the observing times, but the reduction in receiver bands would result in a permanent loss of science. So at a meeting in October 2005, it was agreed that the best way of making the financial savings required was to reduce the number of antennae from 64 to 50. NRAO placed the order with

[4] Spain was not a member of the ESO at this time. They agreed to join in 2006.

[5] The UK's Rutherford Appleton Laboratory and Onsala later agreed to cover band 5 (163–211 GHz) and IRAM agreed to develop the receiver for band 7.

[6] This was to ensure that the bidders for supplying the production antennae were not aware of the test results of their competitor's antenna, see Jeffrey G. Mangum, et al., Evaluation of the ALMA Prototype Antennas, *Publications of the Astronomical Society of the Pacific*, 118, 2006, pp. 1257–1301.

Figure 24.1. Atacama Compact Array, otherwise called the Morita Array, at an altitude of about 5,000 m (16,400 ft) on Llano de Chajnantor in Chile's Atacama Desert. (Courtesy ALMA(ESO/NAOJ/NRAO).)

VertexRSI for their 25 antennae plus an option for 7 more, whilst ESO chose a new consortium of AEM, consisting of Alcatel, EIE and MT Aerospace (of Germany) for their supply of 25 antennae plus an option for an additional 7.

In the meantime Japan, through the National Astronomical Observatory of Japan, had joined ALMA by an agreement signed in September 2004[7] and amended in July 2006. In addition, Taiwan had joined ALMA in September 2005 working through Japan. But instead of providing one third of the project cost Japan had decided to enhance the main Atacama array by building what was called the Atacama Compact Array (ACA) (see Figure 24.1).[8] This consisted of four 12 m antennae and twelve 7 m antennae and a correlator which partly balanced the loss of the fourteen 12 m antennae in the main array. In addition Japan agreed to provide receivers for bands 4, 8 and 10 (125–163, 385–500, and 787–950 GHz, respectively) for the main ALMA array and to pay the other ALMA partners to provide front and back end components for their ACA antennae, including receivers that were identical to those used with the main ALMA antennae.

The addition of the ACA to ALMA produced what was called the Enhanced ALMA, or the Atacama Large Millimeter/Submillimeter Array, with the same acronym of ALMA. Europe and North America (USA and Canada) each agreed to pay 37.5% of the costs of the Enhanced ALMA system whilst the remaining 25% would be paid by East Asia (Japan and Taiwan).

[7] The agreement was signed by ESO, the NSF and the National Institutes for Natural Sciences (NINS) of Japan.

[8] ACA was renamed the Morita Array in March 2013 in memory of Kohichiro Morita who had helped to design ACA but who had died suddenly in Santiago on May 7 2012. See *ALMA Press Release 31*, May 7 2013, Final ACA Antenna Delivered with the new name 'Morita Array'.

The first 12 m antenna to arrive at Chajnantor was not part of the ALMA array but one to take over the work of the 15 m SEST antenna that was to be decommissioned at La Silla in 2003 (see Section 23.2).(8) In June 2001, the Max-Planck-Institut für Radioastronomie, the Astronomisches Institut der Ruhr-Universität Bochum, the Onsala Space Observatory and ESO had agreed to locate a single 12 m antenna at Chajnantor which they purchased from Vertex. It was of a very similar design to the prototype ALMA antenna purchased by NRAO. Clearly, the early availability of this antenna at Chajnantor would not only allow the Europeans to undertake interesting scientific observations, but it would enable them to get early experience of observing near the ALMA site.[9] As a result, this new project was given the name of APEX, or the Atacama Pathfinder Experiment. Telescope installation began in 2003 and the first scientific observations were carried out two years later.

Early APEX observations were generally undertaken with instruments transferred from SEST as instruments specially designed for APEX were mostly not available when the telescope had been completed. However in 2007, a new instrument called LABOCA[10] was fitted to the APEX antenna. It was an innovative 300 element bolometer camera developed in Germany by the Max-Planck-Institut, Bochum and the Friedrich-Schiller-Universität Jena. This camera operated at a wavelength of 0.87 mm, with detectors cooled to about 0.3K, and covered a relatively wide field-of-view of 11 arcmin. LABOCA was, at the time, the largest such camera in the world.

Returning to the main ALMA project, in March 2007 the first fringes were detected between two ALMA antennae at the antenna test facility at the New Mexico VLA site. This was followed a month later by the arrival of the first antenna in Chile. The two antenna transporters arrived in Chile early the following year and in May 2009 the first interferometry was carried out with two antennae at the OSF. Four months later the first antenna was carried to the Chajnantor observation site by a transporter and by the end of the year the first three 12 m antennae were able to operate together at Chajnantor for the first time. The first scientific observations were undertaken in 2011, with the final ALMA antenna being delivered two years later (see Figure 24.2). Array configurations with a maximum baseline of 16 km produced a maximum spatial resolution of better than 10 milliarcseconds (mas). In total, the project had cost over $1.3 billion.

The ALMA project now has one regional centre in each of Europe, North America and Japan to act as the interface between the ALMA user community and the observatory. In addition, nodes have been created in Europe in Bonn-Cologne (Germany), Bologna (Italy), Ondřejov (Czech Republic), Onsala (Sweden), IRAM

[9] APEX was located at about the same altitude as ALMA, but was about 2 km away, see Govert Schilling and Lars Lindberg Christensen, *Europe to the Stars; ESO's First 50 Years of Exploring the Southern Sky*, ESO Wiley-VCH, 2012, p. 221.

[10] LABOCA was an acronym of Large APEX BOlometer CAmera.

Figure 24.2. The final ALMA antenna arriving at Chajnantor to complete the complement of 66 antennae. (Courtesy A.Marinkovic/X-Cam/ALMA(ESO/NAOJ/NRAO).)

(Grenoble, France), Leiden (the Netherlands) and JBCA (Manchester, UK). The European nodes were coordinated from ESO in Garching.

The Llano de Chajnantor and the area nearby have, in recent years, become home to a number of other telescopes. It is impracticable to go into their history here but, for completeness, the most important are listed in Table 24.1.

24.2 South Pole

The South Pole, at an altitude of 2,850 m (9,300 ft) above sea level, is not only very high but extremely cold. The combination of these factors produces an exceptionally low amount of water vapour in its atmosphere, particularly in winter when the average precipitable amount is just 0.2 mm.(9) As a result, it is the ideal place, from a scientific point of view, to make microwave and infrared astronomical observations. On the other hand, its environment is exceptionally harsh for both people and equipment, and it is a considerable distance from civilisation. So astronomical observation campaigns have been undertaken there only relatively recently.

A French experiment, EMILIE (Emission Millimétrique), was the first to make submillimetre astronomical observations from the South Pole. This was during the

Table 24.1. *Llano de Chajnantor Telescopes (in addition to ALMA and APEX)*

Average altitude 5,000m

Telescope name	Operational Frequency	Year Construction Completed on Site	Comments
Cosmic Background Imager (CBI)	Ten bands between 26 and 36 GHz	1999	A radio interferometer consisting of thirteen 0.9 m diameter antennae on a common platform designed to measure the intensity and polarisation of the Cosmic Microwave Background (CMB) with a resolution better than 6 arcmin. Built by Caltech for an international collaborative project. In 2006 the 0.9 m antennae were replaced by 1.4 m antennae. Closed down 2008.
Q/U Imaging ExperimenT (QUIET)	43 and 94 GHz	2008, replacing the CBI	Radio telescope array designed to measure the anisotropy in the polarisation of the CMB. Project led by the University of Chicago.
Pampa La Bola *Average altitude 4,800 m*			
Atacama Submillimeter Telescope Equipment (ASTE)	Up to 500 GHz by 2015	2002	10 m antenna built by Mitsubishi Electric in 2000. Was an early prototype for the Japanese ALMA antennae.
NANTEN2 Observatory	110 to 880 GHz	2004	4 m diameter telescope to observe submillimetre spectra. International collaborative project led by Japan. Moved from Las Campanas Observatory where it had been located since 1995.

austral summer of 1984–1985. Relocated from Mauna Kea because of sky noise, EMILIE measured diffuse radiation from the Milky Way at a wavelength of 0.9 mm. It used a single pixel bolometer which was fed by a 45 cm off-axis mirror.(10)

Over the next few years a number of astronomers set up equipment at or near the pole to try to measure the anisotropy of the Cosmic Microwave Background (CMB). The first attempt to measure the CMB from the polar plateau was undertaken in 1986 by the Radio Physics Research Group at Bell Laboratories. They could find no consistent anisotropy but they were able to measure sky noise and opacity. Unfortunately, an enormous amount of effort was required to ship the equipment to the South Pole where it was only operated through the short, relatively moist summer season, before it all had to be dismantled and removed. Clearly, a permanent facility was required.

AST/RO (Antarctic Submillimeter Telescope and Remote Observatory), which was first used in 1989, was the first radio telescope to operate all year round at the South Pole. It was a general purpose 1.7 m diameter off-axis Gregorian radio telescope designed to map the Milky Way in the spectral lines of atomic carbon. AST/RO, which had receivers operating in the atmospheric windows at 230 GHz (λ 1.3 mm), 460 to 492 GHz, 810 GHz and later 1.46 THz (λ 0.205 mm), was mounted on a dedicated permanent observatory building. Two years later AST/ RO became part of the newly founded American National Science Foundation's (NSF's) Center for Astrophysical Research in Antarctica (CARA).

CARA assembled a number of telescopes, some of which, like earlier instruments, also undertook CMB research. By 1997 they, as a group, had given the first indication that the spectrum of spatial anisotropy in the CMB was consistent with a flat cosmology. More precise results were then provided by DASI (Degree Angular Scale Interferometer) starting in 1999–2000. DASI was a compact interferometer consisting of an array of thirteen 20 cm diameter horns on a fully steerable mount. It operated from 26 to 36 GHz (λ 1.15 to 0.85 cm) and provided baselines from 25 to 121 cm.(11)

Viper, which began observations in 1998, consisted of a 2.1 m off-axis radio telescope designed to measure low contrast millimetre-wave sources. It was used with a variety of instruments including ACBAR (Arcminute Cosmology Bolometer Array Receiver) which was a sixteen element bolometer array that operated at 0.3 K from 2001 to 2005. It made high quality maps of the Sunyaev-Zel'dovich effect in nearby clusters of galaxies and made measurements of CMB anisotropy on a range of angular scales.

The success of these various facilities persuaded the NSF to fund a much larger radio telescope at the South Pole. This was the South Pole Telescope (SPT) (see Figure 24.3) which was a 10 m diameter, altazimuth-mounted[11] off-axis Gregorian

[11] At the south pole an altazimuth-mount is, of course, effectively the same as an equatorial.

Figure 24.3. The 10 m diameter South Pole Telescope. (Courtesy Amble via Wikimedia Commons.)

radio telescope. Its primary purpose was to detect clusters of galaxies using the Sunyaev-Zel'dovich effect over an area of ~4,000 square degrees observing at frequencies of 95 GHz (λ 3 mm), 150 GHz and 220 GHz (λ 1.3 mm). The SPT's primary mirror was made of 60 cm thick aluminium with machined pockets to reduce weight whilst leaving a 3 mm thick reflective surface. This mirror was fitted with a de-icing system that ran continuously to maintain its surface temperature about 1 to 2 K above ambient which kept the surface free of ice except during storms. However running the system for about a day after such a storm was enough to clear any accumulations. The SPT's secondary mirror was 1 m diameter and made of 50 mm thick aluminium. It was cooled to about 10 K.

The SPT, which was installed in 2006–2007, initially used the SPT-SZ (South Pole Telescope-Sunyaev-Zel'dovich) state-of-the-art camera that had a 966 element bolometer array receiver.(12). The camera's focal plane was divided into six 60° wedges that could operate at different frequencies. For much of its lifetime, the camera's focal plane had one wedge operating at 95 GHz, four at 150 GHz and one at 220 GHz. The SPT-SZ camera was replaced in 2011 by the SPTpol camera which, as its name implies, could measure polarisation. More sensitive than its predecessor, its 780 polarisation-sensitive pixels were divided between those sensitive to 90 GHz and those sensitive to 150 GHz radiation. It was designed to measure the 'B-mode' or 'curl' component of the polarised CMB which provides constraints on the growth of structure in the universe. In 2013 it became the first experiment to detect this B-mode.(13)

References

1. Baars, Jacob W. M., *International Radio Telescope Projects; A Life Among Their Designers, Builders and Users*, Jacob W. M. Baars, 2013, p. 128.
2. Shaver, P. A., and Booth, R., A Report on the SEST Users Meeting and Workshop on Millimetre-Wave Interferometry, *ESO Messenger*, No. 65, September 1991, pp. 14–15.
3. Booth, R. S., The LSA/MMA Project, in Ossenkopf, V., Stutzki, J., and Winnewisser, G., (eds.), *The Physics and Chemistry of the Interstellar Medium, Proceedings of a Symposium at Zermatt, September 1998*, GCA-Verlag Herdecke, 1999, pp. 435–439.
4. Madsen, Claus, *The Jewel on the Mountaintop; The European Southern Observatory through Fifty Years*, ESO and Wiley-VCH, 2012, pp. 371–372.
5. Grewing, Michael, and Guilloteau, Stéphane, The LSA Study, *IRAM Newsletter*, No. 32, July 1997.
6. Otárola, A., et al., European Site Testing at Chajnantor: A Step Towards the Large Southern Array, *ESO Messenger*, No. 94, December 1998, pp. 13–20.
7. Baars, Jacob W. M., *International Radio Telescope Projects; A life among their designers, builders and users*, Jacob W. M. Baars, 2013, p. 131.
8. Madsen, Claus, *The Jewel on the Mountaintop; The European Southern Observatory through Fifty Years*, ESO and Wiley-VCH, 2012, pp. 389–391.
9. Wilson, Gregory, Astronomy at the South Pole, *Optics and Photonics News*, January 2004, pp. 34–39.
10. Wilson, Robert W., and Stark, Antony A., Cosmology from Antarctica, *Proceedings of the Smithsonian at the Poles Symposium 2007*, Smithsonian Institution Scholarly Press, 2009, pp. 359–367.
11. Halverson, N. W., et al., Degree Angular Scale Interferometer First Results: A Measurement of the Cosmic Microwave Background Angular Power Spectrum, *Astrophysical Journal*, 568, 2002, pp. 38–45.
12. Carlstrom, J. E., et al., The 10 Meter South Pole Telescope, *Publications of the Astronomical Society of the Pacific*, 123, 2011, pp. 568–581.
13. Hanson, D., et al., Detection of B-mode Polarization in the Cosmic Microwave Background with Data from the South Pole Telescope, *Physical Review Letters*, 111, Issue 14, October 4 2013.

Name Index

Optical/Infrared Observatory and Telescope Index

Bold type indicates the main reference or references when a number of pages are listed.

Optical/Infrared Observatories

Optical/Infrared Telescopes and Interferometers

Radio Observatory and Telescope Index

Bold type indicates the main reference or references when a number of pages are listed.

Millimetre-Wave/Submillimetre Observatories and Telescopes

Radio Observatories and Telescopes

General Index